Lecture Notes in Mathematics

A collection of informal reports and seminars
Edited by A. Dold, Heidelberg and B. Eckmann, Zürich

298

Proceedings of the Second Conference on Compact Transformation Groups

University of Massachusetts, Amherst, 1971

Part I

Springer-Verlag
Berlin · Heidelberg · New York 1972

AMS Subject Classifications (1970): 57D85, 57Exx

ISBN 3-540-06077-4 Springer-Verlag Berlin · Heidelberg · New York
ISBN 0-387-06077-4 Springer-Verlag New York · Heidelberg · Berlin

Offsetdruck: Julius Beltz, Hemsbach/Bergstr.

COMMENTS BY THE EDITORS

The Second Conference on Compact Transformation Groups was held at the University of Massachusetts, Amherst from June 7 to June 18, 1971 under the sponsorship of the Advanced Science Education Program of the National Science Foundation. There were a total of 70 participants at the conference.

As was the case at the first conference at Tulane University in 1967, the emphasis continued to be on differentiable transformation groups. In this connection there was a continued application of surgery typified by the lectures of Browder, Shaneson, and Yang (joint work with Montgomery). A new feature was the applications of the Atiyah-Singer Index Theorem to differentiable transformation groups typified by the lectures of Hinrichsen, Petrie, and Rothenberg. In connection with topological and algebraic methods significant innovations were made by Raymond (joint work with Conner) in the construction of manifolds admitting no effective finite group action, by R. Lee in studying free actions of finite groups on spheres using ideas and methods derived from algebraic K-theory and by Su (joint work with W. Y. Hsiang) in applying the notion of geometric weight systems developed recently by W. Y. Hsiang. There were several lectures on algebraic varieties by Michael Davis, Peter Orlik, and Philip Wagreich. Interest in this area arose from the application several years ago of Brieskorn varieties to the study of actions on homotopy spheres.

These Proceedings contain not only material presented at the conference but also articles received by the editors up to the summer of 1972. We have divided the articles into two volumes; the first volume is devoted to smooth techniques while the second to non-smooth techniques. While the proper assignment of a few papers was not obvious, the editors felt that this classification offered, in general, the most natural division of the material.

H. T. Ku

L. N. Mann

J. L. Sicks

J. C. Su

Amherst, Mass., July 1972

CONFERENCE PARTICIPANTS

E. M. Bloomberg	University of Virginia
J. M. Boardman	Johns Hopkins University
G. Bredon	Rutgers University
W. Browder	Princeton University
H. Cohen	University of Massachusetts
F. Connolly	Notre Dame University
Bruce Conrad	Temple University
W. D. Curtis	Kansas State University
Michael Davis	Princeton University
Murray Eisenberg	University of Massachusetts
Dieter Erle	University of Dortmund
I. Fary	University of California
L. A. Feldman	Stanislaus State College
Gary Ford	Radford College
V. Giambalvo	University of Connecticut
David Gibbs	University of Virginia
R. Goldstein	SUNY at Albany
M. Goto	University of Pennsylvania
E. Grove	University of Rhode Island
Stephen Halperin	University of Toronto
Gary Hamrick	Institute for Advanced Study
Douglas Hertz	University of Massachusetts
John Hinrichsen	University of Virginia
Edward Hood	M.I.T.
Norman Hurt	University of Massachusetts
Sören Illman	Princeton University
Stephen Jones	University of Massachusetts
Louis Kauffman	University of Illinois at Chicago
K. Kawakubo	Osaka University
S. K. Kim	University of Connecticut
Larry King	University of Massachusetts
S. Kinoshita	Florida State University
Robert Koch	Louisiana State University
H. T. Ku	University of Massachusetts
Mei Ku	University of Massachusetts
K. W. Kwun	Michigan State University
Timothy Lance	Princeton University
Connor Lazarov	Lehman College
R. Lee	Yale University

L. Lininger	SUNY at Albany
R. Livesay	Cornell University
S. López de Medrano	University of Mexico
Larry Mann	University of Massachusetts
Deane Montgomery	Institute for Advanced Study
P. Orlik	University of Wisconsin
E. Ossa	University of Bonn
J. Pak	Wayne State University
Ted Petrie	Rutgers University
F. Raymond	University of Michigan
Richard Resch	University of Connecticut
Robert Rinne	Sandia Laboratory, Livermore, California
M. Rothenberg	University of Chicago
Loretta J. Rubeo	University of Virginia
H. Samelson	Stanford University
James Schafer	University of Maryland
V. P. Schneider	S. W. Louisiana University
Reinhard Schultz	Purdue University
J. Shaneson	Princeton University
Jon Sicks	University of Massachusetts
J. C. Su	University of Massachusetts
Richard Summerhill	Institute for Advanced Study
Mel Thornton	University of Nebraska
Philip Wagreich	University of Pennsylvania
Shirley Wakin	University of Massachusetts
H. C. Wang	Cornell University
Kai Wang	University of Chicago
A. Wasserman	University of Michigan
Steven Weintraub	Princeton University
J. E. West	Cornell University
C. T. Yang	University of Pennsylvania

CONFERENCE LECTURE TITLES

J. Boardman	Cobordism of Involutions Revisited
G. Bredon	Strange Circle Actions on Products of Spheres, and Rational Homotopy
W. Browder	Equivariant Differential Topology
M. Davis	Actions on Exotic Stiefel Manifolds
D. Erle	On Unitary and Symplectic Knot Manifolds
I. Fary	Group Action and Betti Sheaf
L. Feldman	Reducing Bundles in Differentiable G-Spaces
V. Giambalvo	Cobordism of Line Bundles with Restricted Characteristic Class
R. Goldstein	Free Differentiable Circle Actions on 6-Manifolds
M. Goto	Orbits of One-Parameter Groups
E. Grove	Classical Group Actions on Manifolds with Vanishing First and Second Integral Pontrjagin Classes
J. Hinrichsen	Operators Elliptic Relative to Group Actions
S. Illman	Equivariant Singular Homology
L. Kauffman	Cyclic Branched Covering Spaces and O(n)-Actions
K. Kawakubo	Invariants for Certain Semi-Free S^1-Actions
	Topological S^1 and Z_{2n} Actions on Spheres
	The Index of Manifolds with Toral Actions
S. Kinoshita	On Infinite Cyclic Covering Transformation Groups on Contractible Open 3-Manifolds
H. T. Ku	Characteristic Invariants of Free Differentiable Actions of S^1 and S^3 on Homotopy Spheres
K. W. Kwun	Transfer Homomorphisms of Whitehead Groups of Some Cyclic Groups
R. Lee	Semi-Characteristic Classes
	The Topological Period of Periodic Groups
S. López de Medrano	Cobordism of Diffeomorphisms of (k-1)-Connected 2k-Manifolds
L. Mann	Degree of Symmetry of Compact Manifolds
D. Montgomery	Introductory Remarks
P. Orlik	The Picard-Lefschetz Monodromy for Certain Singularities
	Actions of the Torus on 4-Manifolds
E. Ossa	Complex Bordism of Isometries and Monogenic Groups
T. Petrie	Applications of the Index Theorem to Smooth Actions on Compact Manifolds
	Applications of the Index Theorem to Smooth Actions on Compact Manifolds II

INTRODUCTORY REMARKS

The subject of transformation groups is in an active period and it is good for all of us interested to meet and exchange ideas at first hand. A generation ago the fewer people then working in a field could manage to keep in touch by correspondence or occasional contacts at general meetings, but this is now more difficult, and specialized conferences of this kind perform an important service not easily achieved in any other way. Transformation groups is an area of topology which has connections with most of the other areas of topology. In the past, progress in any part of topology has often led to progress in transformation groups. This is likely to continue and all of us must keep as well informed as we can about what others are doing at the same time as we are continuing with our own problems. Conversely transformation groups has sometimes contributed to other areas, at the very least by suggesting questions and problems. It is a great pleasure to attend a conference on a very interesting subject under such convenient conditions and congenial surroundings as have been provided here.

Deane Montgomery

TABLE OF CONTENTS

TABLE OF CONTENTS

VOLUME II

INVARIANTS FOR SEMI-FREE S^1-ACTIONS

Katsuo Kawakubo*
Osaka University and Institute for Advanced Study

§1. Introduction

This paper gives the details promised in [10].

In [1] Atiyah and Singer obtained an invariant for certain S^1-actions and in [5] Browder and Petrie used the invariant to distinguish certain semi-free S^1-actions so that they showed the following result.

Theorem 0. For any odd $n \geq 5$, $n = 2k-1$, there are an infinite number of distinct semi-free S^1-actions on the Brieskorn $(2n+1)$-spheres with fixed point set of codimension $4r$, any $r \neq k/2$, $r < (n-1)/2$.

In the present paper, we define some invariants for certain semi-free S^1-actions which are different from that of Atiyah and Singer (see Theorem 1). As an application, we can prove the above theorem of Browder and Petrie without the assumption $r \neq k/2$ (see Corollary 2). Our method is different from that of Atiyah and Singer, and Browder and Petrie. We use the Chern classes due to Borel and Hirzebruch [2] and Grothendieck (see [3]) and the bordism theory. The author wishes to thank Professor F. Uchida who kindly enlightened him about the structure of the normal bundle of fixed point set.

§2. Definitions and statement of results

Definitions: An action (M, φ, G) is called semi-free if G acts freely outside the fixed point set.

We remark the following. Given a semi-free S^1-action (M, φ, S^1) where M may have a boundary, the normal vector bundle of the fixed point set has the unique complex vector bundle structure such that the induced action of S^1 is the scalar multiplication when we regard S^1 as $\{z \mid z \in \mathbb{C}, |z| = 1\}$. By complex bundle we mean that such a complex structure is taken. Therefore a fiber of the normal bundle of the fixed point set has a natural orientation and we denote by $2k$ the codimension of the fixed point set.

Let (M, φ, S^1) be a semi-free S^1-action on a closed manifold M. We suppose that

* Supported in part by National Science Foundation grant.

(i) The fixed point set $F(S^1, M)$ is a homology sphere,

(ii) (M^n, φ, S^1) extends to a semi-free S^1-action (W^{n+1}, Φ, S^1) ($\partial W^{n+1} = M^n$ as S^1-manifold) such that the fixed point set $F(S^1, W)$ is connected and the normal complex bundle of the fixed point set has the trivial Chern classes.

Orient the fixed point set $F(S^1, M)$ so that the orientation class of $F(S^1, M)$ times the orientation class of the oriented normal fiber is the orientation of M^n. We denote the orientation class by $[F(S^1, M)] (\in H_{n-2k}(F(S^1, M)))$. Let

$$\phi : H_{n+1-2k}(F(S^1, W)/F(S^1, M)) \longrightarrow H_{n-2k}(F(S^1, M))$$

be the isomorphism defined by the composition of the following isomorphisms

$$\begin{aligned} & H_{n+1-2k}(F(S^1, W)/F(S^1, M)) \\ & \cong H_{n+1-2k}(F(S^1, W), F(S^1, M)) \\ & \cong H_{n-2k}(F(S^1, M)). \end{aligned}$$

Then we define a generalized orientation class $[F(S^1, W)/F(S^1, M)]$ of $F(S^1, W)/F(S^1, M)$ by the class $\phi^{-1}[F(S^1, M)]$.

First we define Stiefel-Whitney numbers $W_I[F(S^1, W)]$ as follows where I denotes a partition of $\dim F(S^1, W) = n+1-2k$. Let $W_i(F(S^1, W))$ be the i-th Stiefel-Whitney class of $F(S^1, W)$. Since $F(S^1, M)$ $(= \partial F(S^1, W))$ is a homology sphere, the homomorphism

$$j^* : H^i(F(S^1, W)/F(S^1, M) : Z_2) \longrightarrow H^i(F(S^1, W) : Z_2)$$

is an isomorphism for $i \leqq n - 2k$.

Definition: For a nontrivial partition $I = (i_1, \ldots, i_\ell)$ of $\dim F(S^1, W) = n + 1 - 2k$, the Stiefel-Whitney number $W_I[F(S^1, W)]$ is defined by the number

$$\langle j^{*-1} W_{i_1}(F(S^1, W)) \cdots j^{*-1} W_{i_\ell}(F(S^1, W)), [F(S^1, W)/F(S^1, M)] \rangle \in Z_2$$

where nontrivial means $I \neq \dim F(S^1, W)$ and $\langle\ ,\ \rangle$ denotes the Kronecker index. For the trivial partition $I = \dim F(S^1, W)$, $W_I[F(S^1, W)]$ is defined by the number

$$\chi(F(S^1, W)) + 1 \bmod 2 ,$$

where $\chi(F(S^1, W))$ denotes the Euler number of $F(S^1, W)$.

When $\dim F(S^1, W) = n+1-2k \equiv 0 \pmod 4$, we define Pontrjagin numbers and index of $F(S^1, W)$ as follows. Let $P_i(F(S^1, W))$ be the i-th Pontrjagin class of $F(S^1, W)$. Since $F(S^1, M)$ ($= \partial F(S^1, W)$) is a homology sphere, the homomorphism

$$j^* : H^i(F(S^1, W)/F(S^1, M) : Z) \longrightarrow H^i(F(S^1, W) : Z)$$

is an isomorphism for $i \leqq n-2k$.

Definition: For a non trivial partition $I' = (i_1, \dots, i_s)$ of $\dim F(S^1, W)/4 = (n+1-2k)/4$, the Pontrjagin number $P_{I'}[F(S^1, W)]$ of $F(S^1, W)$ is defined by the number

$$\langle j^{*-1}P_{i_1}(F(S^1, W)) \cdots j^{*-1}P_{i_s}(F(S^1, W)), [F(S^1, W)/F(S^1, M)]\rangle \in Z.$$

Definition: The index $\sigma(F(S^1, W))$ of $F(S^1, W)$ is defined by the signature of the cup product pairing

$$H^v(F(S^1, W), F(S^1, M) : \mathbb{R}) \otimes H^v(F(S^1, W), F(S^1, M) : \mathbb{R})$$
$$\longrightarrow H^{2v}(F(S^1, W), F(S^1, M) : \mathbb{R})$$

where $v = \dim F(S^1, W)/2 = (n+1-2k)/2$.

Then we shall have

Theorem 1. Stiefel-Whitney numbers $W_I[F(S^1, W)]$ depend only upon (M, φ, S^1). If $\dim F(S^1, W) \equiv 0 \pmod 4$, Pontrjagin numbers $P_{I'}[F(S^1, W)]$ and the index $\sigma(F(S^1, W))$ depend only upon (M, φ, S^1).

As an application, we shall have

Corollary 2. For any odd, $n \geqq 5$, there are an infinite number of distinct semi-free S^1-actions on the Brieskorn $(2n+1)$-spheres with fixed point set of codimension $4r$, any $r < (n-1)/2$.

§3. Proof of Theorem 1

First we introduce some notations. For each complex vector bundle ξ over an oriented closed smooth manifold X, let $B(\xi)$, $S(\xi)$ and $\mathbb{CP}(\xi)$ denote the total space of the disk bundle, the total space of the sphere bundle and the total space of the complex projective space bundle associated to ξ, respectively. Then the orientation of $S(\xi)$ and $\mathbb{CP}(\xi)$ are induced by those of X and ξ. For an oriented manifold M, we denote by $[M]$ the orientation class and by $\tau(M)$

the tangent vector bundle of M.

Suppose that a semi-free S^1-action (M, φ, S^1) satisfies (i) and (ii) above. Let (W_1, Φ_1, S^1) and (W_2, Φ_2, S^1) be two extensions of (M, φ, S^1) satisfying the condition (ii). Denote by ξ_1 and ξ_2 the normal complex bundles of the fixed point sets $F(S^1, W_1)$ and $F(S^1, W_2)$ respectively. By combining the two actions, we have the action $(W, \Phi, S^1) = (W_1 \underset{id}{\cup} (-W_2), \Phi_1 \cup \Phi_2, S^1)$ where $-W_2$ is W_2 with the opposite orientation. It follows from the uniqueness of the complex structure that the normal bundle of the fixed point set $F = F(S^1, W_1) \cup (-F(S^1, W_2))$ of the action (W, Φ, S^1) has the complex vector bundle structure ξ whose restrictions to $F(S^1, W_1)$ and $F(S^1, W_2)$ are isomorphic to ξ_1 and ξ_2 respectively as complex vector bundles. Let $i_1 : F(S^1, W_1) \longrightarrow F$ and $i_2 : F(S^1, W_2) \longrightarrow F$ be the inclusions, then the i-th Chern class $c_i(\xi)$ satisfies $i_1^* c_i(\xi) = c_i(\xi_1)$ and $i_2^* c_i(\xi) = c_i(\xi_2)$ and the Mayer-Vietoris exact sequence

$$\longrightarrow H^{i-1}(F(S^1, M)) \xrightarrow{\delta} H^i(F) \xrightarrow{i_1^* \oplus i_2^*}$$
$$H^i(F(S^1, W_1)) \oplus H^i(F(S^1, W_2)) \longrightarrow H^i(F(S^1, M)) \longrightarrow$$

shows that the homomorphism

$$i_1^* \oplus i_2^* : H^i(F) \longrightarrow H^i(F(S^1, W_1)) \oplus H^i(F(S^1, W_2))$$

is an isomorphism for $0 < i \leqq n - 2k$. Hence $c_i(\xi) = (i_1^* \oplus i_2^*)^{-1}(c_i(\xi_1) \oplus c_i(\xi_2)) = 0$ for $0 < i < (n+1-2k)/2$. Next we show that $c_i(\xi)$ is zero also for $i \geqq (n+1-2k)/2$. Since it is trivial to prove in the case where $n+1-2k$ is odd, we assume that $n+1-2k$ is even. The i-th Chern class $c_i(\xi)$ is trivially zero for $i \geqq (n+1-2k)/2+1$. If $n+1-2k > 2k$, $c_{(n+1-2k)/2}(\xi)$ is zero by definition. Therefore we have only to prove that $c_{(n+1-2k)/2}(\xi) = 0$ when $n+1-2k \leqq 2k$. Let $t \in H^2(\mathbb{CP}(\xi))$ be the first Chern class of the canonical line bundle over $\mathbb{CP}(\xi)$ and $p : \mathbb{CP}(\xi) \longrightarrow F$ be the projection map. Via the induced homomorphism p^*, $H^*(\mathbb{CP}(\xi))$ is a free graded $H^*(F)$-module with base $1, t, \ldots, t^{k-1}$, [7]. Borel and Hirzebruch [2] and Grothendieck (see [3]) showed

Lemma 3. t and $c_i(\xi)$ satisfy the following relation

$$t^k + \sum_{i=1}^{k} p^* c_i(\xi) \cdot t^{k-i} = 0 .$$

Therefore the Chern class $c_{(n+1-2k)/2}(\xi)$ must satisfy the following relation

$$(*) \qquad t^{(n-1)/2} + p^* c_{(n+1-2k)/2} \cdot t^{k-1} = 0.$$

Let $f : \mathbb{CP}(\xi) \longrightarrow \mathbb{CP}^\infty$ be the classifying map of the fibration

$$S^1 \longrightarrow S(\xi) \longrightarrow \mathbb{CP}(\xi).$$

Then $(\mathbb{CP}(\xi), f)$ defines an element of $\Omega_{n-1}(\mathbb{CP}^\infty)$. As a matter of fact, $(\mathbb{CP}(\xi), f)$ represents the zero element of $\Omega_{n-1}(\mathbb{CP}^\infty)$. This will be seen as follows. We now denote also by $B(\xi)$ an invariant tubular neighborhood of the fixed point set F. Then we have an invariant submanifold $W - \text{Int } B(\xi)$ with boundary $S(\xi)$ and by restricting the action (W, Φ, S^1) to $W - \text{Int } B(\xi)$, we have an action $(W - \text{Int } B(\xi), \Phi, S^1)$. Let

$$\bar{f} : (W - \text{Int } B(\xi))/S^1 \longrightarrow \mathbb{CP}^\infty$$

be the classifying map of the fibration

$$S^1 \longrightarrow (W - \text{Int } B(\xi)) \longrightarrow (W - \text{Int } B(\xi))/S^1.$$

Then the restriction to the boundary

$$\bar{f} \,|\, \partial(W - \text{Int } B(\xi))/S^1 : \partial(W - \text{Int } B(\xi))/S^1 \longrightarrow \mathbb{CP}^\infty$$

is nothing but $f : \mathbb{CP}(\xi) \longrightarrow \mathbb{CP}^\infty$. It follows that $(\mathbb{CP}(\xi), f)$ represents the zero element of $\Omega_{n-1}(\mathbb{CP}^\infty)$.

The map $f : \mathbb{CP}(\xi) \longrightarrow \mathbb{CP}^\infty$ can be written as the composition

$$\mathbb{CP}(\xi) \xrightarrow{\;i\;} (W - \text{Int } B(\xi))/S^1 \xrightarrow{\;\bar{f}\;} \mathbb{CP}^\infty,$$

where i is the inclusion map of the boundary. Let $t_0 \in H^2(\mathbb{CP}^\infty)$ be the first Chern class of the canonical line bundle over $\mathbb{CP}^\infty$. Then $t = f^* t_0$ and

$$t^{(n-1)/2} = f^* t_0^{(n-1)/2} = i^* \bar{f}^* t_0^{(n-1)/2} = 0,$$

since i^* is the zero map. It follows from the cohomology ring structure of $\mathbb{CP}(\xi)$ and the relation (*) above that $c_{(n+1-2k)/2}(\xi) = 0$. Thus all the Chern classes vanish.

Let $h : F \longrightarrow BU(k)$ be the classifying map of the bundle ξ and let $h_0 : F \longrightarrow BU(k)$ be the constant map. Then we have

Lemma 4. (F, h) and (F, h_0) represent the same element of

$\Omega_{n+1-2k}(BU(k))$.

Proof of Lemma 4. As is well-known $H^*(BU(k))$ is the polynomial ring generated by $c_1(\gamma(k)), \ldots, c_k(\gamma(k))$ where $\gamma(k)$ denotes the universal bundle for the unitary group $U(k)$ [11]. Since $h^*(c_i(\gamma(k))) = c_i(\xi) = 0$, the induced homomorphism $h^* : H^i(BU(k)) \longrightarrow H^i(F)$ is the zero map for $i > 0$. Clearly $h_0^* : H^i(BU(k)) \longrightarrow H^i(F)$ is the zero map for $i > 0$. Hence Lemma 4 follows from Theorem 17.5 of Conner-Floyd [6]. This completes the proof of Lemma 4.

It follows from Lemma 4 that there exist a manifold $\bar{F}$ and a map $H : \bar{F} \longrightarrow BU(k)$ such that

$$\partial\bar{F} = F \bigcup_{disj} (-F) \quad \text{and} \quad H|\partial\bar{F} = h \cup h_0$$

where $\bigcup_{disj}$ means the disjoint union. Denote by $\bar{\xi}$ the induced bundle $H^!(\gamma(k))$. Now we construct an oriented closed manifold

$$W' = (W - \mathrm{Int}\, B(\xi)) \cup S(\bar{\xi}) \cup F \times D^{2k}$$

where the attaching maps are the obvious ones, since $\bar{\xi}|\partial\bar{F} = \xi \bigcup_{disj} F \times \mathbb{C}^k$. Naturally we can define a semi-free S^1-action (W', Φ', S^1) on W' as follows:

(1) $\Phi'|W - \mathrm{Int}\, B(\xi) = \Phi$,

(2) $\Phi'|S(\bar{\xi})$ is the restriction of the action defined by the complex scalar multiplication of the complex vector bundle $\bar{\xi}$,

(3) $\Phi'|F \times D^{2k}$ is the restriction of the action defined by the complex scalar multiplication when we regard D^{2k} as $\{z \in \mathbb{C}^k \,|\, |z| \leq 1\}$.

Obviously Φ' is well-defined. The action (W', Φ', S^1) has the fixed point set F whose normal complex bundle is trivial. Hence Theorem 1 follows from the following Lemma 5 and Lemma 8.

Lemma 5. *Let (M^n, φ, S^1) be a semi-free S^1-action on an oriented closed manifold M with connected fixed point set F^m. If the normal complex bundle ξ of the fixed point set F is trivial, then F represents the zero element of the oriented cobordism group Ω_m.*

Proof of Lemma 5. It follows from the assumption that

$$S(\xi) = F \times S^{2k-1} \quad \text{and} \quad \mathbb{CP}(\xi) = F \times \mathbb{CP}^{k-1},$$

where $2k = n - m$. Let $f : F \times \mathbb{CP}^{k-1} \longrightarrow \mathbb{CP}^\infty$ be the classifying map of the fibration $S^1 \longrightarrow F \times S^{2k-1} \longrightarrow F \times \mathbb{CP}^{k-1}$. As we have shown above,

$(F \times \mathbb{CP}^{k-1}, f)$ represents the zero element of $\Omega_{n-1}(\mathbb{CP}^{\infty})$. Let $p_1 : F \times \mathbb{CP}^{k-1} \longrightarrow F$ and $p_2 : F \times \mathbb{CP}^{k-1} \longrightarrow \mathbb{CP}^{k-1}$ be the projection maps. Then we have an isomorphism

$$\tau(F \times \mathbb{CP}^{k-1}) \cong p_1^! \tau(F) \oplus p_2^! \tau(\mathbb{CP}^{k-1}).$$

It follows that the total Stiefel-Whitney class $W(F \times \mathbb{CP}^{k-1})$ can be written as

$$W(F \times \mathbb{CP}^{k-1}) = p_1^* W(F) \cdot p_2^* W(\mathbb{CP}^{k-1}).$$

By the Künneth formula:

$$H^*(F \times \mathbb{CP}^{k-1} : Z_2) \cong H^*(F : Z_2) \otimes H^*(\mathbb{CP}^{k-1} : Z_2),$$

we may write

$$W(F \times \mathbb{CP}^{k-1}) = W(F) \otimes W(\mathbb{CP}^{k-1}).$$

Let $I = (i_1, \ldots, i_\ell)$ be a partition of $\dim F$. Then it is easy to see the following.

Lemma 6. $W_I(F \times \mathbb{CP}^{k-1}) = W_I(F) \otimes 1 + a$ where a denotes the sum of the terms each of which involves a non zero element of $H^i(\mathbb{CP}^{k-1} : Z_2)$ for some $i > 0$.

Let $t_0' \in H^2(\mathbb{CP}^{\infty} : Z_2)$ and $t' \in H^2(\mathbb{CP}^{k-1} : Z_2)$ be the generators. Then consider a Whitney number of the map f associated with $t_0'^{k-1}$ (see page 45 of [6]):

$$\begin{aligned}
&\langle W_I(F \times \mathbb{CP}^{k-1}) f^*(t_0'^{k-1}), [F \times \mathbb{CP}^{k-1}]\rangle \\
&= \langle (W_I(F) \otimes 1 + a)(1 \otimes t'^{k-1}), [F \times \mathbb{CP}^{k-1}]\rangle \\
&= \langle W_I(F) \otimes t'^{k-1}, [F] \otimes [\mathbb{CP}^{k-1}]\rangle \\
&= \langle W_I(F), [F]\rangle \cdot \langle t'^{k-1}, [\mathbb{CP}^{k-1}]\rangle \\
&= \langle W_I(F), [F]\rangle .
\end{aligned}$$

According to Theorem 17.5 of [6], a Whitney number of the map f associated with $t_0'^{k-1}$ must vanish. Therefore we have proved that every Whitney number of F vanishes.

Next we prove that every Pontrjagin number of F vanishes when $\dim F \equiv 0 \pmod 4$. The total Pontrjagin class $P(F \times \mathbb{CP}^{k-1})$ can be written as

$$P(F \times \mathbb{CP}^{k-1}) = p_1^* P(F) \cdot p_2^* P(\mathbb{CP}^{k-1}) + \beta$$

where β denotes a sum of elements of order 2 [11]. Since $H^*(\mathbb{CP}^{k-1})$ is torsion free, we have the isomorphism

$$H^*(F \times \mathbb{CP}^{k-1}) \cong H^*(F) \otimes H^*(\mathbb{CP}^{k-1}).$$

Hence we may write

$$P(F \times \mathbb{CP}^{k-1}) = P(F) \otimes P(\mathbb{CP}^{k-1}) + \beta.$$

Let $I' = (i_1, \ldots, i_s)$ be a partition of $\dim F/4$, then it is easy to see the following

Lemma 7. $P_{I'}(F \times \mathbb{CP}^{k-1}) = P_{I'}(F) \otimes 1 + \alpha + \beta$ where α denotes the sum of the terms each of which involves a non zero element of $H^i(\mathbb{CP}^{k-1})$ for some $i > 0$.

Let $t_0 \in H^2(\mathbb{CP}^\infty)$ and $t \in H^2(\mathbb{CP}^{k-1})$ be the first Chern classes of the canonical line bundles over $\mathbb{CP}^\infty$ and $\mathbb{CP}^{k-1}$ respectively. Then consider a Pontrjagin number of the map f associated with t_0^{k-1} (see page 48 of [6]).

$$\begin{aligned}
&\langle P_{I'}(F \times \mathbb{CP}^{k-1}) f^*(t_0^{k-1}), [F \times \mathbb{CP}^{k-1}]\rangle \\
&= \langle (P_{I'}(F) \otimes 1 + \alpha + \beta)(1 \otimes t^{k-1}), [F \times \mathbb{CP}^{k-1}]\rangle \\
&= \langle P_{I'}(F) \otimes t^{k-1}, [F] \otimes [\mathbb{CP}^{k-1}]\rangle \\
&= \langle P_{I'}(F), [F]\rangle \cdot \langle t^{k-1}, [\mathbb{CP}^{k-1}]\rangle \\
&= \pm \langle P_{I'}(F), [F]\rangle.
\end{aligned}$$

According to Theorem 17.5 of [6], a Pontrjagin number of the map f associated with t_0^{k-1} must vanish. Thus we have shown that every Pontrjagin number of F vanishes. It follows from [12], [13] that F represents the zero element of Ω_m. This completes the proof of Lemma 5.

Lemma 8. Let F_1^m, F_2^m be two oriented compact connected manifolds such that $\partial F_1 = \partial F_2$ and ∂F_1 is a homology sphere. By attaching the two manifolds along their boundaries, we obtain an oriented closed manifold $F = F_1 \cup (-F_2)$. Suppose that F represents the zero element of Ω_m. Then we have

$$w_I[F_1] = w_I[F_2] \text{ for each partition } I \text{ of } m.$$

When $m \equiv 0 \pmod 4$, we have

$P_{I'}[F_1] = P_{I'}[F_2]$ for each non trivial partition I' of $m/4$,

and

$$\sigma(F_1) = \sigma(F_2).$$

Proof of Lemma 8. First we prove in the case of the trivial partition $I = \dim F = m$. It is easy to see that

$$\chi(F_1 \cap F_2) + \chi(F_1 \cup (-F_2)) = \chi(F_1) + \chi(F_2).$$

Since $F_1 \cap F_2$ $(= \partial F_1)$ is a homology sphere, we have

$$\chi(F_1 \cap F_2) \equiv 0 \pmod 2.$$

As is well-known, the mod 2 Euler characteristic is a cobordism invariant, hence we get,

$$\chi(F_1 \cup (-F_2)) \equiv 0 \mod 2.$$

Thus we have proved in the case of the trivial partition $I = \dim F = m$ that

$$W_m[F_1] \equiv \chi(F_1) + 1 \equiv \chi(F_2) + 1 \equiv W_m[F_2] \mod 2.$$

Next we prove in the case of a non trivial partition $I = (i_1, \ldots, i_\ell)$ of m. We introduce some notations. Let $i_a : F_a \longrightarrow F_1 \cup (-F_2)$ $(a = 1, 2)$ be inclusion and let $\pi : F_1 \cup (-F_2) \longrightarrow F_1 \cup (-F_2)/\partial F_1$ be the map obtained by collapsing ∂F_1 to a point. Let $j_a : F_a \longrightarrow F_a/\partial F_a$ $(a = 1, 2)$ be the map obtained by collapsing ∂F_a to a point and $\pi_a : F_1 \cup (-F_2)/\partial F_1 \longrightarrow F_a/\partial F_a$ $(a = 1, 2)$ be the map obtained by collapsing F_x to a point where $x = \{3 - (-1)^a\}/2$. Since

$$j_a^* : H^i(F_a/\partial F_a : Z_2) \longrightarrow H^i(F_a : Z_2)$$

is an isomorphism for $i \leqq m-1$, there exists the unique class ${}_a\hat{W}_i \in H^i(F_a/\partial F_a : Z_2)$ $(a = 1, 2)$ such that $j_a^* {}_a\hat{W}_i = W_i(F_a)$ for $i \leqq m-1$. We shall now show that $\pi^*\{\pi_1^*({}_1\hat{W}_i) + \pi_2^*({}_2\hat{W}_i)\}$ is equal to $W_i(F_1 \cup (-F_2))$ for $i \leqq m-1$. Since

$$i_1^* \oplus i_2^* : H^i(F_1 \cup (-F_2) : Z_2) \longrightarrow H^i(F_1 : Z_2) \oplus H^i(F_2 : Z_2)$$

is an isomorphism for $0 < i \leqq m-1$, an element $x \in H^i(F_1 \cup (-F_2) : Z_2)$ satisfying $i_a^* x = W_i(F_a)$ $(a = 1, 2)$ is nothing but $W_i(F_1 \cup (-F_2))$. We have

$$i_a^* \pi^* \{\pi_1^*({}_1\hat{W}_i) + \pi_2^*({}_2\hat{W}_i)\}$$

$$= i_a^* \pi^* \pi_1^*({}_1\hat{W}_i) + i_a^* \pi^* \pi_2^*({}_2\hat{W}_i)$$

$$= j_a^*({}_a\hat{W}_i)$$

$$= W_i(F_a) \quad \text{for} \quad 0 < i \leqq m-1,$$

since

$$i_a^* \pi^* \pi_{a'}^* = \begin{cases} j_a^* & \text{if} \quad a = a' \\ 0 & \text{if} \quad a \neq a' \end{cases}$$

Therefore we have shown that

$$\pi^* \{\pi_1^*({}_1\hat{W}_i) + \pi_2^*({}_2\hat{W}_i)\} = W_i(F_1 \cup (-F_2)).$$

Let $I = (i_1, \ldots, i_\ell)$ be a non trivial partition of m, then

$$W_I(F_1 \cup (-F_2))$$

$$= \pi^* \{\pi_1^*({}_1\hat{W}_{i_1}) + \pi_2^*({}_2\hat{W}_{i_1})\} \cdots \{\pi_1^*({}_1\hat{W}_{i_\ell}) + \pi_2^*({}_2\hat{W}_{i_\ell})\}$$

$$= \pi^* \{\pi_1^*({}_1\hat{W}_I) + \pi_2^*({}_2\hat{W}_I)\},$$

where ${}_a\hat{W}_I$ $(a = 1, 2)$ means ${}_a\hat{W}_{i_1} \cdots {}_a\hat{W}_{i_\ell}$, since $\pi_a^*({}_a\hat{W}_i) \cdot \pi_{a'}^*({}_{a'}\hat{W}_{i'}) = 0$ for $a \neq a'$. Hence we have

$$\langle W_I(F_1 \cup (-F_2)), [F_1 \cup (-F_2)] \rangle$$

$$= \langle \pi^* \{\pi_1^*({}_1\hat{W}_I) + \pi_2^*({}_2\hat{W}_I)\}, [F_1 \cup (-F_2)] \rangle$$

$$= \langle \pi_1^*({}_1\hat{W}_I) + \pi_2^*({}_2\hat{W}_I), \pi_*[F_1 \cup (-F_2)] \rangle$$

$$= \langle \pi_1^*({}_1\hat{W}_I), \pi_*[F_1 \cup (-F_2)] \rangle + \langle \pi_2^*({}_2\hat{W}_I), \pi_*[F_1 \cup (-F_2)] \rangle$$

$$= \langle {}_1\hat{W}_I, \pi_{1*}\pi_*[F_1 \cup (-F_2)] \rangle + \langle {}_2\hat{W}_I, \pi_{2*}\pi_*[F_1 \cup (-F_2)] \rangle$$

$$= \langle {}_1\hat{W}_I, [F_1/\partial F_1] \rangle + \langle {}_2\hat{W}_I, [-F_2/\partial F_2] \rangle$$

$$= W_I[F_1] - W_I[F_2].$$

According to Thom [12], a Whitney number $\langle W_I(F_1 \cup (-F_2)), [F_1 \cup (-F_2)] \rangle$ must vanish, hence we have shown that

$$W_I[F_1] = W_I[F_2].$$

When $m \equiv 0 \pmod 4$, quite a similar argument shows that

$P_{I'}[F_1] = P_{I'}[F_2]$ for each non trivial partition I' of $m/4$.

The formula

$$\sigma(F_1) = \sigma(F_2)$$

follows from the next well-known facts

$$\sigma(F_1 \cup (-F_2)) = \sigma(F_1) - \sigma(F_2)$$

and

$$\sigma(F_1 \cup -F_2) = 0.$$

(See for examples [1], [8].) This completes the proof of Lemma 8 and simultaneously completes the proof of Theorem 1.

§4. Proof of Corollary 2

Let us recall the explicit description of homotopy spheres in $\Theta_{4p-1+4r}(\partial\pi)$ given by Brieskorn and Hirzebruch [4], [9]:

$$\Sigma^{4p-1+4r}_{3,6k-1} = \{(z_1, \ldots, z_{2p+2r+1}) \in \mathbb{C}^{2p+2r+1} \mid z_1^3 + z_2^{6k-1} + z_3^2 + \ldots + z_{2p+2r+1}^2 = \varepsilon, \ |z_1|^2 + \ldots + |z_{2p+2r+1}|^2 = 1\}$$

where ε is real. Let $\Sigma^{4p-1}_{3,6k-1} \subset \Sigma^{4p-1+4r}_{3,6k-1}$ be the imbedding defined by

$$(z_1, \ldots, z_{2p+1}) \longmapsto (z_1, \ldots, z_{2p+1}, 0, \ldots, 0).$$

Consider the action of S^1 on the last $2r$ variables of $\Sigma^{4p-1+4r}_{3,6k-1}$ defined as follows. Let $A : S^1 \longrightarrow SO(2)$ be the representation defined by

$$A(e^{i\theta}) = \begin{pmatrix} \cos\theta & -\sin\theta \\ \sin\theta & \cos\theta \end{pmatrix}$$

and let $\varphi : S^1 \longrightarrow SO(2r)$ be the representation defined by

$$\varphi(e^{i\theta}) = \begin{pmatrix} A(e^{i\theta}) & & & 0 \\ & A(e^{i\theta}) & & \\ & & \ddots & \\ 0 & & & A(e^{i\theta}) \end{pmatrix}$$

Then S^1 acts on the last $2r$ variables of $\Sigma^{4p-1+4r}_{3,6k-1}$ by means of the representation φ. It is obvious that this action is semi-free and the fixed point set is $\Sigma^{4p-1}_{3,\,6k-1}$. Let us denote this action by $(\Sigma^{4p-1+4r}_{3,\,6k-1}, \varphi_{p,k}, S^1)$. We shall now show that if $k \neq k'$, then $(\Sigma^{4p-1+4r}_{3,\,6k-1}, \varphi_{p,k}, S^1)$ is not equivalent to $(\Sigma^{4p-1+4r}_{3,\,6k'-1}, \varphi_{p,k'}, S^1)$. The manifold

$$W^{4p+4r}_{3,\,6k-1} = \{(z_1, \ldots, z_{2p+2r+1}) \in \mathbb{C}^{2p+2r+1} \mid z_1^3 + z_2^{6k-1} + z_3^2 + \ldots + z^2_{2p+2r+1} = \varepsilon,\ |z_1|^2 + \ldots + |z_{2p+2r+1}|^2 \leqq 1\}$$

admits a semi-free S^1-action defined similarly as above. We denote it by $(W^{4p+4r}_{3,\,6k-1}, \Phi_{p,k}, S^1)$. Then the restriction $\partial(W^{4p+4r}_{3,\,6k-1}, \Phi_{p,k}, S^1)$ of the action $(W^{4p+4r}_{3,\,6k-1}, \Phi_{p,k}, S^1)$ to the boundary $\partial W^{4p+4r}_{3,\,6k-1}$ is nothing but $(\Sigma^{4p-1+4r}_{3,\,6k-1}, \varphi_{p,k}, S^1)$. The fixed point set of the action $(W^{4p+4r}_{3,\,6k-1}, \Phi_{p,k}, S^1)$ is $W^{4p}_{3,\,6k-1}$ and the normal complex bundle of the fixed point set is trivial. Therefore these satisfy the conditions (i) and (ii) in §2. On page 14 of [4] there is the following theorem:

Theorem (Brieskorn)

$$\sigma(W^{4p}_{3,\,6k-1}) = (-1)^p 8k.$$

It follows from Theorem 1 that $(\Sigma^{4p-1+4r}_{3,\,6k-1}, \varphi_{p,k}, S^1)$ is not equivalent to $(\Sigma^{4p-1+4r}_{3,\,6k'-1}, \varphi_{p,k'}, S^1)$ if $k \neq k'$. This completes the proof of Corollary 2.

References

[1] Atiyah, M. F. and Singer, I. M., Index of elliptic operators III, Ann. of Math. 87 (1968), 546-604.

[2] Borel, A. and Hirzebruch, F., Characteristic classes and homogeneous spaces, I, Amer. J. Math. 80 (1958), 458-538.

[3] Bott, R., Notes on K-theory. Harvard University, Cambridge, Mass., 1962.

[4] Brieskorn, E., Beispiele zur Differentiallopologie von Singularitäten, Inventiones Math. 2 (1966), 1-14.

[5] Browder, W. and Petrie, T., Semi-free and quasi-free S^1-actions on homotopy spheres, Memoires dédiés à George DeRham, Springer-Verlag (1970), 136-146.

[6] Conner, P. E. and Floyd, E. E., Differentiable periodic maps, Springer-Verlag, 1964.

[7] Dold, A., Relations between ordinary and extraordinary homology, Colloquium on Algebraic Topology, Aarhus Universitet, 1962.

[8] Hirzebruch, F., Neue topologische Methoden in der algebraischen Geometrie, Springer-Verlag, 1956.

[9] ————, Singularities and exotic spheres, Séminaire Bourbaki, 1966/67, No. 314.

[10] Kawakubo, K., Invariants for semi-free S^1-actions, Bull. A.M.S. (to appear).

[11] Milnor, J. W., Lectures on Characteristic Classes, mimeographed notes. Princeton, 1958.

[12] Thom, R., Quelque propriétés globales des variétés différentiables, Commentarii Math. Helv. 28 (1954), 17-86.

[13] Wall, C. T. C., Determination of the cobordism ring, Ann. of Math. 72 (1960), 292-311.

TOPOLOGICAL S^1 AND Z_{2k} ACTIONS ON SPHERES

K. Kawakubo*
Osaka University and Institute for Advanced Study

In this note we shall study some aspects of topological actions on homotopy spheres. Let X be a finite Poincaré complex. Then a homotopy equivalence $f : M \longrightarrow X$, where M is a compact smooth (or PL or topological respectively) manifold, will be said to define a smooth (or PL or topological respectively) structure on X. The map $f' : M' \longrightarrow X$ defines the same structure if there is a diffeomorphism (or PL-homeomorphism or homeomorphism respectively) $h : M \longrightarrow M'$ such that $f \simeq f' \circ h$. Thus a structure is an equivalence class of pairs (M,f). We denote the set of structures on X by $\mathcal{S}_{Diff}(X)$ (or $\mathcal{S}_{PL}(X)$ or $\mathcal{S}_{Top}(X)$ respectively). Obviously there are natural maps:

$$\mathcal{S}_{Diff}(X) \xrightarrow{\varphi} \mathcal{S}_{PL}(X) \xrightarrow{\psi} \mathcal{S}_{Top}(X).$$

We denote by $\mathcal{S}^C_{Top}(X)$ the set $\mathcal{S}_{Top}(X) - \{\text{Image } \psi\}$.

First we consider the case where X is a complex projective space $\mathbb{CP}^n$. In this case, φ is neither surjective nor injective in general (see [5],[1]). Sullivan showed that the map

$$\mathcal{S}_{PL}(\mathbb{CP}^n) \xrightarrow{\psi} \mathcal{S}_{Top}(\mathbb{CP}^n)$$

is injective for $n > 2$ and classified $\mathcal{S}_{PL}(\mathbb{CP}^n)$ by the characteristic variety theorem [5]. On the other hand, Top transversality and Top surgery due to Kirby-Siebenmann [3][4] enable us to show

$$\mathcal{S}_{Top}(\mathbb{CP}^n) \xleftrightarrow{\text{bijective}} [\mathbb{CP}^{n-1}, F/Top]$$

by the same way as in [5]. Therefore by making use of the following exact sequence

*Supported in part by an NSF grant.

$$\longrightarrow [\mathbb{CP}^{n-1}, Top/PL] \longrightarrow [\mathbb{CP}^{n-1}, F/PL]$$

$$\longrightarrow [\mathbb{CP}^{n-1}, F/Top] \longrightarrow [\mathbb{CP}^{n-1}, B\, Top/PL],$$

it is not hard to show

Theorem 1. *We can give an abelian group structure on* $\mathcal{S}_{Top}(\mathbb{CP}^n)$ $(n > 2)$ *such that the following is an exact sequence of abelian groups*

$$0 \longrightarrow \mathcal{S}_{PL}(\mathbb{CP}^n) \xrightarrow{\psi} \mathcal{S}_{Top}(\mathbb{CP}^n) \longrightarrow Z_2 \longrightarrow 0.$$

Next we consider the case where X is a lense space $L^{2n+1}(2k) = S^{2n+1}/Z_{2k}$. Let $f : \widetilde{\mathbb{CP}}^n \longrightarrow \mathbb{CP}^n$ be an element of $\mathcal{S}_{Top}(\mathbb{CP}^n)$ and let $\eta_k : S^1 \longrightarrow L^{2n+1}(2k) \xrightarrow{\pi} \mathbb{CP}^n$ be the natural fibration. Consider the following commutative diagram

$$\begin{array}{ccc}
S^1 & & S^1 \\
\downarrow & & \downarrow \\
\tilde{L}^{2n+1}(2k) & \xrightarrow{\tilde{f}} & L^{2n+1}(2k) \\
\tilde{\pi}\downarrow \xi_k & & \pi \downarrow \eta_k \\
\widetilde{\mathbb{CP}}^n & \xrightarrow{f} & \mathbb{CP}^n
\end{array}$$

where ξ_k is the induced bundle $f^!\eta_k$ and $\tilde{L}^{2n+1}(2k)$ denotes the total space of ξ_k and $\tilde{f}$ denotes the bundle map and $\tilde{\pi}$ is the projection map. Since a topological manifold $\tilde{L}^{2n+1}(2k)$ has a homotopy type of a CW-complex, $\tilde{f}$ gives a homotopy equivalence by [6]. Then the map

$$\alpha_k : \mathcal{S}_{Top}(\mathbb{CP}^n) \longrightarrow \mathcal{S}_{Top}(L^{2n+1}(2k))$$

is defined by

$$\alpha_k([\widetilde{\mathbb{CP}}^n, f]) = [\tilde{L}^{2n+1}(2k), \tilde{f}].$$

It is easy to prove that α_k is well-defined.

Theorem 2. $\alpha_k(\mathcal{S}^C_{Top}(\mathbb{CP}^n)) \subset \mathcal{S}^C_{Top}(L^{2n+1}(2k))$ *for* $n > 2$ *and* $k > 0$.

By combining Theorem 1 and Theorem 2, we have

Corollary. *There exist non-triangulable homotopy lense spaces* $\tilde{L}^{2n+1}(2k)$ *for* $n > 2$ *and* $k > 0$.

Proof of Theorem 2. First we consider the case $k = 1$. Let $[\tilde{L}^{2n+1}(2), \tilde{f}]$ be $\alpha_1([\widetilde{\mathbb{CP}}^n, f])$ where $[\widetilde{\mathbb{CP}}^n, f]$ represents an element of $\mathcal{S}^C_{Top}(\mathbb{CP}^n)$. Denote by $\tau(M)$, the tangent microbundle of M. Then we shall show that the tangent microbundle $\tau(\tilde{L}^{2n+1}(2))$ cannot reduce to a PL-microbundle. The tangent microbundle $\tau(\tilde{L}^{2n+1}(2))$ satisfies

$$\tau(\tilde{L}^{2n+1}(2)) \underset{\text{stably}}{\cong} \tilde{\pi}^{!}(\xi_1) \oplus \tilde{\pi}^{!}(\tau(\widetilde{\mathbb{CP}}^n)).$$

According to Kirby [2], the non zero element $\mathcal{O} \in H^4(\mathbb{CP}^n, Z_2)$ represents the obstruction to reducing $\tau(\widetilde{\mathbb{CP}}^n)$ to a PL-microbundle. The Gysin exact sequence

$$\longrightarrow H^{i-2}(\widetilde{\mathbb{CP}}^n; Z_2) \longrightarrow H^i(\widetilde{\mathbb{CP}}^n; Z_2) \xrightarrow{\tilde{\pi}^*} H^i(\tilde{L}^{2n+1}(2); Z_2)$$
$$\longrightarrow H^{i-1}(\widetilde{\mathbb{CP}}^n; Z_2) \longrightarrow$$

shows that

$$\tilde{\pi}^* : H^4(\widetilde{\mathbb{CP}}^n; Z_2) \longrightarrow H^4(\tilde{L}^{2n+1}(2); Z_2)$$

is an isomorphism. Therefore, the obstruction $\tilde{\pi}^*(\mathcal{O})$ to reducing $\tilde{\pi}^{!}(\tau(\widetilde{\mathbb{CP}}^n))$ to a PL-microbundle is non zero. Since $\tilde{\pi}^{!}(\xi_1)$ is a vector bundle, we may conclude that $\tau(\tilde{L}^{2n+1}(2))$ cannot reduce to a PL-microbundle, that is, the manifold $\tilde{L}^{2n+1}(2)$ has no PL-structure [2].

Secondly, we prove in general. Let $[\tilde{L}^{2n+1}(2k), \tilde{f}]$ be $\alpha_k([\widetilde{\mathbb{CP}}^n, f])$ where $[\widetilde{\mathbb{CP}}^n, f]$ represents an element of $\mathcal{S}^C_{Top}(\mathbb{CP}^n)$. Then it is easy to see that $\tilde{L}^{2n+1}(2)$ admits a free Z_k-action whose orbit space $\tilde{L}^{2n+1}(2)/Z_k$ is homeomorphic to $\tilde{L}^{2n+1}(2k)$. Hence $\tilde{L}^{2n+1}(2k)$ has no PL-structure, since $\tilde{L}^{2n+1}(2)$ has no PL-structure.

Added in proof. Professor W. Browder and Professor S. Lopez de Medrano kindly informed me that Siebenmann has classified $\mathcal{S}^S_{Top}(\mathbb{RP}^n)$ where $\mathcal{S}^S_{Top}$ is a similar set as $\mathcal{S}_{Top}(\mathbb{RP}^n)$ but an element of $\mathcal{S}^S_{Top}(\mathbb{RP}^n)$ is a "simple" homotopy equivalence $f : M \longrightarrow \mathbb{RP}^n$.

REFERENCES

[1] Kawakubo, K. Inertia groups of low dimensional complex projective spaces and some free differentiable actions on spheres. I. Proc. Japan Acad. 44 (1968), 873-875.

[2] Kirby, R. C., Lectures on triangulation of manifolds, Mimeo. Univ. of Calif., Los Angeles, 1969.

[3] Kirby, R. C., and Siebenmann, L. C. Foundations of topology, Notices Amer. Math. Soc., 16, 1969, 698.

[4] ———————————— to appear.

[5] Sullivan, D., Triangulating and smoothing homotopy equivalences and homeomorphisms. (seminar notes) Princeton Univ., 1967.

[6] Whitehead, J. H. C., Combinatorial homotopy I, Bull. Amer. Math. Soc. 55 (1949), 213-245.

CHARACTERISTIC INVARIANTS OF FREE DIFFERENTIABLE ACTIONS OF S^1 AND S^3 ON HOMOTOPY SPHERES

Hsu-Tung Ku and Mei-Chin Ku

University of Massachusetts, Amherst, Mass.

1. Introduction

The purpose of this paper is to study free differentiable actions of S^1 and S^3 on homotopy spheres by investigating the characteristic invariants and the existence of characteristic homotopy spheres.

We shall denote by (S^1, Σ^{2n+1}) (resp. (S^3, Σ^{4n+3})) a free differentiable action of S^1 (resp. S^3) on homotopy $(2n+1)$-sphere Σ^{2n+1} (resp. $(4n+3)$-sphere Σ^{4n+3}), and $M = \Sigma^{2n+1}/S^1$ (resp. $N = \Sigma^{4n+3}/S^3$) the orbit space.

Definition 1.1. Let (S^1, Σ^{2n+1}) (resp. (S^3, Σ^{4n+3})) be a free differentiable action. There is a homotopy equivalence

$$f:\ M \to CP^n \text{ (resp. } f:\ N \to QP^n),$$

which is transverse regular on the submanifold CP^{n-k} (resp. QP^{n-k}) with $n-k > 2$ (resp. $n-k > 1$) and let $M' = f^{-1}(CP^{n-k})$ (resp. $f^{-1}(QP^{n-k})$). By a characteristic homotopy $(2n-2k+1)$-sphere (resp. $(4n-4k+3)$-sphere) of Σ^{2n+1} (resp. Σ^{4n+3}) we mean a homotopy sphere $\Sigma^{2n-2k+1}$ (resp. $\Sigma^{4n-4k+3}$) which is S^1 (resp. S^3) invariant, and such that $M' = \Sigma^{2n-2k+1}/S^1$ (resp. $M' = \Sigma^{4n-4k+3}/S^3$) with $f|M'$ is a homotopy equivalence. We shall assume that $\dim M' = 0 \pmod 4$ throughout the paper. The characteristic invariants $I_{2k}(S^1, \Sigma^{2n+1})$ (resp. $I_{4k}(S^3, \Sigma^{4n+3})$) of the free differentiable action (S^1, Σ^{2n+1}) (resp. (S^3, Σ^{4n+3})) is defined by

$$I_{2k}(S^1, \Sigma^{2n+1}) = \sigma(M') - \sigma(CP^{n-k}),$$

$$(\text{resp. } I_{4k}(S^3, \Sigma^{4n+3}) = \sigma(M') - \sigma(QP^{n-k})),$$

During the preparation of this paper, both authors were partially supported by NSF Grant NSF-GP-19854.

which lies in the group 8 Z, where σ (X) denotes the index of the smooth manifold X, and Z the ring of integers.

The characteristic invariant $I_{2k}(S^1, \Sigma^{2n+1})$ (resp. $I_{4k}(S^3, \Sigma^{4n+3})$) is the obstruction of the free differentiable S^1 (resp. S^3) action on Σ^{2n+1} (resp. Σ^{4n+3}) having codimension 2k (resp. 4k) characteristic homotopy sphere.

As to notation we shall denote by ρ and $\bar{\rho}$ the canonical bundles over $M = \Sigma^{2n+1}/S^1$ and $N = \Sigma^{4n+3}/S^3$ associated to the principal bundles $S^1 \to \Sigma^{2n+1} \to M$ and $S^3 \to \Sigma^{4n+3} \to N$ respectively, ρ^{-1} and $\bar{\rho}^{-1}$ the additive inverse of ρ and $\bar{\rho}$, $\alpha \in H^2(M,Z)$, $\bar{\alpha} \in H^4(N,Z)$ the generators such that $\langle \alpha^n, [M] \rangle = 1$ and $\langle \bar{\alpha}^n, [N] \rangle = 1$. The L-genus and $\hat{\mathfrak{A}}$-genus of $\xi \in (KO)^\sim(X)$ are the genera belonging to the characteristic power series $x(\tanh x)^{-1}$ and $(x/2)(\sinh x/2)^{-1}$ respectively (cf. [8]):

$$L: (KO)^\sim(X) \to \sum_{i \geq 0} H^{4i}(X;Q),$$

$$\hat{\mathfrak{A}}: (KO)^\sim(X) \to \sum_{i \geq 0} H^{4i}(X;Q).$$

Theorem 1.2 (Browder [4] [14]). *Let (S^1, Σ^{2n+1}) or (S^3, Σ^{4n+3}) be a free differentiable action. Then*

$$I_{2k}(S^1, \Sigma^{2n+1}) = \langle \alpha^k L(\tau_M) L(\rho^{-1})^k, [M] \rangle - 1,$$

$$I_{4k}(S^3, \Sigma^{4n+3}) = \langle \bar{\alpha}^k L(\tau_N) L(\bar{\rho}^{-1})^k, [N] \rangle - \sigma(QP^{n-k}),$$

where τ_M and τ_N denote the tangent bundles of M and N respectively. In particular, the action has a characteristic homotopy (2n+1-2k)-sphere, or (4n+3-4k)-sphere if and only if $\langle \alpha^k L(\tau_M) L(\rho^{-1})^k, [M] \rangle = 1$, or $\langle \bar{\alpha}^k L(\tau_N) L(\bar{\rho}^{-1})^k, [N] \rangle = \sigma(QP^{n-k})$ respectively.

Proposition 1.3 (cf. [14] [16]). *There are infinitely many topologically distinct free differentiable actions of S^1 (resp. S^3) on homotopy (2n+1)-spheres (resp. (4n+3)-spheres) with codimension 2 (resp. 4)* *characteristic homotopy spheres for every* $n \geq 5$.

Proposition 1.4 (cf. [14] [16]). *There are infinitely many topologically distinct free differentiable* S^1 (*resp.* S^3) *actions on homotopy* $(2n+1)$-*spheres* (*resp.* $(4n+3)$-*spheres*), $n \geq 5$, *so that none of them has a codimension* 2 (*resp.* 4) *characteristic homotopy spheres.*

2. Characteristic Invariants

Let us outline briefly Sullivan's point of view on the simply-connected surgery (cf. [21], [22]). Given a smooth m-manifold M^m. A *homotopy smoothing* on M is a pair (L,f), where L is a smooth m-manifold and $f: (L, \partial L) \to (M \partial M)$, is a homotopy equivalence. Two homotopy smoothings $f_i: (L_i, \partial L_i) \to (M, \partial M)$, $i = 0, 1$ are *equivalent* if there is a diffeomorphism $c: (L_o, \partial L_o) \to (L_1, \partial L_1)$ so that f_o is homotopic to $f_1 c$. The set of equivalence classes of homotopy smoothings of M is denoted by hS(M). Let B_O and B_G be the classifying spaces for stable vector bundles and the stable spherical fibration modulo fibre homotopy equivalence, respectively. Define the space G/O to be the fibre of the natural map $B_O \to B_G$. A G/O-*bundle* over X is a pair (ξ,t) consisting of a stable vector bundle ξ over X and a fibre homotopy trivialization t of the associated sphere bundle $S(\xi)$. Two G/O-bundles (ξ_o,t_o) and (ξ_1,t_1) are *equivalent* if there is an G/O bundle (ξ,t) over $X \times I$ and a bundle isomorphism $f_j:\xi_j \cong \xi|X \times j$, $j = 1, 2$, such that $tf_j \sim t_j$. We may identify the equivalence classes of G/O-bundles with $[X,G/O]$.

Now let $i: CP^{n-k} \to CP^n$ (resp. $QP^{n-k} \to QP^n$) be the inclusion. Then i induces

$$i^*: [CP^n, G/O] \to [CP^{n-k}, G/O]$$

$$(\text{resp. } i^*: [QP^n, G/O] \to [QP^{n-k}, G/O]).$$

We have the following exact sequences of pointed sets

$$0 \to hS\,(CP^n) \xrightarrow{\theta} [CP^n, G/O] \xrightarrow{s_{2n}} P_{2n},$$

and

$$0 \to hS\,(QP^n) \xrightarrow{\theta} [QP^n,\ G/O] \xrightarrow{s_{4n}} P_{4n}.$$

Theorem 2.1. [17]. *Let* (S^1, Σ^{2n+1}) *or* (S^3, Σ^{4n+3}) *be a free differentiable action so that* $[M,f] \in hS\,(CP^n)$ *or* $[N,f] \in hS\,(QP^n)$.

Then

$$I_{2k}\,(S^1, \Sigma^{2n+1}) = 8\, s_{2(n-k)}\; i^*\, \theta\, [M,f],$$

$$I_{4k}\,(S^3, \Sigma^{4n+3}) = 8\, s_{4(n-k)}\; i^*\, \theta\, [N,f].$$

We assume that $I_o\,(S^1, \Sigma^{2n+1}) = s_{2n}\,\theta\,[M,f] = 0$, *and* $I_o\,(S^3, \Sigma^{4n+3}) = s_{4n}\,\theta\,[N,f] = 0$.

Proof: We only give the proof for S^1 case. Let $f: N \to CP^n$ be the homotopy equivalence which is transverse regular on CP^{n-k}, and $M' = f^{-1}\,(CP^{n-k})$. Denote $f' = f|M'$. Then $\bar{f}i = i\bar{f}'$, where $i: M' \to M$, and $i: CP^{n-k} \to CP^n$ are inclusions and $\bar{f}$, $\bar{f}'$ are homotopy inverse of f and f' respectively. By definition $I_{2k}\,(S^1, \Sigma^{2n+1}) = \sigma\,(M') - \sigma\,(CP^{n-k})$. But $\theta\,[M,f] = (\bar{f}^*\,\tau_M - \tau_{CP^n},\ t)$

and

$$i^*\,(\bar{f}^*\,\tau_M - \tau_{CP^n}) = \bar{f}'^*\,\tau_{M'} - \tau_{CP^{n-k}}\,.$$

Hence the result is an easy consequence of the formula

$$s_{2(n-k)}\,(\xi,t) = (1/8)\,\langle L(CP^{n-k})\cdot(L(\xi)-1),\ [CP^{n-k}]\rangle.$$

We compute some examples in dimensions 11 and 15:

Proposition 2.2. *Let* (S^1, Σ^{11}) *be a free differentiable action. Then*

$$I_2\,(S^1, \Sigma^{11}) = 32\,(18i^2 + 2i + 7j)$$

for some integers i *and* j.

Proof: Since $k\,\Sigma_M^{11}$, k odd, admits no free differentiable S^1-action, where Σ_M^{11} is the Milnor sphere (cf. [5]), the result follows from [13].

Proposition 2.3. [13]. *Let* (S^3, Σ^{11}) *be a free differentiable action and* S^1 *is a subgroup of* S^3. *Then either* $I_2\ (S^1, \Sigma^{11}) = 224k$, *or* $I_2\ (S^1, \Sigma^{11}) = 224k + 64$ *for some integer* k.

Proposition 2.4. (1) *Let* (S^3, Σ^{15}) *be a free differentiable action. Then*

$$I_4\ (S^3, \Sigma^{15}) = 64\ (7j - 16i^2 - 9i)/45,$$

for some integers i *and* j *(where* i *and* j *are both even or odd). Thus*

$$I_4\ (S^3, \Sigma^{15}) \equiv 0 \bmod 128.$$

(2) *Let* $\beta: \theta_{15} = Z_2 \oplus bP_{16} \to bP_{16}$ *be the projection, and* $\Sigma^{15} \in \theta_{15}$ *be such that* $\beta\ (\Sigma^{15}) = k\ \Sigma^{15}_M$, *where* Σ^{15}_M *is the Milnor sphere. If* $k \not\equiv 0 \bmod 8$, *then* Σ^{15} *admits no free differentiable* S^3*-actions.*

Proof: Let $\bar{\alpha} \in H^4\ (\Sigma^{15}/S^3, Z)$, $\alpha \in H^2\ (\Sigma^{15}/S^1, Z)$ be generators as in section 1. We can show that

$$p_1\ (\Sigma^{15}/S^3) = (32i + 4)\ \bar{\alpha},$$

$$p_2\ (\Sigma^{15}/S^3) = (64j + 12)\ \bar{\alpha}^{\,2},$$

and

$$p_3\ (\Sigma^{15}/S^3) = (64k + 8)\ \bar{\alpha}^{\,3},$$

for some integers i, j and k with i, j and k are all even or odd by a tedious computation from the following:

(3) $$I_4(S^3, \Sigma^{15}) \equiv 0 \bmod 8,$$

(4) $$I_6(S^1, \Sigma^{15}) \equiv 0 \bmod 8,$$

(5) $$\sigma\ (\Sigma^{15}/S^3) = 0,$$

(6) $\hat{\mathfrak{A}}_3\ (\Sigma^{15}/S^3)\ [\Sigma^{15}/S^3]$ is an even integer,

(7) $\langle e^{2\alpha} \hat{\pi} (\Sigma^{15}/S^1), [\Sigma^{15}/S^1]\rangle - 2 \langle e^{\alpha} \hat{\pi} (\Sigma^{15}/S^1), [\Sigma^{15}/S^1]\rangle$ is an integer.

Thus we can compute $I_4 (S^3, \Sigma^{15})$ easily.

For the proof of (2), we can see that the Eells-Kuiper μ-invariant (cf. [7]) is given by

$$\mu (\Sigma^{15}) \equiv n/8.127 \bmod 1,$$

for some integer n. This completes the proof.

3. An identity for characteristic invariants

Theorem 3.1. Let S^3 act freely and differentiably on a homotopy (4n+3)-sphere Σ^{4n+3} ($n \geq 3$), and let S^1 be a subgroup of S^3. Then

$$I_{4k+2} (S^1, \Sigma^{4n+3}) = I_{4k} (S^3, \Sigma^{4n+3}) + I_{4k+4} (S^3, \Sigma^{4n+3}),$$

for $0 \leq k < n-2$.

Proof. There is a fibration

$$\eta: S^2 \to M \xrightarrow{\pi} N.$$

It is not difficult to prove that

$$\langle x, [N]\rangle = \langle \alpha\, \pi^* x, [M]\rangle$$

for $x \in H^* (N,Z)$. It is known that the total Pontrjagin classes of ρ and η are given as follows: $p (\rho) = 1+\alpha^2$, and $p (\eta) = 1 + 4 \bar{\alpha}$. Thus

$$L (\rho) = \alpha/\tanh \alpha,$$

$$\pi^* L (\eta^{-1}) = \tanh 2\alpha/2\alpha = \tanh \alpha/\alpha \, (1 + \tanh^2 \alpha),$$

and

$$\pi^* L (\eta^{-1}) \cdot L (\rho) \{1 + \alpha^2 L (\rho^{-1})^2\} = 1.$$

Applying Theorem 1.2, we have

$$I_{4k}(S^3, \Sigma^{4n+3}) + I_{4k+4}(S^3, \Sigma^{4n+3})$$
$$= \langle \bar{\alpha}^{k} L(\tau_N) L(\bar{\rho}^{-1})^{k}, [N]\rangle + \langle \bar{\alpha}^{k+1} L(\tau_N) L(\bar{\rho}^{-1})^{k+1}, [N]\rangle - 1$$
$$= \langle \bar{\alpha}^{k} L(\tau_N) L(\bar{\rho}^{-1})^{k} \{1 + \bar{\alpha} L(\bar{\rho}^{-1})\}, [N]\rangle - 1$$
$$= \langle \alpha^{2k+1} \pi^{*} L(\tau_N) L(\bar{\rho}^{-1})^{k} \{1 + \bar{\alpha} L(\bar{\rho}^{-1})\}, [M]\rangle - 1$$
$$= \langle \alpha^{2k+1} L(\tau_M) L(\rho^{-1})^{2k+1} \cdot \pi^{*} L(\eta^{-1}) \cdot L(\rho) \{1 + \alpha^2 L(\rho^{-1})\}, [M]\rangle - 1$$
$$= \langle \alpha^{2k+1} L(\tau_M) L(\rho^{-1})^{2k+1}, [M]\rangle - 1$$
$$= I_{4k+2}(S^1, \Sigma^{4n+3}).$$

Remark. A different proof is outlined in [13]. The theorem also can be proved by using the Hirzebruch virtual index (cf. [6], [8]).

Corollary 3.2. *Suppose that* S^3 *acts freely and differentiably on a homotopy* $(4n+3)$*-sphere* Σ^{4n+3} $(n \geq 3)$, *and* $Z_2 \subset S^1 \subset S^3$. *Then*

$$I_4(S^3, \Sigma^{4n+3}) = I_2(S^1, \Sigma^{4n+3}) = I(Z_2, \Sigma^{4n+3}),$$

where $I(Z_2, \Sigma^{4n+3})$ *denotes the Browder-Livesay invariant.*

Proof. The second equality is a result of Montgomery-Yang (cf. [19]).

Corollary 3.3. *Let* Z_2 *act freely and differentiably on a homotopy* 11-*sphere* Σ^{11}. *If* $I(Z_2, \Sigma^{11}) \not\equiv 0 \bmod 32$, *then this action cannot be extended to the free action of* S^1.

Proof. By Proposition 2.2.

Corollary 3.4. *Let* Z_2 *act freely and differentiably on a homotopy* 15-*sphere* Σ^{15}. *If* $I(Z_2, \Sigma^{15}) \not\equiv 0 \bmod 128$, *then this action can not be extended to the free differentiable action of* S^3.

Proof. By Proposition 2.4.

Conjecture. $I_{2k}\,(S^1, \Sigma^{2n+1}) \equiv 0 \bmod 16$, $(n \geq 6)$ and

$I_{4k}\,(S^3, \Sigma^{4n+3}) \equiv 0 \bmod 16$, $(n \geq 4)$ for all k.

4. Characteristic Invariants and Characteristic Homotopy Spheres

Following Brumfiel (cf. [5]) we define the maps α: $[CP^n, G/O] \to \theta_{2n+1}$ as follows: Let CP_o^{n+1} be CP^{n+1} with a disk D^{2n+2} removed. Then CP_o^{n+1} is the total space of the D^2 bundle H over CP^n, $H: CP_o^{n+1} \to CP^n$. Define α, by the compositions

$$\alpha: [CP^n, G/O] \underset{\approx}{\xrightarrow{H^*}} [CP_o^{n+1}, G/O] \underset{\approx}{\xrightarrow{\theta}} hS\,(CP_o^{n+1}) \xrightarrow{j} \theta_{2n+1},$$

where j is the map which assigns to a homotopy smoothing of CP_o^{n+1} its boundary, which is a homotopy sphere. Geometrically, if (S^1, Σ^{2n+1}) is a free differentiable action, $\alpha\,\theta\,[\Sigma^{2n+1}/S^1, f] = \Sigma^{2n+1}$ (cf. [5], Lemma 1.1).

Lemma 4.1 [5]. *Let $\xi \in [CP^{2k-1}, G/O]$ with $\xi = i^*(\hat{\xi})$, $\hat{\xi} \in [CP^{2k}, G/O]$, Then $\alpha\,(\xi) \equiv s_{4k}\,(\hat{\xi}) \in Z \bmod |b\,P_{4k}|Z \in b\,P_{4k}$, where $|b\,P_{4k}|$ denotes the order of the group $b\,P_{4k}$.*

Theorem 4.2. (1). *Let S^1 act freely and differentiably on a homotopy $(2n+1)$-sphere Σ^{2n+1} with characteristic homotopy $(4k+3)$-sphere Σ^{4k+3}. Then*

$$\Sigma^{4k+3} \equiv (1/8)\,I_{2(n-2k-2)}\,(S^1, \Sigma^{2n+1}) \bmod |b\,P_{4k+4}|Z \in b\,P_{4k+4}.$$

(2). *Let S^3 act freely and differentiably on homotopy $(4n+3)$-sphere Σ^{4n+3} with characteristic homotopy $(4k+3)$-sphere Σ^{4k+3}. Then*

$$\Sigma^{4k+3} \equiv (1/8)\,I_{4(n-k-1)}\,(S^3, \Sigma^{4n+3}) \bmod |b\,P_{4k+4}|Z \in b\,P_{4k+4}.$$

Proof. (1). Let $(P^{2n}, f) \in hS\,(CP^n)$ be corresponding to the free action (S^1, Σ^{2n+1}) so that it has characteristic homotopy $(4k+3)$-sphere Σ^{4k+3}. Let $P^{4k+2} = \Sigma^{4k+3}/S^1$. Then the following diagram commutes:

$$\begin{array}{ccc} P^{2n} & \xrightarrow{\quad f \quad} & CP^{n} \\ i \uparrow & & \uparrow i \\ P^{4k+2} & \xrightarrow[f'=f|P^{4k+2}]{} & CP^{2k+1} \end{array}$$

The class $[P^{4k+2}, f']$ in $hS\ (CP^{2k+1})$ is uniquely determined by $[P^{2n}, f]$. Thus we may define $hS^{n-2k}\ (CP^n)$ as the subset of $hS(CP^n)$ consisting of actions with codimension 2 (n-2k) characteristic homotopy spheres. Clearly the diagram below is commutative:

$$\begin{array}{ccc} hS^{n-2k}(CP^n) & \xrightarrow{\ \theta\ } & [CP^n, G/O] \\ i^* \downarrow & & \downarrow i^* \\ hS(CP^{2k+1}) & \xrightarrow{\ \theta\ } & [CP^{2k+1}, G/O] \end{array}$$

Let $\hat{\xi} = \theta\ [P^{2n}, f]$, and $\xi = \theta\ i^*\ [P^{2n}, f]$, then $\xi = i^*\ (\hat{\xi})$

Consider the maps

$$[CP^n, G/O] \xrightarrow{i^*} [CP^{2k+2}, G/O] \xrightarrow{i^*} [CP^{2k+1}, G/O].$$

Set $\hat{\xi}' = i^*\ (\hat{\xi})$ and $\xi = i^*\ (\hat{\xi}')$, where $\hat{\xi}' \in [CP^{2k+2}, G/O]$ and $\xi \in [CP^{2n+1}, G/O]$. Then

$$\Sigma^{4k+3} = \alpha\ (\theta[P^{4k+2}, f']) = \alpha\ (\xi) = \alpha\ i^*\ (\hat{\xi}')$$

$$\equiv s_{4k+4}\ i^*\ \theta\ [P^{2n}, f] \bmod |b\ P_{4k+4}|Z$$

$$\equiv (1/8) I_{2(n-2k-2)}\ (S^1, \Sigma^{2n+1}) \bmod |b\ P_{4k+4}|Z \in b\ P_{4k+4}$$

by Theorem 2.1 and Lemma 4.1.

(2). From the hypotheses, we have $I_{4(n-k)}\ (S^3, \Sigma^{4n+3}) = 0$. Let S^1 be a subgroup of S^3. Then by Theorem 3.1, we have

$$I_{4(n-k-1)}\ (S^3, \Sigma^{4n+3}) = I_{4(n-k)-2}\ (S^1, \Sigma^{4n+3}).$$

Thus by (1),

$$\Sigma^{4k+3} \equiv (1/8) I_{4(n-k)-2} (S^1, \Sigma^{4n+3}) \mod |b\, P_{4k+4}| Z$$

$$\equiv (1/8) I_{4(n-k-1)} (S^3, \Sigma^{4n+3}) \mod |b\, P_{4k+4}| Z.$$

Corollary 4.3. *Let (S^3, Σ^{4n+3}) be a free differentiable action with codimensions 4 and 8 S^3-characteristic homotopy spheres. Then there is a codimension 6 S^1-characteristic homotopy sphere which is standard.*

Corollary 4.4 (cf. [15], [17]). *For any free differentiable S^3 (resp. S^1) action on homotopy $(4n+3)$-sphere Σ^{4n+3}, the codimension 4 (resp. 2) characteristic homotopy sphere is always standard.*

In fact, Montgomery and Yang has shown that the codimension 2 S^1-characteristic homotopy spheres are always standard (cf. [18]).

Corollary 4.5 (cf. [15], [17]). *There are infinitely many topologically distinct free differentiable actions of S^3 (resp. S^1) on standard sphere S^{4n+3} (resp. S^{2n+1}) for every $n \geq 4$ (resp. $n \geq 9$).*

Proof. We only give the proof for S^3 actions. By Proposition 1.3, there are infinitely many topologically distinct free differentiable S^3 actions on some homotopy $(4n+7)$-spheres Σ^{4n+7}, $n \geq 4$, with codimension 4 characteristic homotopy spheres. The codimension 4 characteristic homotopy spheres are always standard by Corollary 4.4. The restricted actions to the codimension 4 characteristic sphere S^{4n+3} have different rational Pontrjagin classes. The result follows.

For S^1 actions on S^{4n+3}, the result can be improved. (cf. Corollary 4.12).

Corollary 4.6. (1). *Suppose that (S^3, Σ^{4n+3}) is a free differentiable action with characteristic homotopy 11-sphere. Then*

$$I_{4(n-3)} (S^3, \Sigma^{4n+3}) \equiv 0 \mod 256.$$

(2). Let (S^3, Σ^{4n+3}) be a free differentiable action such that $I_{4(n-3)}(S^3, \Sigma^{4n+3}) = 0$. Then

$$I_{4(n-4)}(S^3, \Sigma^{4n+3}) \equiv 0 \bmod 64.$$

(3). Let (S^1, Σ^{2n+1}) be a free differentiable action such that $I_{2(n-5)}(S^1, \Sigma^{2n+1}) = 0$. Then

$$I_{2(n-6)}(S^1, \Sigma^{2n+1}) \equiv 0 \bmod 16.$$

Proof. Suppose that $\Sigma^{11} \in \theta_{11}$ admits a free differentiable S^3 action. Then $\Sigma^{11} \in 32\,\theta_{11}$ (cf. [12]). Thus

$$\Sigma^{11} \equiv (1/8) I_{4(n-3)}(S^3, \Sigma^{4n+3}) \bmod |b\,P_{12}|Z \in 32\,\theta_{11}.$$

This implies (1). For the proofs of (2) and (3) we use Proposition 2.3 and the fact that if Σ^{11} admits a free differentiable S^1 action, then $\Sigma^{11} \in 2\,\theta_{11}$.

Corollary 4.7. (1). Let (S^3, Σ^{4n+3}) be a free differentiable action with characteristic homotopy 15-sphere Σ^{15}, that is, $I_{4(n-3)}(S^3, \Sigma^{4n+3}) = 0$. Then

$$I_{4(n-2)}(S^3, \Sigma^{4n+3}) = I_4(S^3, \Sigma^{15}) \equiv 0 \bmod 128.$$

(2). Let (S^1, Σ^{2n+1}) be a free differentiable action with characteristic homotopy 11-sphere Σ^{11}, that is, $I_{2(n-5)}(S^1, \Sigma^{2n+1}) = 0$. Then

$$I_{2n-8}(S^1, \Sigma^{2n+1}) = I_2(S^1, \Sigma^{11}) \equiv 0 \bmod 32.$$

Definition 4.8. Let S^i ($i = 1$ or 3) act freely (and topologically) on m-sphere S^m. We call the action piecewise linear (PL) if the orbit space S^m/S^i is triangulable. By an h-triangulation on KP^n ($K = C$ or Q) is a pair (L,g), where L is a PL-manifold with $\dim L = \dim KP^n$, and $g: L \to KP^n$ is a homotopy equivalence. We shall say that the free PL-action of S^i on S^m is concordant to a smooth action if the h-triangulation $f: S^m/S^1 \to KP^n$ is concordant to an h-smoothing, that is, there is an h-smoothing $f': M \to KP^n$ and a piecewise differentiable homeomorphism $c: S^m/S^1 \to M$ such that $f'c$ is homotopic to f.

By using PL-surgery (cf. [22]), the characteristic invariants are still defined for free PL-action of S^1 on S^m. The characteristic invariants in PL-case can take any integers divisible by 8 (cf. [21]) by Sullivan's classification up to PL-isomorphism of PL-complex, or quaternionic projective spaces.

As an immediate consequence of Corollary 4.6 or Corollary 4.7, we obtain

Corollary 4.9. *There exists infinitely many topologically distinct free PL-actions of S^1 (resp. S^3) on S^{2n+1} (resp. S^{4n+3}), $n \geq 5$ (resp. $n \geq 2$), which can not concordant to smooth actions.*

The case $n = 2$ follows from Proposition 2.3.

Corollary 4.10. *The natural maps*

$$[CP^n, G/O] \to [CP^n, G/PL],$$

and

$$[QP^n, G/O] \to [QP^n, G/PL]$$

are not onto for $n \geq 5$ and $n \geq 2$ respectively.

Another application of Theorem 4.2 is the following:

Proposition 4.11 (cf. [17]). *For $n \geq 5$, there exists at least an exotic $(4n+3)$-sphere Σ^{4n+3} in $b P_{4n+4}$ which admits infinitely many topologically distinct free differentiable actions of S^1 and S^3.*

Proof. It suffices to consider S^3-actions. Let

$$J: (KO)^{\sim} (QP^{n+2}) \to J (QP^{n+2})$$

be the J-homomorphism and $\tau = \tau_{QP^{n+2}}$. We can show that there exist infinitely many $\xi \in (KO)^{\sim} (QP^{n+2})$ with different rational Pontrjagin classes satisfying the following conditions:

(1) $L_{n+2}(\xi) + L_{n-1}(\xi) \cdot L_1(\tau) + \ldots + L_1(\xi) \cdot L_{n+1}(\tau) = 0,$

(2) $J(\xi) = 0,$

(3) $L_n(\xi) \cdot L_1(\tau \oplus \bar{\rho}^{-1}) + \ldots + L_1(\xi) \cdot L_n(\tau \oplus \bar{\rho}^{-1}) = 0,$

(4) $L_n(\xi) + L_{n-1}(\xi) \cdot L_1(\tau \oplus 2\bar{\rho}^{-1}) + \ldots + L_1(\xi) \cdot L_{n-1}(\tau \oplus 2\bar{\rho}^{-1})$ $= 0,$

(5) $\langle L_{n+1}(\xi), x_{n+1} \rangle \not\equiv 0 \bmod 8|b\, P_{4n+4}|$, where x_{n+1} is the generator of $H_{4n+4}(QP^{n+2}, Z)$, and . denotes the cup product in $H^*(QP^{n+2}, Q)$.

(1) and (2) imply that there are infinitely many smooth manifolds $X(\xi)$ of the same homotopy type of QP^{n+2} with stable tangent bundles of the form $f^*(\xi \oplus \tau)$, where $f: X(\xi) \to CP^{n+2}$ are the homotopy equivalence (cf. [4]). It is easy to see from (4), (5) and (6) that

$I_4(S^3, \Sigma^{4n+11}) = \langle L_{n+1}(\xi), X_{n+1} \rangle \not\equiv 0 \bmod 8|b\, P_{4n+4}|$, and

$I_8(S^3, \Sigma^{4n+11}) = 0$

Thus we have infinitely many topologically distinct differentiable actions of S^3 on Σ^{4n+11} so that the restricted actions to codimension 8 characteristic homotopy spheres are still distinct. By Theorem 4.2, the codimension 8 characteristic homotopy spheres, say $\Sigma^{4n+3}(\xi)$, are given by

$$\Sigma^{4n+3}(\xi) \equiv (1/8)\, I_4(S^3, \Sigma^{4n+11}) \in Z \bmod |b\, P_{4n+4}|Z$$

$$\equiv (1/8)\, \langle L_{n+1}(\xi), x_{n+1} \rangle \in Z \bmod |b\, P_{4n+4}|Z$$

$$\not\equiv 0 \text{ in } b\, P_{4n+4} \text{ by (5).}$$

Hence $\Sigma^{4n+3}(\xi)$ are all exotic. Since $b P_{4n+4}$ is finite. This obviously completes the proof of the theorem.

It is interesting to know which integers in $8Z$ may be realized as the characteristic invariants.

Let us construct some particular examples (the construction was given to us by Professor C. T. Yang). Given a free differentiable action (S^1, Σ^{4n+3}), $n > 1$, and an integer m with $m \equiv 0 \mod 8|b P_{4n}|$. Let $S^2 \subset M = \Sigma^{4n+3}/S^1$ represent a generator of $\pi_2(M)$. The normal bundle is trivial, hence there is an imbedding $S^2 \times D^{4n} \subset M$. We may assume that $M' \cap (S^2 \times D^{4n}) = \{y\} \times D^{4n}$, $(y \in S^2)$, where $M' = f^{-1}(CP^{2n})$ as in Definition 1.1. By an unpublished result of Montgomery and Yang, there is a diffeomorphism $\varphi: S^2 \times S^{4n-1} \to S^2 \times S^{4n-1}$ such that $\varphi(y \times S^{4n-1}) \approx S^{4n-1}$ bounds in $S^2 \times D^{4n}$ a compact paralleizable 4n-manifold W of index m. Define $M_m = (M - \text{int}(S^2 \times D^{4n})) \underset{\varphi}{\cup} (S^2 \times D^{4n})$. There is a homotopy equivalence $\psi: M_m \to CP^{2n+1}$, and a free S^1 action (S^1, Σ_m^{4n+3}) with $\Sigma_m^{4n+3}/S^1 = M_m$. The characteristic invariant $I_2(S^1, \Sigma_m^{4n+3})$ is given by

$$I_2(S^1, \Sigma_m^{4n+3}) = I_2(S^1, \Sigma^{4n+3}) + m.$$

By repeating this construction we have the following:

Corollary 4.12. *Let (S^1, Σ^{4n+3}) be a free differentiable action $(n > 1)$. If m is an integer divisible by $8|bP_{4n}||bP_{4n+4}|$, then there exists a free differentiable action (S^1, Σ_m^{4n+3}) with $\Sigma_m^{4n+3} \approx \Sigma^{4n+3}$ and*

$$I_2(S^1, \Sigma_m^{4n+3}) = I_2(S^1, \Sigma^{4n+3}) + m.$$

Thus if Σ^{4n+3} admits a free differentiable S^1 action, then it admits infinitely many topologically distinct free differentiable S^1-actions.

5. Characteristic Invariants and Degree of Symmetry

The degree of symmetry $N(X)$ of a compact connected smooth manifold

X is the maximum of the dimensions of the compact Lie groups which can act effectively and differentiably on M.

Theorem 5.1. Let (S^3, Σ^{4n+3}), $n \geq 5$, be a free differentiable action such that $I_4 (S^3, \Sigma^{4n+3}) \neq 0$, and $I_{4k} (S^3, \Sigma^{4n+3}) = 0$, for $k > 1$. Then $N (\Sigma^{4n+3}/S^3) = 0$.

Proof. Let $p_i (QP^n) = r_i \beta^i$, $\beta \in H^4 (QP^n)$ a generator. If $I_{4k} (S^3, \Sigma^{4n+3}) = 0$ for $k > 1$, then we can show that (cf. [13])

$$p_i (N) = r_i \bar{\alpha}^i, \quad i = 1, \ldots, n-2.$$

Suppose $N (N) > 0$. Then by [1], we have $\hat{\mathfrak{A}} (N) [N] = 0$, and $\hat{\mathfrak{A}} (QP^n) [QP^n] = 0$. Moreover, $L (N) [N] = L (QP^n) [QP^n]$. So we can prove by using [8] that

$$p_i (N) = r_i \bar{\alpha}^i, \quad i = n-1, n.$$

Hence

$$I_4 (S^3, \Sigma^{4n+3}) = 0.$$

Corollary 5.2. There exist infinitely many topologically distinct homotopy quaternionic projective n-spaces with degree of symmetry zero for every $n \geq 5$, and $n = 2$.

Proof. For $n \geq 5$, we have proved in [14] that there exist infinitely many topologically distinct free differentiable action (S^3, Σ^{4n+3}) $(n \geq 5)$ with different rational Pontrjagin classes of the orbit spaces and $I_4 (S^3, \Sigma^{4n+3}) \neq 0$ and $I_{4k} (S^3, \Sigma^{4n+3}) = 0$ for $k > 1$. In case $n = 2$, if we compute directly from $\sigma (N) = 0$, and $\hat{\mathfrak{A}} (N) [N] = 0$, we get $p_i (N) = r_i \bar{\alpha}^i$, $i = 1, 2$. But there exist infinitely many HQP^2 with $p_i (HPQ^2) \neq r_i \bar{\alpha}^i$, $i = 1, 2$.

Lemma 5.3. Let (S^1, Σ^{2n+1}) be a free differentiable action. Then

$I_{2(n-2k)}(S^1, \Sigma^{2n+1}) = 0$ <u>if and only if there exists a</u> $(4k+1)$-<u>dimensional invariant submanifold</u> $\Sigma_Q^{4k+1} \subset \Sigma^{2n+1}$ <u>such that the embedding</u> $i\colon \Sigma_Q^{4k+1}/S^1 \to \Sigma^{2n+1}/S^1$ <u>with normal bundle</u> ν <u>induces isomorphisms</u>

$$i_Q^*\colon H^j(\Sigma^{2n+1}/S^1, Q) \cong H^j(\Sigma_Q^{4k+1}/S^1, Q), \quad o \leq j \leq 4k$$

<u>and</u>

$$p(\nu) = i^*(1 + \alpha^2)^{n-2k},$$

<u>where</u> $\quad i^*\colon H^*(M, Z) \to H^*(\Sigma_Q^{4k+1}/S^1, Z)$.

<u>Proof.</u> Denote the generator of $H_{4k}(M)$ by X_{4k} which satisfies $\langle \alpha^{2k}, X_{4k}\rangle = 1$. According to Theorem 1.2, we have

$$I_{2(n-2k)}(S^1, \Sigma^{2n+1}) = \langle L(\tau_M) \cdot L(\rho^{-1})^{n-2k}, X_{4k}\rangle - 1$$

From hypotheses,

$$i_*[\Sigma_Q^{4k+1}/S^1] = t\, X_{4k}, \quad (t \in Z,\ t \neq 0), \text{ and}$$

$$\begin{aligned} 1 = \sigma(\Sigma_Q^{4k+1}/S^1) &= \langle L(\Sigma_Q^{4k+1}/S^1), [\Sigma_Q^{4k+1}/S^1]\rangle \\ &= \langle i^* L(\tau_M) \cdot L(\nu^{-1}), [\Sigma_Q^{4k+1}/S^1]\rangle \\ &= \langle L(\tau_M) \cdot L(\rho^{-1})^{n-2k}, i_*[\Sigma_Q^{4k+1}/S^1]\rangle \\ &= t\,\langle L(\tau_M) \cdot L(\rho^{-1})^{n-2k}, X_{4k}\rangle \end{aligned}$$

Thus $\quad \langle L(\tau_M) \cdot L(\rho^{-1})^{n-2k}, X_{4k}\rangle = 1$, and $I_{2(n-2k)}(S^1, \Sigma^{2n+1}) = 0$.

Similarly, we have

<u>Lemma 5.4.</u> <u>Let</u> (S^3, Σ^{4n+3}) <u>be a free differentiable action. Then</u> $I_{4(n-k)}(S^3, \Sigma^{4n+3}) = 0$ <u>if and only if there exists a</u> $(4k+3)$-<u>dimensional invariant submanifold</u> $\Sigma_Q^{4k+3} \subset \Sigma^{4n+3}$ <u>such that the embedding</u> $i\colon \Sigma_Q^{4k+3}/S^3 \to \Sigma^{4n+3}/S^3$ <u>with normal bundle</u> ν <u>induces isomorphisms</u>

$$i_Q^*: H^j (\Sigma^{4n+3}/S^3, Q) \cong H^j (\Sigma_Q^{4k+3}/S^3, Q), \; 0 \le j \le 4k$$

and

$$p(\nu) = i^* (1 + 2\bar{\alpha} + \bar{\alpha}^2)^{n-k},$$

where

$$i^*: H^* (\Sigma^{4n+3}/S^3, Z) \to H^* (\Sigma_Q^{4k+3}/S^3, Z)$$

Theorem 5.5. *Let S^1 act freely and differentiably on a homotopy $(4n+3)$-sphere Σ^{4n+3} $(n \ge 2)$ such that the orbit space Σ^{4n+3}/S^1 admits an S^1 action with a component of the fixed point set of codimension 2. Then*

$$I_2 (S^1, \Sigma^{4n+3}) = 0.$$

Proof. Let F be the component of the fixed point set of codimension 2. Then the inclusion $i: F \to \Sigma^{4n+3}/S^1$ induces isomorphisms

$$i^*: H^j (\Sigma^{4n+3}/S^1) \cong H^j (F), \; 0 \le j \le 4n,$$ (cf. [2], [20]), and

so

$$p(\nu) = i^* (1 + \alpha^2)$$

by [4], where ν is the normal bundle of the embedding i. Thus we may apply Lemma 5.3 to complete the proof of the theorem.

Corollary 5.6. *There exist infinitely many topologically distinct homotopy complex projective $(2n+1)$-spaces $(n \ge 5)$ which do not admit differentiable S^1 actions with a component of the fixed point set of codimension two.*

Proof. By Proposition 1.4.

Theorem 5.7. *Let (S^3, Σ^{4n+3}) be a free differentiable action with $n \ge 3$, and S^1 acts differentiably on Σ^{4n+3}/S^3 with a component of the fixed point set of codimension 4. Then $I_4 (S^3, \Sigma^{4n+3}) = 0$.*

Proof. Let ν be the normal bundle of the embedding $i: F \to \Sigma^{4n+3}/S^3$.

It is not difficult to see from [2] that

$$i^*: H^j(N) \cong H^j(F), \quad 0 \leq j \leq 4n-4.$$

Hence the Euler class $e(\nu)$ is given by $e(\nu) = \pm\, i^* \bar{\alpha}$ (cf. [4]). The structure group of the bundle ν can be reduced to $U(2)$, so that the bundle ν is classifying by a classifying map $f: F \to B_{U(2)}$ and $\nu = f^*(E_{U(2)} \times_{U(2)} A)$ for a $U(2)$-module A of (complex) dimension two where $E_{U(2)}$ is the universal space of the principal $U(2)$-bundles. Let the weights of A be w_1, w_2 which may consider as elements of $H^2(B_{S^1 \times S^1})$, where $S^1 \times S^1$ is a maximal torus of $U(2) = S^1 \times S^3$. Then $e(\nu) = f^*(w_1 w_2)$. As $f^*(w_1 + w_2) = 0$, we have $f^*(w_1^2 + w_2^2) = -2 f^*(w_1 w_2) = 2\, i^* \bar{\alpha}$ for appropriate choice of orientation. Thus $p(\nu) = f^*(1 + w_1^2)(1 + w_2^2) = i^*(1 + 2\bar{\alpha} + \bar{\alpha}^2)$. Hence our assertion follows from Lemma 5.4.

Corollary 5.8. *There exist infinitely many topologically distinct homotopy quaternionic projective n-spaces ($n \geq 3$, $n \neq 4$) which do not admit differentiable S^1-actions with a component of the fixed point set of codimension 4.*

Proof. For $n \geq 5$, it follows from Proposition 1.4. If $n = 3$, then from Proposition 2.4 and $\hat{\mathfrak{U}}_3(N)[N] = 0$, we see that $p(N) = p(QP^3)$ (cf. [13]).

As a by product we have

Corollary 5.9. *Let (S^3, Σ^{15}) be a free differentiable action such that S^1 acts differentiably on Σ^{15}/S^3 with a component of the fixed point set of codimension 4. Then $\beta(\Sigma^{15}) = S^{15}$, where $\beta: \theta_{15} \to b P_{16}$ is the projection.*

Corollary 5.10. There exist at most a finite number of homotopy quaternionic projective 3-spaces which admit differentiable S^1-actions with a component of the fixed point set of codimension 4.

Proof. Apply [21].

Remark 5.11 Corollary 5.2 is also true for $n = 3,4$. The proofs are similar to the case $n = 2$ by assuming $p_1(HQP^3) = p_1(QP^3)$, $p_i(HQP^4)$, $i = 1,2$ respectively. In case $n = 3$, we also can show by direct computation that if $N(\Sigma^{15}/S^3) > 0$, then

$$p_1(\Sigma^{15}/S^3) = 4(8i + 1)\ \alpha,$$

$$p_2(\Sigma^{15}/S^3) = 12(8i + 1)^2\ \alpha^2,$$

$$p_3(\Sigma^{15}/S^3) = 8(8i + 1)^3\ \alpha^3,$$

for some integer i. But there exist infinitely many homotopy quaternionic projective 3-spaces HQP^3 with $p_1(HQP^3) = 4\ \alpha$, and $p_2(HQP^3)$, $p_3(HQP^3)$ different from $12\ \alpha^2$ and $8\ \alpha^2$ respectively (see [13]).

References

1. Atiyah, M.F., and Hirzebruch, F., Spin-manifolds and group actions, Essays on Topology and Related Topics, Memoires dédiés Georges de Rham, Springer-Verlag, Berlin, New York, 18-28 (1970).

2. Bredon, G.E., The cohomology ring structure of a fixed point set, Ann. of Math. (2) 80, 524-537 (1964).

3. Browder, W., and Livesay, G.R., Fixed point free involutions on homotopy spheres, Bull. Amer. Math. Soc. 73, 242-245 (1967).

4. Browder, W., Surgery and the theory of differentiable transformation groups, Proc. Conference on Transformation Groups, Springer-Verlag, Berlin, New York, 1-46 (1968).

5. Brumfiel, G., Differentiable S^1 actions on homotopy spheres, (mimeographed), Univ. of California, Berkeley (1970).

6. Conrad, B., Extending free circle actions on spheres to S^3 actions, Proc. Amer. Math. Soc. 27, 168-174 (1971).

7. Eells, J., and Kuiper, N.H., An invariant for certain smooth manifolds, Annali di Math. 60, 93-110 (1962).

8. Hirzebruch, F., Topological Methods in Algebraic Geometry, Springer-Verlag, Berlin, New York (1966).

9. Hsiang, W.C., and Hsiang, W.Y., Some free differentiable actions on 11-spheres, Quart. J. Math. (Oxford) 15, 371-374 (1964).

10. Hsiang, W.C., A note on free differentiable actions of S^1 and S^3 on homotopy spheres, Ann. of Math. (2) 83, 266-272 (1966).

11. Kervaire, M., and Milnor, J., Groups of homotopy spheres, I, Ann. of Math. (2) 77, 504-537 (1963).

12. Ku, H.T., and Ku, M.C., Free differentiable actions of S^1 and S^3 on homotopy 11-spheres, Trans. Amer. Math. Soc. 138, 223-228 (1969).

13. Ku, H.T., and Ku, M.C., Characteristic spheres of free differentiable actions of S^1 and S^3 on homotopy spheres, Trans. Amer. Math. Soc. 156, 493-504 (1971).

14. Ku, H.T., and Ku, M.C., Free differentiable actions of S^1 and S^3 on homotopy spheres, Proc. Amer. Math. Soc. 25, 864-869 (1970).

15. Ku, H.T., and Ku, M.C., Exotic free differentiable actions of S^1 and S^3 on homotopy spheres, Notices of Amer. Math. Soc. 17, 682 (1970).

16. Ku, H.T., and Ku, M.C., Invariant and characteristic homotopy spheres of differentiable actions on homotopy spheres, Studies and Essays presented to Yu-Why Chen on His 60th Birthday, 91-106, Math. Research Center, National Taiwan University, Taiwan (1970).

17. Ku, H.T., and Ku, M.C., Free differentiable actions of S^1 and S^3 on homotopy spheres II, (mimeographed), Univ. of Massachusetts, Amherst (1970).

18. Montgomery, D., and Yang, C.T., Differentiable transformation groups on homotopy spheres, Michigan Math. J. 14, 33-46 (1967).

19. Montgomery, D., and Yang, C.T., Free differentiable actions on homotopy spheres, Proc. Conference on Transformation Groups, Springer-Verlag, Berlin, New York 175-192 (1968).

20. Su, J.C., Transformation groups on cohomology projective spaces, Trans. Amer. Math. Soc. 106, 305-318 (1963).

21. Sullivan, D. P., Triangulating and smoothing homotopy equivalences

and homeomorphisms, Geometric Seminar Notes, Princeton University (1967).

22. Wall, C.T.C., Surgery of compact manifolds (to appear).

23. Wang, K., Free smooth actions of S^1 and S^3 on homotopy spheres, Notices of Amer. Math. Soc. 18, 383 (1971).

DIFFERENTIABLE PSEUDO-FREE CIRCLE ACTIONS ON HOMOTOPY SEVEN SPHERES

Deane Montgomery, Institute for Advanced Study
and
C. T. Yang, University of Pennsylvania*

1. INTRODUCTION

In this paper we study effective differentiable circle actions on homotopy spheres such that all orbits are 1-dimensional and with the further property that exceptional orbits are isolated. An exceptional orbit in this setting is an orbit for which the isotropy group is a finite subgroup different from the identity. Such an action is called pseudo-free or free according to whether there is at least one exceptional orbit or none. Since free actions have been studied extensively, we concern ourselves with those which are pseudo-free.

Let G be the circle group consisting of complex numbers of absolute value 1 and let S^{2n+1} be the unit $(2n+1)$-sphere $|z_0|^2 + \cdots + |z_n|^2 = 1$ in the unitary $(n+1)$-space. Then for any positive integers $q_0,\ldots,q_n$, not all 1 and relatively prime to one another, the orthogonal action of G on S^{2n+1} given by

$$g(z_0,\ldots,z_n) = (g^{q_0} z_0,\ldots,g^{q_n} z_n)$$

is pseudo-free and the number of exceptional orbits is equal to that of the q_i's which are greater than 1. Since any orthogonal pseudo-free circle action on S^{2n+1} may be given this way, we infer that any orthogonal pseudo-free circle action on S^{2n+1} has no more than $n+1$ exceptional orbits. Therefore it is natural to ask whether any differentiable pseudo-free circle action on a homotopy $(2n+1)$-sphere has no more than $n+1$ exceptional orbits. For $n = 1$, the answer to this question is positive. It is known [1] that any pseudo-free circle action on a homotopy 3-sphere is equivalent to an orthogonal action. In this paper, we shall show that for $n = 3$, the answer is negative. In fact, we classify all differentiable pseudo-free circle actions on homotopy 7-spheres. As a consequence of the classification, we know the

*The second-named author was supported in part by the U. S. Army Research Office--Durham and by the National Science Foundation when the paper was under preparation.

existence of a differentiable pseudo-free circle action on a homotopy 7-sphere with exactly k exceptional orbits for any preassigned positive integer k.

For the classification of differentiable pseudo-free circle actions on homotopy 7-spheres, we find a way to present all such actions. This is done in three steps. First, any homotopy 7-sphere on which there is a differentiable pseudo-free circle action is decomposed into finite standard pieces. Second, those pieces are studied and classified. Third, given a finite number of such pieces, it is determined whether a differentiable pseudo-free circle action on a homotopy 7-sphere can be obtained by pasting them together, and how many distinct ones may be obtained by varying the pasting. Details are carried out in later sections.

2. PRELIMINARIES

We begin with notations and elementary results which are needed later.

Throughout this paper, we confine ourselves in the differentiable category. Therefore, unless it is said to the contrary, we take for granted that manifolds (with or without boundary), submanifolds, maps, immersions, imbeddings, group actions and so on are assumed to be differentiable of class C^∞.

By a <u>circle action</u>, we mean an action of the circle group G which consists of complex numbers of absolute value 1. For any integer $q > 1$, we let $\mathbb{Z}q$ be the subgroup of G of order q. Sometimes $\mathbb{Z}q$ also denotes the group of integers modulo q. The group of integers is denoted by $\mathbb{Z}$.

For any integer $m \geqq 0$, $\mathbb{R}^m$ denotes the euclidean m-space, D^m denotes the unit closed m-disk $x_1^2 + \cdots + x_m^2 \leqq 1$ in $\mathbb{R}^m$ and S^{m-1} denotes the boundary ∂D^m of D^m, namely the unit $(m-1)$-sphere $x_1^2 + \cdots + x_m^2 = 1$ in $\mathbb{R}^m$. In case that m is even, say $m = 2n$, D^m and S^{m-1} are also regarded as the unit closed $(2n)$-disk $|z_1|^2 + \cdots + |z_n|^2 \leqq 1$ and the unit $(2n-1)$-sphere $|z_1|^2 + \cdots + |z_n|^2 = 1$ in the unitary n-space with the understanding that $z_i = x_{2i-1} + x_{2i}\sqrt{-1}$, $i = 1,\ldots,n$. Therefore we may naturally identify G with S^1. $\mathbb{R}^m$, D^m and S^{m-1} will be oriented in the natural way whenever an orientation is needed.

Whenever M' is an oriented m'-manifold and M'' is an oriented m''-manifold, we let $M' \times M''$ be so oriented that, if $f'' = \mathbb{R}^{m'} \longrightarrow M'$ and $f'' = \mathbb{R}^{m''} \longrightarrow M''$ are (+)-imbeddings, then $f' \times f'' : \mathbb{R}^{m'} \times \mathbb{R}^{m''} \longrightarrow M' \times M''$ is a (+)-imbedding, where (+)

stands for orientation-preserving. Similarly, (-) will stand for orientation-reversing.

Let M be a manifold on which there is an effective action of the circle group G. For each $x \in M$, $G_x = \{g \in G | gx = x\}$ is called the isotropy group at x and $Gx = \{gx | g \in G\}$ is called the orbit of x. The set of all the orbits Gx with $x \in M$ together with the quotient topology is called the orbit space of the action of G on M and is denoted by M/G. Whenever $A \subset M$, the image of A in M/G is denoted by A^*. An orbit Gx is called principal if the isotropy group G_x is trivial and Gx is called exceptional if G_x is finite but not trivial. If all orbits are principal, the action of G on M is called free. If the action is not free and if all orbits are 1-dimensional and exceptional orbits are isolated and are contained in $M - \partial M$ (∂M denotes the boundary of M), the action is called pseudo-free.

Let M and M' be manifolds on which G acts. A map $f : M \longrightarrow M'$ is called equivariant if for any $g \in G$ and $x \in M$, $f(gx) = gf(x)$. Clearly an equivariant map $f : M \longrightarrow M'$ induces a map $f : M^* \longrightarrow M'^*$ given by $f(Gx) = Gf(x)$. For the sake of simplicity, we let an equivalence mean an equivariant diffeomorphism or the homeomorphism induced by an equivariant diffeomorphism.

Suppose that a free circle action on an m-manifold M is given. Then M^* is an (m-1)-manifold and the projection $\pi : M \longrightarrow M^*$ is a circle fibration. Therefore, if M is oriented, there is a natural orientation for M^* such that, if $f : S^1 \times \mathbf{R}^{m-1} \longrightarrow M$ is a (+)-imbedding with $f(g,x) = gf(1,x)$ for all $(g,x) \in S^1 \times \mathbf{R}^{m-1}$, then $f : \mathbf{R}^{m-1} \longrightarrow M^*$ given by $f(x) = \pi f(1,x) = Gf(1,x)$ is a (+)-imbedding. We note that, if M and M' are manifolds on which G acts freely, then any equivalence of M^* onto M'^* is actually a diffeomorphism, and any (+)-equivalence of M onto M' induces a (+)-equivalence and conversely.

Unless it is explicitly stated to the contrary, the group $\mathbf{Z}$ of integers is used as the coefficient group for both homology and cohomology. Moreover, the induced homomorphism of any map f will also be denoted by f.

Suppose that M is a connected simply connected manifold with a trivial $H_2(M)$ and that there is a free circle action on M. Then M^* is connected and simply connected and $H_2(M^*)$ is an infinite cyclic group having a preferred generator β given as follows. If $\varphi : D^2 \longrightarrow M$ is a map such that for any $g \in S^1$, $\varphi(g) = g\varphi(1)$, then φ induces a map of $D^2/\partial D^2$ into M^* representing β.

Let M be an oriented compact $(2n+1)$-manifold, $n \geqq 3$, which is diffeomorphic to $S^3 \times D^{2n-2}$ and on which there is an effective circle action. If the action is free, we call M a free G-manifold. If the action is pseudo-free, we call M a prime G-manifold or a composite G-manifold according as there is exactly one or at least one exceptional orbit.

Let F_1 be a copy of $S^3 \times D^{2n-2}$ on which there is a circle action such that for any $g \in G$, $(u_1,u_2) \in S^3$ and $(v_1,\ldots,v_{n-1}) \in D^{2n-2}$,

$$g(u_1,u_2;v_1,\ldots,v_{n-1}) = (gu_1,gu_2;v_1,\ldots,v_{n-1}) .$$

Then F_1 is a free G-manifold. Similarly there is a free G-manifold F_2 which is also a copy of $S^3 \times D^{2n-2}$ but on which the circle action is given by

$$g(u_1,u_2;v_1,v_2,\ldots,v_{n-1}) = (gu_1,gu_2;gv_1,v_2,\ldots,v_{n-1}) .$$

(2.1) The constructed free G-manifolds F_1 and F_2 are not equivalent (that means, there does not exist an equivalence of F_1 onto F_2). However, given any free G-manifold F, there is a (+)-equivalence of F onto one of F_1 and F_2. Hence, up to an equivalence or up to a (+)-equivalence, there are exactly two free G-manifolds, namely F_1 and F_2.

Proof. It can be seen that $w_2(F_1^*)$, i.e., the second Stiefel-Whitney class of F_1^*, vanishes but $w_2(F_2^*)$ does not. Therefore F_1^* and F_2^* are not diffeomorphic and hence F_1 and F_2 are not equivalent.

Let F be a given free G-manifold. Then F^* is a compact $(2n)$-manifold which is connected and simply connected and has the integral homology of S^2. Let S^2 be imbedded into $F^* - \partial F^*$ such that it represents a generator of $H_2(F^*)$. Then S^2 is a deformation retract of F^*. Since ∂F^* is connected and simply connected and since $\dim F^* > 5$, it follows from Smale's theorem [2] that F^* is diffeomorphic to a closed tubular neighborhood of S^2 in F^*. Therefore there is a diffeomorphism $f : F^* \longrightarrow F_i^*$, where $i = 1$ or 2 according as $w_2(F^*)$ vanishes or not.

The diffeomorphism $f : F^* \longrightarrow F_i^*$ is induced by an equivalence $f : F \longrightarrow F_i$. Since for each $i = 1,2$, there is a (-)-equivalence of F_i onto F_i, we may assume that $f : F \longrightarrow F_i$ is a (+)-equivalence. Hence the proof is completed.

Suppose now that M is a manifold on which there is a pseudo-free circle action. Let b be a point of M such that Gb is an exceptional orbit, and let D be a slice at b which is a closed disk of center b contained in $M - \partial M$ and on which G_b acts orthogonally. (For the concept of a slice, see, for example, [3].) Let q be the order of G_b (so that $G_b = \mathbb{Z}q$). Since the exceptional orbit Gb is isolated, $\mathbb{Z}q$ acts freely on ∂D. Therefore, if $\dim M = 2n+1$ with $n \geqq 2$, ∂D^* is a (2n-1)-dimensional lens space of fundamental group $\mathbb{Z}q$ and D^*, which is a closed neighborhood of b^* in M^*, is a cone of vertex b^* over ∂D^* so that M^* is not locally euclidean at b^*.

Let E be the union of all exceptional orbits in M. Then $M^* - E^*$ is a manifold even though M^* may not be. Therefore M^* is called a manifold with <u>singularity set</u> E^*. However, M^* is a rational cohomology manifold and $\pi : M \longrightarrow M^*$ is a Seifert fibration. Therefore, if M is oriented, a natural orientation for M^* may be introduced just as in the case of a free circle action.

By the definition of a pseudo-free action, E^* is a discrete subset of M^* contained in $M^* - \partial M^*$, where $\partial M^* = (\partial M)^*$. Therefore, if M is compact, E^* is finite. If $E^* = \{b_1^*,\ldots,b_k^*\}$, M is also called a manifold with <u>singularities</u> $b_1^*,\ldots,b_k^*$.

Let M and M' be manifolds on which G acts pseudo-freely and h^* an equivalence of M^* onto M'^*. Then h^* maps the singularity set E^* in M^* into the singularity set E'^* in M'^* and maps $M^* - E^*$ diffeomorphically onto $M'^* - \partial E'^*$. Moreover, each $b^* \in E^*$ has a neighborhood U^* such that $h^* : U^* \longrightarrow h^*(U^*)$ is induced by a G_b-equivariant orthogonal map of a disk into a disk.

For the rest of this section, we let M be a compact (2n+1)-manifold on which there is a pseudo-free action of G, and let

$$Gb_1,\ldots,Gb_k$$

be distinct exceptional orbits in M such that G acts freely on $M - \cup_{i=1}^{k} Gb_i$. Moreover, we let q_i be the order of the isotropy group at b_i, $i = 1,\ldots,k$, and let

$$q = q_1 \cdots q_k .$$

(2.2) *If* M *has the integral homology of a sphere, then* $q_1,\ldots,q_k$ *are relatively prime to one another.*

This follows from Smith's theorem (see, for example, [4; p. 43]).

Let D_i be a slice at b_i which is a closed $(2n)$-disk of center b_i contained in $M - \partial M$ and on which the isotropy group at b_i acts orthogonally. Clearly $D_1,\ldots,D_k$ can be so chosen that $GD_1,\ldots,GD_k$ are mutually disjoint. Let

$$X = M - \cup_{i=1}^{k} G(D_i - \partial D_i) .$$

Then G acts freely on X so that

$$X^* = M^* - \cup_{i=1}^{k}(D_i^* - \partial D_i^*)$$

is a compact $(2n)$-manifold and its boundary is the disjoint union of $\partial D_1^*,\ldots,\partial D_k^*$ and ∂M^*.

(2.3) *If* M *has the integral homology of a* $(2\ell+1)$-*sphere, then*

$$H^{2\ell}(X^*, \cup_{i=1}^{k}\partial D_i^*) \cong \mathbb{Z}$$

and there is, for each $r = 1,\ldots,\min\{\ell, n-1\}$, *a commutative diagram*

$$\begin{array}{ccccccccc}
0 & \longrightarrow & \mathbb{Z} & \longrightarrow & \mathbb{Z} & \longrightarrow & \mathbb{Z}_q & \longrightarrow & 0 \\
 & & \uparrow & & \uparrow & & \uparrow & & \\
0 & \longrightarrow & H^{2r}(X^*, \cup_{i=1}^{k}\partial D_i^*) & \longrightarrow & H^{2r}(X^*) & \longrightarrow & H^{2r}(\cup_{i=1}^{k}\partial D_i^*) & \longrightarrow & 0
\end{array}$$

where the lower row is a part of the exact integral cohomology sequence of $(X^*, \cup_{i=1}^{k}\partial D_i^*)$ *and vertical arrows are isomorphisms.*

(2.4) *If* M *is a connected simply connected oriented compact* $(2n+1)$-*manifold having the integral homology of a* $(2\ell+1)$-*sphere, where* $\ell \geq 1$ *and* $n \geq 3$, *then each*

Gb_i <u>is contained in a</u> (2n+1)-<u>dimensional prime</u> G-<u>manifold</u> W_i <u>in</u> $M - \partial M$ <u>such that</u> $H_2(\partial W_i^*) \longrightarrow H_2(M^* - \{b_1^*, \ldots, b_k^*\})$ <u>induced by the inclusion map is an isomorphism and the inclusion map of</u> W_i <u>into</u> M <u>is a</u> (+)-<u>imbedding</u>. <u>Moreover, there is a composite</u> G-<u>manifold</u> K <u>such that</u> $\cup_{i=1}^k W_i \subset K \subset M - \partial M$, $H_2(\partial K^*) \longrightarrow$ $\longrightarrow H_2(M^* - \{b_1^*, \ldots, b_k^*\})$ <u>induced by the inclusion map is an isomorphism and the inclusion map of</u> K <u>into</u> M <u>is a</u> (+)-<u>imbedding</u>. <u>Furthermore</u>, $W_1, \ldots, W_k$ <u>and</u> K <u>are unique up to an equivariant isotopy of</u> M.

Proof. Let $D_1, \ldots, D_k$ and X be as before. Then for each $i = 1, \ldots, k$, there is an imbedding

$$\varphi_i^* : D^2 \times D^{2n-2} \longrightarrow X^* - \partial M^*$$

such that (i) $\varphi_i^*(S^1 \times D^{2n-2}) = \varphi_i^*(D^2 \times D^{2n-2}) \cap \partial X^* \subset \partial D_i^*$ and (ii) the image of the preferred generator of $H_2(X^*)$ in $H_2(X^*, \partial D_i^*)$ is equal to q_i times the element represented by $\varphi_i^* | (D^2 \times \{0\}, S^1 \times \{0\})$. Let

$$W_i^* = \varphi_i^*(D^2 \times D^{2n-2}) \cup D_i^*$$

with the corner along $\varphi_i^*(S^1 \times S^{2n-3})$ rounded, and let W_i be the invariant subset of M with $W_i/G = W_i^*$. It is easily seen that W_i is a compact (2n+1)-submanifold of $M - \partial M$ which is connected and simply connected. Since W_i^* has the integral homology of S^2, W_i has the integral homology of S^3. Therefore we may imbed S^3 into $W_i - \partial W_i$ such that it is a deformation retract of W_i. By Smale's theorem, W_i is diffeomorphic to a closed tubular neighborhood of S^3 so that it is diffeomorphic to $S^3 \times D^{2n-2}$. Hence W_i is a prime G-manifold.

Let W_i' be a second (2n+1)-dimensional prime G-manifold in $M - \partial M$ just like W_i. Then we may let D_i and φ_i^* be so constructed that

$$W_i \subset W_i' - \partial W_i' .$$

It can be shown that $W_i'^* - (W_i^* - \partial W_i^*)$ is a simply connected h-cobordism. Therefore there is an equivariant isotopy of M mapping W_i into W_i'. Hence W_i is unique up to an equivariant isotopy of M.

It is easily seen that $\varphi_1^*, \ldots, \varphi_k^*$ may be so chosen that $W_1, \ldots, W_k$ are mutually disjoint. Let

$$Y = M - \cup_{i=1}^{k}(W_i - \partial W_i) \quad .$$

Then Y^* is a connected simply connected compact (2n)-manifold and its boundary is the disjoint union of $\partial W_1^*, \ldots, \partial W_k^*$ and ∂M^*. Moreover, for each $i = 1, \ldots, k$, $H_2(\partial W_i^*) \longrightarrow H_2(Y^*) \longrightarrow H_2(X^*) \longrightarrow H_2(M^* - \{b_1^*, \ldots, b_k^*\})$ induced by inclusion maps are isomorphisms. Since $\dim Y^* \geqq 6$, there is, for each $i = 1, \ldots, k-1$, an immersion $\psi_i^* : S^2 \times [0,1] \longrightarrow Y^* - \partial M^*$ such that (i) for each $j = 0,1$, $\psi_i^*(S^2 \times \{j\}) \subset \partial W_{i+j}^*$ and $\psi_i^*|S^2 \times \{j\}$ is an imbedding representing the preferred generator of $H_2(\partial W_{i+j}^*)$ and (ii) $\psi_i^*(S^2 \times [0,1]) \cap \partial Y^* = \psi_i^*(S^2 \times \{0,1\})$ and $\psi_i^*(S^2 \times [0,1])$ intersects ∂Y^* transversally at $\psi_i^*(S^2 \times \{0,1\})$. Making use of the technique developed in Whitney [5], we are able to eliminate any self-intersection of $\psi_i^*(S^2 \times [0,1])$. Therefore we may assume that $\psi_i^* : S^2 \times [0,1] \longrightarrow Y^* - \partial M^*$ is an imbedding. Let E be the total space of a closed (2n-3)-disk bundle over S^2 which is trivial or non-trivial according as $w_2(Y^*)$ vanishes or not, and let S^2 be regarded as the 0-section in E. Then there is an imbedding $\psi_i^* : E \times [0,1] \longrightarrow Y^* - \partial M^*$ which is a thickening of $\psi_i^* : S^2 \times [0,1] \longrightarrow Y^* - \partial M^*$ and such that for each $j = 0,1$, $\psi_i^*(E \times \{j\}) = \psi_i^*(E \times [0,1]) \cap W_{i+j}^* \subset \partial W_{i+j}^*$. We may let $\psi_1^*, \ldots, \psi_{k-1}^*$ be so chosen that $\psi_1^*(E \times [0,1]), \ldots, \psi_{k-1}^*(E \times [0,1])$ are mutually disjoint. Let

$$K^* = \cup_{i=1}^{k} W_i^* \cup \cup_{i=1}^{k-1} \psi_i^*(E \times [0,1])$$

with the corner along $\cup_{i=1}^{k-1} \psi_i^*(\partial E \times \{0,1\})$ rounded and let K be the invariant subset of M with $K/G = K^*$. Then K is a compact (2n+1)-submanifold of $M - \partial M$ containing $\cup_{i=1}^{k} Gb_i$. It can be seen that K is connected and simply connected. Making use of (2.2), we can see that K^* has the integral homology of S^2 so that K has the integral homology of S^3. Hence as W_i, K is diffeomorphic to $S^3 \times D^{2n-2}$. This shows that K is a composite G-manifold in $M - \partial M$ containing all exceptional orbits in M.

That K is unique up to an equivariant isotopy of M can be proved by the same argument as done for W_i.

In concluding this section, let us describe the pasting technique which will be used in later sections. Let M and N be manifolds or manifolds with singularities and let L be a compact submanifold of ∂M such that $\dim M = \dim N = 1 + \dim L$. Then for any imbedding $f : L \longrightarrow \partial N$, we can use f to paste together M and N obtaining a new manifold

$$M \cup_f N .$$

Precisely speaking, $M \cup_f N$ is the manifold obtained from the disjoint union of M and N by identifying every $x \in L$ with $f(x)$ and then having the corner along ∂L rounded. If there are circle actions on M and N and if f is equivariant, then we have a natural circle action on $M \cup_f N$ by combining those on M and N. If M and N are oriented and if f is a (-)-imbedding, then we have a natural orientation for $M \cup_f N$ such that the inclusion map of either M or N into $M \cup_f N$ is a (+)-imbedding. The construction of $M \cup_f N$ and any of these additional structures on $M \cup_f N$ will be taken for granted without further explanation.

3. A DECOMPOSITION

In this section we shall show that any homotopy (2n+1)-sphere, $n \geqq 3$, on which there is a pseudo-free circle action, can be decomposed into finitely many compact (2n+1)-submanifolds of which one is diffeomorphic to $S^{2n-3} \times D^4$ and contains no exceptional orbits and all others are prime G-manifolds. This decomposition is the foundation of the construction of all pseudo-free circle actions on homotopy 7-spheres given in section 8.

Let Σ be a homotopy (2n+1)-sphere, $n \geqq 3$, on which there is a pseudo-free circle action, and let $Gb_1,\dots,Gb_k$ be distinct exceptional orbits in Σ such that G acts freely on $\Sigma - \cup_{i=1}^k Gb_i$. By (2.4), each Gb_i is contained in a (2n+1)-dimensional prime G-manifold W_i in Σ such that $H_2(\partial W_i^*) \longrightarrow H_2(\Sigma^* - \{b_1^*,\dots,b_k^*\})$ induced by the inclusion map is an isomorphism. Moreover, $W_1,\dots,W_k$ can be so chosen that (i) for any $i,j = 1,\dots,k$ with $|i-j| > 1$, $W_i \cap W_j = \emptyset$ and (ii) for any $i = 1,\dots,k-1$, $W_i^* \cap W_{i+1}^* = \partial W_i^* \cap \partial W_{i+1}^*$ is a closed tubular neighborhood of a 2-sphere in ∂W_i^* representing a generator of $H_2(\partial W_i^*)$. Then

$$K = W_1 \cup \cdots \cup W_k$$

(with the corner along $\cup_{i=1}^{k-1} \partial(W_i \cap W_{i+1})$ rounded) is a (2n+1)-dimensioned composite G-manifold such that $H_2(\partial K^*) \longrightarrow H_2(\Sigma^*-\{b_1^*,\ldots,b_k^*\})$ is an isomorphism. Let

$$F = \text{closure } (\Sigma - K) .$$

Then F is a connected simply connected oriented compact (2n+1)-manifold on which G acts freely. Since K is diffeomorphic to $S^3 \times D^{2n-2}$, $n \geqq 3$, F is diffeomorphic to $S^{2n-3} \times D^4$. Hence for $n = 3$, F is a (2n+1)-dimensional free G-manifold.

By (2.4), K is unique up to an equivariant isotopy of Σ. Then so is F. Hence we have

(3.1) Theorem. *Let Σ be a homotopy (2n+1)-sphere, $n \geqq 3$, on which there is a pseudo-free circle action. Then there is a decomposition*

$$\Sigma = F \cup K$$

where K is a (2n+1)-dimensional composite G-manifold containing all exceptional orbits in Σ and F is diffeomorphic to $S^{2n-3} \times D^4$ and such that $F \cap K = \partial F = \partial K$. Moreover, F and K are unique up to an equivariant isotopy of Σ. Furthermore, there are (2n+1)-dimensional prime G-manifolds $W_1,\ldots,W_k$ in K such that (i) $K = W_1 \cup \cdots \cup W_k$, (ii) for any $i,j = 1,\ldots,k$ with $|i-j| > 1$, $W_i \cap W_j = \emptyset$ and (iii) for $i = 1,\ldots,k-1$, $W_i^* \cap W_{i+1}^* = \partial W_i^* \cap \partial W_{i+1}^*$ is a closed tubular neighborhood of a 2-sphere in ∂W_i^* representing a generator of $H_2(\partial W_i^*)$. The set $\{W_1,\ldots,W_k\}$ is determined by K and conversely.

By (3.1), we know how a pseudo-free circle action on a homotopy (2n+1)-sphere is to be constructed. First, take a finite number of (2n+1)-dimensional prime G-manifolds $W_1,\ldots,W_k$ and paste them together to obtain a composite G-manifold K as said in (3.1). Then find a free circle action on $F = S^{2n-3} \times D^4$ such that there is an equivariant diffeomorphism $f : \partial F \longrightarrow \partial K$ with $F \cup_f K$ being a homotopy 7-sphere. As we shall see in section 8, for $n = 3$, such an F always exists and there are infinitely many distinct $F \cup_f K$ by varying the equivariant diffeomorphism f. However, we do not know whether a similar result holds for $n > 3$.

4. NORMAL BUNDLE OF AN IMMERSED 2-SPHERE

Let M be a compact $(2n+1)$-manifold on which there is a pseudo-free circle action and let E be the union of all exceptional orbits in M. We have said that $M^* - E^*$ is a $(2n)$-manifold. Therefore for any immersed (or imbedded) 2-sphere S in $M^* - E^*$, there is a normal bundle ν of S. It is known that for $n \geqq 3$, the normal bundle ν is trivial iff the value of $w_2(M^*-E^*)$ at the mod 2 homology class represented by S vanishes.

(4.1) Lemma. <u>Let Σ be a homotopy $(2n+1)$-sphere, $n \geqq 2$, on which there is a pseudo-free circle action, and let $Gb_1,\ldots,Gb_k$ be distinct exceptional orbits in Σ such that G acts freely on $\Sigma - \cup_{i=1}^k Gb_i$. Let q_i be the order of the isotropy group at b_i, $i = 1,\ldots,k$, and let $q = q_1 \cdots q_k$. Then Σ^* is a $(2n)$-manifold with singularities $b_1^*,\ldots,b_k^*$ such that the second Stiefel-Whitney class $w_2(\Sigma^*-\{b_1^*,\ldots,b_k^*\})$ vanishes iff $q + n$ is even.</u>

Proof. Let us use Σ in place of M in the discussion given in section 2. It is obvious that our assertion is equivalent to the statement that the second Stiefel-Whitney class $w_2(X^*)$ vanishes iff $q + n$ is even.

Let β be a generator of $H^2(X^*)$. Using Gysin's cohomology sequence of the circle fibration $X \longrightarrow X^*$, we can see that β^r is a generator of $H^{2r}(X^*)$ for $r = 1,\ldots,n-1$. By (2.3), there is a generator α_r of $H^{2r}(X^*,\partial X^*)$ having $q\beta^r$ as its image, $r = 1,\ldots,n-1$. For any $r = 1,\ldots,n-2$, it is clear that

$$\alpha_r \cup \beta = \alpha_{r+1}, \qquad \alpha_r \cup \alpha_1 = q\alpha_{r+1} .$$

By Poincaré duality, $\alpha_{n-1} \cup \beta$ is a generator of $H^{2n}(X^*,\partial X^*)$ which we denote by α_n. Then

$$\alpha_{n-1} \cup \beta = \alpha_n, \qquad \alpha_{n-1} \cup \alpha_1 = q\alpha_n .$$

Let

$$\bar{\alpha}_r = \alpha_r \bmod 2, \quad \bar{\beta} = \beta \bmod 2$$

be elements in the corresponding mod 2 cohomology groups. Then for $r = 1,\ldots,n-1$,

$$\bar{\alpha}_r \cup \bar{\beta} = \bar{\alpha}_{r+1},\ \bar{\alpha}_r \cup \bar{\alpha}_1 = q\bar{\alpha}_{r+1}\ .$$

By induction on r, one can show that for $r = 1,\dots,n-1$,

$$Sq^2\bar{\alpha}_r = (q+r-1)\bar{\alpha}_{r+1}\ .$$

Therefore, if we let $\bar{v} = (q+n-2)\bar{\beta}$, we have

$$Sq^2\bar{\alpha}_{n-1} = \bar{\alpha}_{n-1} \cup \bar{v}\ .$$

Applying Wu's formula (see, for example, [6]), we obtain

$$w_2(X^*) = \bar{v} = (q+n)\bar{\beta}\ .$$

Hence our assertion follow.

(4.2) Lemma. <u>Under the hypothesis of (4.1), if S is an immersed 2-sphere in $\Sigma^* - \{b_1^*,\dots,b_k^*\}$ representing a generator of $H_2(\Sigma^*-\{b_1^*,\dots,b_k^*\})$, then the normal bundle of S is stably trivial iff $q + n$ is even. Hence, for $n > 2$, the normal bundle of S is trivial iff $q + n$ is even.</u>

Notice that if S is an arbitrary immersed 2-sphere in $\Sigma^* - \{b_1^*,\dots,b_k^*\}$ instead, then the normal bundle of S is non-trivial iff $q + n$ is odd and S represents the non-zero element of the second mod 2 homology group of $\Sigma^* - \{b_1^*,\dots,b_k^*\}$.

5. PRIME G-MANIFOLDS

This section studies and classifies 7-dimensional prime G-manifolds, and together with the following two sections prepares for the construction of all pseudo-free circle actions on homotopy 7-spheres to be given in (8.6) of section 8.

For simplicity, every prime G-manifold below is assumed to be 7-dimensional even though many results can be generalized to (2n+1)-dimensional prime G-manifolds for any $n \geqq 3$.

Let W be a prime G-manifold and Gb be the exceptional orbit in W. As in section 2, a prime G-manifold $W' \subset W - \partial W$ may be constructed as follows. Let D be a slice at b which is a closed 6-disk of center b contained in $W - \partial W$ and on which G_b acts orthogonally. If q is the order of G_b, then $G_b = Z_q$ and

$L = \partial D^*$ is a 5-dimensional lens space of fundamental group Z_q. Let

$$\xi : S^1 \longrightarrow L$$

be an imbedding which is the image of a closed path in $\partial(GD)$ homotopic to the exceptional orbit in GD. Then there is a (+)-imbedding

$$\varphi^* : D^2 \times D^4 \longrightarrow W^* - \partial W^*$$

such that (i) $\varphi^*(D^2 \times D^4) \cap D^* = \varphi^*(S^1 \times D^4)$, (ii) $\varphi^*(u;0,0) = \xi(u)$ for all $u \in S^1$ and (iii) $\varphi^*|(D^2 \times \{0\}, S^1 \times \{0\})$ represents a generator of $H_2(W^*,D^*)$. Therefore we have a prime G-manifold W' with

$$W'^* = \varphi^*(D^2 \times D^4) \cup D^* .$$

Since $H_2(\partial W'^*) \longrightarrow H_2(W^*-\{b^*\})$ is an isomorphism, it follows from (2.4) that W' is unique up to an equivariant isotopy of W and the closure of $W^* - W'^*$ is a simply connected h-cobordism. Hence there is a (+)-equivalence of W onto W'.

Since the action of $G_b = \mathbb{Z}q$ on D is orthogonal and is free on ∂D, there are integers r_1, r_2, r_3 such that $(q, r_1r_2r_3) = 1$ and there is a (+) $\mathbb{Z}q$-equivalence of D onto $D(q;r_1,r_2,r_3)$, where $D(q;r_1,r_2,r_3)$ denotes D^6 together with an orthogonal $\mathbb{Z}q$-action given by

$$g(z_1,z_2,z_3) = (g^{r_1}z_1, g^{r_2}z_2, g^{r_3}z_3) .$$

Let G act on $S^1 \times D^6$ such that

$$g(z;z_1,z_2,z_3) = (g^q z; g^{r_1}z_1, g^{r_2}z_2, g^{r_3}z_3) ,$$

and let $V(q;r_1,r_2,r_3)$ denote $S^1 \times D^6$ together with this circle action. Then we may regard $D(q;r_1,r_2,r_3)$ a slice for $V(q;r_1,r_2,r_3)$ at $(1;0,0,0)$ by setting $(1;z_1,z_2,z_3) = (z_1,z_2,z_3)$. Therefore the (+) $\mathbb{Z}q$-equivalence of D onto $D(q;r_1,r_2,r_3)$ can be extended to a (+)-equivalence λ of GD onto $V(q;r_1,r_2,r_3)$.

Let G act on $S^1 \times (D^2 \times D^4)$ such that

$$g(z;u;v_1,v_2) = (gz;u;v_1,v_2) .$$

Then there is an equivariant imbedding φ of $S^1 \times (D^2 \times D^4)$ into $W - \partial W$ which is a lifting of φ^*, and

$$f = \lambda\varphi : S^1 \times (S^1 \times D^4) \longrightarrow \partial V(q;r_1,r_2,r_3)$$

is an equivariant imbedding such that λ can be extended to a (+)-equivalence

$$\lambda : W' \longrightarrow S^1 \times (D^2 \times D^4) \cup_f V(q;r_1,r_2,r_3) .$$

Hence there is a (+)-equivalence of W onto $S^1 \times (D^2 \times D^4) \cup_f V(q;r_1,r_2,r_3)$. This shows that every prime G-manifold, up to a (+)-equivalence, may be constructed as follows.

Let q, r_1, r_2, r_3 be any integers such that

$$q > 1, \quad (q,r_1r_2r_3) = 1 .$$

Let $V(q;r_1,r_2,r_3)$ be the oriented compact 7-manifold $S^1 \times D^6$ on which there is a pseudo-free action of G given by

$$g(z;z_1,z_2,z_3) = (g^q z;g^{r_1}z_1,g^{r_2}z_2,g^{r_3}z_3) ,$$

and let ξ be an imbedding of S^1 into the lens space $L(q;r_1,r_2,r_3) = \partial V(q;r_1,r_2,r_3)^*$ which is the image of a closed path in $V(q;r_1,r_2,r_3)$ homotopic to the exceptional orbit $S^1 \times \{0\}$. Since the normal bundle of $\xi(S^1)$ in $L(q;r_1,r_2,r_3)$ is trivial, there exists a (-)-imbedding $f^* : S^1 \times D^4 \longrightarrow L(q;r_1,r_2,r_3)$ such that $f(u;0) = \xi(u)$ for all $u \in S^1$. Let G act freely on $S^1 \times (D^2 \times D^4)$ such that

$$g(z;u;v_1,v_2) = (gz;u;v_1,v_2) .$$

Then f^* can be lifted to a (-)-equivariant imbedding $f : S^1 \times (S^1 \times D^4) \longrightarrow$ $\longrightarrow \partial V(q;r_1,r_2,r_3)$. Hence

$$W = S^1 \times (D^2 \times D^4) \cup_f V(q;r_1,r_2,r_3)$$

is an oriented compact connected 7-manifold on which there is a pseudo-free circle action with exactly one exceptional orbit.

(5.1) <u>For any integers</u> q, r_1, r_2, r_3 <u>such that</u>

$$q > 1, \quad (q, r_1 r_2 r_3) = 1,$$

<u>the constructed</u>

$$W = S^1 \times (D^2 \times D^4) \cup_f V(q; r_1, r_2, r_3)$$

<u>is a prime</u> G-<u>manifold, which, up to a</u> (+)-<u>equivalence, depends only on the integers</u> q, r_1, r_2, r_3 <u>and the regular homotopy class of the</u> (-)-<u>imbedding</u> $f^* : S^1 \times D^4 \longrightarrow L(q; r_1, r_2, r_3)$.

Proof. Since f^* is diffeotopic to an imbedding $k^* : S^1 \times D^4 \longrightarrow L(q; r_1, r_2, r_3)$ such that for any $u \in S^1$,

$$k^*(u; 0, 0) = G(\bar{u}; 1/\sqrt{3}, 1/\sqrt{3}, 1/\sqrt{3}) ,$$

f is equivariantly diffeotopic to an equivariant imbedding $k : S^1 \times (S^1 \times D^4) \longrightarrow \partial V(q; r_1, r_2, r_3)$ such that for any $u \in S^1$,

$$k(1; u; 0, 0) = (u; 1/\sqrt{3}, 1/\sqrt{3}, 1/\sqrt{3}) .$$

Therefore $f|\{1\} \times (S^1 \times \{0\})$ which is homotopic to $k|\{1\} \times (S^1 \times \{0\})$ represents a generator of $\pi_1(S^1 \times D^6)$.

Using this fact and van Kampen's theorem, we can verify that W is simply connected. Using the same fact and the exactness of Mayer-Vietoris sequence, we can also verify that W has the integral homology of S^3. Hence, as in the proof of (2.4), it follows from Smale's theorem that W is diffeomorphic to $S^3 \times D^4$. This proves that W is a prime G-manifold.

Suppose that $f' : S^1 \times (S^1 \times D^4) \longrightarrow \partial V(q; r_1, r_2, r_3)$ is a second (-)-equivariant imbedding such that $f'^* : S^1 \times D^4 \longrightarrow L(q; r_1, r_2, r_3)$ is regularly homotopic to f^*. Then there is a regular homotopy

$$F^* : S^1 \times D^4 \times [0,1] \longrightarrow L(q; r_1, r_2, r_3) \times [0,1]$$

connecting f^* and f'^*. Since $2(1 + \dim(S^1 \times \{0\})) < \dim L(q; r_1, r_2, r_3)$, we may assume that $F^*|S^1 \times \{0\} \times [0,1]$ is an imbedding, as F^* can be approximated by a regular homotopy satisfying this condition. Now F^* restricted to a neighborhood of $S^1 \times \{0\} \times [0,1]$ is an imbedding. Therefore we may assume that F^* is an imbedding. By lifting F^* we have an equivariant diffeotopy connecting f and f'. Therefore there is a (+)-equivalence of W onto $S^1 \times (D^2 \times D^4) \cup_{f'} V(q;r_1,r_2,r_3)$ and hence the proof of (5.1) is completed.

Let q, r_1, r_2, r_3 be given and let ξ be an imbedding of S^1 into $L(q;r_1,r_2,r_3)$ such that it is the image of a closed path in $V(q;r_1,r_2,r_3)$ homotopic to the exceptional orbit $S^1 \times \{0\}$. Since $\pi_1(SO(4)) \cong \mathbb{Z}_2$, there are, up to a homotopy, only two trivializations of the normal bundle of $\xi(S^1)$, each of which determines a regular homotopy class of a (-)-imbedding $f^* : S^1 \times D^4 \longrightarrow L(q;r_1,r_2,r_3)$ such that $f^*(u;0,0) = \xi(u)$ for all $u \in S^1$. Hence the construction above gives us, up to a (+)-equivalence, at most two prime G-manifolds.

Since ξ is regularly homotopic to the immersion $\eta : S^1 \longrightarrow L(q;r_1,r_2,r_3)$ given by

$$\eta(u) = G(\bar{u};1,0,0) ,$$

we have an imbedding f^* regularly homotopic to the immersion $h^* : S^1 \times D^4 \longrightarrow$ $\longrightarrow L(q;r_1,r_2,r_3)$ given by

$$h^*(u;v_1,v_2) = G(\bar{u};1/\rho,v_1/\rho,v_2/\rho)$$

with $\rho = (1+|v_1|^2+|v_2|^2)^{1/2}$. With this f^*, the prime G-manifold W is uniquely determined (up to a (+)-equivalence). Denote it by

$$W(q;r_1,r_2,r_3) .$$

We have the following

(5.2) <u>Theorem</u>. <u>For any integers</u> q, r_1, r_2, r_3 <u>such that</u>

$$q > 1, \quad (q,r_1r_2r_3) = 1 ,$$

<u>there is a unique prime</u> G-<u>manifold</u>

$$W(q;r_1,r_2,r_3) = S^1 \times (D^2 \times D^4) \cup_f V(q;r_1,r_2,r_3)$$

<u>where</u> $f : S^1 \times (S^1 \times D^4) \longrightarrow \partial V(q;r_1,r_2,r_3)$ <u>is a</u> (-)-<u>equivariant imbedding such</u>

that the imbedding $f^* : S^1 \times D^4 \longrightarrow L(q;r_1,r_2,r_3)$ *induced by* f *is regularly homotopic to the immersion* h^* *given by*

$$h^*(u;v_1,v_2) = G(\bar{u};1/\rho,v_1/\rho,v_2/\rho)$$

with $\rho = (1+|v_1|^2+|v_2|^2)^{1/2}$. *Up to a* (+)-*equivalence, every prime* G-*manifold may be constructed this way.*

Proof. The first part is a consequence of (5.1) and the construction of $W(q;r_1,r_2,r_3)$.

Let W be an arbitrary given prime G-manifold. As seen at the beginning of this section, there are integers q, r_1, r_2, r_3 with $q > 1$ and $(q,r_1r_2r_3) = 1$, and there is a (-)-equivariant imbedding $f : S^1 \times (D^2 \times D^4) \longrightarrow \partial V(q;r_1,r_2,r_3)$ such that (i) if $f^* : S^1 \times D^4 \longrightarrow L(q;r_1,r_2,r_3)$ is the imbedding induced by f, then $f^*|S^1 \times \{0\}$ is the image of a closed path in $V(q;r_1,r_2,r_3)$ homotopic to the exceptional orbit $S^1 \times \{0\}$, and (ii) there is a (+)-equivalence of W onto $S^1 \times (D^2 \times D^4) \cup_f V(q;r_1,r_2,r_3)$.

If f^* is regularly homotopic to the immersion h^* given above, then our assertion follows. Hence we have only to consider the case that f^* is not regularly homotopic to h^*. In this case f^* is regularly homotopic to the immersion $h'^* : S^1 \times D^4 \longrightarrow L(q;r_1,r_2,r_3)$ given by

$$h'^*(u;v_1,v_2) = G(\bar{u};1/\rho,v_1/\rho,\bar{u}v_2/\rho)$$

with $\rho = (1+|v_1|^2+|v_2|^2)^{1/2}$. Let $\lambda : V(q;r_1,r_2,r_3) \longrightarrow V(q;r_1,r_2,r_3-q)$ be the (+)-equivalence given by

$$\lambda(z;z_1,z_2,z_3) = (z;z_1,z_2,z^{-1}z_3)$$

and $\lambda^* : L(q;r_1,r_2,r_3) \longrightarrow L(q;r_1,r_2,r_3-q)$ the diffeomorphism induced by λ. Then $\lambda^*h'^* : S^1 \times D^4 \longrightarrow L(q;r_1,r_2,r_3-q)$ is given by $\lambda^*h'^*(u;v_1,v_2) = G(\bar{u};1/\rho,v_1/\rho,v_2/\rho)$. Since $\lambda^*h'^*$ is regularly homotopic to λ^*f^*, it follows that $S^1 \times (D^2 \times D^4) \cup_{\lambda f} V(q;r_1,r_2,r_3-q) = W(q;r_1,r_2,r_3-q)$. Clearly λ can be extended to a (+)-equivalence of $S^1 \times (D^2 \times D^4) \cup_f V(q;r_1,r_2,r_3)$ onto $S^1 \times (D^2 \times D^4) \cup_{\lambda f} V(q;r_1,r_2,r_3-q)$. We infer that there is a (+)-equivalence of W onto $W(q;r_1,r_2,r_3-q)$. Hence the proof is completed.

Now we are in a position to classify all prime G-manifolds up to a (+)-equivalence. By (5.2), it is sufficient to classify all $W(q;r_1,r_2,r_3)$'s.

(5.3) Let q, r_1, r_2, r_3 be integers such that $q > 1$ and $(q,r_1r_2r_3) = 1$ and let r_1', r_2', r_3' be integers such that $r_i' \equiv r_i \bmod q$, $i = 1,2,3$, and $r_1' + r_2' + r_3' \equiv r_1 + r_2 + r_3 \bmod 2q$. Then $(q,r_1'r_2'r_3') = 1$ and there is a (+)-equivalence of $W(q;r_1,r_2,r_3)$ onto $W(q;r_1',r_2',r_3')$.

Proof. By hypothesis, there are integers S_1, S_2, S_3 such that $r_i' = r_i + S_iq$, $i = 1,2,3$, and $S_1 + S_2 + S_3$ is even.

Assume first that $S_1 = 0$. Then $r_1' = r_1$ and $S_2 + S_3$ is even. Let $h^* : S^1 \times D^4 \longrightarrow L(q;r_1,r_2,r_3)$ be the immersion used in the construction of $W(q;r_1,r_2,r_3)$ and $h'^* : S^1 \times D^4 \longrightarrow L(q;r_1,r_2',r_3')$ the analogs for $W(q;r_1,r_2',r_3')$. Let

$$\lambda : V(q;r_1,r_2,r_3) \longrightarrow V(q;r_1,r_2',r_3')$$

be the (+)-equivalence given by

$$\lambda(z;z_1,z_2,z_3) = (z;z_1,z^{S_2}z_2,z^{S_3}z_3)$$

and $\lambda^* : L(q;r_1,r_2,r_3) \longrightarrow L(q;r_1,r_2',r_3')$ the diffeomorphism induced by λ. Let $\mu^* : S^1 \times D^4 \longrightarrow S^1 \times D^4$ be the diffeomorphism given by $\mu^*(u;v_1,v_2) = (u;u^{S_2}v_1,u^{S_3}v_2)$. Then $\lambda^*h^* = h'^*\mu^*$. Since $S_2 + S_3$ is even, it follows from $\pi_1(SO(4)) \cong \mathbb{Z}_2$ that μ^* is isotopic to the identity. Therefore λ^*h^* is regularly homotopic to h'^* and hence λ can be extended to a (+)-equivalence of $W(q;r_1,r_2,r_3)$ onto $W(q;r_1,r_2',r_3')$.

Next we assert that h^* is regularly homotopic to the immersion $k^* : S^1 \times D^4 \longrightarrow L(q;r_1,r_2,r_3)$ given by

$$k^*(u;v_1,v_2) = G(\bar{u};v_1/\rho,1/\rho,v_2/\rho) .$$

Let $H : S^1 \times D^4 \times [0,1] \to L(q;r_1,r_2,r_3) \times [0,1]$ be given by

$$H(u;v_1,v_2;t) = (G(\bar{u};1/\rho,(t+v_1)/\rho,(t+v_2)/\rho);t)$$

with $\rho = (1+|t+v_1|^2+|t+v_2|^2)^{1/2}$. Then H is an immersion so that h^* is regularly homotopic to the immersion $h_1^* : S^1 \times D^4 \longrightarrow L(q;r_1,r_2,r_3)$ given by

$$h_1^*(u;v_1,v_2) = G(\bar{u};1/\rho,(1+v_1)/\rho,(1+v_2)/\rho) .$$

Similarly k^* is regularly homotopic to the immersion k_1^* given by

$$k_1^*(u;v_1,v_2) = G(\bar{u};(1+v_1)/\rho,1/\rho,(1+v_2)/\rho) .$$

We have shown that r_2 may be replaced by any integer r_2' with $r_2' \equiv r_2$ mod 2q. Therefore we may assume that $r_1 r_2 > 0$. With this assumption, it can be shown that the map $H' : S^1 \times D^4 \times [0,1] \longrightarrow L(q;r_1,r_2,r_3) \times [0,1]$ given by

$$H'(u;v_1,v_2;t) = (G(\bar{u};(1+tv_1)/\rho,(1+(t-1)v_1)/\rho;(1+v_2)/\rho),t)$$

with $\rho = (|1+tv_1|^2+|1+(t-1)v_1|^2+|1+v_2|^2)^{1/2}$ is regular at $S^1 \times \{0\} \times [0,1]$ so that its restriction to a small neighborhood of $S^1 \times \{0\} \times [0,1]$ is an immersion. Hence k_1^* is regularly homotopic to the immersion k_2^* given by

$$k_2^*(u;v_1,v_2) = G(\bar{u};1/\rho,(1-v_1)/\rho,(1+v_2)/\rho) .$$

Since the diffeomorphism of $S^1 \times D^4$ onto $S^1 \times D^4$ which maps every $(u;v_1,v_2)$ into $(u;-v_1,v_2)$ is isotopic to the identity, h_1^* and k_2^* are regularly homotopic. Hence our assertion follows.

Now we are able to prove the case $S_2 = 0$ just like the case $S_1 = 0$ using k^* in place of h^*. The general case is a consequence of these two special cases.

(5.4) _Let_ q, r_1, r_2, r_3 _be integers such that_ $q > 1$ _and_ $(q,r_1r_2r_3) = 1$. _Then there exist integers_ r_1', r_2', r_3' _such that_ q, r_1', r_2', r_3' _are relatively prime to one another_, $r_i' \equiv r_i$ mod q, $i = 1,2,3$, _and_ $r_1' + r_2' + r_3' \equiv r_1 + r_2 + r_3$ mod 2q.

Proof. There are integers r_1'', r_2'', r_3'' such that $r_i'' \equiv r_i$ mod q, $i = 1,2,3$, $r_1'' + r_2'' + r_3'' = r_1 + r_2 + r_3$ and r_2'' and r_3'' are odd. In fact, if q is even, it follows from $(q,r_1r_2r_3) = 1$ that r_2 and r_3 are odd. Hence in this case we may let $r_i'' = r_i$, $i = 1,2,3$. If q is odd, we may let

$$r_1'' = r_1 - \varepsilon_2 q - \varepsilon_3 q, \quad r_2'' = r_2 + \varepsilon_2 q, \quad r_3'' = r_3 + \varepsilon_3 q ,$$

where ε_i is 0 or 1 according as r_i is odd or even, $i = 2,3$.

Since $(q, r_1 r_2 r_3) = 1$ and r_2'' is an odd integer with $r_2'' \equiv r_2 \bmod q$, we infer that $(r_2'', 2q) = 1$. Let d be the largest divisor of r_1'' with $(d, r_2'') = 1$ and let

$$r_1' = r_1'', \quad r_2' = r_2'' + 2qd .$$

Next, let d' be the largest divisor of $r_1' r_2'$ with $(d', r_3'') = 1$ and let

$$r_3' = r_3'' + 2qd' .$$

Then r_1', r_2' and r_3' are as desired.

(5.5) Proposition. Given any prime G-manifold W, there is an orthogonal pseudo-free circle action on S^7 such that W can be canonically imbedded into S^7. Precisely speaking, if $W = W(q; r_1, r_2, r_3)$, where q, r_1, r_2, r_3 are relatively prime to one another, and G acts on S^7 such that

$$g(z_0, z_1, z_2, z_3) = (g^q z_0, g^{r_1} z_1, g^{r_2} z_2, g^{r_3} z_3)$$

then there is a (+)-equivariant imbedding $\lambda : W \longrightarrow S^7$ such that $\lambda(W)$ is a closed tubular neighborhood of the 3-sphere

$$\{(g^q \cos\theta, g'^{r_1} \sin\theta/\sqrt{3}, g'^{r_2} \sin\theta/\sqrt{3}, g'^{r_3} \sin\theta/\sqrt{3}) \mid g, g' \in G \text{ and } 0 \leqq \theta \leqq \pi/2\} .$$

Proof. By (5.2), we may let $W = W(q; r_1, r_2, r_3)$, where q, r_1, r_2, r_3 are integers with $q > 1$ and $(q, r_1 r_2 r_3) = 1$. Because of (5.3) and (5.4), we may assume that q, r_1, r_2, r_3 are relatively prime to one another. Then there is an orthogonal pseudo-free circle action on S^7 given by

$$g(z_0, z_1, z_2, z_3) = (g^q z_0, g^{r_1} z_1, g^{r_2} z_2, g^{r_3} z_3) .$$

Let

$$\lambda' : V(q;r_1,r_2,r_3) \longrightarrow S^7$$

be the (+)-equivariant imbedding given by

$$\lambda'(z;z_1,z_2,z_3) = (z/\rho', z_1/\rho', z_2/\rho', z_3/\rho')$$

with $\rho' = (1+|z_1|^2+|z_2|^2+|z_3|^2)^{1/2}$. We assert that λ' can be extended to a desired imbedding $\lambda : W \longrightarrow S^7$.

As seen in the proof of (5.3), we may assume

$$W = S^1 \times (D^2 \times D^4) \cup_f V(q;r_1,r_2,r_3)$$

where $f : S^1 \times (S^1 \times D^4) \longrightarrow \partial V(q;r_1,r_2,r_3)$ is given by

$$f(z;u;v_1,v_2) = (z^q\bar{u}; z^{r_1}/\rho, z^{r_2}(1+\delta v_1)/\rho, z^{r_3}(1+\delta v_2)/\rho)$$

with $\rho = (1+|1+\delta v_1|^2+|1+\delta v_2|^2)^{1/2}$ and δ being a small positive number. Let δ be so small that there is a (+)-equivariant imbedding $\lambda'' : S^1 \times (D^2 \times D^4) \longrightarrow S^7$ given by

$$\lambda''(z;u;v_1,v_2) = (z^q\bar{u}/\rho'', z^{r_1}/\rho\rho'', z^{r_2}(1+\delta v_1)/\rho\rho'', z^{r_3}(1+\delta v_2)/\rho\rho''),$$

where ρ is as above and $\rho'' = (|u|^2+1)^{1/2}$. Since $\lambda''|S^1 \times (S^1 \times D^4) = \lambda' f$, there is a map

$$\lambda : W \longrightarrow S^7$$

such that $\lambda|V(q;r_1,r_2,r_3) = \lambda'$ and $\lambda|S^1 \times (D^2 \times D^4) = \lambda''$. It is not hard to show that λ is a desired imbedding.

(5.6) Let q, r_1, r_2, r_3 be integers such that $q > 1$ and $(q, r_1r_2r_3) = 1$. If r_1', r_2', r_3' are integers such that there is a permutation σ of $\{1,2,3\}$ and there are $\varepsilon_1, \varepsilon_2, \varepsilon_3 \in \{1,-1\}$ such that $\varepsilon_1\varepsilon_2\varepsilon_3 = 1$ and $\varepsilon_i r_i' = r_{\sigma(i)}$, $i = 1,2,3$, then $(q, r_1'r_2'r_3') = 1$ and there is a (+)-equivalence of $W(q;r_1,r_2,r_3)$ onto $W(q;r_1',r_2',r_3')$.

Proof. By (5.5), there is an orthogonal pseudo-free action of G on S^7 such that $W(q;r_1,r_2,r_3)$ can be canonically imbedded into S^7. Let

$$\lambda : S^7 \longrightarrow S^7$$

be the orthogonal map such that $\lambda(z_0,z_1,z_2,z_3) = (z_0',z_1',z_2',z_3')$ implies that $z_0' = z_0$ and $z_i' = z_{\sigma(i)}$ or $\bar{z}_{\sigma(i)}$ according as ε_i is 1 or -1, $i = 1,2,3$. Since $\varepsilon_1\varepsilon_2\varepsilon_3 = 1$, λ is a (+)-equivalence.

Denote by S'^7 the unit 7-sphere together with the pseudo-free orthogonal action of G such that λ, as a map of S^7 into S'^7, is equivariant. Then there is a canonical imbedding of $W(q';r_1',r_2',r_3')$ into S'^7 with $\lambda W(q;r_1,r_2,r_3)$ as the image. Hence λ defines a (+)-equivalence of $W(q;r_1,r_2,r_3)$ onto $W(q;r_1',r_2',r_3')$.

Combining (5.3) and (5.6), we have

(5.7) Theorem. *Let q, r_1, r_2, r_3 be integers such that $q > 1$ and $(q,r_1r_2r_3) = 1$. If q', r_1', r_2', r_3' are integers such that $q' = q$ and there is a permutation σ of $\{1,2,3\}$ and there are $\varepsilon_1,\varepsilon_2,\varepsilon_3 \in \{1,-1\}$ such that $\varepsilon_1\varepsilon_2\varepsilon_3 = 1$, $\varepsilon_i r_i' \equiv r_{\sigma(i)} \bmod q$, $i = 1,2,3$, and $\varepsilon_1 r_1' + \varepsilon_2 r_2' + \varepsilon_3 r_3' \equiv r_1 + r_2 + r_3 \bmod 2q$, then $q' > 1$ and $(q',r_1'r_2'r_3') = 1$ and there is a (+)-equivalence of $W(q;r_1,r_2,r_3)$ onto $W(q';r_1',r_2',r_3')$.*

The converse of (5.7) is also true. In fact, we have

(5.8) Theorem. *Let q, r_1, r_2, r_3 and q', r_1', r_2', r_3' be integers such that $q > 1$, $(q,r_1r_2r_3) = 1$ and $q' > 1$, $(q',r_1'r_2'r_3') = 1$ and there is a (+)-equivalence of $W(q;r_1,r_2,r_3)$ onto $W(q';r_1',r_2',r_3')$. Then $q = q'$ and there is a permutation σ of $\{1,2,3\}$ and there are $\varepsilon_1,\varepsilon_2,\varepsilon_3 \in \{1,-1\}$ such that $\varepsilon_1\varepsilon_2\varepsilon_3 = 1$, $\varepsilon_i r_i' \equiv r_{\sigma(i)} \bmod q$, $i = 1,2,3$, and $\varepsilon_1 r_1' + \varepsilon_2 r_2' + \varepsilon_3 r_3' \equiv r_1 + r_2 + r_3 \bmod 2q$.*

Proof. Let λ be a (+)-equivalence of $W = W(q;r_1,r_2,r_3)$ onto $W' = W(q';r_1',r_2',r_3')$. Then λ maps the exceptional orbit $S^1 \times \{0\}$ in W into that in W'. Therefore the isotropy groups associated with these two exceptional orbits are of the same order. Hence $q' = q$.

Let $b = (1;0,0,0) \in V(q;r_1,r_2,r_3) \subset W(q;r_1,r_2,r_3)$ and let, for any $\tau \in (0,1]$,

$$D_\tau = \{1\} \times \tau D^6 \subset V(q;r_1,r_2,r_3) \subset W(q;r_1,r_2,r_3) .$$

Clearly D_τ is a slice at b on which $\mathbb{Z}q = G_b$ acts orthogonally. Let b' and D'_τ be the analogs of b and D_τ for $W(q';r'_1,r'_2,r'_3)$. Replacing λ by one equivariantly isotopy to λ if necessary, we may assume that for some $\tau \in (0,1)$, $\lambda(D_\tau) = D'_\tau$ and $\lambda' : D_\tau \longrightarrow D'_\tau$ defined by λ is an orthogonal map such that for some permutation σ of $\{1,2,3\}$, $\lambda'(1;z_1,z_2,z_3) = (1;z'_1,z'_2,z'_3)$ implies $z'_i = z_{\sigma(i)}$ or $\bar{z}_{\sigma(i)}$, $i = 1,2,3$. Let ε_i be 1 or -1 according as $z'_i = z_{\sigma(i)}$ or $\bar{z}_{\sigma(i)}$. Since λ is orientation-preserving, so is λ'. Hence $\varepsilon_1\varepsilon_2\varepsilon_3 = 1$. Since λ is G-equivariant, λ' is $\mathbb{Z}q$-equivariant. Hence $\varepsilon_i r'_i \equiv r_{\sigma(i)} \bmod q$, $i = 1,2,3$.

By (5.7), we may assume that σ is the identity, $\varepsilon_1 = \varepsilon_2 = \varepsilon_3 = 1$ and $r'_1 = r_1$. Then we may assume that $\lambda' : D_\tau \longrightarrow D'_\tau$ is the identity.

Whenever φ is a map of D^2 into W such that for any $u \in S'$, $\varphi(u) = (u;0,0,0) \in V(q;r_1,r_2,r_3)$, φ followed by the projection of W onto W^* represents a generator α_1 of $H_2(W^*)$ which is independent of the choice of φ. Let

$$X^* = W^* - (D^*_\tau - \partial D^*_\tau) .$$

Since there is a natural isomorphism of $H_2(W^*)$ onto $H_2(X^*,\partial D^*_\tau)$, we have a generator α of $H_2(X^*,\partial D^*_\tau)$ corresponding to α_1. From the construction of $W(q;r_1,r_2,r_3)$, it is easily seen that there is an immersion $\psi^* : D^2 \times D^4 \longrightarrow X^*$ such that for any $(u;v_1,v_2) \in S^1 \times D^4$, $\psi^*(u;v_1,v_2) = G(\bar{u};\tau/\rho,\tau v_1/\rho,\tau v_2/\rho)$ with $\rho = (1+|v_1|^2+|v_2|^2)^{1/2}$, and $\psi^*|(D^2 \times \{0\}, S^1 \times \{0\})$ represents α. Let φ', α'_1, X'^*, α', ψ'^* be the analogs of φ, α_1, X^*, α, ψ^* for W'. Since we may let $\lambda\varphi = \varphi'$, $\lambda(\alpha_1) = \alpha'_1$ so that $\lambda(\alpha) = \alpha'$. By assumption $r'_1 = r_1$ so that $\lambda^*\psi^*|S^1 \times \{0\} = \psi'^*|S^1 \times \{0\}$. Therefore it follows from $\lambda(\alpha) = \alpha'$ that $\lambda^*\psi^*|D^2 \times \{0\}$ and $\psi'^*|D^2 \times \{0\}$ are regularly homotopic relative to $S^1 \times \{0\}$. Hence the immersions

$$\lambda^*\psi^*, \psi'^* : S^1 \times D^4 \longrightarrow \partial D'^*_\tau$$

are regularly homotopic relative to $S^1 \times \{0\}$.

Let

$$\mu^* : S^1 \times D^4 \longrightarrow S^1 \times D^4$$

be the diffeomorphism given by

$$\mu^*(u;v_1,v_2) = (u;\bar{u}^{(r_2'-r_2)/q}v_1,\bar{u}^{(r_3'-r_3)/q}v_2) .$$

Then $\lambda^*\psi^* = \psi'^*\mu^*$ so that μ^* is isotopic to the identity relative to $S^1 \times \{0\}$. Therefore

$$(r_2'-r_2)/q + (r_3'-r_3)/q$$

is even and hence $r_2' + r_3' \equiv r_2 + r_3 \mod 2q$, completing our proof.

Remark. Combining (5.2), (5.7) and (5.8), we have a classification theorem for prime G-manifolds up to a (+)-equivalence. Notice that, if prime G-manifolds are not required to be oriented, we can still establish (5.2), (5.7) and (5.8) by dropping orientation throughout as well as $\varepsilon_1\varepsilon_2\varepsilon_3 = 1$ from (5.7) and (5.8). This gives a classification theorem for non-oriented prime G-manifolds up to an equivalence.

Let W be any given prime G-manifold. By (5.2), we may let $W = W(q;r_1,r_2,r_3)$. Then ∂W^* is a closed 5-manifold obtained from the lens space $L(q;r_1,r_2,r_3)$ by performing a surgery associated with the attaching map f^* in the construction of $W(q;r_1,r_2,r_3)$. Therefore it is connected and simply connected and has the integral homology of $S^2 \times S^3$. By Barden's theorem [7], there is a fibration

$$\pi : \partial W^* \longrightarrow S^2$$

of fibre S^3 and structural group $SO(4)$. Up to an isomorphism, there are only two 3-sphere bundles over S^2 with structural group $SO(4)$ of which one is trivial and the other is non-trivial. Moreover, such a bundle is trivial iff the second Stiefel-Whitney class of the total space vanishes. Hence we call a prime G-manifold W _trivial_ or _non-trivial_ according as $w_2(\partial W^*)$ vanishes or not.

(5.9) _The prime G-manifold_ $W(q;r_1,r_2,r_3)$ _is trivial iff_ $q + r_1 + r_2 + r_3$ _is even._

Proof. By (5.4), there are integers r_1', r_2', r_3' such that q, r_1', r_2', r_3' are relatively prime to one another, $r_i' \equiv r_i \bmod q$, $i = 1,2,3$, and $r_1' + r_2' + r_3' \equiv$ $\equiv r_1 + r_2 + r_3 \bmod 2q$. Let G act on S^7 such that

$$g(z_0,z_1,z_2,z_3) = (g^q z_0, g^{r_1'} z_1, g^{r_2'} z_2, g^{r_3'} z_3) .$$

Then the action is pseudo-free and as seen in the proof of (5.5), $W(q;r_1',r_2',r_3')$ can be imbedded naturally into S^7. By (4.1), S^7/G with singularities removed has a vanishing second Stiefel-Whitney class iff $qr_1'r_2'r_3' + 3$ is even. Since q, r_1', r_2', r_3' are relatively prime to one another, $qr_1'r_2'r_3'$ is odd iff $q + r_1' + r_2' + r_3'$ is even or iff $q + r_1 + r_2 + r_3$ is even. Hence $W(q;r_1,r_2,r_3)$ is trivial iff $q + r_1 + r_2 + r_3$ is even.

Let M be an oriented compact 7-manifold which is diffeomorphic to $S^3 \times D^4$ and on which there is a free circle action or a pseudo-free circle action. Then $H_2(\partial M^*)$ has a preferred generator β in the sense of section 2. Let S^3 be an oriented invariant 3-sphere in ∂M such that S^3/G represents β. Then the element $\tilde{\beta}$ of $H_3(\partial M)$ represented by S^3 is independent of the choice of S^3 and will be referred to as the <u>preferred element</u> of $H_3(\partial M)$.

A basis $\{\mu,\nu\}$ of $H_3(\partial M)$ is called a <u>preferred basis</u> if (i) the image of μ in $H_3(M)$ is a generator of $H_3(M)$, (ii) the image of ν in $H_3(M)$ is 0 and (iii) the intersection numbers $\tilde{\beta}\cdot\nu$ and $\mu\cdot\nu$ are not negative, where $\tilde{\beta}$ is the preferred element of $H_3(\partial M)$ and the orientation on ∂M is induced by that on M. Therefore for any preferred basis $\{\mu,\nu\}$ of $H_3(\partial M)$,

$$\mu\cdot\mu = 0, \quad \mu\cdot\nu = 1, \quad \nu\cdot\nu = 0$$

and hence

$$\tilde{\beta} = (\tilde{\beta}\cdot\nu)\mu + (\mu\cdot\tilde{\beta})\nu .$$

Moreover, if $\{\mu,\nu\}$ is a preferred basis of $H_3(\partial M)$, then $\{\mu',\nu'\}$ is a preferred basis of $H_3(\partial M)$ iff for some integer m,

$$\mu' = \mu + m\nu, \quad \nu' = \nu .$$

Hence a preferred basis $\{\mu,\nu\}$ of $H_3(\partial M)$ is completely determined by $\mu\cdot\tilde{\beta}$, called the *index* of $\{\mu,\nu\}$.

(5.10) *If* F *is a free G-manifold,* $\tilde{\beta}$ *is the preferred element of* $H_3(\partial F)$ *and* $\{\xi,\eta\}$ *is a preferred basis of* $H_3(\partial F)$, *then* $\tilde{\beta}\cdot\eta = 1$. *Moreover, there is a unique preferred basis* $\{\xi,\eta\}$ *of* $H_3(\partial F)$ *with* $\xi = \tilde{\beta}$.

Proof. Since G acts freely on F, the image of the preferred generator of $H_2(\partial F^*)$ is that of $H_2(F^*)$ so that the image of $\tilde{\beta}$ in $H_3(F)$ is a generator of $H_3(F)$. Therefore $\tilde{\beta}\cdot\eta = \pm 1$. On the other hand, we know from the definition of a preferred basis that $\tilde{\beta}\cdot\eta \geqq 0$. Hence $\tilde{\beta}\cdot\eta = 1$.

If we require $\xi = \tilde{\beta}$, then $\xi\cdot\tilde{\beta} = 0$ and hence $\{\xi,\eta\}$ is completely determined.

(5.11) *If* W *is the prime G-manifold* $W(q;r_1,r_2,r_3)$, $\tilde{\beta}$ *is the preferred element of* $H_3(\partial W)$ *and* $\{\mu,\nu\}$ *is a preferred basis of* $H_3(\partial W)$, *then* $\tilde{\beta}\cdot\nu = q$ *and* $\mu\cdot\tilde{\beta} \equiv r_1 r_2 r_3 \bmod q$. *Hence for any integer* r *with* $r \equiv r_1 r_2 r_3 \bmod q$, *there is a unique preferred basis* $\{\mu,\nu\}$ *of* $H_3(\partial W)$ *of index* r, *that means,* $\tilde{\beta} = q\mu + r\nu$.

Proof. By (5.3) and (5.4), we may assume that q, r_1, r_2, r_3 are relatively prime to one another. Then, by (5.5), we may let W be canonically imbedded into S^7, where there is a pseudo-free orthogonal action of G on S^7 given by

$$g(z_0,z_1,z_2,z_3) = (g^q z_0, g^{r_1} z_1, g^{r_2} z_2, g^{r_3} z_3) \ .$$

Let

$$a_i = (\delta_{i0},\delta_{i1},\delta_{i2},\delta_{i3}), \qquad i = 0,1,2,3,$$

where δ_{ij} is the Kronecker index. Then S^7 can be naturally regarded as the join of Ga_0, Ga_1, Ga_2, Ga_3 so that we may let

$$S^7 = Ga_0 * Ga_1 * Ga_2 * Ga_3 \ .$$

Moreover, the action of G on S^7 may be regarded as the natural extension of the action of G on $Ga_0 \cup Ga_1 \cup Ga_2 \cup Ga_3$. Let

$$a = (0,1/\sqrt{3},1/\sqrt{3},1/\sqrt{3}), \qquad a' = (0,0,1/\sqrt{2},1/\sqrt{2}) \ .$$

Then W may be regarded as an invariant closed tubular neighborhood of $Ga_0 * Ga$ in $Ga_0 * (Ga_1*Ga_2*Ga_3)$. Notice that under the usual differentiable structure on S^7, $Ga_0 * Ga$ is not a differentiable submanifold. However, it is easy to have a new differentiable structure on S^7 under which $Ga_0 * Ga$ is a differentiable submanifold and then has a closed tubular neighborhood. This is why $Ga_0 * Ga$ may be treated as a differentiable submanifold of S^7 even though it is not. Similar understanding will be needed when we make arguments below.

If x is a point moving along a path in $(Ga_1*Ga_2*Ga_3) - (Ga_1 \cup Ga_2 \cup Ga_3)$ from a to a', $Ga_0 * Gx$ gives an isotopy from $Ga_0 * Ga$ to $Ga_0 * Ga'$. Hence we may also regard W as an invariant closed tubular neighborhood of $Ga_0 * Ga'$. Let U be an invariant closed tubular neighborhood of Ga_0 in $Ga_0 * Ga_1$ and U' an invariant closed tubular neighborhood of Ga' in $Ga_2 * Ga_3$. Then $U * U' \subset (Ga_0*Ga_1) * * (Ga_2*Ga_3)$ is an invariant closed tubular neighborhood of $Ga_0 * Ga'$. Hence we may set

$$W = U * U' .$$

Let $c \in \partial U$ and $c' \in \partial U'$. Then $Gc * Gc'$ is an invariant 3-sphere in ∂W which together with an appropriate orientation represents the preferred element $\tilde{\beta}$ of $H_3(\partial W)$. Let c move inside U toward a_0 and let c' move inside U' toward a'. Then Gc covers Ga_0 q times and Gc' covers Ga' once so that $Gc * Gc'$ covers $Ga_0 * Ga'$ q times. Let $Ga_0 * Ga'$ be so oriented that, if $\tilde{\alpha}$ is the element of $H_3(W)$ represented by $Ga_0 * Ga'$, then the image of $\tilde{\beta}$ in $H_3(W)$ is $q\tilde{\alpha}$. Notice that $\tilde{\alpha}$ is a generator of $H_3(W)$.

Since W is a closed tubular neighborhood of $Ga_0 * Ga'$, there is an imbedding $\varphi : D^4 \longrightarrow S^7$ such that (i) $\varphi(S^3) = Ga_0 * Ga'$ and $\varphi : S^3 \longrightarrow Ga_0 * Ga'$ is orientation-preserving and (ii) $\varphi(D^4) \cap W = \varphi([1/2,1]S^3)$. Let the 3-sphere $\varphi(D^4) \cap \partial W = \varphi(\frac{1}{2} S^3)$ be oriented like S^3 and denote by μ' the element of $H_3(\partial W)$ represented by $\varphi(D^4) \cap \partial W$. Clearly the image of μ' in $H_3(W)$ is $\tilde{\alpha}$.

Let N be the kernel of the natural homomorphism of $H_3(\partial W)$ into $H_3(W)$ and let ν' be the generator of N with $\mu' \cdot \nu' = 1$. Then $\{\mu', \nu'\}$ is a preferred basis of $H_3(\partial W)$ and

$$\tilde{\beta}\cdot\nu' = (q\mu')\cdot\nu' = q .$$

In order to compute $\mu'\cdot\tilde{\beta}$, we first observe that it is the intersection number of $\varphi(D^4) \cap \partial W$ with $Gc * Gc'$ which is equal to the linking number of $Ga_0 * Ga'$ with $Gc * Gc'$. Let c move inside $(Ga_0 * Ga_1) - Ga_0$ toward a_1, let c' move inside $(Ga_2 * Ga_3) - Ga_3$ toward a_2 and afterwards let a' move inside $(Ga_2 * Ga_3) - Ga_2$ toward a_3. Then $Ga_0 * Ga'$ covers $Ga_0 * Ga_3$ r_3 times and $Gc * Gc'$ covers $Ga_1 * Ga_2$ $r_1 r_2$ times. Since $Ga_0 * Ga'$ and $Gc * Gc'$ never intersect during the motion and since the linking numbers of $Ga_0 * Ga_3$ with $Ga_1 * Ga_2$ is 1, it follows that $\mu'\cdot\tilde{\beta} = r_1 r_2 r_3$. Hence

$$\tilde{\beta} = q\mu' + (r_1 r_2 r_3)\nu' .$$

Given any preferred basis $\{\mu,\nu\}$ of $H_3(\partial W)$, there is an integer m such that $\mu = \mu' + m\nu'$ and $\nu = \nu'$. Hence $\tilde{\beta}\cdot\nu = \tilde{\beta}\cdot\nu' = q$ and $\mu\cdot\tilde{\beta} = r_1 r_2 r_3 + mq \equiv r_1 r_2 r_3$ mod q. Moreover, the preferred basis $\{\mu,\nu\}$ of $H_3(\partial W)$ is completely determined if we require that $\mu\cdot\tilde{\beta} = r$, where r is an integer with $r \equiv r_1 r_2 r_3$ mod q. In fact, it follows from the requirement that $m = (r - r_1 r_2 r_3)/q$.

6. COMPOSITE G-MANIFOLDS

As prime G-manifolds in last section, all composite G-manifolds in this section are assumed to be 7-dimensional. We shall see below, every composite G-manifold is determined by certain prime G-manifolds it contains so that we may use the classification of prime G-manifolds to obtain a classification for composite G-manifolds. Moreover (5.11) is to be generalized for composite G-manifolds.

Like (3.1), the following is a consequence of (2.4).

(6.1) Let K be a composite G-manifold and let $Gb_1,\ldots,Gb_k$ be distinct exceptional orbits in K such that G acts freely on $K - \bigcup_{i=1}^k Gb_i$. Then each Gb_i is contained in a prime G-manifold W_i in $K - \partial K$ such that $H_2(\partial W_i^*) \longrightarrow H_2(K^* - \{b_1^*,\ldots,b_k^*\})$ induced by the inclusion map is an isomorphism, and such a prime G-manifold W_i is unique up to an equivariant isotopy of K. Hence $\{W_1,\ldots,W_k\}$ is uniquely determined by K. Moreoever, $\{W_1,\ldots,W_k\}$ may be so chosen that (i) for any $i,j = 1,\ldots,k$ with $|i-j| > 1$, $W_i \cap W_j = \emptyset$ and (ii) for *any* $i = 1,\dots,k-1$, $W_i^* \cap W_{i+1}^* = \partial W_i^* \cap \partial W_{i+1}^*$ *is a closed tubular neighborhood of a 2-sphere in* ∂W_i^* *representing a generator of* $H_2(\partial W_i^*)$. *Then there is a* (+)-*equivalence of* K *onto* $\cup_{i=1}^k W_i$. *Hence* $\{W_1,\dots,W_k\}$ *also determines* K.

Like prime G-manifolds, a composite G-manifold K is called *trivial* or *non-trivial* according as $w_2(\partial K^*)$ vanishes or not. It is clear that, if K is a composite G-manifold determined by prime G-manifolds $W_1,\dots,W_k$, then either $K, W_1,\dots,W_k$ are all trivial or they are all non-trivial. Hence, by (2.2), (5.7) and (5.9), we have

(6.2) *Up to a* (+)-*equivalence, every composite* G-*manifold is determined by prime* G-*manifolds*

$$W(q_1;r_{11},r_{12},r_{13}),\dots,W(q_k;r_{k1},r_{k2},r_{k3})$$

for some integers $k;q_1,r_{11},r_{12},r_{13};\dots;q_k;r_{k1},r_{k2},r_{k3}$ *such that*

(i) $k \geqq 1$,

(ii) $q_1,\dots,q_k$ *are integers* > 1 *relatively prime to one another*,

(iii) *for each* $i = 1,\dots,k$, *at most one of* r_{i1}, r_{i2}, r_{i3} *is even and* $(q_i, r_{i1}r_{i2}r_{i3}) = 1$,

(iv) $q_i + r_{i1} + r_{i2} + r_{i3}$, $i = 1,\dots,k$, *are either all even or all odd*.

Notice that making use of (6.1), (6.2), (5.7) and (5.8), we have a classification for composite G-manifolds up to a (+)-equivalence.

(6.3) *Let* $q_1,\dots,q_k$ *be integers* > 1 *relatively prime to one another and* $q = q_1 \dots q_k$. *Then for any integers* $r_1,\dots,r_k$, *there is an integer* r *such that*

$$r(q/q_i) \equiv r_i \bmod q_i, \qquad i = 1,\dots,k .$$

Moreover, r *is unique up to a congruence modulo* q.

Proof. For $k = 1$, the assertion is trivial.

Assume that $k = 2$. Since q/q_i and q_i are relatively prime, there are integers A_i and B_i such that $A_i(q/q_i) + B_iq_i = r_i$, $i = 1,2$. Let C_1 and C_2 be integers such that $C_1q_1 - C_2q_2 = A_1 - A_2$, and let

$$r = A_1 - C_1q_1 = A_2 - C_2q_2, \quad D_i = B_i + C_i(q/q_i), \qquad i = 1,2.$$

Then $r(q/q_i) + D_iq_i = r_i$ and hence

$$r(q/q_i) \equiv r_i \bmod q_i, \qquad i = 1,2.$$

Now we proceed by induction on k and assume that $k > 2$. By induction hypothesis, there exists an integer s such that for each $i = k-1, k$,

$$s(q_{k-1}q_k/q_i) \equiv r_i \bmod q_i$$

and there is an integer r such that

$$r(q/q_i) \equiv r_i \bmod q_i, \qquad i = 1,\ldots,k-2,$$

$$r(q/q_{k-1}q_k) \equiv s \bmod q_{k-1}q_k \,.$$

Therefore for $i = k-1$ or k,

$$r(q/q_i) = r(q/q_{k-1}q_k)(q_{k-1}q_k/q_i) \equiv s(q_{k-1}q_k/q_i) \equiv r_i \bmod q_i \,.$$

Hence the induction is completed.

If r' is a second integer such that $r'(q/q_i) \equiv r_i \bmod q_i$, $i = 1,\ldots,k$, then $r'(q/q_i) \equiv r(q/q_i) \bmod q_i$ so that $r' \equiv r \bmod q_i$, $i = 1,\ldots,k$. Hence $r' \equiv r \bmod q$.

(6.4) _In_ (6.3), _if_ $r_1,\ldots,r_k$ _are integers such that_ $q_1r_1,\ldots,q_kr_k$ _are either all odd or all even, then there is an integer_ r _such that_

$$r(q/q_i) \equiv r_i \bmod 2q_i, \qquad i = 1,\ldots,k \,.$$

Moreover, r _is unique up to a congruence_ modulo $2q$.

Proof. By (6.3), there is an integer r' such that $r'(q/q_i) \equiv r_i \bmod q_i$, $i = 1,\ldots,k$. If q is even, then exactly one of $q_1,\ldots,q_k$ is even. In this case, we let q_1 be even. Then regardless of whether q is odd or even, exactly one of r' and $r' + q$ is a solution for r in $r(q/q_1) \equiv r_1 \bmod 2q_1$.

If q is even, then $q_1r_1,\ldots,q_kr_k$ are all even. Therefore for any $i = 2,\ldots,k$, both $r(q/q_i)$ and r_i are even so that $r(q/q_i) \equiv r_i \bmod 2q_i$.

If q is odd, then r and r_1 are both odd or both even. Since

$q_1r_1,\dots,q_kr_k$ are either all odd or all even, $r_1,\dots,r_k$ are either all odd or all even. Hence for any $i = 2,\dots,k$, $r(q/q_i) \equiv r_i \bmod 2q_i$.

That r is unique up to a congruence modulo $2q$ is clear.

(6.5) *Let* K *be the composite* G-*manifold determined by prime* G-*manifolds*

$$W(q_1;r_{11},r_{12},r_{13}),\dots,W(q_k;r_{k1},r_{k2},r_{k3})$$

as seen in (6.2). *Let* $q = q_1 \cdots q_k$ *and let* r *be an integer such that*

$$r(q/q_i) \equiv r_{i1}r_{i2}r_{i3} \bmod 2q_i, \qquad i = 1,\dots,k ,$$

as provided by (6.4). *Then* $H_3(\partial K)$ *has a unique preferred basis* $\{\mu,\nu\}$ *of index* r.

Proof. For $k = 1$, (6.5) reduces to (5.11) so that it is true.

Assume that $k = 2$. Let

$$W_i = W(q_i;r_{i1},r_{i2},r_{i3}), \qquad i = 1,2,$$

and let

$$K = W_1 \cup W_2, \quad A = W_1 \cap W_2$$

as seen in (6.1). Then

$$\partial K = (\partial W_1 \cup \partial W_2) - (A-\partial A).$$

Let $\tilde{\beta}_i$ be the preferred element of $H_3(\partial W_i)$. By (5.11), there is a unique preferred basis $\{\mu_i,\nu_i\}$ of $H_3(\partial W_i)$ such that $\tilde{\beta}_i = q_i\mu_i + r(q/q_i)\nu_i$.

Whenever $(h_1,h_2) \in H_3(\partial W_1) \times H_3(\partial W_2)$ with $h_1\cdot\tilde{\beta}_1 = h_2\cdot\tilde{\beta}_2$, there is a unique $h \in H_3(\partial K)$ such that for any $i,j = 1,2$ with $i \neq j$, the image of h in $H_3(\partial K,\partial K \cap \partial W_j)$ is corresponding to the image of h_i in $H_3(\partial W_i,A)$ under the isomorphisms

$$H_3(\partial K,\partial K \cap \partial W_j) \longleftarrow H_3(\partial K \cap \partial W_i,\partial A) \longrightarrow H_3(\partial W_i,A)$$

provided by the excision theorem. In fact, since the orientation for A inherited from ∂W_1 is opposite to that inherited from ∂W_2, it follows from $h_1\cdot\tilde{\beta}_1 = h_2\cdot\tilde{\beta}_2$

that the image of h_1 in $H_3(\partial W_1 \cup \partial W_2, \partial K)$ is equal to the negative of the image of h_2. Therefore, if we denote by h_i' the image of h_i in $H_3(\partial W_1 \cup \partial W_2)$, then the image of $h_1' + h_2'$ in $H_3(\partial W_1 \cup \partial W_2, \partial K)$ is 0. Using the exactness of the homology sequence of $(\partial W_1 \cup \partial W_2, \partial K)$, we obtain an $h \in H_3(\partial K)$ having $h_1' + h_2'$ as its image. It is easily seen that h is as desired and that the uniqueness of h follows from the fact that $H_4(\partial W_1 \cup \partial W_2, \partial K) \cong H_4(A, \partial A) = 0$.

Now we use the result of the preceding paragraph to obtain $\mu, \nu \in H_3(\partial K)$ as follows. Since $q_2\nu_1 \cdot \tilde{\beta}_1 = -q_1 q_2 = q_1 \nu_2 \cdot \tilde{\beta}_2$, it follows that there is an element ν of $H_3(\partial K)$ determined by $(q_2\nu_1, q_1\nu_2) \in H_3(\partial W_1) \times H_3(\partial W_2)$. Let t_1 and t_2 be integers such that $t_2 q_1 + t_1 q_2 = 1$. Then

$$(t_1\mu_1 - t_2 r\nu_1)\cdot\tilde{\beta}_1 = t_1 r q_2 + t_2 r q_1 = (t_2\mu_2 - t_1 r \nu_2)\cdot\tilde{\beta}_2 \ .$$

Therefore there is an element μ of $H_3(\partial K)$ determined by $(t_1\mu_1 - t_2 r\nu_1, t_2\mu_2 - t_1 r\nu_2) \in H_3(\partial W_1) \times H_3(\partial W_2)$. Using the commutative diagram

$$\begin{array}{ccccccccc}
0 & \longleftarrow & H_3(K) & \longleftarrow & H_3(W_1) \oplus H_3(W_2) & \longleftarrow & H_3(A) & \longleftarrow & 0 \\
 & & \uparrow & & & & & & \\
 & & H_3(\partial K) & & \uparrow & & \uparrow & & \\
 & & \downarrow & & & & & & \\
0 & \longleftarrow & H_3(\partial W_1 \cup \partial W_2) & \longleftarrow & H_3(\partial W_1) \oplus H_3(\partial W)_2 & \longleftarrow & H_3(A) & \longleftarrow & 0
\end{array}$$

where rows are parts of exact Mayer-Vietoris sequences, we can easily verify that $\{\mu, \nu\}$ is a preferred basis of $H_3(\partial K)$. Since any imbedded invariant 3-sphere in $\partial K - \partial W_2$ represents $\tilde{\beta}_1$ as well as the preferred generator $\tilde{\beta}$ of $H_3(\partial K)$, it follows that

$$\mu\cdot\tilde{\beta} = (t_1\mu_1 - t_2 r\nu_1)\cdot\tilde{\beta}_1 = t_1 r q_2 + t_2 r q_1 = r,$$

$$\tilde{\beta}\cdot\nu = \tilde{\beta}_1\cdot(q_2\nu_1) = q_1 q_2 = q \ .$$

Hence $\tilde{\beta} = q\mu + r\nu$, completing the proof for the case $k = 2$.

For the general case, we proceed by induction on k and assume that $k > 2$. Let K' be the composite G-manifold determined by $\{W(q_1; r_{11}, r_{12}, r_{13}), \ldots, W(q_{k-1}; r_{k-1,1}, r_{k-1,2}, r_{k-1,3})\}$. By induction hypothesis, our assertion holds for K'. Since K is determined by K' and W_k just as a composite G-manifold determined by two prime G-manifolds, our assertion for K follows from the case $k = 2$.

7. ON DIFFEOMORPHISMS AND HOMOTOPY EQUIVALENCES

Let F be a (7-dimensional) free G-manifold. As seen in the proof of (2.1), there is a fibration

$$\pi : F^* \longrightarrow S^2$$

of fibre D^4 and structural group $SO(4)$. Denote ∂F^* by E. Then E is a connected simply connected closed 5-manifold having the integral homology of $S^2 \times S^3$ and

$$\pi : E \longrightarrow S^2$$

is a 3-sphere bundle with structural group $SO(4)$ such that it is trivial iff $w_2(E) = 0$. In this section, we study diffeomorphisms and homotopy equivalences of E onto itself.

Let $\{\xi, \eta\}$ be the preferred basis of $H_3(\partial F)$ such that ξ is the preferred element, and let S be a fibre of $\pi : E \longrightarrow S^2$. Then S can be so oriented that it represents the image of η in $H_3(E)$. Whenever $f : E \longrightarrow E$ is a map, there is a unique integer $m(f)$ which is the Hopf invariant of $\pi f : S \longrightarrow S^2$. If $f : E \longrightarrow E$ is a map such that $f : H_*(E) \longrightarrow H_*(E)$ is the identity and $\tilde{f} : \partial F \longrightarrow \partial F$ is an equivariant map covering f, then

$$\tilde{f}(\xi) = \xi, \quad \tilde{f}(\eta) = m(f)\xi + \eta .$$

Notice that $m(f)$ is a homotopy invariant of f.

(7.1) <u>Assume that $\pi : E \longrightarrow S^2$ is a trivial bundle. Then the following hold.</u>

(1) <u>Whenever m is an integer, there is a diffeomorphism $h_m : E \longrightarrow E$ such that $h_m : H_*(E) \longrightarrow H_*(E)$ is the identity and $m(h_m) = m$.</u>

(2) <u>Whenever $f : E \longrightarrow E$ is a homotopy equivalence, there is a diffeomorphism $h : E \longrightarrow E$ such that, if $\tilde{f}, \tilde{h} : \partial F \longrightarrow \partial F$ are equivariant maps covering f and h, then $\tilde{f}, \tilde{h} : H_*(\partial F) \longrightarrow H_*(\partial F)$ coincide.</u>

Proof. Let us regard D^4 as the unit closed 4-disk in the quaternion field. Then we may let $F = S^3 \times D^4$ on which the action of G is given by

$$g(u;v) = (gu;v) .$$

Let G act on S^3 such that the projection $\tilde{\pi} : F \longrightarrow S^3$ is equivariant. Then we may let $S^2 = S^3/G$ and $\pi : E \longrightarrow S^2$ the map induced by $\tilde{\pi}$. Therefore for any integer m,

$$h_m : E \longrightarrow E$$

given by

$$h_m(Gu,v) = (Guv^{-m},v)$$

is a diffeomorphism such that $h_m : H_*(E) \longrightarrow H_*(E)$ is the identity and $m(h_m) = m$.

Let $f : E \longrightarrow E$ be any given homotopy equivalence. It is easily seen that there is a diffeomorphism $h' : E \longrightarrow E$ such that $h'f : H_*(E) \longrightarrow H_*(E)$ is the identity and thus we have an integer $m = m(h'f)$. If $\tilde{f}, \tilde{h}', \tilde{h}_m : \partial F \longrightarrow \partial F$ are equivariant maps covering f, h', h_m, then $\tilde{h}'\tilde{f}, \tilde{h}_m : H_*(\partial F) \longrightarrow H_*(\partial F)$ coincide so that $\tilde{f}, \tilde{h}'^{-1}\tilde{h}_m : H_*(\partial F) \longrightarrow H_*(\partial F)$ coincide. Hence $h = h'^{-1}h_m : E \longrightarrow E$ is a desired diffeomorphism.

(7.2) _Let_ S _be a fibre of_ $\pi : E \longrightarrow S^2$ _and_ S' _a cross-section. If_ S'' _is a 3-sphere in_ E _such that_ $S'' \cap S$ _contains only one point and_ S'' _intersects_ S' _transversally at that point, then there is a diffeomorphism_ $h : E \longrightarrow E$ _such that_ $h(S') = S'$, $h(S) = S''$ _and_ $h : H_*(E) \longrightarrow H_*(E)$ _is the identity._

Proof. Let S, S', S'' and E be so oriented that the intersection numbers $S \cdot S'$ and $S'' \cdot S'$ are equal. Let U be a closed tubular neighborhood of S such that $\pi^{-1}\pi(U) = U$ and $\pi(U)$ is a closed 2-cell, and let U' be a closed tubular neighborhood of S' such that for each $y \in S^2$, $\pi^{-1}(y) \cap U'$ is a closed 3-cell. We can easily construct a diffeomorphism h' of $U \cup U'$ onto a closed neighborhood V of $S'' \cup S'$ such that $h'(S) = S''$, $h'(S') = S'$ and $h' : S \longrightarrow S''$ and $h' : S' \longrightarrow S'$ are orientation-preserving.

It is obvious that the closure of $E - (U \cup U')$ is diffeomorphic to $D^2 \times D^3$. Since E, $h'(U)$ and $\partial h'(U)$ are simply connected, it follows from van Kampen's theorem that the closure of $E - h'(U)$ is simply connected. Therefore the closure of $E - V$ is simply connected. Since the closure of $E - V$ is homologically trivial and has a boundary diffeomorphic to $\partial(D^2 \times D^4)$, we infer from a characterization of D^5 [8] that the closure of $E - V$ is diffeomorphic to $D^2 \times D^3$. Hence h' can be extended to a diffeomorphism $h : E \longrightarrow E$ which is as desired.

(7.3) <u>Assume that</u> $\pi : E \longrightarrow S^2$ <u>is non-trivial. Then for any even integer</u> m, <u>there is a diffeomorphism</u> $f_m : E \longrightarrow E$ <u>such that</u> $f_m : H_*(E) \longrightarrow H_*(E)$ <u>is the identity and</u> $m(f_m) = m$.

Proof. Let $X_1 = X_2 = D^2$ and let

$$\varphi : \partial X_1 \times S^3 \longrightarrow \partial X_2 \times S^3$$

be the diffeomorphism given by

$$\varphi(u;v_1,v_2) = (u;uv_1,v_2) .$$

Let $\varphi' : \partial X_1 \longrightarrow \partial X_2$ be the identity. Then we may let

$$S^2 = X_1 \cup_{\varphi'} X_2 ,$$

$$E = X_1 \times S^3 \cup_{\varphi} X_2 \times S^3$$

and $\pi : E \longrightarrow S^2$ the projection. Let S be the 3-sphere $\{0\} \times S^3$ in $X_2 \times S^3$, S^1 the 2-sphere in E which is the union of $D^2 \times \{(0,1)\}$ in $X_1 \times S^3$ and that in $X_2 \times S^3$, and S'' the 3-sphere in E which is the union of $\{(v_1^{-2}v_2^{1};v_1,v_2) | (v_1,v_2) \in S^3$ and $|v_1| \geqq |v_2|\}$ in $X_1 \times S^3$ and $\{(v_1^{2}v_2^{-1};v_1,v_2) | (v_1,v_2) \in S^3$ and $|v_1| \leqq |v_2|\}$ in $X_2 \times S^3$. Then S, S', S'' and E can be so oriented that $\pi : S'' \longrightarrow S^2$ is of Hopf invariant 2 and the intersection numbers $S \cdot S'$ and $S'' \cdot S'$ are equal. By (7.2), there is a diffeomorphism

$$h_2 : E \longrightarrow E$$

such that $h_2(S') = S'$, $h_2(S) = S''$ and $h_2 : H_*(E) \longrightarrow H_*(E)$ is the identity. Since $\pi : S'' \longrightarrow S^2$ is of Hopf invariant 2, $m(h_2) = 2$.

For any even integer m,

$$h_m = (h_2)^{m/2} : E \longrightarrow E$$

is a diffeomorphism such that $h_m : H_*(E) \longrightarrow H_*(E)$ is the identity and $m(h_m) = m$.

For any integer $n \geqq 2$, we regard D^n as a subset of D^{n+1} by setting $x = (x,0) \in \mathbf{R}^n \times \mathbf{R} = \mathbf{R}^{n+1}$ for all $x \in D^n$, and let

$$\varphi : S^1 \times D^n \longrightarrow S^1 \times D^n$$

be the diffeomorphism such that for any $(u;v_1,v_2) \in S^1 \times D^n$ with $v_1 \in D^2$ and $v_2 \in D^{n-2}$,

$$\varphi(u;v_1,v_2) = (u;uv_1,v_2) .$$

Then $D^2 \times S^{n-1} \cup_\varphi S^1 \times D^n$ is diffeomorphic to S^{n+1} so that we may set

$$S^{n+1} = D^2 \times S^{n-1} \cup_\varphi S^1 \times D^n .$$

Let $*$ be a base point of S^2. Whenever $\alpha : (D^n,\partial D^n) \longrightarrow (S^2,*)$ is a map, we let $\alpha' : S^{n+1} \longrightarrow S^2$ be the map such that

$$\alpha'(u,v) = \begin{cases} * & \text{if } (u,v) \in D^2 \times S^{n-1} , \\ \alpha(v) & \text{if } (u,v) \in S^1 \times D^n . \end{cases}$$

Then the homotopy class $[\alpha']$ of α' depends only on the homotopy class $[\alpha]$ of α and the function

$$\lambda : \pi_n(S^2) \longrightarrow \pi_{n+1}(S^2)$$

given by $\lambda[\alpha] = [\alpha']$ is a homomorphism.

(7.4) <u>The homomorphism</u> $\lambda : \pi_n(S^2) \longrightarrow \pi_{n+1}(S^2)$ <u>is onto for</u> $n = 2,3,4$.

Proof. Let G act on $S^3 = D^2 \times S^1 \cup_\varphi S^1 \times D^2$ such that

$$g(u,v) = (gu,g^{-1}v) \qquad \text{for } (u,v) \in D^2 \times S^1 ,$$
$$g(u,v) = (gu,v) \qquad \text{for } (u,v) \in S^1 \times D^2 .$$

Then the action is free so that we may set $S^2 = \{Gx \mid x \in S^3\}$ and the projection

$$p : S^3 \longrightarrow S^2$$

given by $p(x) = Gx$ is of Hopf invariant 1. If $\alpha : (D^2, \partial D^2) \longrightarrow (S^2, *)$ is of degree 1, it follows from the definition of α' that $\alpha' = \alpha^* p$ with $\alpha^* : S^2 \longrightarrow S^2$ of degree 1. Hence $\lambda : \pi_2(S^2) \longrightarrow \pi_3(S^2)$ is onto.

We may set

$$S^{n+2} = D^2 \times \Sigma S^{n-1} \underset{\varphi}{\cup} S^1 \times \Sigma D^n ,$$

where Σ is the suspension. Then one may show directly that the diagram

$$\begin{array}{ccccc} \pi_{n+1}(S^2) & \xrightarrow{\Sigma} & \pi_{n+2}(S^3) & \xrightarrow{p} & \pi_{n+2}(S^2) \\ \uparrow \lambda & & & & \uparrow \lambda \\ \pi_n(S^2) & \xrightarrow{\Sigma} & \pi_{n+1}(S^3) & \xrightarrow{p} & \pi_{n+1}(S^2) \end{array}$$

is commutative, where Σ is the suspension homomorphism. Since for $n = 2,3,4$, $\Sigma : \pi_n(S^2) \longrightarrow \pi_{n+1}(S^3)$ and $p : \pi_{n+1}(S^3) \longrightarrow \pi_{n+1}(S^2)$ are onto and since $\lambda : \pi_2(S^2) \longrightarrow \pi_3(S^2)$ has been shown to be onto, it follows that $\lambda : \pi_n(S^2) \longrightarrow$ $\longrightarrow \pi_{n+1}(S^2)$ is onto for $n = 3,4$.

(7.5) <u>Let</u> $n > 2$ <u>and</u> $\tau : (D^2, \partial D^2) \longrightarrow (S^2, *)$ <u>a given map. Then for any map</u> $\alpha : (D^n, \partial D^n) \longrightarrow (S^2, *)$ <u>the map</u> $\alpha'' : S^{n+1} \longrightarrow S^2$ <u>given by</u>

$$\alpha''(u,v) = \begin{cases} \tau(u) & \text{if } (u,v) \in D^2 \times S^{n-1} , \\ \alpha(v) & \text{if } (u,v) \in S^1 \times D^n , \end{cases}$$

<u>is in</u> $\lambda[\alpha]$.

Proof. We may let S^3 be the unit 3-sphere in the quaternion field, $S^2 = \{G_x | x \in S^3\}$, $p : S^3 \longrightarrow S^2$ the projection and $* = p(1)$.

Let $\tilde{\alpha} : (D^n, \partial D^n) \longrightarrow (S^3, G)$ be a map covering α. Since $n > 2$, we may assume that $\tilde{\alpha}(\partial D^n) = 1$. Let $\tilde{\tau} : (D^2, \partial D^2) \longrightarrow (S^3, G)$ be a map covering τ and let $F : D^2 \times [0,1] \longrightarrow S^3$ be a homotopy such that for any $u \in D^2$, $F(u,0) = \tilde{\tau}(u)$ and $F(u,1) = 1$. Then there is a homotopy

$$H : S^{n+1} \times [0,1] \longrightarrow S^2$$

given by

$$H(u,v;t) = pF(u,t) \qquad \text{if} \quad (u,v) \in D^2 \times S^{n-1} ,$$

$$H(u,v;t) = p(F(u,t)\tilde{\alpha}(v)) \qquad \text{if} \quad (u,v) \in S^1 \times D^n .$$

It is easily seen that for any $(u,v) \in S^{n+1}$,

$$H(u,v;0) = \alpha''(u,v), \quad H(u,v;1) = \alpha'(u,v) .$$

Hence α'' is in $\lambda[\alpha]$.

Let $\pi : E \longrightarrow S^2$ be a non-trivial 3-sphere bundle (of structural group SO(4)). As in the proof of (7.3) we let $X_1 = X_2 = D^2$,

$$S^2 = X_1 \cup_{\varphi'} X_2, \quad E = X_1 \times S^3 \cup_{\varphi} X_2 \times S^3$$

and

$$\pi : E \longrightarrow S^2$$

the projection. Regard S^3 as a subset of E such that for each $v \in S^3$, we set $v = (0;v) \in X_2 \times S^3$, and regard S^2 as a subset of E such that for any $u \in X_i$, we set $u = (u;0,\sqrt{-1}) \in X_i \times S^3$, $i = 1,2$. Then S^3 is a fibre and S^2 is a cross-section for $\pi : E \longrightarrow S^2$ and they intersect at a single point, namely $(0;0,\sqrt{-1})$ which we denote by $*$.

Let

$$D = \{(u;v_1,v_2) \in X_1 \times S^3 \,|\, v_2 - \bar{v}_2 = r\sqrt{-1} \text{ with } r \leqq 0\} ,$$

and N the closure of $E - D$. Then $\partial N = \partial D$, $\partial N \cap X_1 \times S^3 = X_1 \times S^2$ and $\partial N \cap X_2 \times S^3 = \partial X_2 \times D^3_-$, where S^2 is the equator in S^3 and D^3_- is the lower hemisphere in S^3. Therefore there is a natural homeomorphism

$$\beta : S^4 \longrightarrow \partial N$$

with $\beta(D^2 \times S^2) = X_1 \times S^2$ and $\beta(S^1 \times D^3) = \partial X_2 \times D^3_-$, where $S^4 = D^2 \times S^2 \cup_{\varphi} S^1 \times D^3$. Using a natural deformation retraction of N onto $S^3 \vee S^2$, we can show that β, as a map of S^4 into N, is homotopic to a map

$$\beta' : S^4 \longrightarrow S^3 \vee S^2$$

such that for some map $\tau : (D^2, \partial D^2) \longrightarrow (S^2, *)$ and $\tilde{\alpha} : (D^3, \partial D^3) \longrightarrow (S^3, *)$ of degree 1,

$$\beta'(u,v) = \begin{cases} \tau(u) & \text{if } (u,v) \in D^2 \times S^2 , \\ \tilde{\alpha}(v) & \text{if } (u,v) \in S^1 \times D^3 . \end{cases}$$

If $p : S^3 \vee S^2 \longrightarrow S^2$ is a map such that $p|S^3$ is of Hopf invariant 1 and $p|S^2$ is the identity, it follows from (7.4) and (7.5) that $[p\beta'] = \lambda[p\dot{\tilde{\alpha}}]$ is the generator of $\pi_4(S^2)$.

(7.6) <u>Assume that</u> $\pi : E \longrightarrow S^2$ <u>is non-trivial. Then for any homotopy equivalence</u> $f : E \longrightarrow E$, $m(f)$ <u>is even.</u>

Proof. Let $f : E \longrightarrow E$ be a homotopy equivalence. Since there is a diffeomorphism $h : E \longrightarrow E$ such that $m(h) = 0$ and $fh : H_*(E) \longrightarrow H_*(E)$ is the identity, we may assume that $f : H_*(E) \longrightarrow H_*(E)$ is the identity as f can be replaced by fh.

It can be seen that f is homotopic to a homotopy equivalence $f_1 : E \longrightarrow E$ such that $f_1^{-1}(D)$ has a finite number of components, say $C_1, C_2, \ldots, C_m$, each of which is mapped homeomorphically onto D. Since f_1 is of degree 1, m is odd and $C_1, C_2, \ldots, C_m$ may be so arranged that $f_1 : C_i \longrightarrow D$ is orientation-preserving or orientation-reversing according as i is odd or even. Applying a homotopy to f_1 relative to $f_1^{-1}(D)$ if it is necessary, we may assume that there is a topological imbedding $\xi : D^4 \times [-2,2] \longrightarrow E$ such that (i) $\xi(D^4 \times [-2,2]) \cap f_1^{-1}(D) = C_1 \cup C_2$ with $C_1 = \xi(D^4 \times [-2,-1])$ and $C_2 = \xi(D^4 \times [1,2])$ and (ii) for any $(u,t) \in D^4 \times [-2,2]$, $f_1\xi(u,t) = f_1\xi(u,-t)$. Then f_1 is homotopic to an equivalence $f_2 : E \longrightarrow E$ such that for some neighborhood U of $\xi(D^4 \times [-2,2])$, $C_3 \cup \cdots \cup C_m \subset U$, $f_2|E - U = f_1|E - U$ and $f_2(U) \cap D = \emptyset$. Repeating this argument, we obtain a homotopy equivalence $f_3 : E \longrightarrow E$ homotopic to f such that $f_3 : f_3^{-1}(D) \longrightarrow D$ is a homeomorphism.

Since we may use f_3 in place of f, we may assume that $f : f^{-1}(D) \longrightarrow D$ is a homeomorphism. Then f defines a homotopy equivalence

$$f' : N \longrightarrow N .$$

Since $S^3 \vee S^2$ is a deformation retract of N, f' induces a homotopy equivalence

$$f'' : S^3 \vee S^2 \longrightarrow S^3 \vee S^2 .$$

Let β, β', $\tilde{\alpha}$, p be as above. Then $\tilde{\alpha}$ represents a generator b_1 of $\pi_3(S^3)$, $p\tilde{\alpha}$ represents a generator b_2 of $\pi_3(S^2)$ and $p[\beta'] = \lambda p(b_1) = \lambda(b_2)$ is the generator of $\pi_4(S^2)$. Since $f\beta = \beta$, it follows that β' and $f''\beta'$ are homotopic so that

$$\lambda p(b_1) = \lambda p f''(b_1) .$$

Since $f : H_*(E) \longrightarrow H_*(E)$ is the identity, it follows that

$$f''(b_1) = b_1 + mb_2, \quad f''(b_2) = b_2 ,$$

where $m = m(f)$. Therefore

$$\lambda p(b_1) = (m+1)\lambda p(b_1) .$$

Hence m is even.

As a consequence of (7.3) and (7.6), we have

(7.7) <u>The statement</u> (2) <u>of</u> (7.1) <u>is also valid when</u> $\pi : E \longrightarrow S^2$ <u>is non-trivial.</u>

Regardless of whether $\pi : E \longrightarrow S^2$ is trivial or not, we denote by $\mathcal{D}(E)$ the set of diffeomorphisms $h : E \longrightarrow E$ such that $h : H_*(E) \longrightarrow H_*(E)$ is the identity and $m(h) = 0$. Let $\mathcal{P}(E)$ be the set of pseudo-isotopy classes in $\mathcal{D}(E)$. We shall show below that

(7.8) <u>There is a natural one-to-one correspondence between</u> $\mathcal{P}(E)$ <u>and</u> $\mathbb{Z}$.

Assume first that $\pi : E \longrightarrow S^2$ is trivial. Let $F_1^* = F_2^* = F^*$ and let $\xi : \partial F_1^* \longrightarrow \partial F_2^*$ be a (-)-diffeomorphism such that $M = F_1^* \cup_\xi F_2^*$ is diffeomorphic to the complex projective 3-space. For each $h \in \mathcal{D}(E)$, we let $M_h = F_1^* \cup_{\xi h} F_2^*$. Then M_h is a homotopy complex projective 3-space which, up to a diffeomorphism, depends only on the pseudo-isotopy class $[h]$ of h. Therefore there is a function ρ' of

$\mathcal{P}(E)$ into the set of diffeomorphism classes of homotopy complex projective 3-spaces. As seen in [9], ρ' is one-to-one and onto and there is a natural one-to-one correspondence between diffeomorphism classes of homotopy projective 3-spaces and integers. Hence we have a one-to-one correspondence

$$\rho : \mathcal{P}(E) \longrightarrow \mathbf{Z} .$$

This proves (7.8) for the case that $\pi : E \longrightarrow S^2$ is trivial.

For any $h \in \mathcal{D}(E)$, $\rho[h]$ is determined by the first Pontrjagin class $p_1(M_h)$. In fact, we have

$$p_1(M_h) = (24\rho[h]+4)\alpha^2$$

with α being a generator of $H^2(M_h)$. For details, see [10].

Homologically, all the h's in $\mathcal{D}(E)$ resemble the identity, but when $\pi : E \longrightarrow S^2$ is trivial, only half of them are homotopic to the identity. In fact, we have

(7.9) Proposition. When $\pi : E \longrightarrow S^2$ is trivial, any $h \in \mathcal{D}(E)$ is homotopic to the identity iff $\rho[h]$ is even.

Proof. As before, we let $F_1^* = F_2^* = F^*$ and $\xi : \partial F_1^* \longrightarrow \partial F_2^*$ a (-)-diffeomorphism such that $M = F_1^* \cup_\xi F_2^*$ is diffeomorphic to the complex projective 3-space. Let $h \in \mathcal{D}(E)$ and $M_h = F_1^* \cup_{\xi h} F_2^*$. If h is homotopic to the identity, then there is a (+)-homotopy equivalence $f : M_h \longrightarrow M$ given as follows. Let the fibres of $\pi : F_1^* \longrightarrow S^2$ be regarded as unit closed 4-disks and let $H : E \times [1/2,1] \longrightarrow E$ be a homotopy such that for any $x \in E$, $H(x,1/2) = h(x)$ and $H(x,1) = x$. Then $f : M_h \longrightarrow M$ given by

$$f(x) = \begin{cases} x & \text{for } x \in F_1^* \text{ with } |x| \leqq 1/2 , \\ |x|h^{-1}H(x/|x|, |x|) & \text{for } x \in F_1^* \text{ with } |x| \geqq 1/2 , \\ x & \text{for } x \in F_2^* \end{cases}$$

is a homotopy equivalence which leaves every point of F_2^* and every point of a closed tubular neighborhood of the 0-section of $\pi : F_1^* \longrightarrow S^2$ fixed.

Denote by CP^i the complex projective i-space. By Sullivan's theory [11],

$f : M_h \longrightarrow M$ determines an element $a_h \in [CP^2, F/O] \cong [CP^2, F/PL]$. It is known that $[CP^2, F/PL] \cong \mathbb{Z}$ and that $\rho[h]$ is the integer corresponding to a_h. Since $f^{-1}(F_2^*) = F_2^*$ and $f : F_2^* \longrightarrow F_2^*$ is the identity, it follows that the image of a_h in $[CP^1, F/PL] \cong \mathbb{Z}_2$ is 0. Hence $\rho[h]$ is even.

That $\rho[h]$ is even can also be seen as follows. Let N be a submanifold of M as CP^2 in CP^3 such that $N \cap F_1^*$ is a closed 4-disk. By means of frame surgery, we can show that there is a homotopy equivalence $f' : M_h \longrightarrow M$ such that f' is homotopic to h, $f'|F_2^* = f|F_2^*$, f' is transverse regular at N and $f'^{-1}(N)$ is connected and simply connected. Then $f'^{-1}(N) \cap F_1^*$ is a connected simply connected compact parallelizable 4-manifold bounding a 3-sphere so that its index is a multiple of 16, say 16m. Therefore $p_1(M_h) = (48m+4)\alpha^2$ with α being a generator of $H^2(M_h)$. This shows that $\rho[h] = 2m$ which is even.

Conversely let h be an element of $\mathcal{D}(E)$ such that $\rho[h]$ is even. Since $M_h = F_1^* \cup_{\xi h} F_2^*$ is a homotopy complex projective 3-space, there is a (+)-homotopy equivalence. $f' : M_h \longrightarrow M$ such that its restriction to the 0-section S of $\pi : F_2^* \longrightarrow S^2$ is homotopic to the inclusion map. Let a_h be the element of $[CP^2, F/O]$ determined by f'. Since $\rho[h]$ is even, the image of a_h in $[CP^1, F/O] \cong [CP^1, F/PL] \cong \mathbb{Z}_2$ is 0. Therefore we can use frame surgery to show that f' is homotopic to a homotopy equivalence $f : M_h \longrightarrow M$ such that f is transverse regular at S and $f^{-1}(S)$ is a 2-sphere. Since f is orientation-preserving and $f|S$ is homotopic to the inclusion map, we may assume that $f^{-1}(F_2^*) = F_2^*$ and $f : F_2^* \longrightarrow F_2^*$ is the identity, as F_2^* is a closed tubular neighborhood of S. Similarly we may also assume that f leaves every point of a closed tubular neighborhood of the 0-section of $\pi : F_1^* \longrightarrow S^2$ fixed. Therefore f behaves as the homotopy equivalence in the first part and hence we can recover a homotopy $H : E \times [1/2, 1] \longrightarrow E$ such that for any $x \in E$, $H(x, 1/2) = h(x)$ and $H(x, 1) = x$.

Assume now that $\pi : E \longrightarrow S^2$ is non-trivial. We recall that there is a non-trivial closed 4-disk bundle $\pi : F^* \longrightarrow S^2$ with $\partial F^* = E$. Let $F_1^* = F_2^* = F^*$ and $\xi : \partial F_1^* \longrightarrow \partial F_2^*$ a (-)-diffeomorphism with $\pi\xi = \pi$. Then for any $h \in \mathcal{D}(E)$,

$$M_h = F_1^* \cup_{\xi h} F_2^*$$

is a connected simply connected oriented closed 6-manifold having the integral homology of $S^2 \times S^4$. Let us denote by M the manifold M_h when h is the identity. Clearly M is the total space of a non-trivial 4-sphere bundle over S^2.

(7.10) <u>For any</u> $h \in \mathcal{D}(E)$, <u>there is a</u> (+)-<u>homotopy equivalence</u> $\mu : M_h \longrightarrow M$ <u>such that</u> $\mu(F_i^*) = F_i^*$, $i = 1,2$, <u>and</u> $\mu : F_2^* \longrightarrow F_2^*$ <u>is the identity.</u>

Proof. Fix a fibre D of $\pi : F_1^* \longrightarrow S^2$. For any $h \in \mathcal{D}(E)$, we let

$$N_h = D \cup_{\xi h|\partial D} F_2^* .$$

Also we let N denote N_h when h is the identity. We first assert that there is a homotopy equivalence

$$\lambda : N_h \longrightarrow N .$$

Since $h : H_*(E) \longrightarrow H_*(E)$ is the identity and $m(h) = 0$, we infer that $h|\partial D : \partial D \longrightarrow E$ is homotopic to the inclusion map. Let

$$H : \partial D \times [0,1] \longrightarrow E$$

be a homotopy such that for any $x \in \partial D$,

$$H(x,0) = x \qquad H(x,1) = h(x) .$$

Then there is a map $\lambda : N_h \longrightarrow N$ such that $\lambda|F_2^*$ is the inclusion map and for any $x \in D$ (with D being regarded as the unit closed 4-disk),

$$\lambda(x) = \begin{cases} 2x & \text{if } |x| \leqq 1/2 , \\ \xi H(x/|x|, 2|x|-1) & \text{if } |x| \geqq 1/2 . \end{cases}$$

It is easily seen that $\lambda : H_*(N_h) \longrightarrow H_*(N)$ is an isomorphism. Since N_h and N are connected and simply connected, we infer that λ is a homotopy equivalence.

We next assert that the homotopy H can be so chosen that λ can be extended to a map $\mu : M_h \longrightarrow M$ such that $\mu(F_1^*) = F_1^*$. If such a μ exists, it is obvious that $\mu : H_*(M_h) \longrightarrow H_*(M)$ is an isomorphism so that $\mu : M_h \longrightarrow M$ is a (+)-homology equivalence.

Let us regard F_1^* as the compact 6-manifold obtained from $D^2 \times D^4$ such that two points $(u;v_1,v_2)$ and $(u';v_1',v_2')$ are identified iff $|u| = |u'| = 1$, $uv_1 = u'v_1'$ and $v_2 = v_2'$. Let S^2 be regarded as the closed 2-manifold obtained from D^2 by identifying all points in ∂D^2, and let $p : D^2 \times D^4 \longrightarrow F_1^*$ and $p : D^2 \longrightarrow S^2$ be the projections. Then we may let $\pi : F_1^* \longrightarrow S^2$ be induced by the projection $\pi : D^2 \times D^4 \longrightarrow D^2$ and let $D = p(S^1 \times D^4)$ by setting $v = p(1,v)$.

Let

$$S^5 = D^2 \times S^3 \cup_{\varphi} S^1 \times D^4$$

and define

$$\sigma : S^5 \longrightarrow F_1^*$$

as follows.

(1) For any $(u,v) \in D^2 \times S^3$,

$$\sigma(u,v) = p(u,v) .$$

(2) For any $(u,v) \in S^1 \times D^4$,

$$\sigma(u,v) = \begin{cases} 2v & \text{if } |v| \leqq 1/2 , \\ H(v/|v|, 2(1-|v|)) & \text{if } |v| \geqq 1/2 . \end{cases}$$

Then H can be so chosen that σ is null-homotopic. In fact, if σ is not null-homotopic, we replace H by a new homotopy $H' : \partial D \times [0,1] \longrightarrow E$ such that H and H' coincide everywhere except in the interior of a closed 4-disk D' and $H(D') \cup H'(D')$ as the image of a map of S^4 into E represents the generator of $\pi_4(F_1^*)$. (Notice that $\pi_4(F_1^*) \cong \mathbf{Z}_2$ and that the natural homomorphism $\pi_4(E) \longrightarrow \pi_4(F_1^*)$ is onto.) Let $\sigma' : S^5 \longrightarrow F_1^*$ be the analogs of σ using H' in place of H. By (7.4), σ' is not homotopic to σ and hence is null-homotopic.

Let $\gamma : \partial(D^2 \times D^4) \longrightarrow S^5$ be the homeomorphism such that $\gamma : D^2 \times S^3 \longrightarrow D^2 \times S^3$ is the identity and $\gamma : S^1 \times D^4 \longrightarrow S^1 \times D^4$ is equal to φ. Since σ is null-homotopic, $\sigma\gamma$ can be extended to a map $\tau : D^2 \times D^4 \longrightarrow F_1^*$. Now it is easily seen that τ induces a map $\lambda' : F_1^* \longrightarrow F_1^*$ with $\tau = \lambda' p$ and there is a map $\mu : M_h \longrightarrow M$ which is an extension of λ and such that $\mu | F_1^* = \lambda'$. Since μ is of degree 1, it is a (+)-homotopy equivalence.

Contrary to (7.9), we have

(7.11) Proposition. Whenever $\pi : E \longrightarrow S^2$ is non-trivial, a diffeomorphism $h : E \longrightarrow E$ is in $\mathcal{D}(E)$ iff it is homotopic to the identity.

Proof. It is obvious that any diffeomorphism $h : E \longrightarrow E$ homotopic to the identity is in $\mathcal{D}(E)$.

Let $h \in \mathcal{D}(E)$. By (7.10), there is a (+)-homotopy equivalence $\mu : M_h \longrightarrow M$ such that $\mu(F_i^*) = F_i^*$, $i = 1,2$, and $\mu : F_2^* \longrightarrow F_2^*$ is the identity. Then, as in the last part of the proof of (7.9), we can proceed to show that h is homotopic to the identity.

Now we are in a position to prove (7.8) for the case that $\pi : E \longrightarrow S^2$ is non-trivial. Whenever $h \in \mathcal{D}(E)$, there is, by (7.10), a homotopy equivalence $\mu : M_h \longrightarrow M$ such that $\mu(F_i^*) = F_i^*$, $i = 1,2$, and $\mu : F_2^* \longrightarrow F_2^*$ is the identity. Let M_0 be the compact 6-manifold obtained from M by removing an open 6-disk. By Sullivan's theory, μ determines an element a of $[M_0, F/O]$. Since $[M_0, F/O] \cong$ $\cong [M_0, F/PL] \cong [S^2 \vee S^4, F/PL] \cong \mathbf{Z}_2 \oplus \mathbf{Z}$, there is an element $(a_1, a_2) \in \mathbf{Z}_2 \oplus \mathbf{Z}$ corresponding to a. Since μ is transverse regular at the 0-section S of $\pi : F_2^* \longrightarrow S^2$ and $\mu^{-1}(S) = S$, it follows that the associated Arf invariant a_1 vanishes. The integer a_2 depends only on the pseudo-isotopy class $[h]$ of h so that we have a function

$$\rho : \mathcal{P}(E) \longrightarrow \mathbf{Z}$$

given by $\rho[h] = a_2$.

Given any $b_2 \in \mathbf{Z}$, there is an element b of $[M_0, F/O]$ corresponding to $(0, b_2) \in \mathbf{Z}_2 \oplus \mathbf{Z}$. Let b be represented by a closed m-disk bundle $\tau : B \longrightarrow M_0$ and a trivialization map $t : B \longrightarrow D^m$, where m is large. Then t, up to a homotopy, can be so chosen that it is transverse regular at 0 and $t^{-1}(0)$ is a compact 6-manifold such that $\tau : \partial t^{-1}(0) \longrightarrow \partial M_0$ is a diffeomorphism and $\tau : (t^{-1}(0), \partial t^{-1}(0)) \longrightarrow (M_0, \partial M_0)$ is a homotopy equivalence. Let M' be the closed 6-manifold obtained at attaching a closed 6-disk to $t^{-1}(0)$. Then $\tau : t^{-1}(0) \longrightarrow M_0$ can be naturally extended to a homotopy equivalence $\mu : M' \longrightarrow M$. Since the first coordinate of $(0,b_2)$ is 0, we may assume that

$$\mu : \mu^{-1}(F_2^*) \longrightarrow F_2^*$$

is a diffeomorphism. By Smale's theorem, there is a diffeomorphism λ of F_1^* onto the closure of $M' - \mu^{-1}(F_2^*)$. Let λ be so chosen that

$$\mu\lambda : \partial F_1^* \longrightarrow \partial F_2^*$$

is a diffeomorphism equal to ξh for some diffeomorphism $h : E \longrightarrow E$ with $h : H_*(E) \longrightarrow H_*(E)$ being the identity. Now it is not hard to show that $h \in \mathfrak{D}(E)$ and $\rho[h] = b_2$. Hence ρ is onto.

If h and h' are elements of $\mathfrak{D}(E)$ such that $\rho[h] = \rho[h']$, it follows from Sullivan's theory that there is a (+)-diffeomorphism

$$\lambda : M_h \longrightarrow M_{h'} .$$

We can let λ be so chosen that $\lambda(F_i^*) = F_i^*$, $i = 1,2$, and λ is the identity on F_2^* and on a closed tubular neighborhood T of the 0-section of $\pi : F_1^* \longrightarrow S^2$. Then the restriction of λ to the closure of $F_1^* - T$ provides a pseudo-isotopy between h and h'. Hence ρ is 1-1.

8. MAIN THEOREMS

Given any pseudo-free circle action on a homotopy 7-sphere Σ, we have seen in (3.1) that Σ may be regarded as a closed 7-manifold obtained by pasting together a free G-manifold F and a composite G-manifold K via a (-)-equivalence of ∂F onto ∂K. By (2.1) and (6.2), we know how to exhibit all free G-manifolds and all composite G-manifolds. Therefore it is natural to ask when a given free G-manifold and a given composite G-manifold can be pasted together as a pseudo-free circle action on a homotopy 7-sphere. The purpose of this section is to have an answer to this question so that it can be used to exhibit all pseudo-free circle actions on homotopy 7-spheres.

Whenever K is a composite G-manifold, we call the order of $H_2(K^*,\partial K^*)$ the <u>order</u> of K. It is easily seen that the order of K is the product of the orders of the isotropy groups associated with exceptional orbits in K. Therefore, if $\{\mu,\nu\}$

is a preferred basis of $H_3(\partial K)$ and $\mathfrak{F}$ is the preferred element of $H_3(\partial K)$, then $\mathfrak{F}\cdot\nu$ is the order of K while $\mu\cdot\mathfrak{F}$ is the index of $\{\mu,\nu\}$ (see (6.5)).

(8.1) Let K be a composite G-manifold, q the order of K and r the index of a preferred basis of $H_3(\partial K)$. In order that there is a free G-manifold F and a (-)-equivalence $f : \partial F \longrightarrow \partial K$ such that $F \cup_f M$ is a homotopy 7-sphere, it is necessary that

$$r \equiv 1 \quad \text{or} \quad -1 \bmod q .$$

Proof. By (2.1), there is a free G-manifold F such that $w_2(\partial F^*)$ vanishes iff $w_2(\partial K^*)$ vanishes. Then there is a (-)-equivalence f of ∂F onto ∂K. Let $\{\xi,\eta\}$ be the preferred basis of $H_3(\partial F)$ with ξ being the preferred element (see (5.10)) and $\{\mu,\nu\}$ the preferred basis of $H_3(\partial K)$ of index r (see (6.5)). Then the induced homomorphism

$$f : H_3(\partial F) \longrightarrow H_3(\partial K)$$

is given by

$$f(\xi) = q\mu + r\nu ,$$

$$f(\eta) = A\mu + B\nu ,$$

where A and B are integers such that

$$Ar - Bq = 1 .$$

If $F \cup_f K$ is a homotopy 7-sphere, we may use the Mayer-Vietoris sequence of $(F \cup_f K; F, K)$ to show that the image of $f(\eta)$ in $H_3(K)$ is a generator of $H_3(K)$. Therefore $A = 1$ or -1 and hence $\pm r = 1 + Bq \equiv 1 \bmod q$.

The converse of (8.1) is false. In fact, we have

(8.2) Let $W = W(q;r,1,1)$ with $q > 1$ and $r \equiv 1$ or $-1 \bmod q$. If there is a free G-manifold F and a (-)-equivalence $f : \partial F \longrightarrow \partial W$ such that $F \cup_f W$ is a homotopy 7-sphere, then $r \equiv 1$ or $-1 \bmod 2q$.

Proof. Assume first that q is odd. Then, by (4.1) $w_2(\partial W^*) = 0$. Therefore, by (5.9), r is odd. Hence $r \equiv 1$ or $-1 \bmod 2q$.

Assume next that q is even. Let G act on S^7 such that

$$g(z_0,z_1,z_2,z_3) = (g^q z_0, g^r z_1, gz_2, gz_3) .$$

By (5.5), we may set

$$W = \{(z_0,z_1,z_2,z_3) \in S^7 \mid |z_0|^2 + |z_2|^2 \geqq 1/2\} .$$

Let $\{\mu,\nu\}$ be the preferred basis of $H_3(\partial W)$ of index r as constructed in the proof of (5.11). If $r \not\equiv 1$ or $-1 \bmod 2q$, then $q > 2$ and it follows from (5.7) that we may let $r = q+1$ or $q-1$. Let

$$S = \{(su, su, tv^r, tv^{2r-q}) \mid |u| = |v| = 1, s \geqq 0, t \geqq 0, 2(s^2+t^2) = 1\}$$
$$S' = \{(su^q, su^r, tv, tv) \mid |u| = |v| = 1, s \geqq 0, t \geqq 0, 2(s^2+t^2) = 1\}.$$

Then S' is an invariant 3-sphere in ∂W and S is a 3-sphere in ∂W such that S^* is also a 3-sphere. It is easily seen that S and S' can be so oriented that they represent $r\mu + (2r-q)\nu$ and $q\mu + r\nu$ respectively. Since $(q\mu+r\nu)\circ(r\mu+(2r-q)\nu) = -1$, S^* represents a generator of $H_3(\partial W^*)$. Therefore, by (7.2), there is a (-)-equivalence $h : \partial F \longrightarrow \partial W$ such that, if $\{\xi,\eta\}$ is the preferred basis of $H_3(\partial F)$ with ξ being the preferred element, then

$$h(\xi) = q\mu + r\nu, \qquad h(\eta) = r\mu + (2r-q)\nu .$$

Hence, by (7.6), there is an even integer m,

$$\begin{aligned} f(\xi) &= h(\xi) = q\mu + r\nu , \\ f(\eta) &= mh(\xi) + h(\eta) \\ &= (mq+r)\mu + (mr+2r-q)\nu . \end{aligned}$$

Since $F \cup_f M$ is a homotopy 7-sphere, it follows from the Mayer-Vietoris sequence of $(F \cup_f M;F,M)$ that $mq + r = 1$ or -1. This is impossible when $q > 2$, $r = q+1$ or $q-1$ and m is even. Hence $r \equiv 1$ or $-1 \bmod 2q$.

(8.3) <u>Given a prime G-manifold</u> $W(q;r_1,r_2,r_3)$, <u>where at most one of</u> r_1,r_2,r_3 <u>is even, there is an equivariant map</u> $\lambda : W(q;r_1,r_2,r_3) \longrightarrow W(q;r_1r_2r_3,1,1)$ <u>such</u>

that, if we denote $W(q;r_1,r_2,r_3)$ *and* $W(q;r_1r_2r_3,1,1)$ *by* W *and* W' *respectively, then* $\lambda(\partial W) \subset \partial W'$, $\lambda : (W,\partial W) \longrightarrow (W',\partial W')$ *is of degree* 1 *and both* $\lambda^* : W^* \longrightarrow W'^*$ *and* $\lambda^* : \partial W^* \longrightarrow \partial W'^*$ *are homotopy equivalences.*

Proof. Assume that r_2 is odd. Let s and t be integers such that $sr_2 + tq = 1$ and t is even. It is easily seen that there is a $\mathbb{Z}q$-equivariant map

$$\lambda' : \partial D(q;r_1,r_2,r_3) \longrightarrow \partial D(q;r_1,sr_2,r_2r_3)$$

which is of degree 1 and such that

$$\lambda'(z_1,z_2,z_3) = (z_1,z_2^s,z_3^{r_2})$$

when $|z_2|^2 + |z_3|^2$ is small. By the linearity on D^6, λ' can be extended to a $\mathbb{Z}q$-equivariant map $\lambda' : D(q;r_1,r_2,r_3) \longrightarrow D(q;r_1,sr_2,r_2r_3)$ (which may not be differentiable at $0 \in D^6$) and then extended to a G-equivariant map

$$\lambda' : V(q;r_1,r_2,r_3) \longrightarrow V(q;r_1,sr_2,r_2r_3) .$$

Let $h^* : S^1 \times D^4 \longrightarrow L(q;r_1,r_2,r_3)$ and $h'^* : S^1 \times D^4 \longrightarrow L(q;r_1,sr_2,r_2r_3)$ be the immersions which map every $(u;v_1,v_2) \in S^1 \times D^4$ into $G(\bar{u};1/\rho,v_1/\rho,v_2/\rho)$ with $\rho = (1+|v_1|^2+|v_2|^2)^{1/2}$. It is easily seen that λ'^*h^* and h'^* are regularly homotopic. Therefore, we may assume that λ' can be extended to an equivariant map

$$\lambda' : W(q;r_1,r_2,r_3) \longrightarrow W(q;r_1,sr_2,r_2r_3)$$

such that $\lambda'|S^1 \times (D^2 \times D^4)$ is the identity.

Since t is even, $sr_2 \equiv 1 \bmod 2q$ so that, by (5.3), $W(q;r_1,sr_2,r_2r_3) = W(q;r_1,1,r_2r_3)$ which we denote by W''. From the construction of λ', it is clear that $\lambda'(\partial W) \subset \partial W''$ and $\lambda' : (W,\partial W) \longrightarrow (W'',\partial W'')$ is of degree 1. Moreover, $\lambda' : H_*(W^*) \longrightarrow H_*(W'^*)$ and $\lambda' : H_*(\partial W^*) \longrightarrow H_*(\partial W''^*)$ are isomorphisms. Hence both $\lambda'^* : W^* \longrightarrow W''^*$ and $\lambda'^* : \partial W^* \longrightarrow \partial W''^*$ are homotopy equivalences.

Since at most one of r_1, r_2, r_3 is even, it follows from (5.6) that no generality is lost by assuming that r_1 and r_2 are odd. Moreover, we may set $W(q;r_1,1,r_2r_3) = W(q;1,r_1,r_2r_3)$ and $W(q;1,1,r_1r_2r_3) = W(q;r_1r_2r_3,1,1)$. Therefore there is an equivariant map $\lambda'' : W(q;r_1,1,r_2r_3) \longrightarrow W(q;r_1r_2r_3,1,1)$ such that $\lambda''(\partial W'') \subset \partial W'$, $\lambda'' : (W'',\partial W'') \longrightarrow (W',\partial W')$ is of degree 1 and both $\lambda''^* : W''^* \longrightarrow W'^*$ and $\lambda''^* : \partial W''^* \longrightarrow \partial W'^*$ are homotopy equivalences. Hence

$$\lambda = \lambda''\lambda' : W(q;r_1,r_2,r_3) \longrightarrow W(q;r_1r_2r_3,1,1)$$

is a desired equivariant map.

As a generalization of (8.3), we have

(8.4) Let K be the composite G-manifold determined by

$$W(q_1;r_{11},r_{12},r_{13}),\ldots,W(q_k;r_{k1},r_{k2},r_{k3}) ,$$

where $k;q_1,r_{11},r_{12},r_{13};\ldots;q_k,r_{k1},r_{k2},r_{k3}$ are integers satisfying (i) - (iv) of (6.2). Let $q = q_1 \ldots q_k$ and r an integer as provided by (6.5). Then there is an equivariant map λ of K onto $W = W(q;r,1,1)$ such that $\lambda(\partial K) \subset \partial W$, $\lambda : (K,\partial K) \longrightarrow (W,\partial W)$ is of degree 1 and both $\lambda^* : K^* \longrightarrow W^*$ and $\lambda^* : \partial K^* \longrightarrow \partial W^*$ are homotopy equivalences.

Proof. For the sake of convenience, we let $W_i = W(q_i;r_{i1},r_{i2},r_{i3})$ and $W_i' = W(q_i;r_{i1}r_{i2}r_{i3},1,1) = W(q_i;r(q/q_i),1,1)$, $i = 1,\ldots,k$. By (8.3), there is, for each $i = 1,\ldots,k$, an equivariant map $\lambda_i : W_i \longrightarrow W_i'$ such that $\lambda_i(\partial W_i) \subset \partial W_i'$, $\lambda_i : (W_i,\partial W_i) \longrightarrow (W_i',\partial W_i')$ is of degree 1 and both $\lambda_i^* : W_i^* \longrightarrow W_i'^*$ and $\lambda_i^* : \partial W_i^* \longrightarrow \partial W_i'^*$ are homotopy equivalences.

Since $q_1,\ldots,q_k$ are relatively prime to one another, there are integers $s_1,\ldots,s_k$ such that $\Sigma_{i=1}^k s_i(q/q_i) = 1$. Let t_i be the integer such that

$$s_i(q/q_i) + t_iq_i = 1$$

and let

$$\mu_i : W_i' \longrightarrow W$$

be the equivariant map such that for any $(z;z_1,z_2,z_3) \in V(q_i;r(q/q_i),1,1) \subset W_i'$,

$$\mu_i(z;z_1,z_2,z_3) = (z^{q/q_i};z^{rt_i}z_1^{s_i},z_2,z_3) \in V(q;r,1,1) ,$$

and for any $(z;u;v_1,v_2) \in S^1 \times (D^2 \times D^4) \subset W_i'$,

$$\mu_i(z;u;v_1,v_2) = (z^{q/q_i};u^{s_i};v_1,v_2) \in S^1 \times (D^2 \times D^4) .$$

Then $\mu_i(\partial W_i') \subset \partial W$ and $\mu_i : (W_i', \partial W_i') \longrightarrow (W, \partial W)$ is of degree $s_i(q/q_i)$.

As seen in (6.1), we may let

$$K = W_1 \cup \cdots \cup W_k$$

such that (i) for any $i,j = 1,\ldots,k$ with $|i-j| > 1$, $W_i \cap W_j = \emptyset$ and (ii) for any $i = 1,\ldots,k-1$, $W_i^* \cap W_{i+1}^* = \partial W_i^* \cap \partial W_{i+1}^*$ is a closed tubular neighborhood of a 2-sphere in ∂W_i^* representing a generator of $H_2(\partial W_i^*)$. Since $\mu_i^* \lambda_i^* : H_2(W_i^*) \longrightarrow H_2(W^*)$ has a cokernel of order q/q_i, $\mu_i^* \lambda_i^* : H_2(\partial W_i^*) \longrightarrow H_2(\partial W^*)$ is an isomorphism, $i = 1,\ldots,k$. Therefore we can use equivariant homotopy to adjust

$$\mu_i \lambda_i : (W_i, \partial W_i) \longrightarrow (W, \partial W), \qquad i = 1,\ldots,k$$

such that for any $i = 1,\ldots,k-1$, $\mu_i \lambda_i$ and $\mu_{i+1}\lambda_{i+1}$ coincide on $W_i \cap W_{i+1}$. Hence we have an equivariant map

$$\lambda : K \longrightarrow W$$

such that $\lambda | W_i = \mu_i \lambda_i$, $i = 1,\ldots,k$. Clearly $\lambda(\partial K) \subset \partial W$ and the degree of $\lambda : (K, \partial K) \longrightarrow (W, \partial W)$ is $\Sigma_{i=1}^k s_i(q/q_i) = 1$. It is easily seen that $\lambda^* : H_*(K^*) \longrightarrow H_*(W^*)$ is an isomorphism. Hence $\lambda^* : K^* \longrightarrow W^*$ is a homotopy equivalence. Similarly, $\lambda^* : \partial K^* \longrightarrow \partial W^*$ is also a homotopy equivalence.

Now we can strengthen (8.1) so that its converse also holds. In fact, we have

(8.5) Theorem. *Let K be the composite G-manifold determined by*

$$W(q_1;r_{11},r_{12},r_{13}),\ldots,W(q_k;r_{k1},r_{k2},r_{k3})$$

where $k;q_1,r_{11},r_{12},r_{13};\ldots;q_k,r_{k1},r_{k2},r_{k3}$ *are integers satisfying* (i) - (iv) *of* (6.2). *Let* $q = q_1 \ldots q_k$ *and* r *an integer as provided by* (6.5). *If there is a free G-manifold* F *and a* (-)-*equivalence* $f : \partial F \longrightarrow \partial K$ *such that* $F \cup_f K$ *is a homotopy* 7-*sphere, then* $r \equiv 1$ *or* $-1 \bmod 2q$. *The converse is also true.*

Proof. Suppose first that there is a free G-manifold F and a (-)-equivalence

$f : \partial F \longrightarrow \partial K$ such that $F \cup_f K$ is a homotopy 7-sphere. By (8.4), there is an equivariant map

$$\lambda : K \longrightarrow W = W(q;r,1,1)$$

such that $\lambda(\partial K) \subset \partial W$, $\lambda : (K,\partial K) \longrightarrow (W,\partial W)$ is of degree 1 and both $\lambda^* : K^* \longrightarrow W^*$ and $\lambda^* : \partial K^* \longrightarrow \partial W^*$ are homotopy equivalences. Since $\lambda^{-1} f : \partial F \longrightarrow \partial K$ is a (-)-equivariant map such that $(\lambda^{-1} f)^* : \partial F^* \longrightarrow \partial K^*$ is a homotopy equivalence, it follows from (7.1), (2) or (7.7) that there is a (-)-equivalence $f' : \partial F \longrightarrow \partial K$ such that the induced homomorphisms $\lambda^{-1} f, f' : H_*(\partial F) \longrightarrow H_*(\partial W)$ coincide. By hypothesis, $F \cup_f K$ is a homotopy 7-sphere so that the homomorphism $H_3(\partial F) \longrightarrow$ $\longrightarrow H_3(F) \oplus H_3(K)$ in the Mayer-Vietoris sequence of $(F \cup_f K;F,K)$ is an isomorphism. Therefore, the corresponding homomorphism in the Mayer-Vietoris sequence of $(F \cup_{f'} W;F,W)$ is also an isomorphism so that $F \cup_{f'} W$ is a homotopy 7-sphere. Hence, by (8.2), $r \equiv 1$ or $-1 \bmod 2q$.

Suppose conversely that $r \equiv 1$ or $-1 \bmod 2q$. By (5.3), $W = W(q;r,1,1)$ may be naturally identified with $W(q;1,1,1)$ or $W(q;-1,1,1)$. Therefore there is a free G-manifold F and a (-)-equivalence $f' : \partial F \longrightarrow \partial W$ such that $F \cup_{f'} W = S^7$ and the circle action on $F \cup_{f'} W$ is orthogonal. Hence we can use the same argument as above to show that there is a (-)-equivalence $f : \partial F \longrightarrow \partial K$ such that the induced homomorphisms $f, \lambda f' : H_*(\partial F) \longrightarrow H_*(\partial K)$ coincide. This implies that $F \cup_f K$ is a homotopy 7-sphere.

Now we are ready to exhibit all pseudo-free circle actions on homotopy 7-spheres.

(8.6) __Theorem.__ __Let__ $k;q_1,r_{11},r_{12},r_{13};\ldots;r_{k1},r_{k2},r_{k3}$ __be integers such that__

(i) $k \geqq 1$,

(ii) $q_1,\ldots,q_k$ __are integers__ > 1 __relatively prime to one another,__

(iii) __for each__ $i = 1,\ldots,k$, __at most one of__ r_{i1}, r_{i2}, r_{i3} __is even and__ $(q_i, r_{i1}r_{i2}r_{i3}) = 1$,

(iv) $q_i + r_{i1} + r_{i2} + r_{i3}$, $i = 1,\ldots,k$, __are either all even or all odd,__

(v) __if__ $q = q_1 \ldots q_k$, __then for all__ $i = 1,\ldots,k$, $r_{i1}r_{i2}r_{i3} \equiv \varepsilon(q/q_i) \bmod 2q_i$, __where__ $\varepsilon = 1$ __or__ -1 __independent of__ i.

Let K be the composite G-manifold determined by the prime G-manifolds

$$W_i = W(q_i; r_{i1}, r_{i2}, r_{i3}), \qquad i = 1,\ldots,k .$$

(See (5.2) and (6.2).) Then there is a free G-manifold F and a (-)-equivalence $f : \partial F \longrightarrow \partial K$ such that $\Sigma = F \cup_f K$ is a homotopy 7-sphere on which there is a pseudo-free circle action with exactly k exceptional orbits. If F' and $f' : \partial F' \longrightarrow \partial K$ are analogous to F and f, then there is a (+)-equivalence $\lambda : F \longrightarrow F'$ such that $h = f^{-1}f'\lambda : \partial F \longrightarrow \partial F$ is a (+)-equivalence such that $h : H_*(\partial F) \longrightarrow H_*(\partial F)$ is the identity. Conversely, if F and f are as above and $h : \partial F \longrightarrow \partial F$ is a (+)-equivalence such that $h : H_*(\partial F) \longrightarrow H_*(\partial F)$ is the identity then $F \cup_{fh} K$ is a homotopy 7-sphere on which there is a pseudo-free circle action. Moreover, any pseudo-free circle action on a homotopy 7-sphere, up to a (+)-equivalence, may be given this way. Furthermore, if $k'; q'_1, r'_{11}, r'_{12}, r'_{13}; \ldots; q'_{k'}, r'_{k'1}, r'_{k'2}, r'_{k'3}; W'_i = W(q'_i; r'_{i1}, r'_{i2}, r'_{i3})$, $i = 1,\ldots,k'; K', F'$, $f' : \partial F' \longrightarrow \partial K'$ and $\Sigma' = F' \cup_{f'} K'$ are analogous to those above without prime, then there is a (+)-equivalence of Σ onto Σ' iff

a) $k = k'$,

b) for some permutation τ of $\{1,\ldots,k\}$, W_i can be identified with $W'_{\tau(i)}$ by means of a (+)-equivalence (see (5.7) and (5.8)), $i = 1,\ldots,k$,

c) F and F' can be identified by means of a (+)-equivalence (see (2.1)),

d) after the identifications $K = K'$ and $F = F'$, $(f^{-1}f')^* : \partial F^* \longrightarrow \partial F^*$ is pseudo-isotopic to the identity.

After having (2.1), (3.1), (5.2), (5.7), (5.8), (6.2) and (8.5), the proof of (8.6) is easy. Details are omitted. As a consequence of (8.6), we have the following classification theorem.

(8.7) Theorem. Up to an equivalence, distinct pseudo-free circle actions on homotopy 7-spheres are given as follows. Let k be a positive integer and let $q_1 < \cdots < q_k$ be k integers > 1 relatively prime to one another. For each $i = 1,\ldots,k$, let r_{i1}, r_{i2}, r_{i3} be integers such that at most one of them is even *and* $r_{i1}r_{i2}r_{i3} \equiv (q_1 \cdots q_k)/q_i \bmod 2q_i$, *and let* W_i *be the prime* G-*manifold* $W(q_i; r_{i1}, r_{i2}, r_{i3})$. *Let* K *be the composite* G-*manifold determined by* $W_1, \ldots, W_k$ *and let* F *be a free* G-*manifold such that there is a* (-)-*equivalence* $f : \partial F \longrightarrow \partial K$ *with* $F \cup_f K$ *being a homotopy* 7-*sphere. Let* $h : \partial F \longrightarrow \partial F$ *be a* (+)-*equivalence such that* $h : H_*(\partial F) \longrightarrow H_*(\partial F)$ *is the identity. Then* $F \cup_{fh} K$ *is a homotopy* 7-*sphere on which there is a pseudo-free circle action, and the action, up to an equivalence, depends only on the choice of*

a) *the integer* $k > 1$;

b) *the integers* $1 < q_1 < \cdots < q_k$ *relatively prime to one another;*

c) *the* (+)-*equivalence class of*

$$W(q_1; r_{11}, r_{12}, r_{13}), \ldots, W(q_k; r_{k1}, r_{k2}, r_{k3})$$

where r_{i1}, r_{i2}, r_{i3} *are integers such that at most one of them is even and* $r_{i1}r_{i2}r_{i3} \equiv (q_1 \cdots q_k)/q_i \bmod 2q_i$, $i = 1, \ldots, k$;

d) *the pseudo-isotopy class of* $h^* : \partial F^* \longrightarrow \partial F^*$.

Moreover, by making distinct choices in a), b), c), d), *we obtain all distinct pseudo-free circle actions on homotopy* 7-*spheres up to an equivalence.*

Remarks.

(1) The choice of $k, q_1, \ldots, q_k$ in a) and b) is arbitrary. In fact, if k is any integer > 1 and $1 < q_1 < \cdots < q_k$ are any integers relatively prime to one another, then we may let $W_i = W(q_i; (q_1 \cdots q_k)/q_i, 1, 1)$, $i = 1, \ldots, k$, and then use (8.5) to show the existence of a (-)-equivalence $f : \partial F \longrightarrow \partial K$ such that $F \cup_f K$ is a homotopy 7-sphere on which there is a pseudo-free circle action.

(2) If the choice in a), b), c) has been made, it follows from (7.8) that choices in (d) are in one-to-one correspondence with integers.

(3) There is a pseudo-free circle action on a homotopy 7-sphere with exactly one exceptional orbit, which is distinct from any orthogonal circle action on S^7. For example, we may let $K = W(7;5,3,1)$.

9. A HOMOTOPY CLASSIFICATION

In the following the symbol Σ, with or without index, denotes a homotopy 7-sphere on which there is a pseudo-free circle action.

By an *h-equivalence* of Σ_1 into Σ_2, we mean an equivariant map $\lambda : \Sigma_1 \longrightarrow \Sigma_2$ such that $\lambda^* : \Sigma_1^* \longrightarrow \Sigma_2^*$ is a homotopy equivalence. It is obvious that for any Σ, the identity map of Σ into Σ is an h-equivalence and that, if $\lambda_1 : \Sigma_1 \longrightarrow \Sigma_2$ and $\lambda_2 : \Sigma_2 \longrightarrow \Sigma_3$ are h-equivalences, then so is $\lambda_2\lambda_1 : \Sigma_1 \longrightarrow \Sigma_3$. However, it will be seen later that given any h-equivalence $\lambda : \Sigma_1 \longrightarrow \Sigma_2$, there may not exist an h-equivalence of Σ_2 into Σ_1. Therefore it is natural to have a homotopy classification for pseudo-free circle actions on homotopy 7-spheres in the following sense. The actions on Σ_1 and Σ_2 are called *h-equivalent* if for some Σ_3, there exist h-equivalences $\lambda_1 : \Sigma_1 \longrightarrow \Sigma_3$ and $\lambda_2 : \Sigma_2 \longrightarrow \Sigma_3$.

The purpose of this section is to show that any two pseudo-free circle actions on homotopy 7-spheres are h-equivalent iff their orbit spaces have isomorphic integral cohomology rings and that in each equivalence class, there is an orthogonal pseudo-free circle action on S^7 with exactly one exceptional orbit, which is unique up to an orthogonal transformation.

For any integer $q > 1$, we denote by Sq the 7-sphere S^7 together with the orthogonal pseudo-free circle action on S^7 given by

$$g(z_0, z_1, z_2, z_3) = (g^q z_0, gz_1, gz_2, gz_3) \quad .$$

Then the integral cohomology ring of the orbit space Sq^* is given as follows. First,

$$H^i(Sq^*) \cong \begin{cases} \mathbb{Z} & i = 0,2,4,6; \\ 0 & \text{otherwise.} \end{cases}$$

Second, if α_1 is a generator of $H^2(Sq^*)$, then there are generators α_2 and α_3 of $H^4(Sq^*)$ and $H^6(Sq^*)$ such that $\alpha_1^2 = q\alpha_2$ and $\alpha_1^3 = q^2\alpha_3$. Since the orbit spaces of any two h-equivalent pseudo-free circle actions on homotopy 7-spheres must have isomorphic integral cohomology rings, it follows that

(9.1) *Whenever* q *and* q' *are distinct integers* > 1, *the orthogonal pseudo-free circle actions on* Sq *and* Sq' *are not h-equivalent.*

Let

$$W = \{(z_0, z_1, z_2, z_3) \in Sq \,\big|\, |z_0|^2 + |z_1|^2 \geqq 1/2\}$$

and F the closure of $Sq - W$. Then

$$Sq = F \cup W$$

is a decomposition as seen in section 3.

Whenever $f : \partial F \longrightarrow \partial W$ is a (-)-equivalence such that $f : H_*(\partial F) \longrightarrow H_*(\partial W)$ is the identity,

$$\Sigma = F \cup_f W$$

is a homotopy 7-sphere on which there is a natural pseudo-free circle action. If q is even, then W is a non-trivial prime G-manifold. By (7.11), f^* is homotopic to the identity so that f is equivariantly homotopic to the identity. Hence one can easily construct an h-equivalence

$$\lambda : \Sigma \longrightarrow Sq$$

such that $\lambda|W$ is the inclusion map.

If q is odd, it is still true that there exists an h-equivalence $\lambda : \Sigma \longrightarrow Sq$. However, as seen in (7.9), f^* may not be homotopic to the identity so that we need a proof for this case.

Let q be odd and $f : \partial F \longrightarrow \partial W$ a (-)-equivalence such that $f : H_*(\partial F) \longrightarrow H_*(\partial W)$ is the identity and $f^* : \partial F^* \longrightarrow \partial F^*$ is not homotopic to the identity. By (5.5), we may identify W with $W(q;1,1,1)$. Therefore we may set

$$W = S^1 \times (D^2 \times D^4) \cup V(q;1,1,1) .$$

Since $[S^2 \times D^4, F/O] \cong \pi_2(F/O) \cong \mathbf{Z}_2$, it follows from Sullivan's theorem that there is a homotopy equivalence

$$\mu^* : (S^2 \times D^4, S^2 \times S^3) \longrightarrow (S^2 \times D^4, S^2 \times S^3)$$

corresponding to the generator of $\mathbf{Z}_2$ and such that $\mu^* : H_*(S^2 \times S^3) \longrightarrow H_*(S^2 \times S^3)$ is the identity. Moreover, we may let μ^* be such that for some $a \in S^2$, μ^* is transverse regular at $\{a\} \times D^4$ and $\mu^{*-1}(\{a\} \times D^4) = \{a\} \times D^4$ which is mapped identically onto $\{a\} \times D^4$. Let $S^2 = D^2/\partial D^2$ and let the image of ∂D^2 in S^2

be a. Then we have a homotopy equivalence

$$\lambda^* : (W^*, \partial W^*) \longrightarrow (W^*, \partial W^*)$$

such that $\lambda^* : V(q;1,1,1)^* \longrightarrow V(q;1,1,1)^*$ is the identity and $\lambda^* : D^2 \times D^4 \longrightarrow$ $\longrightarrow D^2 \times D^4$ is a map which maps $S^1 \times D^4$ identically onto $S^1 \times D^4$ and induces the map μ^* above. Since q is odd, $\lambda^* : \partial W^* \longrightarrow \partial W^*$ is a homotopy equivalence which is not homotopic to the identity. Therefore we infer from $[S^2 \times D^4, F/O] \cong \mathbb{Z}_2$ that λ^* is homotopic to f^* or to f^{*-1}. Hence λ^* can be extended to a homotopy equivalence

$$\lambda^* : \Sigma^* \longrightarrow S_q^*$$

with $\lambda^*(F^*) = F^*$. By lifting λ^*, we have an h-equivalence

$$\lambda : \Sigma \longrightarrow Sq$$

such that $\lambda(F) = F$ and $\lambda(W) = W$. This proves

(9.2) Let $Sq = F \cup W$ be a decomposition as seen in section 3 and let $f : \partial F \longrightarrow \partial W$ be any (-)-equivalence such that $f : H_*(\partial F) \longrightarrow H_*(\partial W)$ is the identity. Then $\Sigma = F \cup_f W$ is a homotopy 7-sphere on which there is a pseudo-free circle action and there is an h-equivalence $\lambda : \Sigma \longrightarrow Sq$ with $\lambda(F) = F$ and $\lambda(W) = W$.

In general, we have the following result which contains (9.2) as a special case.

(9.3) Theorem. Let Σ be a homotopy 7-sphere on which there is a pseudo-free circle action, and let q be the product of the orders of the isotropy groups associated with exceptional orbits in Σ. Then there is an h-equivalence $\lambda : \Sigma \to Sq$.

Proof. By (8.7), we may let

$$\Sigma = F \cup_f K$$

where K is a composite G-manifold determined by prime G-manifolds

$$W_i = W(q_i; r_{i1}, r_{i2}, r_{i3}), \qquad i = 1, \ldots, k,$$

for some integers $k;q_1,r_{11},r_{12},r_{13};\dots;q_k,r_{k1},r_{k2},r_{k3}$ satisfying the requirements stated in (8.7), F is a free G-manifold and F a (-)-equivalence of ∂F onto ∂K. Then it is clear that

$$q = q_1 \cdots q_k .$$

Let

$$Sq = F' \cup W$$

be a decomposition as seen in section 3, where $W = W(q;1,1,1)$. By (8.4), there is an equivariant map

$$\lambda : K \longrightarrow W$$

such that $\lambda(\partial K) = \partial W$, $\lambda : (K,\partial K) \longrightarrow (W,\partial W)$ is of degree 1 and both $\lambda^* : K^* \longrightarrow W^*$ and $\lambda^* : \partial K^* \longrightarrow \partial W^*$ are homotopy equivalences.

Let $\{\mu,\nu\}$ be the preferred basis of $H_3(\partial K)$ of index 1 and $\{\xi,\eta\}$ the preferred basis of $H_3(\partial F)$ with ξ being the preferred element. Then

$$f(\xi) = q\mu + \nu, \quad f(\eta) = \mu .$$

Therefore, if

$$\mu' = \lambda(\mu),, \qquad \nu' = \lambda(\nu) ;$$
$$\xi' = q\mu' + \nu' , \quad \eta' = \mu' ,$$

$\{\mu',\nu'\}$ is the preferred basis of $H_3(\partial W)$ of index 1 and $\{\xi',\eta'\}$ is the preferred basis of $H_3(\partial F')$ with ξ' being the preferred element of $H_3(\partial F')$. Hence there is an equivariant diffeomorphism

$$\lambda' : F \longrightarrow F'$$

such that λf, $\lambda' : H_*(\partial F) \longrightarrow H_*(\partial F')$ coincide.

Let S^2 be a 2-sphere in $\partial F'^*$ representing a generator of $H_2(\partial F'^*)$. In order that $(\lambda f)^* : \partial F^* \longrightarrow \partial F'^*$ is homotopic to a map which is transverse regular at S^2 and under which the inverse image of S^2 is a 2-sphere, a necessary and sufficient condition is that the associated Arf invariant c vanishes. If this condition is satisfied, then $(\lambda f)^*$ is homotopic to a diffeomorphism so that we have an equivariant diffeomorphism $\lambda' : F \longrightarrow F'$ such that $\lambda f, \lambda' : \partial F \longrightarrow \partial F'$ are equivariantly homotopic. Hence λ can be extended to an h-equivalence of Σ onto Sq mapping F into F'. This shows that it suffices to find a λ such that the associated Arf invariant c for $(\lambda f)^* : \partial F^* \longrightarrow \partial F'^*$ vanishes.

Assume first that q is odd. By (7.9), there is a (-)-equivalence $f_1 : \partial F' \longrightarrow \partial W$ such that $f_1 : H_*(\partial F') \longrightarrow H_*(\partial W)$ is the identity but the associated Arf invariant for $f_1^* : \partial F'^* \longrightarrow \partial F'^*$ does not vanish. Let

$$\lambda_1 : F' \cup_{f_1} W \longrightarrow Sq$$

be an h-equivalence as provided by (9.2) such that $\lambda_1(W) \subset W$ and $\lambda_1(F') \subset F'$. Then the associated Arf invariant for $\lambda_1^* : \partial F'^* \longrightarrow \partial F'^*$ does not vanish. If the associated Arf invariant for $(\lambda f)^* : \partial F^* \longrightarrow \partial F'^*$ does not vanish, then that for $(\lambda_1 \lambda f)^* : \partial F^* \longrightarrow \partial F'^*$ vanishes. In this case, let us use $\lambda_1 \lambda$ in place of λ. Then the associated Arf invariant for $(\lambda f)^* : \partial F^* \longrightarrow \partial F'^*$ always vanishes and hence we have an h-equivalence of Σ onto Sq.

Assume next that q is even. We assert that the associated Arf invariant c for $(\lambda f)^* : \partial F^* \longrightarrow \partial F'^*$ always vanishes. Since $W = W(q;1,1,1)$, we may let

$$W^* = D^2 \times D^4 \cup V^*$$

with $V = V(q;1,1,1)$ (see section 5). Let

$$S^2 = D_1^2 \cup D_2^2 \qquad (D_1^2 = D_2^2 = D^2)$$

and let

$$\varphi : S^2 \longrightarrow W^*$$

be an imbedding which maps D_1^2 diffeomorphically onto $D^2 \times \{0\}$ and maps D_2^2 onto the cone of vertex b^* over $\varphi(\partial D_1^2)$ $(= \varphi(\partial D_2^2))$, where b^* is the singularity in V^* and D_2^2 is regarded as the cone of vertex 0 over ∂D_2^2. Then φ represents a generator α of $H_2(W^*)$ and $q\alpha$ may be represented by an immersion

$$\psi : S^2 \longrightarrow W^*$$

such that for some $0 < \delta < 1$,

$$\psi(u) = \varphi(u^q/|u|^{q-1}) \quad \text{when} \quad u \in D_1^2 \quad \text{with} \quad |u| \geqq \delta ,$$

ψ imbeds the interior of δD_1^2 into $\delta D^2 \times D^4$ and ψ imbeds the interior of D_2^2 into $V^* - \{b^*\}$.

As above, we let

$$W_i^* = D^2 \times D^4 \cup V_i^*$$

with $V_i = V(q_i; r_{i1}, r_{i2}, r_{i3})$. From the construction of λ^* in the proof of (8.4), we know that

$$\lambda^{*-1}(V^* - \partial V^*) = \cup_{i=1}^k (V_i^* - \partial V_i^*)$$

and that λ^* may be assumed to be transverse regular at $\psi(D_2^2)$. Then by means of framed surgery, we can have a λ such that $\lambda^{*-1}\psi(D_2^2) \cap V_i^*$ consists of q/q_i immersed closed 2-disks (which may not be disjoint).

Since $\lambda^* : (K^*, \partial K^*) \longrightarrow (W^*, \partial W^*)$ is of degree 1, we may assume that for some radius L of D_1^2 there is a neighborhood N of $\varphi(L)$ in V^* such that λ^* maps $\lambda^{*-1}(N)$ diffeomorphically onto N, because this can be achieved by applying a homotopy to λ^* if necessary. Now we let δ be so small that $\psi(\delta D_2^2) \in \lambda^{*-1}(N)$. Since $\psi(D_2^2 - \delta D_2^2)$ covers $\varphi(D_2^2 - \delta D_2^2)$ an even number of times, it follows that the associated Arf invariant for λ^* at $\psi(S^2)$, which is an even multiple of that for λ^* at $\varphi(D_2^2 - \delta D_2^2)$, vanishes.

The immersion $\psi : S^2 \longrightarrow W^* - \{b^*\}$ is regularly homotopic to an imbedding of S^2 into ∂W^* representing a generator of $H_2(\partial W^*)$. Therefore the associated Arf invariant of $\lambda^* : \partial K^* \longrightarrow \partial W^*$ vanishes and hence the proof of (9.3) is completed.

Notice that if Σ has more than one exceptional orbit, then there does not exist an h-equivalence of Sq into Σ as the image of an exceptional orbit under an h-equivalence is an exceptional orbit with an isotropy group not of smaller order. This shows that an h-equivalence, not like a homotopy equivalence, may not have an "inverse" in any natural sense.

A summary of these results has appeared in [12].

BIBLIOGRAPHY

1. Jacoby, One-parameter transformation groups of the three-sphere, Proc. Amer. Math. Soc. 7, 131-142 (1956).

2. Smale, On the structure of manifolds, Amer. J. of Math.,84, 387-399 (1962).

3. Montgomery, Samelson, and Yang, Orbits of highest dimension, Trans. Amer. Math. Soc. 87, 284-293 (1958).

4. Borel, Seminar on Transformation Groups, Princeton University Press (1960).

5. Whitney, Differentiable manifolds, Ann. of Math. 37, 645-680 (1936).

6. Milnor, Characteristic Classes, notes by J. Stasheff, mimeographed, Princeton University (1957).

7. Barden, Simply connected five-manifolds, Ann. Math. 82, 365-385 (1965).

8. Milnor, Lectures on the h-Cobordism Theorem, Princeton University Press (1965).

9. Montgomery and Yang, Differentiable actions on homotopy seven spheres, Trans. Amer. Math. Soc. 122, 480-498 (1966).

10. Wall, Classification problems in differential topology V, On certain 6-manifolds, Invent. Math. 1, 355-374 (1966).

11. Sullivan, Triangulating Homotopy Equivalences, Princeton University Thesis, 1966, University Microfilms Inc., Ann Arbor.

12. Montgomery and Yang, Differentiable pseudo-free circle actions, Proc. Nat. Acad. Sci. USA 68, 894-896 (1971).

SEMIFREE CIRCLE ACTIONS WITH TWISTED FIXED POINT SETS

Reinhard Schultz

Purdue University

An action of the group G is called semifree if its only isotropy subgroups are the trivial subgroup and G itself. In this paper we shall consider only the case in which $G = S^1$ and the action is a smooth action on a homotopy sphere; such actions have been extensively studied.

The standard examples of semifree circle actions are given by linear representations consisting of k copies of the standard representation of S^1 as SO_2 plus a trivial representation. In this case it is easy to see that the fixed point set of the induced action on the unit sphere is an unknotted subsphere of codimension 2k. Furthermore, the equivariant normal bundle of the fixed point set is equivariantly trivial (in Browder's terminology, the fixed point set is untwisted [2, §6, p.36]). There are many previously known examples of nonlinear actions whose fixed point sets are ordinary spheres (e.g., see [2]-[4], [6], [8], [9]); in all these examples the fixed point set is untwisted. In this paper we shall construct infinite families of semifree circle actions on the standard sphere for which the fixed point sets are twisted (see Thrm. 3.1). The construction uses Toda's results on the odd-primary homotopy groups of spheres ([14], [15]) and some elementary formulas in surgery theory (related to long

Partially supported by NSF Grant GP-19530.

exact sequence (D) in [4]). The techniques resemble those used in [11], but here we obtain existence rather than non-existence theorems.

This paper is divided into three sections. The first section establishes some necessary results on the odd-primary components of certain homotopy groups, and the second section contains some results on the normal invariants of fiber homotopy trivializations. These are used to construct the examples of the third section.

1. Some homotopy computations

Let G be a compact Lie group acting smoothly, let K be a component of its fixed point set, and let ρ be the local representation on the component K. Then the equivariant normal bundle of K is a canonical reduction of the structure group of the ordinary normal bundle to the centralizer of ρ; If G is a circle group acting semifreely, this centralizer is merely the unitary group. If the manifold acted upon and its fixed point set are both homotopy spheres, it is easy to see that this equivariant normal bundle is equivariantly fiber homotopically trivial [2, Thrm. 5.3, p.32].

In this section we shall produce candidates for nontrivial equivariant normal bundles; these will be nontrivial complex vector bundles over spheres which are equivariantly fiber homotopically trivial. Stated differently, if $E(CP^k)$ is the topological monoid of homotopy self-equivalences of CP^k and

$$j_*: \pi_*(U_{k+1}) \to \pi_*(E(CP^k))$$

is induced by the action of U_{k+1} on CP^k via projective collineations, then we shall produce nonzero elements in the kernel of j_*.

Recall (e.g. [11, §1]) that the filtration of U_{k+1} by the standardly embedded subgroups U_s $(s \leq k+1)$ gives rise to an exact couple and hence a spectral sequence converging to $\pi_{s+t}(U_{k+1})$ with

$$E^1_{s,t} = \begin{cases} \pi_{s+t}(S^{2s-1}) & s \leq k+1 \\ 0 & s > k+1. \end{cases}$$

Let $d^1_{s,t}$ be the first differential in this spectral sequence.

PROPOSITION 1.1. _In the above notation,_ $d^1_{3,t}$ _is zero on elements of odd order._

PROOF. It follows from the definition $d^1 = j_*\partial$ that under the canonical isomorphism from $\pi_{s+t-1}(S^3)$ to $\pi_{s+t}(KP^\infty)$ (KP^∞ = infinite-dimensional quaternionic projective space), d^1 corresponds to an induced homomorphism

$$h_* : \pi_{s+t}(S^5) \to \pi_{s+t}(KP^\infty)$$

for some $h \in \pi_5(KP^\infty)$. Since S^4 is the 7-skeleton of KP^∞, we clearly have $\pi_5(KP^\infty) = Z_2$; the class h is actually the nontrivial element, but this fact is not relevant to our purposes.

If X is a simple space, let $X^\#$ denote its localization away from the prime 2 (see [13, Ch.II] for a discussion of localizations). By the universality properties of $(\ )^\#$, there is a commutative diagram as follows:

$$\begin{array}{ccc}
\pi_{s+t}(S^5) \times \pi_5(KP^\infty) & \longrightarrow & \pi_{s+t}(KP^\infty) \\
\downarrow & & \downarrow \\
[S^{s+t\#}, S^{5\#}] \times [S^{5\#}, KP^{\infty\#}] & \longrightarrow & [S^{s+t\#}, KP^{\infty\#}]
\end{array}$$

Since $[S^{n\#}, Y^{\#}]$ is isomorphic to $\pi_n(Y) \otimes Z[\frac{1}{2}]$ (see [13, Thrm. 2.1]), it is immediate that $h^{\#}$ is trivial. By the naturality of the above isomorphism, it follows that the composition map $(h^{\#})_*$ corresponds to h_* tensored with the identity on $Z[\frac{1}{2}]$, and hence $h_* \otimes id_{Z[\frac{1}{2}]}$ is zero. Since the natural map $A \to A \otimes Z[\frac{1}{2}]$ is a monomorphism on elements of odd order (for any abelian group A), naturality considerations imply that h_* must be zero on all elements of odd order.

Let p denote an odd prime which will remain fixed throughout this paper. According to results of Toda, the p-primary component of $\pi_{4p-3}(S^3)$ is isomorphic to Z_p; if $p \geq 5$, then $\pi_{6p-3}(S^5)$ is also isomorphic to Z_p [14, Thrm. 13.4, p.176].

PROPOSITION 1.2. (i) *The generator of the p-primary component of* $E^1_{2,4p-5}$ $(= Z_p)$ *corresponds to a nonzero element of* $\pi_{4p-3}(U_{k+1})$ *for* $1 \leq k \leq p-1$.

(ii) *If* $p \geq 5$, *the generator of the p-primary component of* $E^1_{3,6p-6}$ $(= Z_p)$ *corresponds to a nonzero element of* $\pi_{6p-3}(U_{k+1})$ *for* $2 \leq k \leq p$.

PROOF. By the results of [14, Thrm. 13.4, p. 176], the p-primary component of $E^1_{s,t}$ vanishes provided $s + t = 4p - 2$, $s \neq 2$, and $k \leq p - 1$. Likewise, if $p \geq 5$ the p-primary component of $E^1_{s,t}$ vanishes provided

$s + t = 6p - 2$, $s \neq 3$, and $k \leq p$. Hence all differentials d^r with $E^r_{2,4p-5}$ and $E^r_{3,6p-6}$ as codomains are zero on p-primary components. On the other hand, the only differential with $E^r_{2,4p-5}$ or $E^r_{3,6p-6}$ as domain and a group with nonzero p-primary component as codomain is d^1 on $E^1_{3,6p-6}$. However, this differential is trivial on the p-primary component by Proposition 1.1. Thus the p-primary components of $E^1_{2,4p-5}$ and $E^1_{3,6p-6}$ survive to E^∞ as claimed.

PROPOSITION 1.3. _The p-primary components of_ $\pi_{4p-3}(E(CP^k))$ _and_ $\pi_{6p-3}(E(CP^k))$ _are zero provided_ $2 \leq k \leq 2p-3$ _in the first case, and_ $3 \leq k \leq 3p-3$ _combined with_ $p \geq 5$ _in the second case._

PROOF. Consider the Federer spectral sequence for $\pi_{s+t}(E(CP^k))$ ([5]; also see [11,1.2]). The nonzero groups in the initial term of this sequence in total degrees $4p-3$ and $6p-3$ (if $p \geq 5$) have the form

$$\pi_{q+2j}(S^{2k+1}),$$

where $q = 4p-3$ or $6p-3$ and $0 \leq j \leq k$. By Toda's results ([14] again), the p-primary components of these homotopy groups vanish provided $2 \leq k \leq 2p-3$ if $q = 4p-3$ and $3 \leq k \leq 3p-3$ if $q = 6p-3$. Clearly the p-primary component of the E^∞ term of the Federer spectral sequence must also be zero in total degree q for the above choices of q and k.

The principal result of this section now follows easily.

THEOREM 1.4. _There is a nonzero element in the p-primary component of_ $\pi_q(U_{k+1})$ _which maps trivially into_ $\pi_q(E(CP^k))$ _via_ j_*, _provided_ $q = 4p - 3$ _and_ $2 \leq k \leq p-1$ _or_ $q = 6p-3$, $p \geq 5$, _and_ $3 \leq k \leq p$.

PROOF. According to Proposition 1.2 there is a nonzero element in the p-primary component of $\pi_q(U_{k+1})$ under the above conditions on q and k. On the other hand, according to Proposition 1.3 the p-primary component of $\pi_q(E(CP^k))$ is zero under these same conditions, and hence the conclusion is immediate.

Remark. The elements in Theorem 1.4 have been constructed to satisfy the following useful property: If $u_\ell \in \pi_q(U_{\ell+1})$, $u_k \in \pi_q(U_{k+1})$, $\ell \leq k$, and $i: U_{\ell+1} \to U_{k+1}$ is the standard inclusion, then $i_* u_\ell = u_k$. This is a consequence of the naturality of the exact couple with respect to standard inclusions.

2. Normal invariants of fiber homotopy trivializations

Let U be a compact Lie group acting smoothly on the closed manifold P, and let $\xi \in \pi_{n-1}(U)$ be an element whose image in $\pi_{n-1}(E(P))$ is trivial. Choose a smooth representative f of ξ, and denote the action map from U to E(P) by j. If $P(\xi)$ is the smooth fiber bundle over S^n with fiber P constructed via the map f, then a contracting homotopy H of jf determines a fiber homotopy (FH) trivialization Φ_H of $P(\xi)$. Explicitly, let $P(\xi)$ be the union of $D^n_- \times P$ and $D^n_+ \times P$ modulo identification of the two copies of $S^{n-1} \times P$ by the clutching function $(x,y) \to (x, f(x)y)$. Then if $x \in D^n_-$, let $\Phi_H(x,y)=(x,y)$, and if $x \in D^n_+$ with representation $x = (z,t) \in R^n \times R = R^{n+1}$, then

$$\Phi_H(z,t;y) = (z,t;H_t(\frac{z}{|z|})y)$$

(the right hand side tends to a limit as $z \to 0$ and $t \to 1$

since H_1 is constant); it is easy to check that this map is consistent with the clutching function.

Different homotopy classes of contracting homotopies yield different homotopy classes of FH trivializations, and it is immediate that the set of homotopy classes of such trivializations is canonically isomorphic to $\pi_n(E(P))$, the FH trivialization group of the trivial bundle. Every homotopy class of FH trivializations determines an equivalence class of homotopy smoothings of $S^n \times P$; let

$$C(\xi) \subseteq [S^n \times P, G/O]$$

denote the set of normal invariants determined by FH trivializations of ξ. The purpose of this section is to establish the following result:

THEOREM 2.1. <u>The set</u> $C(0)$ <u>is a subgroup of</u> $[S^n \times P, G/O]$, <u>the set</u> $C(\xi)$ <u>is a coset of</u> $C(0)$, <u>and</u> $C(\xi + \xi') = C(\xi) + C(\xi')$.

The addition on $[X,G/O]$ is the natural abelian group structure induced by the Whitney sum (compare [1]).

We shall need a fairly standard "sum theorem" to prove the above result. Let M be a closed smooth manifold, and let M_1, M_2 be compact submanifolds with boundary of the same dimension such that:

(i) $M = M_1 \cup M_2$

(ii) $\partial M_1 \subseteq \text{Int}M_2$ and $\partial M_2 \subseteq \text{Int}M_1$.

It is immediate that $M_{12} = M_1 \cap M_2$ is also a compact submanifold of the same dimension and $\partial(M_1 \cap M_2) = \partial M_1 \cup \partial M_2$, a disjoint union. Suppose $f_1\colon X_1 \to M_1$ and $f_2\colon X_2 \to M_2$ are homotopy smoothings which are diffeomorphisms on $f_1^{-1}(M_{12})$ and $f_2^{-1}(M_{12})$ respectively. If

$$\varphi\colon f_2^{-1}(M_{12}) \to f_1^{-1}(M_{12})$$

is a diffeomorphism such that $f_1\varphi = f_2$ on $f_2^{-1}(M_{12})$, then (X_1, f_1) and (X_2, f_2) may be pasted together via φ to yield a homotopy smoothing (X,f) of M which is a diffeomorphism on $f^{-1}(M_{12})$. Now (X_1, f_1), (X_2, f_2), and (X,f) may also be interpreted as relative homotopy smoothings, and as such their normal invariants u_1, u_2, and u lie in $[M_1/M_{12}, G/O]$, $[M_2/M_{12}, G/O]$, and $[M/M_{12}, G/O]$ respectively (compare [10, §6, Relative Sullivan Theory]). Since $M - \mathrm{Int}\, M_1$ and $M - \mathrm{Int}\, M_2$ are disjoint, there is a canonical homeomorphism h from

$$(M_1/M_{12}) \vee (M_2/M_{12})$$

to M/M_{12}.

FORMULA 2.2. (Sum Theorem). The restrictions of h^*u to $[M_1/M_{12}, G/O]$ and $[M_2/M_{12}, G/O]$ are u_1 and u_2 respectively.

This follows immediately from the definitions.

Let ξ and ξ' be two elements of $\pi_{n-1}(U)$, and let f and f' be smooth representatives of these classes. If N and N' are two well disjoint disks in S^n, then the bundle classified by $\xi + \xi'$ may be formed from $[S^n - \mathrm{Int}(N \cup N')] \times P$, $N \times P$, and $N' \times P$ by identifying the two copies of $\partial N \times P$ and $\partial N' \times P$ via f and f' respectively. Given contracting homotopies H and H' of jf and jf', a fiber homotopy trivialization $\Phi_H + \Phi_{H'}$ may be readily constructed on this model for $\xi + \xi'$. To do this, take $\Phi_H + \Phi_{H'}$ to be the identity on $[S^n - \mathrm{Int}(N \cup N')] \times P$, an extension resembling the construction of $\Phi_H | D^n_+ \times P$ on $N \times P$, and an extension resembling the construction of $\Phi_{H'} | D^n_+ \times P$ on $N' \times P$.

PROPOSITION 2.3. *The normal invariants of* FH *trivializations satisfy the following sum formula:*

$$q(P(\xi + \xi'), \Phi_H + \Phi_{H'}) = q(P(\xi), \Phi_H) + q(P(\xi'), \Phi_{H'})$$

PROOF. By construction $\Phi_H + \Phi_{H'}$ is a diffeomorphism on the inverse image of $[S^n - \text{Int } N \cup N'] \times P$. Hence the sum formula of 2.2 is applicable with $M_1 = [S^n - \text{Int}N'] \times P$ and $M_2 = [S^n - \text{Int}N] \times P$. However, it is easy to determine the components of the normal invariant u of $\Phi_H + \Phi_{H'}$ as a relative homotopy smoothing of (M, M_{12}). Namely, the $[M_1/M_{12}, G/O]$ component is the normal invariant u_1 of Φ_H and the $[M_2/M_{12}, G/O]$ component is the normal invariant u_2 of $\Phi_{H'}$. Let

$$q: M \to M/M_{12} = (M_1/M_{12}) \vee (M_2/M_{12})$$
$$q_1: M \to M/M_2 \cong M_1/M_{12}$$
$$q_2: M \to M/M_1 \cong M_2/M_{12}$$

be the collapsing maps. The proof of 2.3 then amounts to showing that $q_1^* u_1 + q_2^* u_2 = q^* u \ \ (= q^*(u_1, u_2))$. But this is a straightforward formal homotopy exercise using the fact that the collapsing map

$$S^n \to S^n/S^n - \text{Int}(N \cup N') \cong S^n \vee S^n$$

is a standard comultiplication map.

PROOF OF THEOREM 2.1. Consider the special case in which $\xi = \xi' = 0$. Then Proposition 2.3 essentially says that the composite

$$\pi_n(E(P)) \to hS(S^n \times P) \to [S^n \times P, G/O]$$

is a homomorphism; hence C(0), its image, is a subgroup (compare [4, sequence (D), p.161]). On the other hand, a few standard homotopic manipulations show that every FH trivialization is homotopic to one having the form Φ_H. Therefore, Proposition 2.3 implies that $C(\xi) + C(\xi') \subseteq C(\xi + \xi')$; the rest of the conclusions of Theorem 2.1 follow quickly from this inclusion (compare [7,§4]).

3. Construction of examples

We shall now prove our main result.

THEOREM 3.1. *Let $k \geq 2$ be a positive integer. Then there exist infinitely many values of n for which S^{2n} has a smooth semifree circle action with $S^{2(n-k-1)}$ as twisted fixed point set.*

Remark. In the cases $k = 0, 1$, the fixed point set is always untwisted. The prime p will be fixed as in Section 1.

LEMMA 3.2. *Let ℓ be an even integer. Then $[S^{\ell}CP^k, G/O]$ has no p-torsion if $2k + \ell \leq 2p^2 - 2p - 4$.*

PROOF OF LEMMA 3.2. Induction on k; if $k = 1$, then $CP^k = S^2$ and the result follows since $\pi_{\ell+2}(G)$ has no p-torsion. Assume the result is true for k. Consider the following Puppe sequence:

$$[S^{\ell+1}CP^k, G/O] \to \pi_{2k+\ell+2}(G/O) \to [S^{\ell}CP^{k+1}, G/O] \to [S^{\ell}CP^k, G/O]$$

If $x \in [S^{\ell}CP^{k+1}, G/O]$ is p-torsion, the nonexistence of p-torsion in $[S^{\ell}CP^k, G/O]$ implies that x comes from an element $y \in \pi_{2k+\ell+2}(G/O)$. Since $px = 0$, the element py must be in the image of $[S^{\ell+1}CP^k, G/O]$. But this is impossible. For $2(k+1) + \ell < 2p^2 - 2p - 2$ implies $\pi_{2k+\ell+2}(G/O)$ has no p-torsion (since the corresponding homotopy group of G has none; compare [11,1.1] or [15,§6]), and hence $x \neq 0$ would imply y had infinite order. Since py also has infinite order, there would have to be an element of $[S^{\ell+1}CP^k, G/O]$ of infinite order. But $S^{\ell+1}CP^k$ has only odd-dimensional cells, and the odd homotopy groups of G/O are all finite. Hence $[S^{\ell+1}CP^k, G/O]$ is also finite, a contradiction which establishes the nonexistence of p-torsion.

PROOF OF THEOREM 3.1. Consider the elements ξ and ζ in the p-primary components of $\pi_{4p-3}(U_{k+1})$ and $\pi_{6p-3}(U_{k+1})$ described in Theorem 1.4. We wish to determine the cosets $C(\xi)$ and $C(\zeta)$. Since $p\xi = 0$ and $p\zeta = 0$, it follows from Theorem 2.1 that $pC(\xi) \subseteq C(0)$ and $pC(\zeta) \subseteq C(0)$. In particular $C(\xi)$ and $C(\zeta)$ are contained in the torsion subgroups of $[S^{4p-2} \times CP^k, G/O]$ and $[S^{6p-2} \times CP^k, G/O]$ respectively.

On the other hand, Lemma 3.2 and the canonical splitting

$$[X \times Y, G/O] = [X, G/O] \oplus [Y, G/O] \oplus [X \wedge Y, G/O]$$

imply that the above groups have no p-torsion, provided $k \leqq p^2 - 3p + 1$ in the first case and $k \leqq p^2 - 4p + 1$ in the second case. If $p \geqq 5$, then these upper bounds exceed $p - 1$ and p respectively. The nonexistence of p-torsion implies that multiplication by p is an automorphism of the torsion subgroups; hence the inverse image of $C(0)$ under p is precisely $C(0)$. In other words, $C(\xi)$ and $C(\zeta)$ must be $C(0)$ themselves. Therefore, any FH trivialization of $C(\xi)$ or $C(\zeta)$ must have the same normal invariant as a trivialization of the trivial bundle. Since the manifolds in question are simply connected and even-dimensional, the normal invariant map is a monomorphism. Thus any FH trivialization of $C(\xi)$ or $C(\zeta)$ is equivalent as a homotopy smoothing to a self-equivalence

$h: S^q \times CP^k \to S^q \times CP^k$ $(q = 4p - 2$ or $6p - 2)$. Hence $C(\xi)$ and $C(\zeta)$ are diffeomorphic to $S^{4p-2} \times CP^k$ and $S^{6p-2} \times CP^k$ respectively. By Browder's classification theorem [2, Thrm. 5.7, p.33], ξ and ζ represent equivariant normal bundles of semifree circle actions on some homotopy $(q+2k+2)$-sphere with S^q as (twisted) fixed point set. The existence of infinitely many such actions follows since we have constructed two for each prime $p \geq \max(5,k)$ if $k \geq 3$ and one if $k = 2$.

It remains to show that such actions exist on the ordinary sphere. The homotopy spheres acted upon have dimensions between $4p + 4$ and $6p - 4$ in the first case (ξ elements) and between $6p + 6$ and $8p - 2$ in the second case (ζ elements). However, it is well-known that Γ_{2m} has no p-torsion for $2m < 2p^2 - 2p - 2$ (compare [11,1.1]); if $p \geq 5$, then $6p - 4$ is always less than this number, while if $p \geq 7$, then $8p - 2$ is also less than this number. Thus if n is the order of the homotopy sphere on which the action exists, then n and p are relatively prime. Take a connected sum of n copies of the action on the exotic sphere along the fixed point set. A smooth semifree action is then obtained on the ordinary sphere, its fixed point set is still an ordinary sphere, and its equivariant normal bundle is $n\xi$ or $n\zeta$. Since $(n,p) = 1$, both of these elements are nonzero, and the proof is complete.

It is reasonable to ask whether these equivariant normal bundles are trivial or nontrivial as ordinary vector bundles. We shall show that the ξ bundles are nontrivial provided $k \leq \frac{1}{2}(p-3)$; further homotopy spectral sequence arguments imply the ζ bundles are always trivial as ordinary vector bundles.

LEMMA 3.3. _Conjugation induces an automorphism of the spectral sequence for $\pi_*(U_{k+1})$ constructed in Section 1. The induced map on $E^1_{s,t}$ is $(-1)^s$ on odd-primary components._

PROOF. The first statement follows because conjugation leaves the standardly embedded U_s subgroups invariant. The induced map on $\pi_{s+t}(S^{2s-1})$ is merely composition with the restriction of conjugation on C^s to the unit sphere; this map certainly has degree $(-1)^s$. Since the suspension of composition is bilinear, $S(\{(-1)^s\}_* x) = (-1)^s S(x)$ is trivial. Since suspension is a monomorphism on the odd-primary components of $\pi_{s+t}(S^{2s-1})$ [12, Ch.IV, §4, p.281] clearly $S(\{(-1)^s\}_* x) = S((-1)^s x)$ implies $\{(-1)^s\}_* x = (-1)^s x$ if x has odd order.

PROPOSITION 3.4. _Let $\xi \in \pi_{4p-3}(U_{k+1})$ be as in Theorem 1.4, and suppose $k \leq \frac{1}{2}(p - 3)$. Then the image of ξ in $\pi_{4p-3}(SO_{2k+2})$ is nontrivial._

PROOF. We shall prove a stronger statement. Namely, the composite of ξ with the canonical maps $U_{k+1} \to SO_{2k+2}$ and $SO_{2k+2} \to U_{2k+2}$ is nontrivial. The composite of these canonical maps is just the standard representation of U_{k+1} plus its conjugate. Hence the image of ξ in $\pi_{4p-3}(U_{2k+2})$ is $i_*(\xi + \bar{\xi})$, where $\bar{\xi}$ is the conjugate of ξ.

But ξ corresponds to an element of $E^1_{2,4p-5}$, and conjugation induces the identity on such elements; since there are no p-primary elements of total degree $4p - 3$ in lower filtrations, it follows that ξ is self-conjugate. Thus the image of ξ in $\pi_{4p-3}(U_{2k+2})$ is twice $i_*\xi$; since $i_*\xi$ is nontrivial by the remark following Theorem 1.4 and has odd order, the image element $2\, i_*\xi$ is also nontrivial.

Remark. It is natural to ask whether semifree circle actions with twisted fixed point sets exist on odd-dimensional spheres. Although the p-primary component of $\pi_{n+k}(S^k)$ is known up to roughly $n = 2p^3 + O(p^2)$ [15, §11], the results are considerably more complicated than in the range $n \leq p^2$. This fact causes several complications in any attempt to find odd-dimensional examples like those of Theorem 3.1, none of which appear in the arguments of this paper.

Purdue University
West Lafayette, Indiana 47907

REFERENCES

1. J.M. Boardman and R.M. Vogt, Homotopy-everything H-spaces, Bull. Amer. Math. Soc. 74 (1968), 1117-1122.

2. W. Browder, Surgery and the theory of differentiable transformation groups, Proceedings of the Conference on Transformation Groups (New Orleans, 1967), 1-46. Springer-Verlag, New York, 1968.

3. ________, and T. Petrie, Semifree and quasifree S^1 actions on homotopy spheres, Essays on Topology and Related Topics (Memoires dediés à G. de Rham), 136-146. Springer-Verlag, New York, 1970.

4. ________, Diffeomorphisms of manifolds and semifree actions of homotopy spheres, Bull. Amer. Math. Soc. 77 (1971), 160-163.

5. H. Federer, A study of function spaces by spectral sequences, Trans. Amer. Math. Soc. 82 (1956), 340-361.

6. K. Kawakubo, Invariants for semifree S^1 actions, mimeographed, Institute for Advanced Study, Princeton, 1971.

7. M. Kervaire and J. Milnor, Groups of homotopy spheres, Ann. of Math. 78 (1963), 514-537.

8. H.-T. Ku, A note on semifree actions of S^1 on homotopy spheres, Proc. Amer. Math. Soc. 22 (1969), 614-617.

9. ________, and M.C. Ku, Semifree differentiable actions of S^1 on homotopy (4k+3)-spheres, Mich. Math. J. 15 (1968), 471-476.

10. C.P. Rourke, The Hauptvermutung according to Sullivan, mimeographed, Institute for Advanced Study, Princeton, 1968.

11. R. Schultz, Semifree circle actions and the degree of symmetry of homotopy spheres, Amer. J. Math., to appear.

12. J.-P. Serre, Groupes d'homotopie et classes des groupes abéliens, Ann. of Math. 58 (1953), 258-294.

13. D. Sullivan, Geometric Topology I: Localization, periodicity, and Galois symmetry, mimeographed, Massachusets Institute of Technology, 1970.

14. H. Toda, Composition Methods in Homotopy Groups of Spheres, Annals of Mathematics Study No. 49. Princeton University Press, Princeton. 1962.

15. ______, On iterated suspensions I, J. Math. Kyoto Univ. 5 (1965-1966), 87-142. Ibid. II, 209-250.

$\underline{\underline{Z}}_2$-TORUS ACTIONS ON HOMOTOPY SPHERES

Reinhard Schultz
Purdue University

One problem appearing in the proceedings of the Tulane conference asks for the maximum rank of those Z_2-tori which can act smoothly, orientation-preservingly, and effectively on an exotic n-sphere [2, Problem 6, p. 235]. Cohomological methods imply that the rank cannot exceed n; on the other hand, Z_2^n acts orthogonally and orientation-preservingly on S^n. In this short note we shall show that every homotopy n-sphere which represents an element of odd order in the group Θ_n also has a smooth Z_2^n action, and this action is (equivariantly) homotopically equivalent to the standard one on S^n (in fact, topologically equivalent if $n \geq 5$). Since Θ_n frequently has elements of odd order (e.g., whenever $n \equiv 3 \bmod 4$, provided $n \neq 3$), the maximum rank for exotic n-spheres is usually n. The existence of Z_2^n actions as described above is a special case of the following easy observation:

Let G be a finite group which acts smoothly and orientation-preservingly on S^n, let (H) be the principal isotropy subgroup type, and let $|G/H|$ denote the number of cosets of H in G. Then every element of Θ_n which is divisible by $|G/H|$ supports a smooth G-action which is homotopically equivalent to the given one.

Partially supported by NSF Grant GP-19530.

To prove this, take an equivariant embedding of G/H in S^n as a principal orbit. Now G/H may be viewed as an equivariantly framed zero-manifold in S^n by the differentiable slice theorem; since $\bar{G}$ acts orientation-preservingly, the framed bordism class of this equivariant framing is $|G/H| \in Z = \pi_n(S^n)$. The observations of [1, p. 444] now imply that if $\sum^n$ is a homotopy n-sphere, then $|G/H|\sum^n$ has a smooth G-action which is homotopically equivalent to the given one.

REFERENCES

1. G.E. Bredon, A π_*-module structure for Θ_* and applications to transformation groups, Ann. of Math. 86 (1967), 434-448.

2. P.S. Mostert (ed.), Proceedings of the Conference on Transformation Groups (New Orleans, 1967). Springer-Verlag, New York, 1968.

FREE AND SEMI-FREE SMOOTH ACTIONS

OF S^1 AND S^3 ON HOMOTOPY SPHERES

KAI WANG

University of Chicago

This paper contains some results on actions of S^1 and S^3 on homotopy spheres. All the actions are smooth.

At first, we consider an old problem, namely, "Do there exist infinitely many inequivalent free S^1-actions on standard spheres ?". The first result on this problem is due to W. C. Hsiang and W. Y. Hsiang that the answer is yes on standard 11-sphere (see [7]). Followed soon by a joint paper of D. Montgomery and C. T. Yang that the answer is yes on standard 7-sphere (see [9][10]).

In the former, the special results of Eells and Kuiper on 8-dimensional manifolds were used, while in the later, special properties of dimension 7 were used to translate the classification of free S^1-actions into a question about knotted 3-spheres in 6-sphere, and then the results of A. Haefliger were applied.

It is very difficult to generlize their methods to higher dimensional spheres. However in 1966, W. C. Hsiang applied surgery to construct infinitely many inequivalent free S^1-actions on homotopy (2n-1)-spheres, for $n > 3$. But he was not able to determine the diffeomorphism type of the total spaces (see [6]).

Now for almost all possible dimensions we can answer the question affirmatively. In fact we will prove a little bit stronger result.

Theorem A. Let Σ^{2n-1} be a homotopy sphere which supports a "decomposable" free S^1-action, then it supports infinitely many inequivalent "decomposable" free S^1-actions, for $n \geq 4$, $n \neq 5$.

A free S^1-action (Σ^{2n-1}, f) is decomposable if it is the equivariant gluing of two standard pieces $(S^{2p-1} \times D^{2q}, A)$ and $(D^{2p} \times S^{2q-1}, A)$ by an equivariant diffeomorphism g of $(S^{2p-1} \times S^{2q-1}, A)$ where the action A on each piece is linear. For a more precise definition see Definition 1.

It is clear that there is a decomposable free S^1-action on standard sphere, namely the linear one. Then as a corollary, we have

Corollary. There are infinitely many inequivalent decomposable free S^1-actions on the standard (2n-1)-spheres, for $n \geq 4$, $n \neq 5$.

Remark. I was informed that Dan Burghelea had proved similar result by a different approach, but I have not seen the proof yet. And H. T. Ku and M. C. Ku had also proved similar result on higher dimensional spheres by considering codimension 2 characteristic subspheres of free S^1-actions on homotopy spheres. It was also shown by G. Brumfield that there are infinitely many inequivalent free S^1-actions on standard 9-sphere (see [5]).

It is also interesting to ask when a free S^1-action is decomposable it turns out that the decomposability is determined by the Pontryagin classes of the orbit space. To be more precise, we have

Let $S^1 = \{g \in C \mid |g| = 1\}$, $S^{2m-1} = \{u = (u_1,\dots,u_m) \in C^m \mid \|u\| = 1\}$ and $D^{2m} = \{u = (u_1,\dots,u_m) \in C^m \mid \|u\| \leq 1\}$. Let $gu = (gu_1,\dots,gu_m)$ for $g \in S^1$ and $u \in C^m$.

Define S^1-actions $(S^{2p-1} \times D^{2q}, A)$, $(S^{2p-1} \times S^{2q-1}, A)$ and $(D^{2p} \times S^{2q-1}, A)$ by the equation

$$A(g; (u,v)) = (gu, gv)$$

where $g \in S^1$, $u \in C^p$, $v \in C^q$.

Let k be an equivariant diffeomorphism of $(S^{2p-1} \times S^{2q-1}, A)$. Then we can define an action $A(k)$ on $\Sigma(k)$, where

$$\Sigma(k) = S^{2p-1} \times D^{2q} \cup_k D^{2p} \times S^{2q-1}$$

so that $A(k)|S^{2p-1} \times D^{2q} = A$ and $A(k)|D^{2p} \times S^{2q-1} = A$, which is clearly a free S^1-action.

Definition 1. A free S^1-action (Σ^{2n-1}, f) on a homotopy sphere Σ^{2n-1} is decomposable if there is an equivariant diffeomorphism k of $(S^{2p-1} \times S^{2q-1}, A)$ such that $\Sigma(k)$ is diffeomorphic to Σ^{2n-1} and the actions (Σ^{2n-1}, f) and $(\Sigma(k), A(k))$ are equivalent.

At first, we have following elementary construction which is useful in our furture applications.

Lemma 2. For $q \leq p \leq 2q$, let $\{k_i\}_{i \in N}$ be an infinite set of orientation preserving diffeomorphisms of $S^p \times S^q$ such that $k_i|x_0 \times S^q$

is homotopy to the inclusion. Then there is an infinite subset $\{k_{i_s}\}_{s\in N}$ of $\{k_i\}_{i\in N}$ such that $\Sigma(k_0 k_{i_0}^{-1} k_{i_s})$ is diffeomorphic to $\Sigma(k_0)$.

Note that $(\Sigma(k), A(k))$ depends only on the equivariant pseudo-isotopy class of k. Hence in order to study the free S^1-actions on homotopy spheres, it is reasonable to study the equivariant pseudo-isotopy classes of $(S^{2p-1} \times S^{2q-1}, A)$.

Let (M, G, f), or (M, f) if G is understood, be an action of compact Lie group on smooth manifold M. Let D(M, f) be the group of equivariant pseudo-isotopy classes of equivariant diffeomorphisms of (M, f) and let $D_0(M, f)$ be the subgroup of those elements which are equivariantly homotopic to identity.

Let W be an oriented connected smooth manifold. Then set $hS(W, \partial W)$ consists of equivalence classes of pairs [M, h] where M is an oriented smooth manifold and $h : (M, \partial M) \dashrightarrow (W, \partial W)$ is a simple homotopy equivalence preserving orientation and $h|\partial M$ is a diffeomorphism. Two pairs $[M_1, h_1]$ and $[M_2, h_2]$ are equivalent if there is a diffeomorphism $k : M_1 \dashrightarrow M_2$ such that $h_2 k$ is homotopic to h_1 rel M_1.

Let B_O and B_G denote the classifying spaces for stable vector bundles and stable spherical fibre spaces respectively. Define G/O to be the fiber of the inclusion $B_O \dashrightarrow B_G$. The inclusion of G/O into B_O is denoted by i. The set of homotopy classes of maps of W to G/O mapping ∂W to a point is denoted by $[W, \partial W; G/O]$. Let $\pi = \pi_1(W)$. If dimW = n. Let $L_n(\pi)$ be the Wall group of π. $L_{n+1}(\pi)$ acts on $hS(W, \partial W)$. There are functions ω, d and σ such that the following sequence of sets is exact if $n \geq 5$.

$$L_{n+1}(\pi) \overset{\omega}{\dashrightarrow} hS(W,\partial W) \overset{d}{\dashrightarrow} [W,\partial W; G/O] \overset{\sigma}{\dashrightarrow} L_n(\pi)$$

If $W = K \times D^q$, where K is a closed manifold, then the set $hS(K \times D^q, K \times S^{q-1})$ has an addition for $q \geq 1$, which makes it an abelian group and ω, d and σ are homomorphisms.

Let K be the orbit space of $(S^{2p-1} \times S^{2q-1}, A)$ and let k be an equivariant diffeomorphism of $(S^{2p-1} \times S^{2q-1}, A)$, then the induced map $\bar{k}$ on K is a diffeomorphism and if k is equivariantly homotopic to identity then the mapping torus $K_{\bar{k}} = K \times I/ (x, 0) \sim (\bar{k}(x), 1)$ is homotopic equivalent to $K \times S^1$ by a homotopy equivalence $\tilde{k}$, hence gives an element $[K_{\bar{k}}, \tilde{k}] \in hS(K \times S^1)$.

Let $[M, l] \in hS(K \times I, \partial)$, i. e., $l : M \dashrightarrow K \times I$ is a homotopy equivalence such that $l|\partial M : \partial M \dashrightarrow \partial(K \times I)$ is a diffeomoephism. Let $W = M/ l^{-1}(x, 0) \sim l^{-1}(x, 1)$, then l induces a homotopy equivalence $\bar{l} : W \dashrightarrow K \times S^1$ and $[W, \bar{l}]$ depends only on the class $[M, l]$. Hence we have a well-defined map $\mu : hS(K \times I,\partial) \dashrightarrow hS(K \times S^1)$ which is 1 - 1 and $[K_{\bar{k}}, \tilde{k}] \in \mathrm{Im}\mu$.

Conversely, every element in $\mathrm{Im}\mu$ gives rise an equivariant diffeomorphism of $(S^{2p-1} \times S^{2q-1}, A)$ which is equivariantly homotopic to identity.

Consider following exact sequence of groups

$$0 \dashrightarrow hS(K \times I, \partial) \otimes Q \xrightarrow{d} [\Sigma K_+, G/O] \otimes Q \dashrightarrow \begin{cases} 0 \text{ if } p+q \text{ is even} \\ Q \text{ if } p+q \text{ is odd} \end{cases}$$

We also have following sequence

$$hS(K \times I, \partial) \otimes Q \xrightarrow{d} [\Sigma K_+, G/O] \otimes Q \xrightarrow{i} [\Sigma K_+, B_O] \otimes Q \xrightarrow{\tilde{L}} \tilde{H}^{4*}(\Sigma K_+, Q)$$

where i_* is an isomorphism because the homotopy group of B_G is finite. $\tilde{L}$ is known to be an isomorphism. Hence

(a) $\tilde{L}i_*d$ is an isomorphism if p+q is even

(b) A computation of σ in terms of the Hirzebruch genus L shows that $\tilde{L}i_*d$ is an isomorphism onto the subspace $\Sigma_{i=1}^{p+q-2} H^{4i}(\Sigma K_+, Q)$ if p+q is odd.

Hence a computation of $H^*(\Sigma K_+, Q)$ shows that for $p > 1$, $q > 1$ if p+q is even and $p \gneq 2$, $q \gneq 2$ if p+q is odd, then $hS(K \times I, \partial)$ is an infinite group.

Let $G = \{[K \times S^1, 1] \in \operatorname{Im}\mu\}$. Then G acts on $hS(K \times S^1)$ by composition and there is a 1 - 1 correspondence between the orbit set of $hS(K \times S^1)$ under G and the set of diffeomorphism classes of manifolds with the same homotopy type as $K \times S^1$. But G is finite, then $\operatorname{Im}\mu/G$ is infinite, this implies $D_O(S^{2p-1} \times S^{2q-1}, A)$ is infinite.

Theorem 3. $D_O(S^{2p-1} \times S^{2q-1}, A)$ is an infinite group for $p > 1$, $q > 1$ if p+q is even and $p > 2$, $q > 2$ if p+q is odd.

On the other hand, we have

Theorem 4. $D(S^{2p-1} \times D^{2q}, A)$ for $p \leq 2q$ and $D(D^{2p} \times S^{2q-1}, A)$ for $q \leq 2p$ are finite groups.

Sketch of the proof: Every equivariant diffeomorphism of $(S^{2p-1} \times D^{2q}, A)$ is equivariantly pseudo-isotopic to an equivariant bundle map and the equivariant bundle map is determined by the equivariant homotopy class of an equivariant map $a : (S^{2p-1}, L) \longrightarrow (SO(2q), C)$ where L is the linear action of S^1 on S^{2p-1} and $C(g, x) = gxg^{-1}$ for $g \in S^1$ which is embedded in $SO(2q)$ along the diagonal and $x \in SO(2q)$. Then the set of equivariant homotopy classes of equivariant maps of (S^{2p-1}, L) to $(SO(2q), C)$ is finite.

The following theorem is easy to prove.

Theorem 5. Let k_i, $i = 1, 2$ be equivariant diffeomorphisms of $(S^{2p-1} \times S^{2q-1}, A)$ so that they define actions $(\Sigma(k_i), A(k_i))$ on homotopy spheres. Then $(\Sigma(k_i), A(k_i))$ are equivalent if and only if there exist equivariant diffeomorphisms η of $(S^{2p-1} \times D^{2q}, A)$ and ρ of $(D^{2p} \times S^{2q-1}, A)$ such that $k_2 = \rho k_1 \eta$.

Altogether, we can prove:

Theorem 6. There are infinitely many inequivalent decomposable free S^1-actions on homotopy $(2n-1)$-spheres for $n \geq 3$, $n \neq 5$.

Proof: Choose p and q satisfying the conditions in (2), (3), (4) with $p+q = n$. Since $D_0(S^{2p-1} \times S^{2q-1}, A)$ is infinite, we can construct infinitely many actions $\mathcal{A} = \{(\Sigma(k_i), A(k_i))\}_{i \in N}$ where k_i are choosen from each class of $D_0(S^{2p-1} \times S^{2q-1}, A)$. It is sufficient to show that each equivalence class of actions in $\mathcal{A}$ is finite. Suppose not, there are infinitely many actions in $\mathcal{A}$ which are equivalent to one another, i. e., we have $(\Sigma(k_{i_s}), A(k_{i_s})) \sim (\Sigma(k_{i_0}), A(k_{i_0}))$ for $s \in N$. By (5) there exist equivariant diffeomorphisms η_s of $(S^{2p-1} \times D^{2q}, A)$ and ρ_s of $(D^{2p} \times S^{2q-1}, A)$ such thar $k_{i_0} = \rho_s k_{i_s} \eta_s$. By (4) for some $s \neq t$, ρ_s is equivariantly pseudo-isotopic to ρ_t and η_s is equivariantly pseudo-isotopic to η_t. Hence $\rho_s k_{i_s} \eta_s = \rho_t k_{i_t} \eta_t$ implies k_{i_s} is equivariantly pseudo-isotopic to k_{i_t}, which contradicts to our choice of k_i.

Now we will give a direct proof of a corollary of theorem A, and the idea of the proof of theorem A is the same but a little bit complicated.

Corollary 7. There exist infinitely many inequivalent decomposable free S^1-actions on standard (2n-1)-spheres for $n > 3$, $n \neq 5$.

Proof: Let k_i, $i = 1, 2, \ldots$ be infinitely many equivariant diffeomorphisms of $(S^{2p-1} \times S^{2q-1}, A)$ which are homotopic to identity and define inequivalent actions. By (2), we can choose an infinite subset $\{k_{i_s}\}_{s \in N}$ of $\{k_i\}_{i \in N}$ such that $\Sigma(k_{i_0}^{-1} k_{i_s})$ is diffeomorphic to $\Sigma(id)$ i.e., S^{2n-1}. Note that $k_{i_0}^{-1} k_{i_s}$ are equivariant diffeomorphisms of $(S^{2p-1} \times S^{2q-1}, A)$. Hence we have an infinite set of actions $\mathcal{A} = \{(\Sigma(k_{i_0}^{-1} k_{i_s}), A(k_{i_0}^{-1} k_{i_s}))\}_{s \in N}$. By the same argument as in (6) one can show that each equivalence class of actions in $\mathcal{A}$ is finite. Hence there are infinitely many inequivalent decomposable free S^1-actions on S^{2n-1}.

Proposition 8. There exist only finitely many inequivalent

decomposable free S^1-actions on homotopy 9-spheres.

Proposition 8'. There exist only finitely many inequivalent decomposable free S^3-actions on homotopy 11-spheres.

Naturally one may ask following question: "When is a free S^1-action on homotopy sphere decomposable?".

We have a simple criterion on the decomposability of free S^1-actions on homotopy spheres, namely

Theorem 9. Let (Σ^{2n-1}, f) be a free S^1-action on homotopy sphere. If (Σ^{2n-1}, f) is equivalent to $(\Sigma(k), A(k))$ for an equivariant diffeomorphism k of $(S^{2p-1} \times S^{2q-1}, A)$ for $p \geq q$, then

$$p(\Sigma^{2n-1}/f) \equiv (1+z^2)^n \pmod{z^p}$$

Conversely, if

$$p(\Sigma^{2n-1}/f) \equiv (1+z^2)^n \pmod{z^{[n+1/2]}}$$

then (Σ^{2n-1}, f) is decomposable, where z is a generator of $H^2(\Sigma^{2n-1}/f, Z)$ and p is the total pontryagin class of the orbit space, for n $\geq$ 4.

Sketch of theproof: This theorem is proved by imbedding a copy of CP^{p-1} in the orbit space $\widetilde{CP}^{n-1}$ of (Σ^{2n-1}, f), and then constructing a homotopy equivalence $CP^{n-1} \dashrightarrow \widetilde{CP}^{n-1}$ which is the identity on CP^{p-1}. Using results of Sanderson (see [11]), Adams (see [1]) and Atiyah (see [2]), one proves that the normal bundles of CP^{p-1} in CP^{n-1} and $\widetilde{CP}^{n-1}$ are equivalent, hence there is an equivariant imbedding of

$(S^{2p-1} \times D^{2q}, A)$ into (Σ^{2n-1}, f). Then using the techiniques of G. R. Livesay and C. B. Thomas (see [8]), one can prove our theorem.

Similarly, we have

Theorem 9'. Let (Σ^{4n-1}, f) be a free S^3-action on homotopy sphere. If (Σ^{4n-1}, f) is equivalent to $(\Sigma(k), A(k))$ for an equivariant diffeomorphism k of $(S^{4p-1} \times S^{4q-1}, A)$ for $p \geqq q$, then

$$p(\Sigma^{4n-1}/f) \equiv (1+z)^{2n}(1+4z)^{-1} \quad (\text{mod } z^p)$$

Conversely, if

$$p(\Sigma^{4n-1}/f) \equiv (1+z)^{2n}(1+4z)^{-1} \quad (\text{mod } z^{[n+1/2]})$$

then (Σ^{4n-1}, f) is decomposable, where z is a generator of $H^4(\Sigma^{4n-1}/f, Z)$ and p is the total Pontryagin class of the orbit space for $n \geqq 3$.

Remark. Put together Hsiang's result and (9), we get an alternative proof of theorem 6.

Corollary 10. All free S^1-actions on homotopy 7-spheres are decomposable.

Contrary to theorem 6. we have:

Theorem 11. There are infinitely many inequivalent non-decomposable free S^1-actions on homotopy (2n-1)-spheres for $n \geqq 5$, and there are infinitely many inequivalent non-decomposable free S^3-actions on homotopy (4n-1)-spheres for $n \geq 3$.

Theorem 12. There exist infinitely many free S^1-actions on homotopy 11-spheres which are not extendable to free S^3-actions.

Proof: Let (Σ^{4n-1}, S^1, f) be a free S^1-action which is the restriction of a free S^3-action (Σ^{4n-1}, S^3, F). Then by (9) and (9'), it is easy to see that (Σ^{4n-1}, S^1, f) is decomposable if and only if (Σ^{4n-1}, S^3, F) is. But there are only finitely many inequivalent decomposable free S^3-actions on homotopy 11-spheres , while there are infinitely many inequivalent decomposable free S^1-actions on homotopy 11-spheres, and using the fact that the restriction of (Σ^{4n-1}, S^3, F) to (Σ^{4n-1}, S^1, f) is essentially unique. So there exist infinitely many inequivalent free S^1-actions on homotopy 11-spheres which are not extendable to free S^3-actions.

REFERENCES

1. J. F. Adams: Vector fields on spheres. Annals of Math. 75 (1966) 602-632.

2. J. F. Adams: On the group J(X) II. Topology 3 (1965) 137-171.

3. M. F. Atiyah: Thom complexes. Proc. London Math. Soc. (3) 11 (1961) 291-310.

4. W. Browder: Surgery and the theory of differentiable transformation groups in Proc. Conference on Transformation groups (New Orleans 1967) Springer 1968, 1-46.

5. G. Brumfield: Differentiable S^1 actions on homotopy spheres. Preprint.

6. W. C. Hsiang: A note on free differentiable actions of S^1 and S^3 on homotopy spheres. Annals of Math. 83 (1966) 266-271.

7. W. C. Hsiang & W. Y. Hsiang: Some free differentiable actions on 11-spheres. Quat. J. Math. (Oxford) 15 (1964) 371-374.

8. G. R. Livesay & C. B. Thomas: Free Z_2 and Z_3 actions on homotopy spheres. Topology 7 (1968) 11-14.

9. D. Montgomery & C. T. Yang: Differentiable actions on homotopy seven spheres (I). Trans. A. M. S. 122 (1966) 480-498.

10. D. Montgomery & C. T. Yang: Differentiable actions on homotopy seven spheres (II), In "Proc. Conference on transformation groups (New Orleans 1967)" Springer 1968 125-134.

11. B. J. Sanderson: Immersions and embeddings of projective spaces. Proc. of London Math. Soc. 14(1964) 137-153.

12. K. Wang: Free and semi-free actions of S^1 and S^3 on homotopy spheres. Preprint.

COBORDISM OF INVOLUTIONS REVISITED

J.M. Boardman

The Johns Hopkins University
Baltimore, Maryland 21218

We present a complete elementary proof of the five halves theorem, which was announced as Theorem 1 of [2].

THEOREM 0. *Let* T *be a smooth involution on a smooth closed* n-*manifold* V, *and let* k *be the fixed-point dimension* (that is, the maximum of the dimensions of the various components of the fixed-point set). *If* V *does not bound* (in the unoriented sense), *then* $n \leq 5k/2$. *If further the class* $[V]$ *is indecomposable in the cobordism ring* N, *then* $n \leq 2k+1$.

This completely answers the question raised by Conner and Floyd in (27.1) of [3]. Examples we construct for the proof and their products show that these bounds are the best possible, apart from the obvious trivial improvements. (If k is odd, we can obviously improve the first bound to $n \leq (5k-1)/2$. If $k+1$ is a power of 2, the second bound can be improved to $n \leq 2k$, because the hypothesis ensures that $n+1$ is not a power of 2.)

In the original (no longer available) preprint version a complicated and obscure computational algebraic proof was given. Later, in some privately circulated notes, Conner offered some key observations that shed much light on the subject. We find that we can reinterpret the ideas geometrically and provide direct geometric proofs. Consequently, we now have a vastly improved and very much shorter proof of the theorem. In particular, we no longer need the cohomology theory $\underline{N}^*(-)$. Our proof is more-or-less self-contained and depends on very little cobordism theory, all of which is found in Conner and Floyd's book [3]. We shall summarize what little we need.

All manifolds are assumed smooth in the C^∞ sense, compact, unoriented, and

The author was partially supported by the National Science Foundation under grant GP-19481.

without boundary unless otherwise indicated. We work entirely mod 2, and $H_*(X)$ denotes the mod 2 homology, $H_*(X;Z_2)$. The symbol]]] signals the end of a proof, or the judged absence of any need for further proof.

§1. The cobordism ring. The unoriented cobordism ring $\underline{N}$ was defined and investigated by Thom [8]. The group $\underline{N}_n$ is the set of cobordism classes $[V]$ of n-manifolds V, with addition by means of disjoint union, $[V] + [W] = [V \cup W]$. The graded group $\underline{N}$ is made into a graded algebra over Z_2 by the cartesian product, $[V][W] = [V \times W]$. Thom determined the structure of this algebra.

In order to discuss cobordism adequately, we need certain well-known polynomials in the Stiefel-Whitney classes rather than the Stiefel-Whitney classes themselves. These are conveniently defined axiomatically, as follows. For each positive integer n there is for each real vector bundle ξ a well-defined characteristic class $\sigma_n(\xi) \in H^n(X)$, where X is the base space of ξ, that satisfies the following axioms:

(a) it is natural, $\sigma_n(f^*\xi) = f^*\sigma_n(\xi)$;

(b) it is additive, $\sigma_n(\xi \oplus \eta) = \sigma_n(\xi) + \sigma_n(\eta)$ for vector bundles ξ and η over the same space;

(c) $\sigma_n(\xi) = w_1(\xi)^n$ when ξ is a line bundle.

In terms of the Stiefel-Whitney classes $w_i(\xi)$, $\sigma_n(\xi)$ is that polynomial which expresses the sum of powers $\Sigma_i\, t_i^n$ in terms of the elementary symmetric functions of the indeterminates t_i (as is clear when ξ is a Whitney sum of line bundles). The axioms for the characteristic classes σ_n enable us to compute, much as with Stiefel-Whitney classes.

If the base space X happens to be a n-manifold with fundamental class $z_X \in H_n(X)$, we define $e_n(\xi) = \langle \sigma_n(\xi), z_X \rangle$, with value 0 or 1. We also define $\sigma_n(X) = \sigma_n(\tau)$ and $e_n(X) = e_n(\tau)$, where τ is the tangent bundle of X. It is clear that if Y is a k-manifold and n and k are both positive, then

$$\sigma_{n+k}(X\times Y) = p^*\sigma_{n+k}(X) + q^*\sigma_{n+k}(Y) = 0$$

for dimensional reasons, where $p: X \times Y \to X$ and $q: X \times Y \to Y$ are the projections; which shows that $e_{n+k}(X \times Y) = 0$.

We can now state Thom's theorem (see Théorème IV.12 of [8]).

THEOREM 1. *The ring $\underline{N}$ is a graded polynomial algebra over Z_2 with one generator in each degree not of the form 2^q-1. The class $[V]$ in $\underline{N}_n$ serves as a polynomial generator in degree n if and only if $e_n(V) = 1$.*

Example: real projective space (compare p. 80-81 of [8]). Let P_n denote real projective n-space, and $\alpha \in H^1(P_n)$ the generator of cohomology. If we add a trivial line bundle to the tangent bundle, we obtain the Whitney sum of $n+1$ copies of the canonical line bundle over P_n. Therefore $\sigma_n(P_n) = (n+1)\alpha^n$, and $e_n(P_n) = n+1$. Thus $[P_n]$ is a polynomial generator if n is even, but not if n is odd.

Example: the hypersurface $H_{m,n}$. This is the surface in $P_m \times P_n$ defined by the equation

$$t_0 u_0 + t_1 u_1 + \ldots + t_m u_m = 0,$$

where $t_0, t_1, \ldots, t_m$ are homogeneous coordinates in P_m and $u_0, u_1, \ldots, u_n$ are homogeneous coordinates in P_n, and we assume $m \leq n$. These very useful manifolds were introduced into cobordism theory by Milnor [5].

Their characteristic classes are particularly easy to compute. Let α and β be the generators of $H^1(P_m \times P_n)$ inherited from P_m and P_n respectively. For the normal line bundle ν to $H_{m,n}$ in $P_m \times P_n$ we have $w_1(\nu) = \alpha + \beta$. Therefore

$$\sigma_{m+n-1}(H_{m,n}) = \sigma_{m+n-1}(P_m) + \sigma_{m+n-1}(P_n) + (\alpha + \beta)^{m+n-1}$$

$$= (m+1)\alpha^{m+n-1} + (n+1)\beta^{m+n-1} + (\alpha + \beta)^{m+n-1}.$$

The first two terms vanish if we assume $2 \leq m \leq n$, because $\alpha^{m+1} = 0$ in $H^*(P_m)$ and $\beta^{n+1} = 0$ in $H^*(P_n)$. The value of the third term on the fundamental class z_H is not completely obvious.

LEMMA 2. For any class $\varphi \in H^*(P_m \times P_n)$ we have

$$\langle j^*\varphi , z_H\rangle = \langle\varphi . (\alpha + \beta), z_{P\times P}\rangle ,$$

where $J:H_{m,n} \subset P_m \times P_n$.

Proof. The one-dimensional cohomology class in $P_m \times P_n$ dual to the submanifold $H_{m,n}$ must be $\alpha + \beta$; in other words, $j_* z_H = z_{P\times P} \cap (\alpha + \beta)$. Then

$$\langle j^*\varphi , z_H\rangle = \langle\varphi , j_* z_H\rangle = \langle\varphi, z_{P\times P} \cap (\alpha + \beta)\rangle = \langle\varphi \cdot (\alpha + \beta), z_{P\times P}\rangle . \qquad]]]$$

We therefore seek the coefficient of $\alpha^m \beta^n$ in $(\alpha + \beta)^{m+n}$, which is by definition the binomial coefficient $\binom{m+n}{m}$, taken mod 2 of course. That is, $e_{m+n-1}(H_{m,n}) = \binom{m+n}{m}$.

If $m = 1$, our computation needs to be modified. Since $H_{1,1}$ is just a circle, let us assume $n \geq 2$. Then $e_n(H_{1,n})$ is the coefficient of $\alpha\beta^n$ in $(n+1)\beta^n(\alpha + \beta) + (\alpha + \beta)^{n+1}$, which is always zero.

The mod 2 properties of binomial coefficients are well known. We have $\binom{m+n}{m} = 1$ if and only if there are no carry digits when we add the numbers m and n in binary notation (see for example Lemma I.2.6 of [7]). In particular, given k not a power of 2, write $k = 2^r(2s+1)$ and take $m = 2^r$ and $n = 2^{r+1}s$, so that $e_{k-1}(H_{m,n}) = 1$, provided $r \geq 1$. If $r = 0$, we cannot use $H_{1,n}$, but k is odd and we already know $e_{k-1}(P_{k-1}) = 1$. So the classes $[P_n]$ and $[H_{m,n}]$ generate the algebra $\underline{N}$.

§2. Bordism theory. In [3], Conner and Floyd developed and exploited the bordism groups $\underline{N}_n(X)$ of a space X. The elements are the bordism classes $[V,f]$ of maps $f:V \to X$, where V is a n-manifold. It is useful to note that because $V\times I$ is a cobordism from V to itself, $[V,f] = [V,g]$ whenever f and g are homotopic. These groups form a generalized homology theory, $\underline{N}_*(-)$. When X is a point, the map f is redundant, and so the coefficient groups of this homology theory are just the cobordism groups $\underline{N}_n$.

There are obvious commutative and associative products in this theory, $\underline{N}_m(X) \times \underline{N}_n(Y) \to \underline{N}_{m+n}(X\times Y)$, given by $[V,f].[W,g] = [V\times W, f\times g]$. In particular, when X is a point, $\underline{N}_*(Y)$ is naturally endowed with a $\underline{N}$-module structure.

The natural transformation $\mu : \underline{N}_*(X) \to H_*(X)$ is defined by $\mu[V,f] = f_* z_V$, where z_V is the fundamental class of V. It preserves the product structure; that is, we have the commutative square

$$\begin{array}{ccc} \underline{N}_*(X) \times \underline{N}_*(Y) & \longrightarrow & \underline{N}_*(X\times Y) \\ \downarrow \mu\times\mu & & \downarrow \mu \\ H_*(X) \times H_*(Y) & \longrightarrow & H_*(X\times Y). \end{array}$$

The structure of the $\underline{N}$-module $\underline{N}_*(X)$ is described by the following theorem, which combines (8.1) and (8.3) of [3].

THEOREM 3. The $\underline{N}$-module $\underline{N}_*(X)$ is free, and μ induces an isomorphism $Z_2 \otimes_{\underline{N}} \underline{N}_*(X) \cong H_*(X)$. More specifically, the elements $z_\alpha \in \underline{N}_*(X)$ form a $\underline{N}$-base of $\underline{N}_*(X)$ if and only if their augmentations μz_α form a Z_2-base of $H_*(X)$.]]]

Remark (addressed mainly to experts). This result depends ultimately on showing that (in the language of Milnor and Moore [6]) the left $\underline{A}$-module coalgebra "$H^*(MO)$" is $\underline{A}$-free, where $\underline{A}$ is the Steenrod algebra. This was established by Thom in his original paper (II.§6 of [8]) in proving our Theorem 1. Although Theorem 3 as stated suffices for our purposes, far more is true. There exists an unnatural natural isomorphism of homology theories $\underline{N}_*(X) \cong H_*(X) \otimes \underline{N}$, that preserves the product structure and is therefore in particular an isomorphism of $\underline{N}$-modules; we call it unnatural because there is no preferred isomorphism in view. This boils down to the existence of an isomorphism of left $\underline{A}$-module coalgebras $H^*(MO) \cong \underline{A} \otimes \underline{C}$, where $\underline{C}$ is the coalgebra dual to $\underline{N}$ and is given the trivial $\underline{A}$-module structure. This was proved in effect also by Thom in [8] (his $\underline{A}$-base for $H^*(MO)$ visibly spans a subcoalgebra over Z_2) and explicitly by Liulevicius [4]. Moreover, this is no accident; any connected commutative associative $\underline{A}$-free left $\underline{A}$-module coalgebra over the

Steenrod algebra $\underline{A}$ splits in the same fashion. (This happens if "g" in the proof of 4.4 in [6] is chosen to be a homomorphism of coalgebras as well as of comodules. As in [4], the special properties of the Steenrod algebra ensure that this can always be done.)

§3. Bordism of vector bundles. Let ξ be a vector bundle with fibre dimension k over a n-manifold V. Let $f: V \to BO(k)$ be a classifying map, which is unique up to homotopy. Then the element $[V,f]$ of $\underline{N}_n(BO(k))$ depends only on ξ. This correspondence (see §25 of [3]) interprets $\underline{N}_n(BO(k))$ as the cobordism group of n-manifolds with vector bundles over them having fibre dimension k.

Alternatively, we may restrict attention to smooth vector bundles over manifolds; that is, vector bundles in which the total space is itself a smooth manifold, in such a way that the vector bundle is smoothly locally trivial. Conner and Floyd show (§25 of [3] again) that we get the same cobordism group.

As we let k vary, we obtain $\underline{N}_*(\coprod_k BO(k))$, where $\coprod$ denotes the disjoint union. This becomes an $\underline{N}$-algebra under the pairings

$$\underline{N}_*(BO(k)) \times \underline{N}_*(BO(m)) \longrightarrow \underline{N}_*(BO(k+m))$$

induced by the cartesian product of vector bundles. Since we can write this pairing as the composite

$$\underline{N}_*(BO(k)) \times \underline{N}_*(BO(m)) \longrightarrow \underline{N}_*(BO(k) \times BO(m)) \longrightarrow \underline{N}_*(BO(k+m))$$

induced by the usual map $BO(k) \times BO(m) \longrightarrow BO(k+m)$, it is carried by the augmentation μ into the corresponding pairing in homology.

To interpret $\underline{N}_*(BO)$ we stabilize, by ignoring the addition of trivial line bundles: we impose the relation $[\xi] = [\xi \oplus \varepsilon] = [\xi] \times [R]$ for any vector bundle ξ. The cartesian product makes $\underline{N}_*(BO)$ into a $\underline{N}$-algebra, whose structure we now determine. The homology $H_*(BO)$ is well known to be a polynomial algebra on the generators $a_i \in H_i(BO(1)) \subset H_i(BO)$, for $i > 0$. (We may take P_∞ as $BO(1)$.) To recognize other systems of polynomial generators of $H_*(BO)$ we again need the characteristic classes σ_i introduced in §1. If ξ is a vector bundle with classifying map $f:X \to BO(k) \to BO$ and $u \in H_n(X)$, the class f_*u serves as one of the polynomial generators of $H_*(BO)$ if and only if $\langle \sigma_n(\xi), u \rangle = 1$. We combine this with Theorem 3.

LEMMA 4. The ring $\underline{N}_*(BO)$ is a polynomial algebra over $\underline{N}$ with one generator in each positive degree. The class $[\xi]$ of a vector bundle ξ over a n-manifold serves as a polynomial generator if and only if $e_n(\xi) = 1$.]]]

We shall also be concerned with $\underline{N}_*(BO)$ regarded as an algebra over Z_2. For convenience let us write $b_i(\xi) = e_i(V)$ whenever ξ is a vector bundle over a manifold V, to detect generators for the subalgebra $\underline{N}$ of $\underline{N}_*(BO)$.

LEMMA 5. Suppose given for each positive n, elements x_n' and x_n'' of $\underline{N}_n(BO)$, with x_n' absent if n has the form 2^q-1. Then these elements form a system of polynomial generators of the ring $\underline{N}_*(BO)$ over Z_2 if and only if for each n the pairs of numbers $(b_n(x_n'), e_n(x_n'))$, $(b_n(x_n''), e_n(x_n''))$ and $(0,0)$ are distinct.

Proof. We study $\underline{N}_*(BO)$ by means of the augmentation homomorphism $\mu:\underline{N}_*(BO) \to H_*(BO)$, the homomorphism $p_*:\underline{N}_*(BO) \to \underline{N}$ induced by the map p from BO to a point, and the inclusion of rings $i_*:\underline{N} \to \underline{N}_*(BO)$ induced by the inclusion of a point into BO. For each n we choose an element $y_n \in \underline{N}_n(BO)$ such that $e_n(y_n) = 1$, so that $\underline{N}_*(BO)$ is a polynomial algebra over $\underline{N}$ on the generators y_n, by Lemma 4. We may assume $b_n(y_n) = 0$, by replacing y_n by $y_n + i_*p_*y_n$ if necessary, because $b_n(y_n) = e_n(p_*y_n)$ by definition. We next choose polynomial generators z_n (for n not of the form 2^q-1) for $\underline{N}$ by Theorem 1, included in $\underline{N}_*(BO)$ by i_*; then $e_n(z_n) = 0$, $b_n(z_n) = 1$, and $\underline{N}_*(BO)$ is indeed a polynomial algebra over Z_2 with the y_n and z_n as generators.

Any element x in $\underline{N}_n(BO)$ can be expressed as a polynomial in the y_r and z_r. Since b_n and e_n both vanish on decomposable elements, we must have

$$x = e_n(x)y_n + b_n(x)z_n + \text{decomposable elements.}$$

The lemma now follows.]]]

§4. Transfer homomorphisms. The reason that the isomorphisms $\underline{N}_*(X) \cong H_*(X) \otimes \underline{N}$ mentioned in §2 do not exhaust the study of the bordism groups $\underline{N}_*(X)$ is that this homology theory, in common with all practical homology theories, has more structure than the Eilenberg-Steenrod axioms suggest. For example, if V is a manifold, $\underline{N}_*(V)$ contains the canonical element $[V,1]$.

More generally, suppose $f:V \longrightarrow W$ is a smooth map of manifolds. It follows from §9 of [3] and transversality theory [8] that we can represent any element of $\underline{N}_*(W)$ by a smooth map $g:Q \longrightarrow W$ that is transverse to f (that is, $f\times g:V\times Q \longrightarrow W\times W$ transverse to the diagonal submanifold ΔW of $W\times W$). The pullback P of f and g is again a manifold, which is equipped with a map $h:P \to V$. We can do the same for cobordisms.

LEMMA 6. We can define a N-module homomorphism $f^!:\underline{N}_*(W) \longrightarrow \underline{N}_*(V)$, called the transfer homomorphism induced by f, by the formula $f^![Q,g] = [P,h]$.]]]

This homomorphism does not preserve degrees. As a special case, the canonical element $[V,1]$ in $\underline{N}_*(V)$ may be described as $p^!1$, where p maps V to a point. Transfer homomorphisms have some obvious properties, such as $(f \circ g)^! = g^! \circ f^!$ and $(f\times g)^!(x\times y) = (f^!x)\times(g^!y)$.

Remark. It is easy to see that none of the natural isomorphisms $\underline{N}_*(X) \cong H_*(X) \otimes \underline{N}$ of §2 expresses the bordism transfer solely in terms of the corresponding homology transfer $f^!:H_*(W) \longrightarrow H_*(V)$. (For consider the homomorphisms $f_* \circ f^!:\underline{N}_*(W) \longrightarrow \underline{N}_*(W)$ as $f:V \longrightarrow W$ runs through all classes of $\underline{N}_*(W)$.)

In fact, we make $\underline{N}_*(V)$ an algebra over $\underline{N}$ (if degrees are ignored) by means of the intersection product, with $[V,1]$ as identity element. Given mutually transverse smooth maps $f_1:Q_1 \longrightarrow V$ and $f_2:Q_2 \longrightarrow V$, we form the pullback $g:P \longrightarrow V$ and define the intersection product by $[Q_1,f_1].[Q_2,f_2] = [P,g]$. This multiplication is well defined by Lemma 6, because we can express it as the composite

$$\underline{N}_*(V) \times \underline{N}_*(V) \longrightarrow \underline{N}_*(V\times V) \xrightarrow[\Delta^!]{} \underline{N}_*(V),$$

where $\Delta: V \to V\times V$ is the diagonal inclusion. It is immediate, categorically or geometrically, that for a smooth map of manifolds $f: V \to W$ the transfer $f^!: \underline{N}_*(W) \to \underline{N}_*(V)$ is a homomorphism with respect to the intersection products.

§5. Manifolds with involution. We are interested in various cobordism groups of manifolds with involution, that is, of manifolds V equipped with a smooth map $T: V \to V$ satisfying $T^2 = 1$. We define the following groups of cobordism classes,

- $\underline{I}_n$, of n-manifolds with involution,
- $\underline{F}_n$, of n-manifolds with free involution,
- $\underline{M}_n$, of n-manifolds with boundary, with an involution that is free on the boundary.

As usual, these are abelian groups by means of disjoint union, and every element has order 2.

The graded groups $\underline{I}$, $\underline{F}$ and $\underline{M}$ possess much algebraic structure. All are clearly $\underline{N}$-modules. The cartesian product makes $\underline{I}$ into a $\underline{N}$-algebra and $\underline{M}$ and $\underline{F}$ into $\underline{I}$-modules. Moreover, $\underline{M}$ too becomes a $\underline{N}$-algebra if we first reinterpret it. Let V be a manifold with boundary and involution. According to Conner and Floyd (§22 of [3]), the set F of fixed points is the disjoint union of various submanifolds F_i of dimension i, and each F_i admits a tubular neighborhood N_i that is isomorphic, as a manifold with involution, to the unit disk bundle of the normal bundle to F_i in V, equipped with the fibrewise antipodal involution that is minus on each fibre. We may assume the N_i disjoint. Outside the union N of the N_i, T acts freely, from which it follows that V is cobordant in $\underline{M}$ to N. But N is determined by the normal bundles to the fixed-point sets F_i; in other words, $\underline{M}_n$ may be regarded as the cobordism group of smooth vector bundles over manifolds, where n is now the total dimension. The cartesian product of vector bundles now makes $\underline{M}$ a $\underline{N}$-algebra, whose structure was determined in §3 as

$$\underline{M}_n = \Sigma_i \, \underline{N}_i(BO(n-i)).$$

The cartesian product multiplication in $\underline{F}$ turns out to be inappropriate, being identically zero. Instead, we make $\underline{F}$ a $\underline{N}$-algebra by defining $[V,T][W,T] = [V \times_T W, T\times 1]$, where $V \times_T W$ denotes the orbit space $V\times W/T\times T$, so that for the orbit spaces we have $(V \times_T W)/(T\times 1) = (V/T) \times (W,T)$. This multiplication also makes $\underline{M}$ and $\underline{I}$ into $\underline{F}$-modules. The structure of $\underline{F}$ was found rather easily by Conner and Floyd, in §19 of [3]. We regard an involution on V as an action of $O(1) = Z_2$, so that there is a classifying map $f:V/T \longrightarrow BO(1)$. This sets up the isomorphism of $\underline{N}$-algebras $\underline{F} \cong \underline{N}_*(BO(1))$, with the multiplication in $\underline{N}_*(BO(1))$ induced by the tensor product of line bundles. The tensor product of vector bundles also describes the $\underline{F}$-module structure of $\underline{M}$.

There are obvious homomorphisms $\underline{F}_n \longrightarrow \underline{I}_n \longrightarrow \underline{M}_n$, defined by merely broadening the equivalence relation. We have also the boundary homomorphism $\partial : \underline{M}_n \longrightarrow \underline{F}_{n-1}$, defined by $\partial[V,T] = [\partial V, T|\partial V]$ where ∂V is the boundary of V, and called the <u>bordism J-homomorphism</u>. (The classical Hopf-Whitehead J-homomorphism $J:\pi_i(BO(n)) \longrightarrow \pi_{i+n-1}(S^n)$ can be regarded as assigning to each vector bundle over the i-sphere S^i with fibre dimension n the fibre homotopy equivalence class of its unit sphere bundle.) These are all homomorphisms of $\underline{I}$-modules, and of $\underline{F}$-modules.

Elementary geometric arguments show that these homomorphisms form an exact sequence, which is, in fact, a special case of the extremely general cobordism exact sequence. (For instance, if $\partial[V,T] = 0$, then ∂V bounds some manifold W with free involution, and V is cobordant in $\underline{M}$ to the closed manifold $V \cup W$, where we glue W to V along ∂V. The other steps are even easier.) In this case, however, we have a splitting $\underline{F} \rightarrow \underline{M}$, because any manifold V with free involution is the unit sphere bundle in the line bundle over V/T classified by the map $V/T \longrightarrow BO(1)$ corresponding to V, and this sphere bundle bounds the unit disk bundle. Therefore the long exact sequence reduces to the exact sequence below, exactly as for (28.1) of [3].

<u>THEOREM 7.</u> <u>We have the split short exact sequence</u>

$$0 \longrightarrow \underline{I}_n \longrightarrow \underline{M}_n \xrightarrow[\partial]{} \underline{F}_{n-1} \longrightarrow 0. \qquad]]]$$

Since we already know $\underline{F}$ and $\underline{M}$ as $\underline{N}$-modules, we can deduce the structure of $\underline{I}$ as $\underline{N}$-module,

$$\underline{I}_n \cong \Sigma_{i\neq 1} \underline{N}_{n-i}(BO(i)).$$

The complete determination of $\underline{I}$ as $\underline{N}$-algebra is exactly the problem of finding the kernel of ∂. We know already that $\underline{I}$ is a subalgebra of $\underline{M}$, which is itself a polynomial algebra over $\underline{N}$. In spite of this, the behavior of ∂ on products is not clear. It is not, as one might hope, a derivation.

§6. The stable bordism J-homomorphism. We find we can compute ∂ on products if we first stabilize. To do this we need the suspension element $s = [D^1, T]$ of $\underline{M}_1$, where T is the antipodal involution on the unit 1-disk D^1. In terms of vector bundles, multiplication by s, or suspension, merely adds a trivial line bundle.

Let x and y be the classes in $\underline{M}$ of typical vector bundles ξ over V and η over W, with fibre dimensions m and n respectively, where V is a i-manifold and W is a j-manifold. Then xy is represented by the cartesian product $\xi\times\eta$ over $V\times W$. By compactness, we can embed ξ in the trivial bundle ε^A over V with fibre R^A, and embed η in the trivial bundle ε^B over W, if A and B are large enough. Then $\xi\times\eta$ embeds in the trivial bundle $\varepsilon^A \times \varepsilon^B$. We consider it as the intersection of $\xi\times\varepsilon^B$ and $\varepsilon^A\times\eta$. When we restrict to the unit sphere bundles and project to the unit sphere S^{A+B-1} in $R^A\times R^B$, we find the pullback square

$$\begin{array}{ccc} S(\xi\times\eta) & \longrightarrow & S(\xi\oplus\varepsilon^B) \\ \downarrow & & \downarrow \\ S(\varepsilon^A\oplus\eta) & \longrightarrow & S^{A+B-1}, \end{array}$$

where $S(\omega)$ denotes the unit sphere bundle of a vector bundle ω. The manifolds with involution $S(\xi\times\eta)$, $S(\xi\oplus\varepsilon^B)$ and $S(\varepsilon^A\oplus\eta)$ represent respectively $\partial(xy)$, $\partial s^B x$, and $\partial s^A y$. When we take orbit spaces we still have a pullback square, and therefore

$$\partial(xy) = (\partial s^B x)\cdot(\partial s^A y) \quad \text{in} \quad \underline{N}_*(P_{A+B-1}) \subset \underline{N}_*(P_\infty) = \underline{N}_*(BO(1))$$

with respect to the intersection product in $\underline{N}_*(P_{A+B-1})$.

We shall evidently need $\partial(s^k xy)$ as well. This is easily deduced from the previous formula by replacing x by $s^k x$, which has the effect of increasing A by k. We find

$$\partial(s^k xy) = (\partial s^{B+k}x).(\partial s^{A+k}y) \quad \text{in} \quad \underline{N}^*(P_{A+B+k-1}).$$

The choice of A and B is unimportant, as long as they are large enough. We easily recover ∂x from ∂sx. With notation as above, sx is represented by $\xi \oplus \varepsilon$, which is contained in $\varepsilon^A \oplus \varepsilon$. To recover $S(\xi)$ from $S(\xi \oplus \varepsilon)$ we intersect with $S(\varepsilon^A)$. In terms of orbit spaces, we have precisely the transfer associated to the inclusion $h:P_{A-1} \subset P_A$ as projective hyperplane. So $\partial x = h^! \partial sx$, which is nothing more than a restatement of Theorem (26.4) of [3]. We can now stabilize.

DEFINITION 8. For any integer n, and A sufficiently large (at least n), we define $\underline{L}_n = \text{invlim}_k \underline{N}_{n+k}(P_{A+k})$, where this inverse limit is formed using the maps $h^!:\underline{N}_{n+(k+1)}(P_{A+(k+1)}) \to \underline{N}_{n+k}(P_{A+k})$, so that the graded group $\underline{L}$ is a $\underline{N}$-module. We make $\underline{L}$ a $\underline{N}$-algebra by means of the multiplication $\underline{L}_m \times \underline{L}_n \to \underline{L}_{m+n}$ induced from the intersection pairings

$$\underline{N}_{m+(B+k)}(P_{A+(B+k)}) \times \underline{N}_{n+(A+k)}(P_{B+(A+k)}) \to \underline{N}_{(m+n)+k}(P_{(A+B)+k})$$

in $\underline{N}_*(P_{A+B+k})$, where we write $\underline{L}_m = \text{invlim}_r \underline{N}_{m+r}(P_{A+r})$, $\underline{L}_n = \text{invlim}_r \underline{N}_{n+r}(P_{B+r})$, and $\underline{L}_{m+n} = \text{invlim}_r \underline{N}_{m+n+r}(P_{A+B+r})$. We define the stable bordism J-homomorphism $J:\underline{M}_{i+n} \to \underline{L}_{i+n}$ on the class $[\xi]$ from the sequence of elements $\partial s^{k+1}[\xi] \in \underline{N}_{i+n+k}(P_{A+k})$.

The intersection products do make $\underline{L}$ an algebra, because they are preserved by the maps $h^!$.

We have an immediate consequence of our product formula and definitions.

THEOREM 9. $J:\underline{M} \to \underline{L}$ is a homomorphism of $\underline{N}$-algebras.]]]

The algebraic structure of $\underline{L}$ is easily determined. Let $\theta_n \in \underline{L}_{-n}$ be the element represented by the inclusions $P_{-n+k} \subset P_{A+k}$, defined for $n \geq -A$. From the definition of multiplication in $\underline{L}$, $\theta_m \theta_n = \theta_{m+n}$ and $\theta_0 = 1$. We therefore write θ for θ_1 from now on, so that $\theta_n = \theta^n$, the nth power, for any integer n. By Theorem 3, the θ^n for $-A \leq n \leq k$ map to a $\underline{N}$-base of $\underline{N}_*(P_{A+k})$, and θ^n maps to zero if $n > k$. As we let k vary, we deduce that $\underline{L}$ is the graded algebra of (homogeneous) Laurent series in θ with coefficients in $\underline{N}$. Because θ has degree -1, there are only finitely many terms with negative powers of θ in any Laurent series.

Example. $Js = \theta^{-1}$. It is represented in $\underline{N}_{k+1}(P_{A+k})$ by the element ∂s^{k+2}, which is the class of the inclusion $P_{k+1} \subset P_{A+k}$.

Denote by $\underline{K}$ the subalgebra of power series in θ.

LEMMA 10. The map $\underline{L}_n \to \underline{N}_{n+k}(P_{A+k})$ is onto, with kernel $\theta^{k+1}\underline{K}_{n+k+1}$.]]]

If we take $k = -1$, this kernel is $\underline{K}_n$, and we recover the original homomorphism ∂ as the composite $\underline{M}_n \xrightarrow{J} \underline{L}_n \to \underline{N}_{n-1}(P_\infty)$. If we combine this with the short exact sequence of Theorem 7, we obtain a description of $\underline{I}$.

THEOREM 11. Let $x \in \underline{M}$. Then $x \in \underline{I}$ if and only if $Jx \in \underline{K}$, that is, the Laurent series Jx is without negative powers of θ.]]]

Theorems 9 and 11 at last provide an algebraic reason why $\underline{I}$ should be a subalgebra of $\underline{M}$.

We now see why it is difficult to compute ∂ on products. In effect, we extract ∂x from Jx by ignoring non-negative powers of θ. Hence to find $\partial(xy)$, unless we can put a bound on the negative powers of θ occurring in Jy, we need arbitrarily many of the coefficients in the Laurent series Jx.

Given an element $[V,T]$ in $\underline{I}$, the constant term in the power series $J[V,T]$ is easily determined. We take $k = 0$ in Lemma 10, so that we project to $\underline{N}_n(P_A)$ with kernel $\theta . \underline{K}_{n+1}$, and only the constant term survives. For the image in $\underline{N}_n(P_A)$ we have

$$\partial s[V,T] = [\partial(V\times D^1), T\times T] = [V\times S^0, T\times T] = [V\times S^0, 1\times T] = [V].1,$$

because the manifolds with involution $(V\times S^0, T\times T)$ and $(V\times S^0, 1\times T)$ are isomorphic. Hence the desired constant term.

THEOREM 12. Let V be a closed manifold with involution T. Then

$$J[V,T] = [V] + \text{terms with positive powers of } \theta. \quad]]]$$

This corresponds to Theorem 2 of [2], apart from some changes in notation, and is a reformulation of Theorem (24.2) of Conner and Floyd [3], with essentially the same proof.

Before we consider the other coefficients in the power series Jx for x in $\underline{I}$, it is useful to compute the composite homomorphism $p_*\partial : \underline{M} \to \underline{N}$, where p maps P_∞ to a point, so that $p_*\partial[V,T] = [\partial V/T]$. We factor $p_*\partial$ as $\underline{M} \xrightarrow{J} \underline{L} \longrightarrow \underline{N}_*(P_\infty) \xrightarrow{p_*} \underline{N}$. By Lemma 10, with $k = -1$, this in turn factors through the quotient $\underline{N}$-module $\underline{L}/\underline{K}$, which admits the elements θ^{-n}, for positive n, as $\underline{N}$-base. But $\theta^{-n} = Js^n$. It follows that we only need to know $p_*\partial s^n$, which is clearly $[P_{n-1}]$.

LEMMA 13. Given x in $\underline{M}$, write $Jx = \Sigma_n c_n \theta^n$. Then $p_*\partial x = \Sigma_{n\geq 1} c_{-n}[P_{n-1}]$.]]]

Actually, we know that $[P_n] = 0$ if n is odd, so that alternate terms in this expression are automatically zero. This fact fails to simplify any of our arguments, however.

Now let V be a closed manifold with involution, and write $d_k = p_*\partial s^{k+1}[V,T]$ for $k \geq 0$, so that as above $d_0 = [V]$. We know $J[V,T]$ has the form $\Sigma_{n\geq 0} c_n \theta^n$, by Theorem 11. Since $Js^{k+1}[V,T] = \Sigma_n c_n \theta^{n-k-1}$ by Theorem 9, Lemma 13 yields $d_k = \Sigma_n c_n [P_{k-n}]$. Let us introduce the power series

$$\pi = 1 + [P_1]\theta + [P_2]\theta^2 + [P_3]\theta^3 + \dots,$$

so that $\Sigma_k d_k \theta^k = J[V,T].\pi$. This formula determines $J[V,T]$, because the power series π is evidently invertible.

THEOREM 14. $J[V,T] = (\Sigma_k d_k \theta^k) \cdot (\Sigma_r [P_r]\theta^r)^{-1}$ *for a closed manifold* V *with involution* T, *where* $d_k = p_* \partial s^{k+1}[V,T]$.]]]

Hence our homomorphism J coincides with the homomorphism called J' in [2], except for multiplication by an irrelevant power of θ .

§7. *Interpretation of Alexander's theorem.* In [1], Alexander gives a $\underline{N}$-base of the $\underline{N}$-algebra $\underline{I}$ in terms of certain explicit elements of $\underline{I}$ and an operation Γ in $\underline{I}$. We recover his result from a different point of view.

The graded algebra $\underline{M}$ is a polynomial algebra over $\underline{N}$. If γ denotes the canonical line bundle over P_∞ , the elements $y_n = [\gamma | P_{n-1}] \in \underline{M}_n$ for positive n form a system of polynomial generators, by Lemma 4. In particular, $y_1 = s$, the suspension element.

The subalgebra $\underline{I}$ of $\underline{M}$ contains a large polynomial subalgebra. Consider the involution T_0 on P_n defined by

$$T_0(t_0,t_1,t_2,\ldots,t_n) = (-t_0,t_1,t_2,\ldots,t_n).$$

Its fixed-point sets are the single point $t_1 = t_2 = \ldots = t_n = 0$, and the subspace P_{n-1} defined by $t_0 = 0$, with normal bundle $\gamma | P_{n-1}$. Therefore $[P_n,T_0] = y_n + y_1^n$, an element of $\underline{M}$. It follows that the elements $[P_n,T_0]$ for $n \geq 2$, along with s, form another system of polynomial generators of $\underline{M}$. Let us write $\underline{Q}$ for the polynomial subalgebra of $\underline{I}$ generated by the $[P_n,T_0]$, so that $\underline{M} = \underline{Q}[s]$.

According to Theorem 11, the subalgebra $\underline{I}$ of $\underline{M}$ consists of those polynomials H in s over $\underline{Q}$ for which JH is a power series. Since $Js = \theta^{-1}$, there is, given any element y in $\underline{M}$, a unique polynomial F in $\underline{N}[s]$, without constant term, such that $J(y+F)$ is a power series and $y + F$ lies in $\underline{I}$. (In other words, we are choosing a different splitting of the short exact sequence of Theorem 7.)

For elements in $s^n\underline{I}$ we can be more precise. Let V be any manifold with involution T. Since $J[V,T] = [V]$ + terms in θ by Theorem 12, $Js([V,T] + [V])$ is again a power series. This shows that $s([V,T] + [V])$ is an element of $\underline{I}$, which we call $\Gamma[V,T]$. Then $J\Gamma[V,T] = \Gamma J[V,T]$, if we denote also by Γ the shift operation on power series that deletes the constant term and divides by θ. There is, moreover, an explicit construction $[V',T']$ for $\Gamma[V,T]$, given by [3] and [1]: V' is the quotient of $V \times S^1$ by the relation $(v,z) = (Tv, -z)$, with involution $T'(v,z) = (Tv, \bar{z})$, where S^1 is the unit circle of complex numbers. For higher powers of s, we simply iterate; $s^n[V,T]$ occurs in $\Gamma^n[V,T]$. We can now deduce what is essentially Alexander's theorem, where $\underline{Q}^+$ denotes the augmentation ideal $\mathrm{Ker}(\underline{Q} \to \underline{N})$ consisting of polynomials over $\underline{N}$ without constant term, a $\underline{N}$-submodule of $\underline{Q}$.

THEOREM 15. The N-module $\underline{I}$ is the direct sum of $\underline{Q}$ and the $\underline{N}$-submodules $\Gamma^n\underline{Q}^+$ for $n > 0$, and Γ^n embeds $\underline{Q}^+$ in $\underline{I}$.]]]

§8. Proof of the main theorem. We prove the five halves theorem by introducing another system of generators of $\underline{M}$. Actually, these will only generate $\underline{M}$ stably. To $\underline{M}$ we adjoin a formal inverse s^{-1} of s, to give $\underline{M}[s^{-1}]$. Then $\underline{M}[s^{-1}] = \underline{N}_*(BO) \otimes Z_2[s,s^{-1}]$, and the effect is to destroy the grading in $\underline{N}_*(BO)$ by allowing addition of elements of different degrees, if we first insert the correct powers of s. Since $Js = \theta^{-1}$ is obviously invertible, the stable J-homomorphism extends uniquely to an algebra homomorphism $J:\underline{M}[s^{-1}] \to \underline{L}$.

It is trivial to see that $\underline{I}[s^{-1}] = \underline{M}[s^{-1}]$. For our main theorem, however, we are interested in the dimensions of the fixed-point sets. Therefore, we *filter* $\underline{M}$, and hence $\underline{M}[s^{-1}]$, according to the dimension of the base manifold of a representative vector bundle. (If the base is disconnected, we take the highest of the dimensions of the various components.) Let us write $\mathrm{fil}(x)$ for the filtration of an element x. We know the structure of $\underline{M}$ as a filtered algebra over Z_2, from §7 or §3. It is a polynomial algebra over Z_2 on generators x_r' and x_r'' of filtration r, for each $r \geq 0$, with x_r' absent if $r+1$ is a power of 2, where the x_r' generate $\underline{N}$ and the x_r'' are polynomial generators for $\underline{M}$ over $\underline{N}$. Polynomials in these generators have the expected filtration:

$fil(xy) = fil(x) + fil(y)$ for any $x \neq 0$, $y \neq 0$;

$fil(x+y) = \max(fil(x), fil(y))$ if the polynomials x and y have no monomials in common.

We need generators on which the effect of J is known. Our proof therefore hinges on the following lemma, which is essentially Theorem 3 of [2].

LEMMA 16. For each n not of the form 2^q-1, we can find elements $x_n \in \underline{I}_n \subset \underline{M}_n$ and $z_n \in \underline{N}_n$ with the properties:

(a) the x_n form a system of polynomial generators of $\underline{M}[s^{-1}]$ over the subalgebra $Z_2[s,s^{-1}]$;

(b) the z_n form a system of polynomial generators of $\underline{N}$;

(c) $Jx_n = z_n$ + terms involving θ;

(d) $fil(x_n) = n/2$ if n is even, or $(n-1)/2$ if n is odd;

(e) the filtration of a polynomial in the x_n is the maximum of the filtrations of its terms.

Proof. We consider the examples of involutions given in [2]. We define the involution T_i on the projective space P_{2i} by

$$T_i(t_0,t_1,t_2,\ldots,t_i,t_1',\ldots,t_i') = (t_0,t_1,t_2,\ldots,t_i,-t_1',\ldots,-t_i'),$$

in terms of homogeneous coordinates $t_0,t_1,\ldots,t_i,t_1',\ldots,t_i'$ on P_{2i}. Its fixed-point sets are the linear subspaces P_i defined by $t_1' = t_2' = \ldots = t_i' = 0$, and P_{i-1} defined by $t_0 = t_1 = \ldots = t_i = 0$. For the normal bundle ν of P_i we have $\sigma_i(\nu) = i\alpha^i$, where α generates the cohomology. For the normal bundle ν of P_{i-1} we have $\sigma_{i-1}(\nu) = (i+1)\alpha^{i-1}$.

The product involution $T_i \times T_j$ on $P_{2i} \times P_{2j}$ maps the hypersurface $H_{2i,2j}$ defined by

$$t_0u_0 + t_1u_1 + \ldots + t_iu_i + t_1'u_1' + \ldots + t_i'u_i' = 0$$

into itself, where for clarity we write $u_0,u_1,\ldots,u_j,u_1',\ldots,u_j'$ for the coordinates on P_{2j} and assume $i \leq j$. We consider the fixed-point sets of $T_i \times T_j | H_{2i,2j}$, which are just the intersections with $H_{2i,2j}$ of the four fixed-point sets of $T_i \times T_j$ on $P_{2i} \times P_{2j}$. Now $P_i \times P_j$ intersects $H_{2i,2j}$ in a copy of $H_{i,j}$. The normal bundle ν_1 of $H_{i,j}$ in $H_{2i,2j}$ is the restriction to $H_{i,j}$ of the normal bundle of $P_i \times P_j$ in $P_{2i} \times P_{2j}$; we therefore find $\sigma_{i+j-1}(\nu_1) = i\alpha^{i+j-1} + j\beta^{i+j-1}$. The sets $P_i \times P_{j-1}$ and $P_{i-1} \times P_j$ are entirely contained in $H_{2i,2j}$. The normal bundle ν_2 of $P_i \times P_{j-1}$ becomes, if we add the normal line bundle of $H_{2i,2j}$ in $P_{2i} \times P_{2j}$, the normal bundle of $P_i \times P_{j-1}$ in $P_{2i} \times P_{2j}$; therefore

$$\sigma_{i+j-1}(\nu_2) = i\alpha^{i+j-1} + (j+1)\beta^{i+j-1} + (\alpha + \beta)^{i+j-1}.$$

Similarly, for the normal bundle ν_3 of $P_{i-1} \times P_j$ in $H_{2i,2j}$ we find

$$\sigma_{i+j-1}(\nu_3) = (i+1)\alpha^{i+j-1} + j\beta^{i+j-1} + (\alpha + \beta)^{i+j-1}.$$

Finally, $P_{i-1} \times P_{j-1}$ intersects $H_{2i,2j}$ in a copy of $H_{i-1,j-1}$, which has dimension $i + j - 3$ and for our purposes may safely be ignored.

We shall list the generators x_n of the Lemma explicitly, according to the form of n. Assertion (c) and Theorem 12 will define the elements z_n, for which assertion (b) follows immediately from Theorem 1 and the examples discussed in §1. Assertion (d) will be obvious. Assertion (e) is a consequence of (a) and (d) and the known structure of $\underline{M}$ by a counting argument, or directly as follows. Since $\underline{M}[s^{-1}] = \underline{N}_*(BO) \otimes Z_2[s,s^{-1}]$, Lemma 5 provides a system of elements y_n of $\underline{M}[s^{-1}]$ satisfying (a), (d) and (e) of Lemma 16. Our hypotheses assure that we can adjust the y_n (by multiplication by powers of s, and replacement by linear combinations) until

$$x_n = y_n + \text{decomposable terms},$$

from which assertion (e) for the x_n is clear. Thus only (a) needs substantial proof. We apply Lemma 5. If x_n has filtration m, we therefore need $b_m(x)$ and $e_m(x)$; in computing these we may ignore terms of lower filtration. The proof of our Lemma will follow from these remarks and the calculations below.

Case n = 4k-2. We take $x_{4k-2} = [P_{4k-2}, T_{2k-1}]$. We need only consider the fixed-point set P_{2k-1}, on which we have to evaluate $(2k-1)\alpha^{2k-1}$. Hence

$e_{2k-1}(x_{4k-2}) = 1$, and $b_{2k-1}(x_{4k-2}) = e_{2k-1}(P_{2k-1}) = 0$.

Case n = 4k-1, k not of the form 2^q-1. As in §1 we can choose positive integers i and j such that $i + j = k$ and the binomial coefficient $\binom{k}{i} = 1$ (mod 2 of course). We take $x_{4k-1} = [H_{4i,4j}, T_{2i}\times T_{2j}|H_{4i,4j}]$. The fixed-point dimension is $2k-1$. To find $e_{2k-1}(x_{4k-1})$ we have to evaluate $\sigma_{2k-1}(\nu)$ on the fundamental classes of the various components and add. For $H_{2i,2j}$ we have $\sigma_{2k-1}(\nu_1) = 2i\alpha^{2k-1} + 2j\beta^{2k-1} = 0$. To evaluate $\sigma_{2k-1}(\nu_2)$ on $P_{2i}\times P_{2j-1}$ we need the coefficient of $\alpha^{2i}\beta^{2j-1}$ in $(\alpha+\beta)^{2k-1}$, which is $\binom{2k-1}{2i}$. Similarly, $P_{2i-1}\times P_{2j}$ contributes $\binom{2k-1}{2j} = \binom{2k-1}{2i-1}$. Hence

$$e_{2k-1}(x_{4k-1}) = \binom{2k-1}{2i} + \binom{2k-1}{2i-1} = \binom{2k}{2i} = \binom{k}{i} = 1,$$

by standard properties of binomial coefficients. Also,

$$b_{2k-1}(x_{4k-1}) = e_{2k-1}(H_{2i,2j}) = 1.$$

Case n = 4k. We take $x_{4k} = [P_{4k}, T_{2k}]$. For the normal bundle of the fixed-point set P_{2k} we have $\sigma_{2k}(\nu) = 2k\alpha^{2k} = 0$, so that $e_{2k}(x_{4k}) = 0$. Also, $b_{2k}(x_{4k}) = e_{2k}(P_{2k}) = 1$.

Case n = 4k+1. We take $x_{4k+1} = [H_{2,4k}, T_1\times T_{2k}|H_{2,4k}]$. For the fixed-point set $H_{1,2k}$ we have $\sigma_{2k}(\nu_1) = \alpha^{2k} + 2k\beta^{2k} = \alpha^{2k}$. To evaluate this on $H_{1,2k}$ we use Lemma 2, which yields zero. For the set $P_1\times P_{2k-1}$ we found $\sigma_{2k}(\nu_2) = \alpha^{2k} + (2k+1)\beta^{2k} + (\alpha+\beta)^{2k}$, in which the coefficient of $\alpha\beta^{2k-1}$ is $\binom{2k}{1} = 0$. For the set $P_0\times P_{2k}$ we found $\sigma_{2k}(\nu_3) = 2\alpha^{2k} + 2k\beta^{2k} + (\alpha+\beta)^{2k}$, which does contain the term β^{2k}. Therefore $e_{2k}(x_{4k+1}) = 1$, and $b_{2k}(x_{4k+1}) = e_{2k}(H_{1,2k}) = 0$.]]]

We give two immediate consequences, as well as the proof of the main theorem.

COROLLARY 17. *$J:\underline{M} \to \underline{L}$ and $J:\underline{M}[s^{-1}] \to \underline{L}$ are monomorphisms.*]]]

COROLLARY 18. *Given x in $\underline{M}$, Jx is a finite Laurent series if and only if x is a polynomial in s with coefficients in $\underline{N}$.*

Proof. Since J is a homomorphism of $\underline{N}$-algebras and $Js = \theta^{-1}$, Corollary 17 shows that Jx is a finite Laurent series in θ if and only if x is a finite Laurent series in s, with coefficients in $\underline{N}$. If x lies in $\underline{M}$, negative powers of s do not occur.]]]

Proof of Theorem 0. Let V be a n-manifold with involution, with fixed-point dimension k, so that $[V,T]$ has filtration at most k. We write $[V,T]$ by Lemma 16 as a polynomial in the x_i and s and s^{-1}; by that Lemma and Theorem 11, this polynomial must take the form

$$[V,T] = \lambda_0 + \lambda_1 s^{-1} + \lambda_2 s^{-2} + \dots,$$

where each λ_i is a polynomial over Z_2 in the x_r. By Theorem 12 and Lemma 16, $[V]$ is the polynomial in the z_r corresponding to λ_0. In case (a), $\lambda_0 \neq 0$. Now for all generators x_r, $\mathrm{fil}(x_r) \geq 2r/5$ (with equality if $r = 5$), and so for dimensional reasons,

$$k \geq \mathrm{fil}([V,T]) \geq \mathrm{fil}(\lambda_0) \geq 2n/5.$$

In case (b), the term x_n actually appears in λ_0, and so

$$k \geq \mathrm{fil}(\lambda_0) \geq \mathrm{fil}(x_n) \geq (n-1)/2. \qquad]]]$$

References

[1] Alexander, J.C., The bordism ring of manifolds with involution, to appear.

[2] Boardman, J.M., On manifolds with involution, Bull. Amer. Math. Soc. 73(1967) 136-138.

[3] Conner, P.E. and Floyd, E.E., Differentiable periodic maps, Ergebnisse der Mathematik 33, Springer-Verlag (Berlin 1964).

[4] Liulevicius, A.L., A proof of Thom's theorem, Comment. Math. Helvet. 37(1962/1963)121-131.

[5] Milnor, J.W., On the Stiefel-Whitney numbers of complex manifolds and of spin manifolds, Topology 3(1965)223-230.

[6] Milnor, J.W. and Moore, J.C., On the structure of Hopf algebras, Annals of Math. 81(1965)211-264.

[7] Steenrod, N.E. and Epstein, D.B.A., Cohomology operations, Annals of Math. Study 50, Princeton U.P. (Princeton, 1962).

[8] Thom, R. Quelques propriétés globales des variétés différentiables, Comment. Math. Helvet. 28(1954)17-86.

BEMERKUNGEN ÜBER ÄQUIVARIANTE EULER-KLASSEN

Tammo tom Dieck

Universität des Saarlandes

1. Euler-Klassen äquivarianter Bündel und Erzeugende für U_G^*-

Sei G eine kompakte kommutative Liescne Gruppe. Sei η das universelle komplexe Geradenbündel über dem unendlichen komplexen projektiven Raum CP^∞. Ist V ein nicht-trivialer irreduzibler komplexer G-Modul, so betrachten wir V auch als G-Vektorraumbündel über einem Punkt. Wir arbeiten mit dem G-Vektorraumbündel $V \otimes \eta$ über CP^∞; dabei operiert G trivial auf η. Die äquivariante unitäre Kobordismen-Theorie, die wir in [5] entwickelt haben, liefert uns eine Euler-Klasse

$$e(V \otimes \eta) \in U_G^2(CP^\infty)$$

des Bündels $V \otimes \eta$. Da G trivial auf CP^∞ operiert, haben wir einen Isomorphismus

$$U_G^*(CP^\infty) \cong U_G^*[[C]]$$

von U_G^*-Algebren; das Element C entspricht dabei der Euler-Klasse $e(\eta) \in U^2(CP^\infty)$. Wir entwickeln $e(V \otimes \eta)$ in eine Potenzreihe

$$e(V \otimes \eta) = a_0(V) + a_1(V)C + a_2(V)C^2 + \dots .$$

Unser Ziel ist die Untersuchung der Elemente

$$a_i(V) \in U_G^{2-2i}.$$

Sei $(V_j \mid j \in J)$ die Familie der nicht-trivialen irreduziblen komplexen G-Moduln.

Satz 1. Die Elemente

$$a_i(V_j), \quad j \in J, \quad i = 0,1,2,\dots$$

sind algebraisch unabhängig über U^*.

Der Beweis von Satz 1 beruht auf dem Lokalisierungssatz aus [5]. Wir zeigen nämlich, daß die fraglichen Elemente sogar in einer geeigneten Lokalisierung von U_G^* unabhängig sind.

Sei $\lambda : U_G^* \to S^{-1}U_G^*$ die Lokalisierung, in der S die multiplikativ abgeschlossene Teilmenge von U_G^* ist, die von den Euler-Klassen der V_j erzeugt wird. Theorem 3.1 aus [5] gibt uns einen Isomorphismus

$$(1) \quad S^{-1}U_G^* \cong U_*(\prod_{j\in J}{}' BU) \otimes Z[eV_j, eV_j^{-1} \mid j \in J].$$

Der Strich am Produkt bedeutet, daß nur die Punkte des Produktes genommen werden, bei denen bis auf endlich viele alle Koordinaten gleich dem Grundprodukt sind. Natürlich bezeichnet eV_j die Euler-Klasse von V_j und eV_j^{-1} das dazu inverse Element. Wir betrachten (1) als Identifikation. Wir setzen

$$b_i(V) := \lambda a_i(V)$$

und bestimmen die $b_i(V)$. Zu diesem Zweck beschreiben wir zunächst einige kanonische Elemente in $S^{-1}U_G^*$.

Wir gehen aus von dem kanonischen Isomorphismus $U^*(BU(1)) \cong U^*[[C]]$, worin C wieder der U^*-Euler-Klasse von η entspricht. Sei $\beta_i \in U_{2i}(BU(1))$ das zu C^i duale Element bezüglich der üblichen Paarung $U^*(BU(1)) \otimes U_*(BU(1)) \to U_*$. Sei

$$h_i \in U_{2i}(BU)$$

das Bild von β_i bei der kanonischen Abbildung $U_*(BU(1)) \to U_*(BU)$. Dann ist $U_*(BU)$ eine Polynomalgebra über U_* in $h_1, h_2, h_3, \ldots$; außerdem ist $h_0 = 1$. Für jedes $j \in J$ haben wir eine Injektion

$$t_j : BU \to \prod_{j\in J}{}' BU.$$

Wir setzen

$$h_i(j) := (t_j)_* h_i.$$

Dann ist $U_*(\Pi' BU)$ eine Polynomalgebra über U_* in den $h_i(j)$. Wir identifizieren $h_i(j)$ auch mit dem nach (1) gegebenen Element $h_i(j) \otimes 1$ von $S^{-1}U_G^*$.

<u>Satz 2.</u> <u>Es gilt</u>

$$a_0(V_j) = e(V_j)$$

<u>und für alle</u> $i \geq 0$ <u>und</u> $j \in J$

$$b_i(V_j) = h_i(j) \otimes e(V_j).$$

Beweis von Satz 2. Wir gehen auf die Definition von $U_G^*(X)$ als ein direkter Limes von äquivarianten Homotopiemengen zurück [5]. Das Element $e(V_j \otimes \eta)$ wird durch eine Abbildung

$$f_j : CP^{\infty+} \to M_G\, U(1)$$

in den Thom-Raum des universellen eindimensionalen komplexen G-Vektorraumbündels repräsentiert. (Die Definition für $U_G^*(X)$ in [5] ist nur für kompakte X vernünftig. Für unsere Zwecke genügt die Definition

$$U_G^*(CP^\infty) = \operatorname{inv} \lim U_G^*(CP^n).)$$

Einschränkung auf die Fixpunktmenge induziert den Isomorphismus (1). Eine repräsentierende Abbildung für $\lambda e(V_j \otimes \eta)$ ist

$$g_j : CP^{\infty+} \underset{k}{\to} MU(0) \wedge BU(1)^+ \underset{v_j}{\to} M_G\, U(1),$$

wobei $MU(0) = S^0$ ist, k eine Homotopieäquivalenz (nämlich im wesentlichen die Identität $CP^\infty = BU(1)$) und v_j die Injektion desjenigen Teiles der Fixpunktmenge, die zum Summanden V_j gehört (vergleiche Lemma 2.1 in [5]).

Wir benutzen die Bezeichnungen aus [5], § 2, und entnehmen der dort angegebenen Konstruktion der Abbildung φ, daß $\lambda e(V_j \otimes \eta_n)$, $\eta_n = \eta|\, CP^n$, ein Element

$$h_n^{(j)} \otimes e(V_j) \in U^*(CP^n;\, \Pi'\, BU) \otimes Z[eV_j, eV_j^{-1} \mid j \in J]$$

ist, wobei $h_n^{(j)}$ durch eine Abbildung

$$CP^n \subset CP^\infty = BU(1) \subset BU \underset{t_j}{\to} \Pi'\, BU$$

repräsentiert wird. Bei dem kanonischen Isomorphismus

$$U^*(BU(1);\, BU) \cong U_*(BU)[[C]]$$

wird das durch $BU(1) \subset BU$ repräsentierte Element auf die Reihe

$$1 + h_1 C + h_2 C^2 + \dots$$

abgebildet. Deshalb liefern die $h_n^{(j)}$ für $n \to \infty$ eine Potenzreihe

$$h(j) \in U^*(CP^\infty;\, \Pi'\, BU) \cong U_*(\Pi'\, BU)[[C]],$$

die gleich

$$1 + h_1(j)C + h_2(j)C^2 + \dots$$

ist. Durch Koeffizientenvergleich folgt die Behauptung über die $b_i(V_j)$.

Die Gleichheit $a_0(V_j) = e(V_j)$ sieht man ein, wenn man $V_j \otimes \eta$ auf einen Punkt einschränkt.

Beweis von Satz 1. Der fragliche Satz ist eine unmittelbare Folge von Satz 2, da man aus (1) und der angegebenen Struktur von $U_*(\Pi' BU)$ sofort entnimmt, daß die Elemente $\lambda a_i(V_j)$ algebraisch unabhängig über U^* sind.

2. Charakteristische Zahlen für unitäre S^1-Mannigfaltigkeiten

Satz 3. Die Bündelungstransformation

$$\alpha : U^*_{S^1} \to U^*(BS^1)$$

ist injektiv.

Satz 4. Charakteristische Zahlen bestimmen die Bordismenklasse einer unitären S^1-Mannigfaltigkeit.

Der Beweis von Satz 3 beruht im wesentlichen auf dem folgenden Satz von P. Löffler; für den Beweis verweisen wir auf [11].

Satz 5. Die Lokalisierungsabbildung

$$\lambda : U^*_{S^1} \to S^{-1}U^*_{S^1}$$

ist injektiv.

Wenden wir die Lokalisierung auf die Bündelungstransformation ([5], Prop.1.2) an, so erhalten wir das kommutative Diagramm

$$\begin{array}{ccc} U^*_{S^1} & \xrightarrow{\ \alpha\ } & U^*(BS^1) \\ \lambda \downarrow & & \downarrow \Lambda \\ S^{-1}U^*_{S^1} & \xrightarrow[S^{-1}\alpha]{} & S^{-1}U^*(BS^1) . \end{array}$$

Λ ist injektiv, weil $U^*(BS^1)$ keine Nullteiler hat. Im nächsten Lemma zeigen wir, daß $S^{-1}\alpha$ injektiv ist. Mit Satz 5 folgt dann die Injektivität von α.

Lemma 1. $S^{-1}\alpha$ ist injektiv.

Beweis. Wir verwenden den Isomorphismus (1) für $G = S^1$ und zeigen, daß für jede endliche Teilmenge P von J der Unterring

$$(2) \quad B := U_*(\prod_{j\in P} BU) \otimes Z[eV_j \mid j\in P]$$

von $S^{-1}U_G^*$ bei $S^{-1}\alpha$ injektiv abgebildet wird. Da jedes Element von $S^{-1}U_G^*$ algebraisch über einem Unterring dieser Form ist, folgt die Injektivität von $S^{-1}\alpha$.

Sei P gegeben. Wir wählen eine Primzahl p so, daß die S^1-Moduln V_j für $j\in P$ bei Beschränkung auf $Z_p \subset S^1$ immer noch paarweise nicht isomorph sind. Wir betrachten dann die V_j für $j\in P$ auch als Z_p-Moduln und sehen (2) als einen Unterring von $S^{-1}U_{Z_p}^*$ an. Der Beweis von Lemma 1 beruht darauf, $S^{-1}\alpha$ für die Gruppen S^1 und Z_p eingeschränkt auf den Unterring (2) zu vergleichen.

Aus Satz 2 entnehmen wir, daß jedes Element von B algebraisch über dem Unterring A der "ganzen Elemente" ist (d.h. der Elemente im Bild von λ). Es genügt deshalb zu zeigen, daß $S^{-1}\alpha | A$ injektiv ist. Da Λ injektiv ist, können wir das Bild von Λ mit $U^*(BS^1)$ identifizieren und $S^{-1}\alpha | A$ als eine Abbildung $A \to U^*(BS^1)$ auffassen. Wir betrachten deshalb die Zusammensetzung

$$(3) \quad A \xrightarrow{S^{-1}\alpha} U^*(BS^1) \longrightarrow U^*(BZ_p) \xrightarrow{\Lambda} S^{-1}U^*(BZ_p),$$

worin die unbezeichnete Abbildung durch $Z_p \subset S^1$ induziert wird und Λ die Lokalisierung für den Fall Z_p bedeutet. (S^{-1} hat natürlich je nach Gruppe eine andere Bedeutung.)

Wir behaupten: (3) ist auch $S^{-1}\alpha$ für die Gruppe Z_p, wenn wir A als Unterring von $S^{-1}U_{Z_p}^*$ auffassen. Diese Behauptung folgt aus einer expliziten Beschreibung von $S^{-1}\alpha$, die man aus Theorem 3.3 von [5] und den daran anschließenden Bemerkungen entnimmt. Wir haben in [8] gezeigt, daß $S^{-1}\alpha$ für Z_p injektiv ist. Damit ist der Beweis von Lemma 1 beendet.

Beweis von Satz 4. Wir haben zu erläutern, welche charakteristischen

Zahlen wir meinen. Sei $\mathfrak{U}_*^G$ für $G = S^1$ der Bordismenring unitärer S^1-Mannigfaltigkeiten. Wir haben Abbildungen

$$\mathfrak{U}_*^G \xrightarrow{i} U_G^* \xrightarrow{\alpha} U^*(BG) \xrightarrow{B} H^*(BG)[[a_1,\ldots]];$$

dabei ist i die Pontrjagin-Thom Konstruktion ([5], p. 349) und B die Boardman-Abbildung, die im vorliegenden Fall bekanntlich injektiv ist (vergleiche Bröcker-tom Dieck [2], p. 154, für eine analoge Aussage). Aus Lemma 1, [5] Prop.4.1 und der Arbeit [10] von Hamrick und Ossa schließen wir, daß αi injektiv ist. Die genaue Formulierung von Satz 4 ist die Aussage "$B\alpha i$ ist injektiv" — und das haben wir gerade bewiesen. Für nähere Einzelheiten und verwandte Resultate sei auf [6], [8] hingewiesen.

3. Über die Struktur der universellen formalen Gruppe

Die universelle formale Gruppe

$$(3)\quad F(X,Y) = X + Y + \sum_{i,j\geq 1} c_{i,j}\, X^i Y^j$$

über dem Lazard-Ring L ist kanonisch isomorph zur formalen Gruppe, die in der komplexen Kobordismen-Theorie auftritt (Quillen [12]). Wir benutzen diesen Isomorphismus als Identifizierung und fassen die Elemente $c_{i,j}$ als Elemente von $U_{2i+2j-2}$ auf.

Wir wollen zeigen, daß eine große Anzahl von Potenzreihen, die aus F(X,Y) abgeleitet werden, algebraisch unabhängig über $L = U_*$ sind. Sei

$$[n] = [n]_F(X)\ ,\quad n \in Z,$$

die Multiplikation mit n in der formalen Gruppe (Fröhlich [9],p.58). Wenn wir [n] in die Reihe $\sum_{i=1}^{\infty} c_{i,j} X^i$ einsetzen, soll das Ergebnis mit $x_j(n)$ bezeichnet werden ($j = 1,2,3,\ldots;\ n \in Z - \{0\}$).

<u>Satz 6.</u> <u>Die Elemente</u>

$$x_j(n)\ ,\ j = 1,2,3,\ldots\ ,\quad n \in Z - \{0\}$$
$$[n]\ ,\qquad n \in Z - \{0\}$$

<u>von</u> $U^*[[C]]$ <u>sind algebraisch unabhängig über</u> U^*.

Beweis. Die Bündelungstransformation α erhält Euler-Klassen. Wir bezeichnen deshalb $\alpha e(V)$ wieder mit $e(V)$. Wir verwenden das kommutative Diagramm

$$\begin{array}{ccc} U^*_{S^1}(CP^\infty) & \xrightarrow{\alpha} & U^*(BS^1 \times CP^\infty) \\ \downarrow \cong & & \downarrow \cong \\ U^*_{S^1}[[C]] & \xrightarrow{\alpha} & U^*(BS^1)[[C]] . \end{array}$$

Wenn wir die Euler-Klasse

(4) $\alpha e(V \otimes \eta) = \alpha a_0(V) + \alpha a_1(V)C + \ldots$

gemäß der formalen Gruppe (3) entwickeln

(5) $\alpha e(V \otimes \eta) = e(V) + C + \Sigma\, c_{i,j}\, e(V)^i C^j$

und in (4) und (5) Koeffizienten vergleichen, erhalten wir

$$\alpha a_0(V) = e(V)$$

$$(6) \qquad \alpha a_1(V) = 1 + \sum_{i=1}^{\infty} c_{i,1}\, e(V)^i$$

$$\alpha a_j(V) = \sum_{i=1}^{\infty} c_{i,j}\, e(V)^i, \quad j \geq 2.$$

Wir können die nicht-trivialen irreduziblen S^1-Moduln so durch Elemente von $Z - \{0\}$ indizieren, daß gilt

$$\alpha e(V_n) = [n] \in U^*(BS^1) = U^*[[C]].$$

Wir setzen das in (6) ein, verwenden die algebraische Unabhängigkeit der $h_i(j)$ in Satz 2 und die Injektivität von $S^{-1}\alpha$ (Lemma 1) und erhalten das gewünschte Resultat.

4. Über eine Konstruktion von Conner und Floyd

Sei in diesem Abschnitt $G = S^1$. Sei $\varepsilon: U^*_G \to U^* \subset U^*_G$ oder $\varepsilon: U^*(BG) \to U^*$ die Augmentation. Sei $e_1 = e(V_1)$ die Euler-Klasse des in Abschnitt 3 eingeführten Standardmoduls V_1. Sei $S3 \subset C^2$ mit der folgenden S^1-

Operation versehen:

$$(t\ (z_1,z_2)) \longmapsto (tz_1,\ t^{-1}z_2).$$

Ist dann X ein S^1-Raum, so machen wir den Quotientraum $(S^3 \times X)/S^1$ wieder zu einem S^1-Raum, indem wir auf den Repräsentanten die S^1-Operation durch

$$(t\ (z_1,z_2,x)) \longmapsto (tz_1,z_2,tx)$$

definieren. Den so aus X konstruierten S^1-Raum $(S^3 \times X)/S^1$ nennen wir b(X). Ist X eine unitäre S^1-Mannigfaltigkeit, so auch b(X) in kanonischer Weise.

Satz 7. Es gibt genau eine U^*-lineare Abbildung

$$\beta : U_G^* \to U_G^* \qquad (G = S^1)$$

vom Grad -2 , so daß für alle $x \in U_G^*$ gilt $\lambda\beta x = e_1^{-1}\ \lambda(x - \varepsilon x)$. Diese Abbildung hat ferner die folgenden Eigenschaften:

(1) Für alle x,y aus U_G^* gilt
$\beta(xy) = \beta(x)y + \varepsilon(x)\ \beta(y) = \beta(x)\ \varepsilon(y) + x\ \beta(y)$.

(2) β ist linksinvers zur Multiplikation mit e_1.

(3) Wird $x \in U_G^*$ durch die unitäre G-Mannigfaltigkeit M repräsentiert (vermöge Pontrjagin-Thom-Konstruktion), so $\beta(x)$ durch b(M).

(4) Ferner läßt sich für ein x wie in (3) $\alpha(x)$ schreiben als
$\varepsilon[M] + \varepsilon[b(M)]C + \varepsilon[b^2(M)]C^2 + \dots$,
wobei $b^i(M)$ die i-fache Iteration der Konstruktion b ist.

Beweis. Sei $\lambda_P : U_G^* \to P^{-1}U_G^*$ die Lokalisierung nach der durch e_1 erzeugten Teilmenge P. Die exakte Lokalisierungssequenz dazu [7] hat die Gestalt

$$(7)\quad 0 \to U_G^* \xrightarrow[\lambda_P]{} P^{-1}U_G^* \xrightarrow[\partial]{} U_*(BG) \longrightarrow 0,$$

da jede freie S^1-Mannigfaltigkeit berandet, wenn beliebige Operationen zugelassen sind. Für $x \in U_G^*$ ist

$$\partial(e_1^{-1}\lambda_P x) = \partial(e_1^{-1}\lambda_P \varepsilon x),$$

wie man zum Beispiel durch Anwenden von α auf (7) erkennt. Deshalb läßt sich β wie gewünscht eindeutig unter der Nebenbedingung $\lambda\beta x = e_1^{-1}\lambda(x - \mathfrak{E}x)$ definieren. Aus dieser Formel folgt unmittelbar (1) und dann wegen $\mathfrak{E}(e_1 x) = 0$ auch (2). Um (3) einzusehen, schaue man sich das Normalenbündel an die Fixpunktmenge von b(M) an. Ist ν das Normalenbündel an die Fixpunktmenge von M, so ist

$$(\nu \otimes 1) - (M \times C \to M)$$

dasjenige an die Fixpunktmenge von b(M); das Minuszeichen bedeutet hier, daß M die zur gegebenen unitären Struktur inverse Struktur bekommt. Verwendet man [5], Prop. 4.1 und Satz 5, so folgt die Behauptung. (Aus einer sogleich vermerkten homotopiemäßigen Konstruktion von β kann man (3) auch einsehen, ohne Satz 5 zu verwenden.) Die in (4) behauptete Formel für $\alpha(x)$ folgt unmittelbar aus der Definitionsgleichung von β unter Beachtung von $\mathfrak{E}\alpha[M] = \mathfrak{E}[M]$.

Wir geben nun eine andere Konstruktion von β an, die (für hier nicht beschriebene) Verallgemeinerungen wichtig ist. Wir erinnern daran, daß U_G^{2n} als direkter Limes über äquivariante Homotopiemengen der Form

$$[V^c, M_{n+|V|}(G)]_G^o$$

definiert ist [5], wobei V^c die Einpunkt-Kompaktifizierung des komplexen G-Moduls V ist und $M_k(G)$ der Thom-Raum des universellen komplexen G-Bündels $\xi_k(G)$ der Dimension k. (Es ist $|V| = \dim_C V$.) Wir definieren Abbildungen (mit der Abkürzung $k = n+|V|$) wie folgt:

$$\begin{array}{ll}
[V^c, M_k(G)]_G^o & \xrightarrow{(1)} \\
[S^{3+} \wedge V^c, S^{3+} \wedge M_k(G)]_G^o & \xrightarrow{(2)} \\
[(S^{3+} \wedge V^c)/G, (S^{3+} \wedge M_k(G))/G]_G^o & \xrightarrow{(3)} \\
[(S^{3+} \wedge V^c)/G, M_k(G)]_G^o & \xrightarrow{(4)} \\
\tilde{U}_G^{2k}((S^{3+} \wedge V^c)/G) & \xrightarrow{(5)} \\
U_G^{2n}(S^3/G) & \xrightarrow{(6)} \\
U_G^{2n-2} \quad . &
\end{array}$$

(1) ist smash-Produkt mit S^{3+}. (2) ist Ausdividieren von S^1 und Auf-

prägen einer neuen S^1-Operation wie in der Definition von b(X). (3) wird induziert durch die klassifizierende Abbildung von $(S^3 \times \xi_k(G))/G$. (4) kommt von der Definition der unitären Kobordismengruppen als direkter Limes. (5) ist ein Thom-Isomorphismus und (6) der Gysin-Homomorphismus zu $S^3/G \to$ Punkt. Die Zusammensetzung der Abbildungen (1) bis (6) werde mit β_V bezeichnet. Man verifiziert, daß die verschiedenen β_V im Limes eine Abbildung

$$\beta : U_G^{2n} \to U_G^{2n-2}$$

induzieren.

Man vergleiche die in Satz 7 gegebene Konstruktion von β mit Conner-Floyd [4], 42., Conner [3], 13., and J.C. Alexander [1].

Literatur

1. Alexander, J. C.: The bordism ring of manifolds with involution. Proc. Amer. Math. Soc. 31, 536-542 (1972).

2. Bröcker, Th., und tom Dieck, T.: Kobordismentheorie. Lecture Notes in Math. 178. Springer-Verlag 1970.

3. Conner, P.E.: Seminar on periodic maps. Lecture Notes in Math. 46. Springer-Verlag 1967.

4. Conner, P.E., and Floyd, E.E.: Differentiable periodic maps. Springer-Verlag 1964.

5. tom Dieck, T.: Bordism of G-manifolds and integrality theorems. Topology 9, 345-358 (1970).

6. ——— : Characteristic numbers of G-manifolds. I. Inventiones math. 13, 213-224 (1971).

7. ——— : Lokalisierung äquivarianter Kohomologie-Theorien. Math. Z. 121, 253-262 (1971).

8. ——— : Periodische Abbildungen unitärer Mannigfaltigkeiten. To appear Math. Z. (1972).

9. Fröhlich, A.: Formal groups. Lecture Notes in Math. 74. Springer-Verlag 1968.

10. Hamrick, G., and Ossa, E.: Unitary bordism of monogenic groups and isometries.

11. Löffler, P.: Dissertation Saarbrücken 1972.

12. Quillen, D.: Elementary proofs of some results of cobordism theory using Steenrod operations. Advances in Math. 7, 29-56 (1971).

EXISTENCE OF FIXED POINTS

Tammo tom Dieck

Universität des Saarlandes

In part I of this note we obtain results about the action of a finite abelian p-group on a unitary manifold. This extends the results in [8] for cyclic groups.

In part II we indicate how unitary cobordism theory can be used to derive existence theorems for fixed points which are usually proved by using P.A. Smith theory.

I. Unitary manifolds

Let p be a prime number, let G be a finite abelian p-group and suppose that G is the product of n cyclic factors. We consider differentiable actions of G on closed manifolds preserving a complex structure on the stable tangent bundle (unitary G-manifold for short). Let U_* be the bordism ring of unitary manifolds. Let m_i be a polynomial generator of U_* in dimension $2(p^i - 1)$ such that all the Chern numbers of m_i are divisible by p $(i \geq 1)$. Put $m_0 = p \in U_0$ and denote by

$$I(p,n)$$

the ideal of U_* generated by $m_0, m_1, \ldots, m_{n-1}$. Let

$$F_n(G,k) \subset U_*/\, I(p,n)$$

be the subring of U_* generated by bordism classes x which are represented by a manifold M admitting a unitary G-action with fixed point set of dimension less than or equal to 2k. Let

$$U_n(r) \subset U_*/\, I(p,n)$$

be the subring generated by $U_0, U_2, \ldots, U_{2r}$. Let $|G|$ denote the order of G.

Theorem 1. (a) *For every* $n \geq 1$ *and every* $k \geq 0$ *we have the equality*

$$F_n(G,k) = U_n((k+1)|G| - 1).$$

(b) *Let M be a unitary G-manifold such that its bordism class (forgetting the G-action) in $U_*/\, I(p,n)$ is indecomposable. Then M has a component of the fixed point set with dimension greater than*

$$\frac{1}{|G|} \dim M - 2.$$

Example. If V is a complex G-module and P(V) the associated projective space considered as unitary G-manifold then it is easy to see that Theorem 1(b) holds and that the lower bound for the dimension of the fixed point set is obtained. So if $\dim_{\mathbb{C}} V \not\equiv 0 \mod p$ we see that arbitrary differentiable unitary actions on P(V) cannot give fixed point sets of lower dimension than those which come from linear actions.

Remark 1. The proof of Theorem 1 will show that there is an analogous theorem for orientation preserving G-actions if p is odd.

The proof of Theorem 1 requires the next lemma which provides us with enough unitary G-manifolds with low dimensional fixed point sets. Let $H(m,n) \subset P(\mathbb{C}^{m+1}) \times P(\mathbb{C}^{n+1})$ for $m \leq n$ be the Milnor manifold defined by

$$H(m,n) = \{([x_i],[y_i]) \mid \sum_{i=0}^{m} x_i y_i = 0\}.$$

Lemma 1. Let $m \leq n$ and $m+n \leq (k+1)|G|$. Then there exists a unitary G-action on H(m,n) with fixed point set of dimension less than or equal to 2k.

Proof. The proof is essentially given in [8], Lemma 1. One only has to replace q by $|G|$ everywhere. (The reader will be able to supply a proof if he takes suitable linear G-actions on $P(\mathbb{C}^{m+1})$ and $P(\mathbb{C}^{n+1})$ such that H(m,n) is a G-invariant subset.)

The following discussion will prepare the proof of Theorem 1. Let $U_G^*(-)$ be the equivariant unitary cobordism theory for the group G and let $\alpha: U_G^*(X) \to U^*(EG \times_G X)$ be the bundling transformation (see [3] for details). Let $S \subset U_G^*$ be the multiplicatively closed subset consisting of 1 and the Euler-classes of complex G-modules without trivial direct summand. The localization $S^{-1}U_G^*$ can be computed. Theorem 3.1 of [3] gives us an isomorphism

$$S^{-1}U_G^* \cong U_*(\prod_{j\in J} BU) \otimes \mathbb{Z}[V_j, V_j^{-1} \mid j\in J].$$

The V_j correspond under this isomorphism to the Euler classes of nontrivial irreducible G-modules. We consider the ungraded ring associated to $S^{-1}U_G^*$. Let

$$F_k \subset S^{-1}U_G^*, \qquad k = 0,1,2,\dots,$$

be the subring generated by

$$\bigoplus_{i \leq k} U_{2i}(\prod_{j \in J} BU) \otimes \mathbb{Z}[V_j^{-1} \mid j \in J].$$

Put $q = |G|$. Then F_k is a polynomial ring over the integers in

$$(k + 1)q - 1$$

indeterminates. We denote by

$$\lambda \colon U_G^* \to S^{-1} U_G^*$$

the localization map. A unitary G-manifold M yields via Pontrjagin-Thom-construction a well defined element $[M]_G \in U_G^*$. The element $\lambda[M]_G$ can be computed from the normal bundle to the fixed point set in M, as explained in [3]. We apply this without further special reference.

Since a suitable linear combination of the manifolds H(m,n) with $m + n - 1 = r$ give a polynomial generator of U_* in dimension 2r, we extract from Lemma 1 the following fact: There exist unitary G-manifolds x_j, $1 \leq j \leq (k+1)q - 1$, with the following properties:

(i) $\dim x_j = 2j$;

(ii) $[x_j]$ is a polynomial generator of U_* ;

(iii) $\lambda[x_j]_G \in F_k$.

<u>Proof of Theorem 1</u>. Let

$$\Lambda : U^*(BG) \to S^{-1} U^*(BG)$$

be the localization. We see from [7], Satz 1, that

$$U^*(BG) \xrightarrow[\varepsilon]{} U^* \to U^*/ I(p,n)$$

factorizes over the image of Λ (here ε denotes the augmentation) and therefore induces a map

$$\beta_1 : \Lambda U^*(BG) \to U^*/ I(p,n).$$

Put

$$D_k = F_k \cap \lambda U_G^* .$$

Then $S^{-1}\alpha$ induces a map

$$\beta_2 : D_k \to \Lambda U^*(BG).$$

Let $\beta = \beta_1\beta_2$ denote the composition. The elements

$$\lambda[x_j]_G =: y_j$$

are contained in D_k. We now use the fact that $\alpha : U_G^* \to U^*(BG)$ maps $[M]_G$ into an element with augmentation $[M]$. So we conclude that the elements

$$\beta y_j \ , \quad j \neq p^k-1, \qquad k = 1,\dots,n-1$$

are algebraically independent over $\mathbb{Z}_p$ and hence the corresponding y_j are algebraically independent over $\mathbb{Z}_p$ in $D_k \otimes \mathbb{Z}_p$. (Note $|G| \geq p^n$.) The $\mathbb{Z}_p$-algebra $D_k \otimes \mathbb{Z}_p$ has transcendence degree at most $(k+1)|G| - 1$, because F_k is a polynomial ring in $(k+1)|G| - 1$ generators. Since we found already $(k+1)|G| - 1 - (n-1)$ independent elements in $D_k \otimes \mathbb{Z}_p$ among the y_j we look for another set of $n-1$ independent elements in the kernel of $\beta \otimes \mathbb{Z}_p$. (This is in fact the only difference of our present proof to the one given in [8]. I did not realize the simple argument that will follow.)

So let us consider the case $n \geq 2$. Let W_i for $2 \leq i \leq n$ be a p-dimensional complex G-module where one summand comes from a non-trivial irreducible representation of the first factor of G and the other p-1 summands come from different non-trivial representations of the i-th factor of G. Let $P(W_i)$ be the associated projective space. Then $P(W_i)$ has only isolated fixed points ; there are p of them. Hence we have

$$z_i := \lambda[P(W_i)]_G \in D_o.$$

The z_i are contained in the kernel of β, because $\beta z_i = [P(\mathbb{C}^p)]$ and $[P(\mathbb{C}^p)]$ is contained in $I(p,n)$ for $n \geq 2$. (We could have taken m_1 to be the class of $P(\mathbb{C}^p)$.) The tangential representations at the fixed points of $P(W_i)$ are build up from tensorproducts of irreducible representations in W_i. If $W_i = A_1 \oplus \dots \oplus A_p$ then

$$z_i = \sum_{j=1}^{p} (A_1^{-1} \oplus \dots \oplus \hat{A}_j^{-1} \oplus \dots \oplus A_p^{-1}) \otimes \bar{A}_j^{-1},$$

where ^ means "leave out" and ‾ means "complex conjugate". So by construction the p summands are not all equal. Therefore $z_i \not\equiv 0 \bmod p$. Moreover we see that the z_i are polynomials in the $V_j^{-1} \in D_o$ built up from <u>disjoint</u> sets of generators V_j^{-1}. Hence $(D_o/\text{ Kernel } \beta) \otimes \mathbb{Z}_p$ has transcendence degree at most $(k+1)|G| - 1 - (n-1)$.

The inclusion

$$F_n(G,k) \supset U_n((k+1)|G| - 1)$$

follows immediately from Lemma 1. Now suppose x is contained in $F_n(G,k)$ but not in $U_n((k+1)|G| - 1)$. Then x is algebraically independent of the elements in $U_n((k+1)|G| - 1)$. But x is contained in the image $\beta(D_k)$, essentially by definition of $F_n(G,k)$. The considerations above show that $\beta(D_k)$ has the same transcendence degree, namely $(k+1)|G| - n$, as $U_n((k+1)|G| - 1)$. Hence x is algebraic over $U_n((k+1)|G| - 1)$; a contradiction. This proves Theorem 1 (a). Part (b) is an immediate corollary.

Remark 2. We have to divide out I(p,n) because I(p,n) is the ideal of manifolds admitting fixed point free actions, as was shown in [4]. Special cases and related theorems are proved in [5], [8]. The first theorems of this kind appeared in Boardman [1] and Conner [2].

II. Compact spaces

It is well known that a torus group or a product $(\mathbb{Z}_p)^n$ of cyclic groups $\mathbb{Z}_p$ of prime order p cannot act on, e.g., a contractible compact space without fixed points (P.A. Smith theory). We show in this note how unitary cobordism theory can be used to give a simple proof of analogous results.

Let G = T x H be a compact abelian Lie group where T is a Torus and H a finite abelian p-group (p prime number). We allow T or H to be trivial. We consider continuous G-actions on compact Hausdorff spaces X.

Let $q : EG \to BG$ be a numerable universal principal G-bundle. If X is a G-space we have the associated bundle with fibre X

$$q_X : EG\times_G X \to BG.$$

Let U*(-) denote unitary cobordism theory. The induced map

$$q_X^* : U^*(BG) \to U^*(EG\times_G X)$$

will give information about the G-action on X. (For the next theorem see also Hsiang [10], Prop. 1.)

Theorem 2. *Let* X *be a compact Hausdorff* G-*space. The following assertions are equivalent:*

(a) X *has a fixed point.*

(b) q_X *has a section.*

(c) q_X^* *is injective.*

Corollaries. (1) If q_X is fibrehomotopy trivial, then X has a fixed point.

(2) If X is contractible, then X has a fixed point.

To prove theorems of this kind we use the method of localization [6]. Let V be a complex G-module without trivial direct summand. Let

$$q_V : EG\times_G V \to BG$$

be the associated vector bundle and let $e(V) \in U^*(BG)$ be its U^*-Euler-class. We localize with respect to the set S of such Euler classes. Since for a pair (X,Y) of G-spaces the cobordism group

$$h^*(X,Y) := U^*(EG\times_G X,\ EG\times_G Y)$$

is a graded module over $U^*(BG)$ we can form the localization $S^{-1}h^*(X,Y)$. The $h^*(X,Y)$ constitute an equivariant cohomology theory.

Basic for our purpose is the following lemma which was proved in [7].

Lemma 2. *Let* $G = T \times H$ *be the product of a torus* T *and a finite abelian p-group* H. *Then*

$$S^{-1}U^*(BG) \neq 0.$$

Proof of Theorem 2. (a) $\Rightarrow$ (b). If $x \in X$ is a fixed point then $s(qe) = [e,x]$ defines a section s of q_X.

(b) $\Rightarrow$ (c). Let s be a section of q_X. Then $s^*q_X^* = \mathrm{id}$. Hence q_X^* is injective.

(c) $\Rightarrow$ (a). Since q_X^* is injective, so is the localization

$$S^{-1}q_X^* : S^{-1}U^*(BG) \to S^{-1}U^*(EG\times_G X).$$

From Lemma 2 we conclude that

$$S^{-1}U^*(EG\times_G X) \neq 0.$$

On the other hand we show that

$$S^{-1}U^*(EG\times_G X) = 0$$

if X has no fixed point. This follows from [6], Satz 1. Note that axiom (K4) of [6], p. 254, is not used; for A we use the fixed point set; but since A is empty X-A is compact and so $U^*(EG\times_G X)$ is annihilated by a suitable element of S ([6], Satz 2 and proof of Satz 1).

Proof of the corollaries. (1) If q_X is fibrehomotopy trivial then q_X has (up to homotopy) a right inverse. Hence q_X^* is injective.
(2) If X is contractible then q_X is fibrehomotopy trivial (Dold [9]).

References

1. Boardman, J.M.: On Manifolds with involution. Bull. Amer. Math. Soc. 73, 136-138 (1967).

2. Conner, P.E.: Seminar on periodic maps. Lecture Notes in Math. 46. Springer-Verlag 1967.

3. tom Dieck, T.: Bordism of G-manifolds and integrality theorems. Topology 9, 345-358 (1970).

4. —— : Actions of finite abelian p-groups without stationary points. Topology 9, 359-366 (1970).

5. —— : Characteristic numbers of G-manifolds I. Inventiones math. 13,213-224 (1971).

6. —— : Lokalisierung äquivarianter Kohomologie-Theorien. Math. Z. 121, 253-262 (1971).

7. —— : Kobordismentheorie klassifizierender Räume und Transformationsgruppen. To appear Math. Z. (1972).

8. —— : Periodische Abbildungen unitärer Mannigfaltigkeiten. To appear Math. Z. (1972).

9. Dold, A.: Partitions of unity in the theory of fibrations. Ann. Math. 78, 223-255 (1963).

10. Hsiang, Wu-Yi: Some fundamental theorems in cohomology theory of topological transformation groups. Bull. Amer. Math. Soc. 77, 1094-1098 (1971).

COBORDISM OF LINE BUNDLES WITH RESTRICTED CHARACTERISTIC CLASS

V. Giambalvo

The University of Connecticut, Storrs

In studying spin manifolds with free orientation preserving involution, the first complication comes from the fact that the quotient space does not have a spin structure. The purpose of this note is to give some results on a cobordism theory which takes into account some of the deviation of the quotient space from being a spin manifold. Details will appear elsewhere. All cohomology will be with coefficients in Z_2.

Define a cobordism theory Λ_* algebraically as follows: Let $f: BSO \times \mathbb{RP}^\infty \to K(Z_2,2)$ by $f^*(\iota) = w_2 \otimes 1 + 1 \otimes t^2$, $t \in H^1(\mathbb{RP}^\infty)$ the generator. Let E be the total space of the fibration induced by f from the path fibration over $K(Z_2,2)$. Denote by ξ the bundle over E obtained by pulling back the universal vector bundle from BSO, and by $M(\xi)$ the Thom space of this bundle. Then $\Lambda_* = \pi_*(M(\xi))$, the stable homotopy of this Thom space.

Geometrically, Λ_* is given by the obvious cobordism relation on triples (M,η,c), where η is a line bundle over the oriented manifold M, and $c: M \to E$ is a lift of $\nu \times \eta: M \to BSO \times \mathbb{RP}^\infty$; i.e., a way of saying $w_2(M) = (w_1(\eta))^2$. Note that for a 2-connected spin manifold N with involution T, N/T has either a spin structure or there is a lift of $\nu_{N/T} \times \rho$ to E, where $\rho: N/T \to \mathbb{RP}^\infty$ classifies the involution.

<u>Theorem</u> 1: There is a module L over the Steenrod algebra $\mathcal{A}$, and a graded Z_2-vector space Y such that $(L \otimes Y) \oplus F$ is isomorphic to $H^*(M(\xi))$, where F is a free $\mathcal{A}$ module and Y is generated by all finite sequences $(i_1,\ldots,i_k)$ of integers greater than 1 with $i_j \leq i_{j+1}$, for all j.

L is obtained from $\mathcal{A}/\mathcal{A}Sq^1 \oplus \bigoplus_{i=1}^\infty \mathcal{A}$ by the relations $Sq^2\alpha_0 = Sq^1\alpha_1$; $Sq^2Sq^3\alpha_i = Sq^1\alpha_{i+1}$, $i > 0$, where α_0 generates $\mathcal{A}/\mathcal{A}Sq^1$, and α_i is the generator of the ith summand.

<u>Theorem</u> 2: Λ_* has only 2-primary torsion. The Adams spectral for Λ_* collapses. For each summand in $H^*(M(\xi))$ isomorphic to L there is one element of infinite order

in each dimension congruent to 0 mod 4, and for each integer $q = 4k + 1 +$ (dimension of an $\mathcal{A}$ generator of L) there is an element of order 2^{2k+2}. There are further elements of order 2 corresponding to F.

<u>Theorem</u> 3: The map $\mu: \Lambda_* \to \Omega_*^{spin}(Z_2)$ given by passing to sphere bundles is a monomorphism modulo torsion.

REFERENCES

1. Giambalvo, V. Cobordism of line bundles with a relation, mimeographed.

2. Stong, R.E. Cobordism Theories. Princeton University Press, 1969.

UNITARY BORDISM OF MONOGENIC GROUPS AND ISOMETRIES

Gary Hamrick and Erich Ossa*

University of Texas at Austin

Friedrich-Wilhelms-Universtät zu Bonn

The Institute for Advanced Study

Introduction

Bordism of manifolds with group action was first studied by Conner and Floyd in [2]. Stong has made a thorough investigation of the bordism of unoriented manifolds with actions of finite groups [7]. The next most tractable case consists of unitary G-manifolds, which are manifolds with G-action commuting with a stable complex structure of the tangent bundle. The bordism groups of such actions, denoted by $U_*(G)$, form a graded module over the unitary bordism ring U_*.

Stong [6] has shown that $U_*(G)$ is a free U_*-module on even dimensional generators when G is a p-primary finite abelian group. Landweber [4] has proven the same result for finite cyclic G.

Let $U_*^{(\ell)}(G)$ be the bordism module of unitary G-manifolds such that the codimension of each of the isotropy subgroups is at least ℓ. In particular, $U_*^{(0)}(G) = U_*(G)$. The main result of this note extends Landweber's theorem by means of the technique used in [5] to study S^1 actions.

Theorem 1: Let G be a compact monogenic Lie group. Then $U_*^{(\ell)}(G)$ is a free U_*-module on generators in dimensions congruent to ℓ mod 2.

It follows from Theorem 1 that every unitary G-manifold on which the identity component G_0 has no fixed points bounds a unitary G-

*First author holds a National Science Foundation postdoctoral fellowship; second author supported in part by National Science Foundation grant GP-7952X2.

manifold. Equivalently, bordism classes in $U_*(G)$ are determined by the equivariant normal bundle to the fixed point set of G_0.

The freeness of $U_*(G)$ quickly leads to a similar result on the bordism of isometries. When we speak of an isometry ϕ on a manifold we shall not have reference to a particular Riemannian metric; we shall mean only that there exists some unspecified metric for which ϕ is an isometry. Let $U_*(I)$ be the bordism module of isometries on unitary manifolds.

Theorem 2: $U_*(I)$ is a free module on even dimensional generators.

1. Bordism of Monogenic Groups

First of all we have to introduce some more notation. Let G' be a compact Lie group, $k_1,\dots,k_m$ non-negative integers, and let F, $F_1,\dots,F_m$ be families of closed subgroups of G'. We look at unitary G'-manifolds M in the sense of the introduction and m-tuples of complex G'-vectorbundles $E_1,\dots,E_m$ over M with the following properties:

(i) all isotropy groups in M are elements of F

(ii) E_i has dimension k_i and all isotropy groups in the spherebundle of E_i are elements of F_i.

We denote the bordism group of such objects $(M,E_1,\dots,E_m)$ by $U_*(G';F;(k_1,F_1),\dots,(k_m,F_m))$. If any of the families $F,F_1,\dots,F_m$ is the family of all subgroups of G', we delete it from the notation. Often we shall simply write (k) instead of $(k_1,\dots,k_m)$.

The most important family of subgroups in this note will be the family $F(G',\ell)$ of all subgroups of G' which have codimension at least ℓ (we always assume $\ell \geq 0$). We shall denote this family by $F(\ell)$ whenever it is obvious which group is to be taken as G'. As in the introduction we write also $U_*^{(\ell)}(G';(k))$ instead of $U_*(G';F(\ell);(k))$.

Now we investigate more closely the case where G' is a product $G' = S^1 \times G$, we have the following subfamilies of $F(\ell) = F(S^1 \times G, \ell)$:

$$F_S(\ell) := \{H \in F(\ell) \mid S^1 \subset H\},$$

$$F_\infty(\ell) := \{H \in F(\ell) \mid S^1 \not\subset H\},$$

$$F_n(\ell) := \{H \in F(\ell) \mid S^1 \cap H \text{ is of order } \leq n\},$$

$$\tilde{F}_n(\ell) := F_n(\ell) - F_{n-1}(\ell) = \{H \in F(\ell) \mid S^1 \cap H \text{ is of order } n\}.$$

Our basic tool is the well-known exact sequence introduced by Conner [1]. Here this sequence takes the following form:

$$(*)\qquad \begin{array}{ccc} U_*(S^1 \times G; F_{n-1}(\ell); (k)) & \xrightarrow{i_n} & U_*(S^1 \times G; F_n(\ell); (k)) \\ \partial_n \nwarrow & & \swarrow j_n \\ & \bigoplus_{k_0} U_*(S^1 \times G; \tilde{F}_n(\ell); (k_0, F_{n-1}(\ell)), (k)) & \end{array}$$

i_n is the obvious forgetful map, j_n takes the normal bundle to the fixed point set of $\mathbb{Z}_n \subset S^1$ and ∂_n restricts to the spherebundle of the k_0-dimensional bundle E_0. The maps i_n and j_n preserve the (total) dimension, whereas ∂_n lowers the dimension by 1.

Lemma 1:

(a) j_n is surjective,

(b) $i: U_*(S^1 \times G; F_\infty(\ell); (k)) \to U_*(S^1 \times G; F(\ell) \cup F_S(\ell-1); (k))$ is the zero map.

We note in particular that, as a consequence of (b), any unitary $S^1 \times G$-manifold on which the S^1-action has no fixed points, is the boundary of a unitary $S^1 \times G$-manifold. It follows that the bordism class of a unitary $S^1 \times G$-manifold is uniquely determined by the equivariant normal bundle to the fixed point set of S^1.

Proof of Lemma 1:

Let $(M,E_0,E_1,\dots,E_m)$ represent an element of $U_*(S^1 \times G;\tilde{F}_n(\ell);(k_0,F_{n-1}(\ell)),(k))$. We shall see that any such object can be viewed as a family of complex G-vector bundles over M/S^1. For later use we note that this interpretation leads to a U_*-isomorphism.

$$U_*(S^1 \times G;\tilde{F}_n(\ell);(k_0,F_{n-1}(\ell)),(k))$$

$$\cong \bigoplus_{r_0,r_1,\dots,r_m} U_*(G;F(\ell-1);1,(k_{ij})_{0 \le i \le m,\ 1 \le j \le n}),$$

where the sum is to be taken over all complex representations $r_i: \mathbb{Z}_n \to U(k_i)$ such that r_0 has no trivial summands and where k_{ij} is the multiplicity of the j-th irreducible representation of $\mathbb{Z}_n$ in the representation r_i. The isomorphism (**) lowers dimensions by 1.

All this can be seen for instance by looking at the fixed point set of $\mathbb{Z}_n$ in the classifying space for $S^1 \times G$-vectorbundles (cf. [8]). We, however, will use a direct geometric construction which can be exploited to prove Lemma 1.

So let us go back to the above $(M,E_0,E_1,\dots,E_m)$. Since the S^1-action has isotropy groups $\mathbb{Z}_n$ in M, we can split the bundles E_i into $S^1 \times G$-subbundles E_{ij}, $1 \le j \le n$, such that an element $t \in \mathbb{Z}_n \subset S^1 \subset \mathbb{C}$ acts on E_{ij} by multiplication with t^j in the fibres.

Now we define a new $S^1 \times G$-action on the E_{ij} as follows:

$$(t,g)\#e := t^{-j}((t,g)e) \quad \text{for } t \in S^1,\ g \in G,\ e \in E_{ij}.$$

With respect to this new action, $\mathbb{Z}_n$ acts trivially and $S^1/\mathbb{Z}_n$ acts freely. Taking the quotient by the $S^1/\mathbb{Z}_n$-action we get complex vectorbundles $\tilde{E}_{ij}$ over $\tilde{M} = M/S^1$. Moreover the projection $M \to \tilde{M}$ defines a one-dimensional complex vectorbundle ξ over $\tilde{M}$. All these bundles inherit a G-action, which we denote also by $\#$. This exhibits the above isomorphism (**), since, as we shall see now, the original $(M,E_0,E_1,\dots,E_m)$ can be recovered from these data. For this, we make the G-vectorbundles $\tilde{E}_{ij}$, ξ into $S^1 \times G$-vectorbundles defining the action by

$$(t,g) \circ e := t^j(g\# e) \quad \text{for } t \in S^1,\ g \in G,\ e \in \tilde{E}_{ij},$$

$$\text{and } (t,g) \circ e := t^n(g\# e) \quad \text{for } t \in S^1,\ g \in G,\ e \in \xi.$$

Setting $\tilde{E}_i = \bigoplus_j \tilde{E}_{ij}$, let $X = S(\xi \oplus \tilde{E}_0)$ be the spherebundle of the bundle $\xi \oplus \tilde{E}_0$ over $\tilde{M}$, and let E_i', $1 \le i \le m$, be the pullback of $\tilde{E}_i$ to X. Then $(X,E_1',\dots,E_m')$ with the $\circ$-action defines an element of $U_*(S^1 \times G;F_n(\ell);(k))$, and it is easily checked that the normal bundle to the fixed point set of $\mathbb{Z}_n$ is precisely $(M,E_0,E_1,\dots,E_m)$.

This proves the surjectivity of j_n. Moreover, the element $(X,E_1',\dots,E_m')$ constructed above is, as a spherebundle, the boundary of the corresponding discbundle. Therefore it maps to zero in $U_*(S^1 \times G;F(\ell) \cup F_S(\ell-1);(k))$. Now an easy induction argument, using the exact sequences (*), shows that any element of $U_*(S^1 \times G;F_n(\ell);(k))$ maps to zero in $U_*(S^1 \times G;F(\ell) \cup F_S(\ell-1);(k))$. Since $U_*(S^1 \times G;F_\infty(\ell);(k)) = \varinjlim U_*(S^1 \times G;F_n(\ell);(k))$ this proves part (b) of Lemma 1.

From the above we obtain easily the following

<u>Corollary of Lemma 1</u>:

Suppose that for any $(k) = (k_1,\dots,k_m)$ the U_*-module $U_*(G;F(\ell-1);(k))$ is free. Then $U_*(S^1 \times G;F_\infty(\ell);(k))$ and $U_*(S^1 \times G;F(\ell) \cup F\ (\ell-1);(k))$

are free U_*-modules.

A special case worth mentioning is the case $\ell = 1$. Since $F(1) \cup F_S(0) = F(0)$ we obtain: if $U_*(G;(k))$ is a free U_*-module for any $(k) = (k_1,\dots,k_m)$, then $U_*(S^1 \times G;(k))$ is also a free U_*-module.

Proof of the corollary:

From the isomorphism (**), the exact sequence (*) and Lemma 1 (a) it follows immediately that $U_*(S^1 \times G;F_\infty(\ell);(k)) = \varinjlim U_*(S^1 \times G;F_n(\ell);(k))$ is a free U_*-module. In analogy to (*) we have now the exact sequence

$$\begin{array}{ccc} U_*(S^1 \times G;F_\infty(\ell);(k)) & \xrightarrow{i} & U_*(S^1 \times G;F(\ell) \cup F_S(\ell-1);(k)) \\ & \nwarrow \partial \qquad \swarrow j & \\ & \bigoplus\limits_{k_0} U_*(S^1 \times G;F_S(\ell-1);(k_0,F_\infty(\ell)),(k)) & \end{array}$$

From Lemma 1(b) we have $i = 0$, and so it is enough to show that $U_*(S^1 \times G;F_S(\ell-1);(k_0,F_\infty(\ell)),(k))$ is a free U_*-module (since then the group in question is a direct summand of a free U_*-module, therefore projective, and therefore free [3]). But the latter is isomorphic to

$$\bigoplus_{r_0,r_1,\dots,r_m} U_*(G;F(\ell-1);(k_{ij})_{0 \le i \le m}) ,$$

where the sum is taken over all representations $r_i\colon S^1 \to U(k_i)$ such that r_0 has no trivial summands and where k_{ij} is the multiplicity of the j-th irreducible representation of S^1 in the representation r_i.

Finally we come now to the case of a monogenic group G.

<u>Theorem 1</u>: Let G be a monogenic group. Then $U_*^{(\ell)}(G;(k)) = U_*(G;F(\ell);(k))$ is a free U_*-module on generators of dimension congruent to ℓ modulo 2.

<u>Proof of Theorem 1</u>:

Let d be the dimension of G. For $d = 0$ the assertion has been proved in [4], so that we may use induction on d.

Let us first consider the case $\ell = d$ (the case $\ell > d$ is trivial). Since we may assume that $d > 0$, we can write $G = S^1 \times G'$ and have

$$U_*(G;F(d);(k)) = U_*(S^1 \times G';F_\infty(d);(k)).$$

Since $U_*(G';F(d-1);(k))$ is free on the right generators by the induction hypothesis, we conclude from the corollary of Lemma 1 that $U_*(G;F(d);(k))$ is a free U_*-module. A quick check through the proof of the corollary shows that the generators have the right dimensions. For the case $0 \leq \ell < d$ we use now downward induction on ℓ. Let T_ℓ be any subtorus of G of codimension ℓ. We denote by $F(T_\ell)$ the family of all subgroups of G which have codimension ℓ and contain T_ℓ. In analogy to the sequences (*) we have the exact sequence

$$\begin{array}{ccc} U_*(G;F(\ell+1);(k)) & \xrightarrow{i} & U_*(G;F(\ell);(k)) \\ {\scriptstyle\partial}\nwarrow & & \swarrow{\scriptstyle j} \\ & \bigoplus_{T_\ell < G} \bigoplus_{k_0} U_*(G;F(T_\ell);(k_0,F(\ell+1)),(k)) & \end{array}$$

First of all we show now that $U_*(G;F(T_\ell);(k_0,F(\ell+1)),(k))$ is free on generators of dimension congruent to ℓ mod 2. But since T_ℓ is a direct factor of G we have, as above, an isomorphism

$$U_*(G;F(T_\ell);(k_0,F(\ell+1)),(k)) \cong$$

$$\cong \bigoplus_{r_0,\dots,r_m} U_*(G/T_\ell;F(\ell);(k_{ij})_{0 \le i \le m}) .$$

Here $r_0,\dots,r_m$ run through all representations $r_i\colon T_\ell \to U(k_i)$ such that r_0 has no trivial summands and k_{ij} is the multiplicity of the j-th irreducible representation of T_ℓ in the representation r_i. Since $\ell < d$, the group G/T_ℓ has dimension $< d$, and the assertion follows from the induction hypothesis.

To prove the theorem it is now enough to show that $i\colon U_*(G;F(\ell+1);(k)) \to U_*(G;F(\ell);(k))$ is the zero mapping, since then $U_*(G;F(\ell);(k))$ is a direct summand of a free U_*-module and hence free. So let us take an element $(M,E_1,\dots,E_m)$ in $U_*(G;F(\ell+1);(k))$. Since $\ell + 1 \ge 1$ there is a $S^1 < G$ which has no fixed points in M. If we write $G = S^1 \times G'$ this means that $(M,E_1,\dots,E_m)$ lies in the image of the obvious map

$$U_*(S^1 \times G';F_\infty(\ell+1);(k)) \to U_*(G;F(\ell+1);(k)).$$

But we have the commutative diagram

$$\begin{array}{ccc} U_*(S^1 \times G';F_\infty(\ell+1);(k)) & \to & U_*(G;F(\ell+1);(k)) \\ \downarrow i & & \downarrow i \\ U_*(S^1 \times G';F(\ell+1) \cup F_S(\ell);(k)) & \to & U_*(G;F(\ell);(k)) \end{array}$$

and Lemma 1(b) states that the left column is zero.

2. Bordism of isometries

We call a G-manifold effective if the G-action on each G-component is effective. Let $U_*(G,ef)$ be the bordism module of effective unitary G-manifolds. We claim that the obvious homomorphism

$$\rho: \bigoplus_{H \triangleleft G} U_*(G/H, ef) \to U_*(G)$$

is an isomorphism. Clearly ρ is epic. To see that ρ is monic, we assume that a disjoint union $\cup \{M_H | H \triangleleft G\}$ of effective G/H-manifolds bounds a G-manifold W and show that M_H bounds a G/H-manifold. Let W_H be the union of those components of W which meet M_H. The equivariant collaring theorem implies that H acts trivially on W_H and thus that W_H is an effective G/H-manifold. Hence M_H bounds W_H.

Let $T^m \times \mathbb{Z}_k$ be the product of the m-torus with the cyclic group of order k. Since $U_*(T^m \times \mathbb{Z}_k, ef)$ is a direct summand of $U_*(T^m \times \mathbb{Z}_k)$, Theorem 1 implies

<u>Proposition 1</u>: $U_*(T^m \times \mathbb{Z}_k, ef)$ is a free U_*-module on even dimensional generators.

Let two generators of $T^m \times \mathbb{Z}_k$ be equivalent if there is a continuous automorphism of $T^m \times \mathbb{Z}_k$ taking one to the other. Let γ be an equivalence class of generators, and let $U_*(T^m \times \mathbb{Z}_k, ef)_\gamma$ simply be a copy of $U_*(T^m \times \mathbb{Z}_k, ef)$ indexed by γ. From each equivalence class γ we select a generator g_γ.

<u>Proposition 2</u>: $U_*(I)$ is isomorphic to $\bigoplus_{m,k,\gamma} U_*(T^m \times \mathbb{Z}_k, ef)_\gamma$.

<u>Proof</u>: If M is a $T^m \times \mathbb{Z}_k$ manifold representing a class in $U_*(T^m \times \mathbb{Z}_k, ef)_\gamma$, there exists a Riemannian metric on M such that $T^m \times \mathbb{Z}_k$ acts by isometries. The action of g_γ on M yields a class in $U_*(I)$. This defines a homomorphism

$$\sigma: \bigoplus_{m,k,\gamma} U_*(T^m \times \mathbb{Z}_k, ef)_\gamma \to U_*(I).$$

We wish to define an inverse τ for σ. Let ϕ be an isometry on M representing a class in $U_*(I)$. We may assume that ϕ acts transitively on the set of components of M. The closure of

$\{\phi^n | n \text{ an integer}\}$ in the group of diffeomorphisms on M is a compact monogenic Lie group acting differentiably on M. Obviously M is an effective G-manifold. Since G is generated by ϕ, there is a unique isomorphism of G to $T^m \times \mathbb{Z}_k$ (for some m and k) carrying ϕ to some g_γ. Let $\tau[M]$ be the corresponding class in $U_*(T^m \times \mathbb{Z}_k, ef)_\gamma$. To check that τ is well defined, one can apply the equivariant collaring theorem as before.

From the preceding propositions we obtain immediately

Theorem 2: $U_*(I)$ is a free module on even dimensional generators.

REFERENCES

1. P. E. Conner: A bordism theory for actions of an abelian group, Bull. Amer. Math. Soc. 69(1963), pp. 244-247.
2. P. E. Conner and E. E. Floyd: Differentiable periodic maps, Springer-Verlag, 1964.
3. P. E. Conner and L. Smith: On the complex bordism of finite complexes, Publ. Math. IHES, 37(1969), pp. 117-221.
4. P. S. Landweber: Equivariant bordism and cyclic groups, to appear in Proc. Amer. Math. Soc.
5. E. Ossa: Fixpunktfreie S^1-Aktionen, Math. Ann. 186(1970), pp.45-52.
6. R. E. Stong: Complex and oriented equivariant bordism; in Topology of Manifolds, Edited by J. C. Cantrell and C. H. Edwards, Jr., Markham Pub,. Co., Chicago, 1970, pp. 291-316.
7. R. E. Stong: Unoriented bordism and actions of finite groups, Memoir no. 103 of the Amer. Math. Soc., 1970.
8. T. tom Dieck: Faserbündel mit Gruppen operationen, Archiv der Math., 20(1969), 136-143.

QUILLEN'S THEOREM FOR MU^*

Connor Lazarov*

Herbert Lehman College
City University of New York

The aim of this paper is to prove an MU^* analogue of Quillen's theorem which relates the Krull dimension of the cohomology of BG, for G a compact connected Lie group to the Krull dimension of the cohomology of BA where A runs over the closed Abelian subgroups of G. Let MU^* be $MU^*(\text{point})$ and let R be a commutative ring with unit which is a flat module over MU^*. Let

$$\Omega(X) = (\bigoplus_0^\infty MU^{2i}(X)) \otimes R$$

where the tensor product is taken over MU^*. We will show,

THEOREM: Let G be a compact connected Lie group, then

Krull dimension $(\Omega(BG))$ = maximum Krull dimension $(\Omega(BA))$

where A runs over the closed Abelian subgroups of G.

While the proof uses many of Quillen's techniques, it differs in one important aspect. Quillen shows ([10] p. 21) that for a compact G-space, $H_G(X)$ (see section 1) is finitely generated over $H_G(\text{point})$. We do not know whether the same result is true for Ω. Instead we use Landweber's result about the flatness of $MU^*(BZ_n)$. Since $H^*(BZ_n)$ is not a flat Z-module, this proof will not yield the result for H^*.

I am indebted to Prof. Quillen and Prof. Landweber for preprints as well as to Prof. Alex Heller for pointing out various facts.

1. Preliminaries

Let A be a commutative ring with unit. Let $p_0 \subsetneq p_1 \subsetneq \cdots \subsetneq p_n$ be a chain of prime ideals. We say that such a chain has length

---

*Research partially supported by NSF grant GP-12639.

n. The Krull dimension of A is the maximum length of a chain of prime ideals in A. Hereafter we will just say dimension instead of Krull dimension. Let A be a subring of B and suppose B is finitely generated as a module over A. Then B is integral over A and dimension (B)=dimension (A). (See [5] sec. 1-6.) If $f: A \to B$ is a homomorphism of commutative rings with unit and B is finitely generated as an A module, then dimension $(A) \geq$ dimension (B). For, let I be the kernel of f and consider the map $A \xrightarrow{\phi} A/I \to 0$. If p is a prime ideal in A/I, $\phi^{-1}(p)$ is a prime ideal in A and so it follows that dimension $(A) \geq$ dimension (A/I). However B is finitely generated over the subring A/I and so dimension (B) = dimension (A/I).

If I is an ideal in A which has the property that for each x in I there is an integer n such that $x^n = 0$, then the ideal I is called a nil ideal. We prove a useful lemma regarding such ideals.

LEMMA (1.1): Let A be a commutative ring with unit and I a nil ideal in A. Then A and A/I have the same dimension.

Proof: Let $\phi: A \to A/I$ be the quotient map. As above dimension $(A) \geq$ dimension (A/I). Now let p be a prime ideal in A. Consider $\phi(p)$. Suppose $\phi(a)\phi(b)$ is in $\phi(p)$. Then $ab - x = z$ for some x in p and some z in I. $z^n = 0$ for some integer n, so $(ab - x)^n = (ab)^n + y = 0$, where y is in p. Thus, either a or b is in p and so $\phi(p)$ is a prime ideal. Suppose now, that p and q are two prime ideals in A with $p \supsetneq q$. Let y be an element in $p - q$ and suppose there is an x in q such that $y - x$ is in I. Then $(y - x)^n = y^n - xt = 0$ for some integer n and some t in A. Since x is in q, y^n is in q which is impossible. Thus $\phi(p) \supsetneq \phi(q)$ and so dimension $(A/I) \geq$ dimension (A).

Now we prove a lemma which will be known to some readers. This proof was shown to me by Alex Heller.

LEMMA (1.2): Let T be an n-dimensional torus and A a closed subgroup of T. Then one can write T as a product $S^1 \times \cdots \times S^1$ of n circles in such a way that $A = \prod_{j=1}^{n} (A \cap S^1_{(j)})$ where $S^1_{(j)}$ is the jth factor.

Proof: Let C be the quotient T/A and consider the character groups $\hat{C}$, $\hat{T}$, $\hat{A}$. Then it is well known that $0 \to \hat{C} \to \hat{T} \to \hat{A} \to 0$ (see [9] theorem 37). However $\hat{T} = Z^n$ and so $\hat{C}$ is a finitely generated free Abelian group. We can choose a basis of Z^n in such a way that multiples of the basis elements form a basis for $\hat{C}$ ([4] theorem 7.1). Then the map $\hat{C} \to \hat{T}$ is a product $\prod_{j=1}^{r} f_j : Z^r \to Z^n$. Now applying Pontrjagin duality ([7] theorem 39), $\hat{\hat{C}} = C$, $\hat{\hat{A}} = A$, $\hat{\hat{T}} = T$ and the map $T \to C$ is the product

$$\prod_{j=1}^{n} g_j : T = \prod_1^n S^1 \to \prod_1^r S^1 = C$$

where $g_j = \hat{f}_j$ $j=1,\ldots,r$ and $g_j = 0$ for $j=r+1,\ldots,n$. Thus, the kernel, A, is the product $\prod_{j=1}^{n}$ kernel (g_j).

Let X be a compact left G-space and $E(G)$ a universal principal G-bundle. As in [10] we define $\Omega_G(X)$ to be $\Omega(E(G) \times_G X)$ and $X_G = E(G) \times_G X$. If X is a point, we denote $\Omega_G(X)$ by Ω_G. Then it is immediate that if A is a subgroup of G, $\Omega_G(G/A) = \Omega_A$. If $X \to Y$ is a G-map, then we have an induced map $X_G \to Y_G$ and so a map $\Omega_G(Y) \to \Omega_G(X)$. If we take X to be G/A and Y to be a point, then Ω_A is a module over Ω_G.

2. Proof of Theorem

We prove the main theorem in two parts.

THEOREM (2.1): Let A be a closed Abelian subgroup of the compact connected Lie group G. Then dimension $\Omega_G \geq$ dimension Ω_A.

Proof: It is enough to show that Ω_A is a finitely generated module over Ω_G. By the Peter-Weyl theorem G has a faithful finite dimensional complex representation. Let $G \subset U(n)$ be the inclusion corresponding to this representation. From the diagram,

$$\begin{array}{ccc} & \nearrow \Omega_{U(n)}(U(n)/G) = \Omega_G & \\ \Omega_{U(n)} & & \downarrow \\ & \searrow \Omega_{U(n)}(U(n)/A) = \Omega_A & \end{array}$$

it is enough, by the remarks in section 1, to show that Ω_A is a finitely generated $\Omega_{U(n)}$ module. Since any complex representation of a compact Abelian Lie group is a sum of one-dimensional representations, A is contained in a maximal torus of $U(n)$. Thus $A \subset T \subset U(n)$. (Of course, if $U(n)$ were replaced by an arbitrary compact connected Lie group, this statement would be false.) Now $MU^*(BT)$ is the power series ring in elements $u_1,\ldots,u_n$ of degree 2 and $MU^*(BU(n))$ maps injectively onto the power series in the elementary symmetric functions in the u's ([8] p. 42). Thus $MU^*(BT)$ is a finitely generated $MU^*(BU(n))$ module with even dimensional elements $\{u_1^{k_1}\ldots u_n^{k_n}\}$ $k_j < j$ as basis. Thus $\Omega(BT)$ is a finitely generated $\Omega(BU(n))$ module. (Further, dimension $\Omega(BT)$ = dimension $\Omega(BU(n))$.) So, it is enough to show that Ω_A is a finitely generated Ω_T module. By (1.2) there is a decomposition $\underbrace{S^1 \times \cdots \times S^1}_{n}$ of T so that $A = \prod_{j=1}^{n} A \cap S^1_{(j)} = \prod_{j=1}^{n} A_j$, and each A_j is either finite cyclic or S^1. From [7] lemma 1, lemma 5, and final remark 4 it follows that $MU^*(BS^1)$ and $MU^*(BZ_k)$ are flat MU^* modules. It is further shown in [7] that if $Z_k \subset S^1$ then the map $MU^*(BS^1) \to MU^*(BZ_k)$ is onto. From this and the decomposition of T it follows that $\Omega(BT) = (\bigotimes_1^n MU^*(BS^1)) \otimes R$, $\Omega(BA) = (\bigotimes_1^n MU^*(BA_j)) \otimes R$ and the map $\Omega(BT) \to \Omega(BA)$ is just the tensor product of the maps $MU^*(BS^1_{(j)}) \to MU^*(BA_j)$ tensored with the identity of R. Since each of these maps is onto, $\Omega(BA) = \Omega_A$ is a finitely generated module over $\Omega(BT) = \Omega_T$. Thus it follows that Ω_A is a finitely generated module over Ω_G and so the result follows.

LEMMA (2.2): Let X be a compact G-space. Suppose $X = \bigcup_1^n X_i$ where each X_i is a compact G-space and $(X_i)_G \to X_G \to X_G/(X_i)_G$ is a cofibration. Then the kernel I of $\Omega_G(X) \to \bigoplus_1^n \Omega_G(X_i)$ is a nil ideal.

Proof: From the exactness of

$$\Omega(X_G,(X_i)_G) \to \tilde{\Omega}(X_G) \to \tilde{\Omega}((X_i)_G)$$

we see that if x is in I there is a y_i in $\Omega(X_G,(X_i)_G)$ which restricts to x. From the commutative diagram,

$$\begin{array}{ccccc} \bigotimes_1^n \Omega(X_G,(X_i)_G) & \to & \tilde{\Omega}(\bigwedge_1^n X_G/(X_i)_G) & \xrightarrow{\Delta^*} & \Omega(X_G, \bigcup_1^n (X_i)_G) = 0 \\ \downarrow & & \downarrow & & \downarrow \\ \bigotimes_1^n \tilde{\Omega}(X_G) & \longrightarrow & \tilde{\Omega}(\bigwedge_1^n X_G) & \xrightarrow{\Delta^*} & \tilde{\Omega}(X_G) \end{array}$$

where Δ is the diagonal map, it follows that $x^n = 0$.

Corollary 2.3: Let X be a compact G-manifold and suppose $X = \bigcup_1^n X_i$ where each X_i is a closed G-invariant tubular neighborhood of an orbit G/A_i where A_i is a closed Abelian subgroup. Then

$$\text{dimension } \Omega_G(X) = \text{maximum dimension } \Omega_{A_i}.$$

Proof: Each X_i has G/A_i as a G-deformation retract. Further, it is easy to see that $(X_i)_G \to X_G \to X_G/(X_i)_G$ is a cofibration. Thus the kernel of

$$\Omega_G(X) \to \bigoplus_1^n \Omega_G(X_i) \cong \bigoplus_1^n \Omega_G(G/A_i) = \bigoplus_1^n \Omega_{A_i}$$

is a nil ideal by the previous lemma. Also, from the proof of (2.1) we have seen that each Ω_{A_i} and hence $\bigoplus_1^n \Omega_{A_i}$ is a finitely generated module over Ω_G. Thus $\bigoplus_1^n \Omega_{A_i}$ is a finitely generated module over $\Omega_G(X)$ and since the kernel is a nil ideal it follows from the preliminary remarks that dimension $\Omega_G(X) =$ dimension $\bigoplus_1^n \Omega_{A_i}$. However it is an easy fact about commutative rings that dimension $\bigoplus_1^n \Omega_{A_i}$ $=$ maximum dimension Ω_{A_i}.

We precede the next argument with a few remarks. If X is a compact manifold, then it follows from [1] that $\varprojlim{}^1 K^*(X_G) = 0$.

Landweber has pointed out ([6]) that it follows from [3] that $\varprojlim^1 K^*(Y) = 0$ implies $\varprojlim^1 MU^*(Y) = 0$. Now, let E be a complex G-vector bundle over the compact G-manifold X and $F(E)$ the unitary flag bundle of E. (See [1] for a treatment of $K^*(F(E))$ and $H^*(F(E))$ and notice that the structure of $MU^*(F(E))$ follows using the same arguments and the cobordism Chern classes.) Then $MU^*(X)$ injects into $MU^*(F(E))$ and $MU^*(F(E))$ is finitely generated over $MU^*(X)$ by even dimensional elements given by Chern classes. It follows from this and the vanishing of $\varprojlim^1$ that $MU^*(F(E)_G)$ is finitely generated over the subring $MU^*(X_G)$ and thus $\Omega_G(F(E))$ is finitely generated over the subring $\Omega_G(X)$. Now we proceed as Quillen does.

<u>THEOREM (2.4)</u>: Let G be a compact connected Lie group. Then

dimension $\Omega_G \leq$ maximum dimension Ω_A.

Proof: As in (2.1) let V be a faithful complex representation of G and let $F(V)$ be the unitary flag bundle of V. $F(V)$ is a compact G-manifold homeomorphic to $U(n)/T$. $F(V)_G$ is the unitary flag bundle of the complex vector bundle $E(G) \underset{G}{\times} V$ over BG and so by the preceding remarks $\Omega_G(F(V))$ is a finitely generated module over the subring Ω_G and so dimension $\Omega_G(F(V))$ = dimension Ω_G. Now a point of $F(V)$ is a family of mutually orthogonal one-dimensional subspaces of V. The isotropy group of a point in $F(V)$ is a closed subgroup of G which leaves each of these one-dimensional subspaces fixed. Thus V is a faithful representation of the isotropy group and splits into a sum of one-dimensional representations. So, the isotropy group of any point in $F(V)$ is a closed Abelian subgroup of G. Thus $F(V) = \bigcup_1^n X_i$ where each X_i is a closed tubular neighborhood of an orbit G/A_i where A_i is a closed Abelian subgroup of G. Then by (2.3) dimension $\Omega(F(V))$ = maximum dimension Ω_{A_i} and so dimension Ω_G = maximum dimension Ω_{A_i}.

Remarks

1. There are lots of examples of rings R which are flat modules over MU^* and which have finite Krull dimension. For example, since MU^* is an infinite polynomial ring over Z with generators X_i in dimension 2i, we could take S to be the multiplicative set consisting of all products of X_i's. Then $S^{-1}MU^*$ is such a ring. For such a ring the corresponding Ω (point) has finite Krull dimension. However, we don't know what dimension $\Omega(BZ_n)$ is.

2. This proof goes through word for word for complex K-theory, since $K^*(BZ_n)$ is a flat module over K^*(point).

BIBLIOGRAPHY

1. Atiyah, M. F., and Segal, G. B.: "Equivariant K-Theory and Completions," J. Diff. Geo. 3 (1969), pp. 1-18.
2. Bott, R.: "Lectures on K(X)," Lecture notes, Harvard Univ. (1963).
3. Buhstaber, V. B., and Miscenko, A. S.: "Elements of Infinite Filtrations in K-Theory," Dokl. Akad. Nauk SSR 178 (1968) 1234-1237 = Soviet Math. Dokl. 9 (1968) 256-259.
4. Eilenberg, S., and Steenrod, N.: "Foundations of Algebraic Topology," Princeton Univ. Press (1952).
5. Kaplansky, I.: "Commutative Rings," Allyn and Bacon (1970).
6. Landweber, P.: "A Note on the Cobordism of Classifying Spaces," to appear.
7. ----------: "Coherence, Flatness, and Cobordism of Classifying Spaces," Proc. Aarhus Inst. Alg. Topology (1970).
8. Novikov, S. P.: "The Methods of Alg. Topology from the Viewpoint of Cobordism Theories," Lecture notes, Aarhus Universitat.
9. Pontrjagin, L. S.: "Topologische Gruppen," vol. 2, B. G. Teubner Verlagsgesellschaft (1958).
10. Quillen, D.: "The Spectrum of an Equivariant Cohomology Ring I," to appear.

EQUIVARIANT CHARACTERISTIC NUMBERS

Chung N. Lee* and Arthur G. Wasserman*

University of Michigan
Ann Arbor, Michigan

It is a classical result of Thom [12] that a closed manifold** M is a boundary iff all the Stiefel-Whitney numbers of M vanish. This result has been generalized to manifolds together with a reduction of the structural group of the normal bundle e.g. oriented or stably almost complex manifolds.

The purpose of this paper is to consider the equivariant version of these results. More specifically, equivariant cohomology theories applied to the classifying space for G vector bundles are used to define equivariant characteristic classes and the associated numbers; then it is shown that in special cases these numbers vanish on the G manifold M iff M is a G boundary.

In part I several examples of unoriented cobordism are presented and a general theorem, which covers all the cases considered, is proved. The difficult part of the theorem is proving an equivariant version of the statement: $\mathfrak{N}_*(BO(K)) \longrightarrow \mathfrak{N}_*(BO(K+1))$ is a monomorphism. The oriented case is considered (for $G = Z_2$) in part 2 using the results of [3] although new geometric proofs are provided for these results.

I

Let G be a compact Lie group and let $B(O, G)_n$ be the classifying space for G vector bundles of dimension n (see [13] or [6]) and let $\mu_n \longrightarrow B(O, G)_n$ be the universal vector bundle. If h^* is an equivariant cohomology theory ([2]) then elements of $h^*(B(O, G)_n)$ are called universal h^* characteristic classes. If $E \longrightarrow X$ is a G vector bundle induced by an equivariant map $f: X \longrightarrow B(O, G)_n$ then $f^*(h^*(B(O, G)_n)) \subset h^*(X)$ is the characteristic subgroup of the bundle E.

* This work was partially supported by the National Science Foundation under grants GP-7952X3 and GP-20038.

** All manifolds considered in this paper will be C^∞ and all group actions are assumed to be smooth.

Note that this subgroup is well defined since f is unique up to equivariant homotopy. In particular, for a G manifold M^n we have the tangent map $\tau_M : M \longrightarrow B(O, G)_n$.

Let $\mathcal{G}$ be a collection of G manifolds e.g. $\mathcal{G}$ = manifolds with free G action or $\mathcal{G}$ = all G manifolds. $\mathfrak{N}_*(\mathcal{G})$ will denote the cobordism group of compact manifolds in $\mathcal{G}$ i.e. equivalence classes of closed manifolds in $\mathcal{G}$ where $M_1^n \sim M_2^n$ if there is a compact manifold $W \in \mathcal{G}$ with $\partial W = M_1 \cup M_2$. ($\mathcal{G}$ must satisfy some mild conditions to make $\sim$ an equivalence relation.) We wish to describe characteristic numbers for $\mathfrak{N}_*(\mathcal{G})$.

Suppose that h^* is given by $H^* \circ A$ where A is a functor from the category of G spaces and equivariant maps to the category of topological spaces and continuous maps and H^* is singular cohomology and let $h_* = H_* \circ A$ denote the associated equivariant homology theory. Let $< \, , \, > : h^*(X) \otimes_{H^*(pt)} h_*(X) \longrightarrow H_*(pt)$ be the Kronecker pairing. Then, if $[\quad]$ assigns to each compact G-manifold W in $\mathcal{G}$ a "top class" $[W, \partial W] \in h_*(W, \partial W)$ satisfying

i) $[W_1 \cup W_2, \partial W_1 \cup \partial W_2] = [W_1, \partial W_1] + [W_2, \partial W_2]$ and

ii) $\partial_*[W, \partial W] = [\partial W]$

then the x characteristic number of M is defined by $x(M) = <\tau_M^* x, [M]> \in H_*(pt)$ where $x \in h^*(B(O, G)_n)$. We then have the classical

Theorem 1 ([9]). The x characteristic number is a $\mathcal{G}$ equivariant cobordism invariant if x is in the image of $j^* : h^*(B(O, G)_{n+1}) \longrightarrow h^*(B(O, G)_n)$ where $j : B(O, G)_n \longrightarrow B(O, G)_{n+1}$ is the map classifying $\mu_n \oplus 1$.

The proof just uses the commutative diagram

$$\begin{array}{ccc} \partial W & \xrightarrow{\tau_{\partial W}} & B(O, G)_n \\ \cap & & \downarrow j \\ W & \xrightarrow{\tau_W} & B(O, G)_{n+1} \end{array}$$

and properties i) and ii) of the top class.

Note 1). The requirement on x is necessary because the tangent map rather than the normal map is being used. In the cases considered j^* will be an epimorphism.

Note 2). If $M \neq 0 \in \mathfrak{N}_*(\mathcal{G})$ but $M = \partial W$, $W \notin \mathcal{G}$ then the h^* characteristic numbers of M will vanish if there is a class $y \in h_*(W, \partial W)$ such that $\partial_* y = [M]$ i.e. if W can be given a "top class". Choosing h_* carefully can sometimes avoid this problem. An alternative approach (not considered here) is to use restricted classifying spaces, B_n say, such that the tangent bundle of W

is <u>not</u> induced from B_{n+1} and hence the diagram used in the proof of Theorem 1 cannot be constructed.

Special Cases

1) Let $\mathscr{G}$ = manifolds with free G action. Then $\mathfrak{N}_*(\mathscr{G}) = \mathfrak{N}_*(BG)$ by the map which sends (M) to (M/G, f) where $f : M/G \longrightarrow BG$ classifies the principal bundle $M \longrightarrow M/G$ ([4]). Let $h^*(X) = H^*(E_G \times_G X; Z_2)$ where E_G is the universal principal G bundle. Then h^* is an equivariant cohomology theory.

Now $\pi : E_G \times_G M \xrightarrow{E_G} M/G$ is a fibration and since E_G is contractible π is a homotopy equivalence with inverse $q : M/G \longrightarrow E_G \times_G M$ given by $q([m]) = [\bar{f}(m), m]$ where $\bar{f} : M \longrightarrow E_G$ is equivariant. Note that $q([gm]) = [\bar{f}(gm), gm] = [g\bar{f}(m), gm] = [\bar{f}(m), m]$ and hence q is well defined. Thus $h_*(M) = H_*(E_G \times_G M; Z_2) = H_*(M/G; Z_2)$ has a top class (in dimension n-dim G) if M is compact since M/G is a compact manifold. Hence h^* characteristic numbers are defined.

<u>Theorem</u> 2. $M = 0 \in \mathfrak{N}_*(\mathscr{G})$ iff all h^* characteristic numbers vanish.

Proof. The universal characteristic classes $h^*(B(O, G)_n)$ may be easily computed as follows: we have an inclusion $i : BO_n \longrightarrow B(O, G)_n$ where BO_n is the classifying space for vector bundles with trivial G action.

<u>Lemma</u> 3. If $L \subset G$ then $B(O, G)_n | L = B(O, L)_n$ i.e. $B(O, G)_n$ thought of as an L space is just $B(O, L)_n$.

Proof. Let $E \longrightarrow X$ be a L vector bundle. Then $G \times_L E \longrightarrow G \times_L X$ is a G vector bundle and hence has a G equivariant classifying map $f : G \times_L X \longrightarrow B(O, G)_n$. Since $X \subset G \times_L X$, $f|X$ is a L equivariant map to $B(O, G)_n | L$ which classifies $E \longrightarrow X$.

Since $B(O, e)_n = BO_n$ we have that $i : BO_n \longrightarrow B(O, G)_n$ is a homotopy equivalence. From the diagram of fibrations

$$\begin{array}{ccccc} BG \times BO_n & \xrightarrow{\approx} & E_G \times_G BO_n & \longrightarrow & E_G \times_G B(O, G)_n \\ \downarrow & & \downarrow {\scriptstyle BO_n} & & \downarrow {\scriptstyle B(O, G)_n} \\ BG & = & BG & = & BG \end{array}$$

we see that $E_G \times_G B(O, G)_n = BG \times BO_n$ (as observed by T. tom Dieck) and hence $h^*(B(O,G)_n) = H^*(BG; Z_2) \otimes_{Z_2} H^*(BO_n; Z_2)$.

To complete the proof we need

Theorem 4 ([4]). Let $[M^n, f] \in \mathfrak{N}_n(X)$. Then $[M, f] = 0$ iff $\langle w_1^{i_1} w_2^{i_2} \cdots w_n^{i_n} \cup f^*(x), [M] \rangle = 0$ for all $x \in H^*(X; Z_2)$ where w_j is the jth Stiefel-Whitney class of M.

We have the maps

$$M/G \xrightarrow{q} E_G \times_G M \xrightarrow{1 \times \tau_M} E_G \times_G B(O,G)_n = BG \times BO_n$$

and will denote the composite by $\ell \times k$ where $\ell : M/G \longrightarrow BG$ and $k : M/G \longrightarrow BO_n$. We can identify ℓ from the diagram of principal fibrations below as the classifying map for the bundle $M \longrightarrow M/G$.

$$\begin{array}{ccccc} M & \xrightarrow{\bar{f} \times id} & E_G \times M & \longrightarrow & E_G \\ \downarrow & & \downarrow & & \downarrow \\ M/G & \xrightarrow{q} & E_G \times_G M & \longrightarrow & BG. \end{array}$$

To identify k note that we have a bundle map $T(M) \longrightarrow T(M)/G$ and hence $\tau_M : M \longrightarrow B(O,G)_n$ factors as $M \xrightarrow{\pi} M/G \xrightarrow{p} BO_n \xrightarrow{i} B(O,G)_n$. Crossing with E_G we get

$$M/G \xrightarrow{q} E_G \times_G M \xrightarrow{\pi \times \pi} B_G \times M/G \xrightarrow{id \times p} BG \times BO_n \xrightarrow{\approx} E_G \times_G B(O,G)_n$$

and since $\pi \circ q$ = identity, we have $k = p$, i.e. k classifies $T(M)/G = T(M/G) \oplus T_F/G$ where T_F is the tangent bundle along the fibre of $\pi : M \longrightarrow M/G$. T_F is induced by $M/G \xrightarrow{\ell} BG \xrightarrow{ad} BO_s$ where s = dimension G and $ad : G \longrightarrow O_s$ is the adjoint representation.

Thus the Stiefel-Whitney classes of M/G may be expressed in terms of the Whitney classes of $T(M)/G$, $k^*w_1, \ldots, k^*w_n$ and the Whitney classes of T_F which are in the image of ℓ^*. Thus all cohomology classes $w_1^{i_1} \cdots w_n^{i_n} \cup f^*x$ are in the image of $\ell \times k$ and h^* numbers vanish iff the numbers required by Theorem 4 vanish.

There is an oriented version of Theorem 3 namely let $\Omega_*(\mathcal{G})$ denote the cobordism group of oriented manifolds with free orientation preserving actions of G. Let $h^*(X) = H^*(E_G \times_G X; Z \oplus Z_2)$. Since M/G is an oriented manifold, there is a top class and h^* characteristic numbers are defined.

Theorem 5. Suppose $H_*(BG; Z)$ has all torsion of order two. Then $M = 0 \in \Omega_*(\mathcal{G})$ iff all h^* characteristic numbers vanish.

The proof uses Theorem 6 instead of Theorem 4 and is otherwise identical.

Theorem 6 [4]. If X is a CW complex with finite skeleta and all torsion in $H_*(X; Z)$ is two torsion then $[M^n, f] = 0 \in \Omega_n(X)$ iff all numbers of the form $\langle w_2^{i_2} \ldots w_n^{i_n} \cup f^*x, [M]\rangle$ (where $x \in H^*(X; Z_2)$ and w_j is the jth Stiefel-Whitney class of M) and of the form $\langle p_1^{i_1} \ldots p_r^{i_r} \cup f^*x, [M]\rangle$ (where $x \in H^*(X; Z)$, $r \leq n/4$ and p_j is the jth Pontrjagin class of M) vanish.

2) Let H be a closed subgroup of G and let $\mathcal{G}$ = compact manifolds M such that G_x, the isotropy group at x, is conjugate to H for all $x \in M$.

Theorem 7. $\mathfrak{N}_*(\mathcal{G})$ is determined by h^* characteristic numbers where $h^*(X) = H^*(E_K \times_K F(X, H); Z_2)$ where $K = \frac{N(H)}{H}$, $F(X, H)$ is the fixed point set of H on X.

The oriented version of this result is more complicated to state and is omitted.

3) $G = Z_2^k$ and $\mathcal{G}$ = all G manifolds. Let $h^*(X) = H^*(F(X, Z_2^k); Z_2)$. Since $F(M, Z_2^k)$ is a manifold, there is a top class.

Theorem 8. $N = 0 \in \mathfrak{N}_*(\mathcal{G})$ iff all h^* characteristic numbers vanish.

The proof uses the result of [10].

4) $G = Z_2^k$. Let $t_1, \ldots, t_k$ be coordinates for G with $t_i = \pm 1$ and let $\mathcal{G}$ = manifolds M such that G_x is given by $t_{i_1} = t_{i_2} = \ldots t_{i_r} = 1$ for $0 \leq r \leq k$. Let h^* be as in 3).

Theorem 9. $M = 0 \in \mathfrak{N}_*(\mathcal{G})$ iff all h^* characteristic numbers vanish.

The proof uses the result of [14].

These special cases may be treated in a uniform manner. If M is a G-manifold and $x \in M$ we have the slice representation $\rho(x) : G_x \longrightarrow O_m$ where m = dimension M - dimension G/G_x and an invariant neighborhood of x is equivariantly diffeomorphic to $G \times_{G_x} V(\rho(x))$ where $V(\rho(x))$ is $\mathbb{R}^m$ with the G action defined by $\rho(x)$. We shall specify a collection, $\mathcal{G}$, of G manifolds by listing the allowable isotropy groups and allowable slice representations.

Let $\mathcal{C}$ be a collection of pairs (H, ρ) where H is a closed subgroup of G and $\rho : H \longrightarrow O(m)$ is a representation. We assume that $\mathcal{C}$ satisfies

i) $(H, \rho) \in \mathcal{C} \Longleftrightarrow (gHg^{-1}, i_{g^{-1}} \circ \rho) \in \mathcal{C}$ where $i_{g^{-1}} : gHg^{-1} \longrightarrow H$ by $i_{g^{-1}}(x) = g^{-1}xg$

ii) $(H, \rho) \in \mathcal{C}$ and $x \in V(\rho) \Longrightarrow (H_x, \rho(x)) \in \mathcal{C}$

iii) $(H, \rho) \in \mathcal{C} \Longleftrightarrow (H, \rho+1) \in \mathcal{C}$ where $\rho+1 : H \xrightarrow{\rho} O(m) \subset O(m+1)$.

Definition. Let $\mathcal{G}(\mathcal{C})$ denote the category of G manifolds M such that $(G_x, \rho(x)) \in \mathcal{C}$ for all $x \in M$ and let $\mathfrak{N}(\mathcal{C})$ denote the associated unoriented co-bordism theory. Condition iii) guarantees $\partial W \in \mathcal{G}(\mathcal{C})$ if $W \in \mathcal{G}(\mathcal{C})$ and that $M \times I \in \mathcal{G}(\mathcal{C})$ if $M \in \mathcal{G}(\mathcal{C})$. i) and ii) are consistency conditions.

Definition. Let $\mathcal{C}' \subset \mathcal{C}$ satisfy i), ii), iii) and suppose that for $(H, \rho) \in \mathcal{C} - \mathcal{C}'$

i) there is no $(H, \rho') \in \mathcal{C}'$

ii) there is no $(H', \rho') \in \mathcal{C}$ and $x \in V(\rho')$ with $(H'_x, \rho(x)) = (H, \rho)$

Then $\mathcal{C}$, $\mathcal{C}'$ are said to be close.

Note 1). See [11] for alternative adjacency conditions.

Note 2). If $\mathcal{C}$, $\mathcal{C}'$ are close, $H \in \mathcal{C} - \mathcal{C}'$, $M \in \mathcal{G}(\mathcal{C})$, then $M_H = \{x \in M | G_x = H\}$ is a closed submanifold of M by ii) and $\frac{N(H)}{H}$ acts freely on M_H. Moreover, $F(M, H) = M_H \cup Y$ where Y is also closed in M.

We have an exact sequence

$$\mathfrak{N}_n(\mathcal{C}') \longrightarrow \mathfrak{N}_n(\mathcal{C}) \longrightarrow \mathfrak{N}_n(\mathcal{C}, \mathcal{C}') \longrightarrow \mathfrak{N}_{n-1}(\mathcal{C}')$$

as usual (see e.g. [5]).

Let $h^*(X) = \sum_i H^*(E_{K_i} \times_{K_i} F(X, H_i); Z_2)$ where the H_i are a complete set of representatives of the conjugacy classes of subgroups in $\mathcal{C} - \mathcal{C}'$ and $K_i = N(H_i)/H_i$. If $M \in \mathcal{G}(\mathcal{C})$, $\partial M \in \mathcal{G}(\mathcal{C}')$ then M_{H_i} is a closed submanifold contained in the interior of M and hence $E_{K_i} \times_{K_i} M_{H_i} \simeq M_{H_i}/K_i$ has the homotopy type of a closed manifold. We define a top class $[M]$ in

$$\sum_i H_*(E_{K_i} \times_{K_i} F(M, K_i); Z_2) = \sum_i H_*(E_{K_i} \times_{K_i} (M_{H_i} \cup Y_i); Z_2)$$

$$= \sum_i H_*(E_{K_i} \times_{K_i} M_{H_i}; Z_2) \oplus \sum_i H_*(E_{K_i} \times_{K_i} Y_i; Z_2)$$

$$= \sum_i H_*(M_{H_i}/K_i; Z_2) \oplus \sum_i H_*(E_{K_i} \times_{K_i} Y_i; Z_2)$$

by $[M] = \sum_i [M_{H_i}/K_i]$.

Theorem 10. $M = 0 \in \mathcal{N}_*(\mathcal{C}, \mathcal{C}')$ iff all h^* characteristic numbers vanish.

Corollary 11. If $\mathcal{N}_*(\mathcal{C}') \longrightarrow \mathcal{N}_*(\mathcal{C})$ is the zero map then $M = 0 \in \mathcal{N}_*(\mathcal{C})$ iff all h^* characteristic numbers vanish.

Note that Theorems 2, 7, 8, 9 are special cases of the corollary.

Proof. We first outline the proof.

Step 1). Reduce the study of $\mathcal{N}(\mathcal{C}, \mathcal{C}')$ to an ordinary unoriented bordism problem as in [4], [5], [8], [15], [11] so that Theorem 4 applies. That is, we construct spaces $B_1, \ldots, B_r$ such that $\mathcal{N}_n(\mathcal{C}, \mathcal{C}') \overset{\phi}{\cong} \sum_i \mathcal{N}_{t_i}(B_i)$ where $\phi(M) = \sum(N_i, f_i)$, $M \in \mathcal{N}_n(\mathcal{C}, \mathcal{C}')$ and $f_i : N_i \longrightarrow B_i$.

Step 2). Show that for $M \in \mathcal{N}(\mathcal{C}, \mathcal{C}')$, $E_{K_i} \times_{K_i} F(M, H_i) = \bigcup N_i$.

Step 3). Show that $E_K \times_K F(B(O, G)_n, H) = \bigcup_i (BO_{s_i} \times B_i')$.

Step 4). Produce maps $\theta_i : B_i \longrightarrow B_i'$ and show that $(\theta_i)_* : \mathcal{N}_*(B_i) \longrightarrow \mathcal{N}_*(B_i')$ is a monomorphism.

Step 5). Show that $\tau : M \longrightarrow B(OG)_n$ which induces $\tilde{\tau} : F(M, H) \times_K E_K \longrightarrow F(B(O, G)_n, H) \times_K E_K$ or equivalently $\tilde{\tau}_i : N_i \longrightarrow BO_{s_i} \times B_i'$ can be expressed as $\tilde{\tau}_i = a_i \times \theta_i \circ f_i$ where $a_i : N_i \longrightarrow BO_{s_i}$ is related to the tangent map for N_i as in Theorem 2.

Step 6). Show that the subring (image τ_i^*) where $\tau_i^* : H^*(BO_{s_i} \times B_i'; Z_2) \longrightarrow H^*(N_i; Z_2)$ contains the Stiefel-Whitney classes of N_i and hence h^* characteristic numbers vanish for M iff $(N_i, \theta_i \circ f_i) = 0 \in \mathcal{N}_*(B_i')$ which implies by step 4) that $(N_i, f_i) = 0 \in \mathcal{N}_*(B_i)$ and hence, by step 1), that $M = 0 \in \mathcal{N}_n(\mathcal{C}, \mathcal{C}')$.

It will become clear during the proof that is sufficient to consider the case where $\mathcal{C} - \mathcal{C}'$ contains only elements of the form $(H, \rho \oplus t)$ and the translates of such elements by elements of G. Here ρ is a representation of H containing no trivial factor and t denotes the trivial representation of dimension t. The more general situation only involves taking direct sums but is notationally awkward.

Hence, we shall do steps 1) - 6) under the assumption that $r = 1$ and drop the subscript i from B, t, N, f, s, B', θ, τ, α.

Step 1). An element of $\mathfrak{N}_n(\mathcal{C}, \mathcal{C}')$ may be represented by a compact manifold $W \in \mathcal{Y}(\mathcal{C})$ such that $\partial W \in \mathcal{Y}(\mathcal{C}')$. Let $W_{(H)} = \{x \in W \mid G_x = gHg^{-1}$ for some $g \in G\}$. $W_{(H)}$ is a closed G invariant submanifold of the interior of W. Let $\nu(W_{(H)}, W)$ denote the normal bundle of $W_{(H)}$ in W and let $D\nu$ (resp. $S\nu$) be the disk (resp. sphere) bundle of $\nu(W_{(H)}, W)$. An elementary argument shows that $D\nu = W \in \mathfrak{N}_n(\mathcal{C}, \mathcal{C}')$. If $x \in W_H \subset W_{(H)}$ then $T(W)_x = T(W_H)_x \oplus \nu(W_H, W_{(H)})_x \oplus \nu(W_{(H)}, W)_x$.

Since H operates trivially on W_H, $T(W_H)_x$ is a trivial representation of H of dimension $t + \dim K$, say. Because $W_{(H)} \xrightarrow{\pi} G/N(H)$ is a fibration with fibre W_H, $\nu(W_H, W_{(H)}) = \pi^*\nu(eN(H), G/N(H)) = W_H \times V$ where V is the representation of $N(H)$ on $T(G/N(H))_{eN(H)}$. We have another decomposition of $T(W)_x = T(Gx)_x \oplus T(Sx)$ where Sx is the slice at x and the slice representation at x, $T(Sx)$ is assumed to be a translate of $\rho \oplus t$. Since $V \subset T(Gx)_x$ and $T(W_H)_x$ is trivial, we have $\nu(W_{(H)}, W)_x$ is a translate of ρ. Hence, we may regard $\mathfrak{N}(\mathcal{C}, \mathcal{C}')$ as the cobordism group of G vector bundles E over manifolds $M = M_{(H)}$ such that the fibres are translates of ρ.

Let $\pi: E \longrightarrow M$ be such a bundle. Then $M = G \times_{N(H)} M_H$ and $E = G \times_{N(H)} E|M_H$; therefore the study of such G vector bundles is equivalent to the study of $N(H)$ vector bundles $E|M_H \longrightarrow M_H$ with fibre a translate of ρ. Now $M_H \longrightarrow M_H/K$ is a K principal bundle and the composite $E/M_H \longrightarrow M_H \longrightarrow M_H/K$ is a fibre bundle with fibre $N(H) \times_H V(\rho)$; if $P(H, \rho)$ denotes the structure group of such bundles then $\mathfrak{N}_n(\mathcal{C}, \mathcal{C}')$ may be identified with $\mathfrak{N}_t(B)$ where $t = n - p - \dim K$, $B = BP(H, \rho)$ is the classifying space of $P(H, \rho)$ and the correspondence sends $E \longrightarrow M$ to (N, f) where $N = M/G = M_H/K$ and $f: N \longrightarrow B$ is the classifying map for $E|M_H \longrightarrow M_H/K$.

Step 2). If $M \in \mathfrak{N}_n(\mathcal{C}, \mathcal{C}')$, $E_K \times_K F(M, H) = M_H/K \cup E_K \times_K Y = N^t \cup E_K \times_K Y$. Since the union is disjoint and Y, M_H are compact the extra piece $E_K \times_K Y$ will not disturb the argument.

Step 3). We must investigate $F(B(O, G)_n, H)$.

Lemma 12. $F(B(O, G)_n, H) = \bigcup_{i,j} F(\delta_{ij})$ where the $F(\delta_{ij})$ are the components of $F(B(O, G)_n, H)$, δ_{ij} are distinct n-dimensional representations of H and for fixed i, j, j' there is an $n \in N(H)$ such that $\delta_{ij'} = \delta_{ij} \circ i_n$ where $i_n: H \longrightarrow H$ is the automorphism of H given by conjugation by n. Furthermore, for each i $\bigcup_j F(\delta_{ij})$ is $N(H)$ invariant.

Proof. The components of $F(B(O, G)_n, H)$ are in 1-1 correspondence with equivariant homotopy classes of maps $f : G/H \longrightarrow B(O, G)_n$, the correspondence being given by $f \longrightarrow$ component of $F(B(O, G)_n, H)$ containing $f(eH)$. However, such homotopy classes are also in 1-1 correspondence with equivalence classes of G vector bundles, $\pi : E \longrightarrow G/H$, of dimension n which, in turn, are classified by the n-dimensional representation of H on $\pi^{-1}(eH)$, hence the first statement. For the second part, note that $N(H)$ acts on $F(B(O, G)_n, H)$, on maps $f : G/H \longrightarrow B(O, G)_n$ by $nf(gH) = f(gnH)$, and on representations of H by $n.\ \delta = \delta \circ i_{n^{-1}}$ and the correspondences defined above are easily seen to be equivariant. Since the identity component of $N(H)$, $N(H)_0$, acts trivially on the components of $F(B(O, G)_n, H)$ and $N(H)$ is compact, for each i there are only a finite number of j's. Note that for any equivariant map $f : G/H \longrightarrow B(O, G)_n$ the image of $N(H)/H$ is contained in $\bigcup_j F(\delta_{ij})$ for some i, hence $\bigcup_j F(\delta_{ij})$ is $N(H)$ invariant.

Lemma 13. $\bigcup_j F(\delta_{ij})$ is characterized up to $N(H)/H$ equivariant homotopy equivalence as the classifying space for $N(H)$ vector bundles over $N(H)/H$ (and hence $N(H)$) spaces with fibres having the H representations δ_{ij}.

Proof. Standard.

Lemma 14. Let δ_{1j}, δ_{2q} be representations as above. Suppose $\mathrm{Hom}_H(V(\delta_{1j}), V(\delta_{2q})) = 0$ for all j, q. Then $\bigcup_j F(\delta_{1j}) \times \bigcup_q F(\delta_{2q}) = \bigcup_{j,q} F(\delta_{1j} \oplus \delta_{2q})$.

Proof. First note that if $(j, q) \neq (j', q')$ then $\delta_{1j} \oplus \delta_{2q} \neq \delta_{1j'} \oplus \delta_{2q'}$ since $\mathrm{Hom}_H(V(\delta_{1j}), V(\delta_{2q})) = 0$ and hence the union on the right is, in fact, disjoint. Whitney sum induces an equivariant map $\psi: \bigcup_j F(\delta_{1j}) \times \bigcup_q F(\delta_{2q}) \longrightarrow \bigcup_{j,q} F(\delta_{1j} \oplus \delta_{2q})$ and we must produce an inverse for ψ. It is sufficient to show that every $N(H)$ vector bundle E over an $N(H)/H$ space with fibres $V(\delta_{1j} \oplus \delta_{2q})$ can be written as a Whitney sum of $N(H)$ vector bundles $E_1 \oplus E_2$ where E_1 has fibres $V(\delta_{1j})$, E_2 has fibres $V(\delta_{2q})$. Let $\gamma_1, \ldots, \gamma_r, \beta_1, \ldots, \beta_s$ be irreducible representations of H such that $\delta_{1j} = \sum_{i=1}^{r} r_{ij}\gamma_i$, $\delta_{2q} = \sum_{i=1}^{s} r_{ij}\beta_j$ and such that for fixed i not all $r_{ij} = 0$. Then $E_1 = \sum_{i=1}^{r} \mathrm{Hom}_H(V(\gamma_i), E) \otimes_{R_i} V(\gamma_i)$ where $R_i = \mathrm{Hom}_H(V(\gamma_i), V(\gamma_i))$, $E_2 = \sum_{i=1}^{s} \mathrm{Hom}_H(V(\beta_i), E) \otimes_{R_i} V(\beta_i)$ and $E = E_1 \oplus E_2$ as H vector bundles [1]. To complete the proof it is only necessary to show that E_1 and E_2 are $N(H)$ vector bundles. Consider the composite $(E_1)_x \subset E_x \xrightarrow{n} E_{nx} = (E_1)_{nx} \oplus (E_2)_{nx} \longrightarrow (E_2)_{nx}$. Now $(E_1)_x \approx V(\delta_{1j})$ for some j, $(E_2)_{nx} = V(\delta_{2q})$ for some q; but $n(E_1)_x = V(\delta_{ij} \circ i_{n^{-1}}) = V(\delta_{ij'})$ since $hnx = n(n^{-1}hn)x$. Thus $n(E_1)_x \longrightarrow E_{nx} \longrightarrow (E_2)_{nx}$ is zero since inclusion and projection are H equivariant and $\mathrm{Hom}_H(V(\delta_{1j'}), V(\delta_{2q})) = 0$. Hence E_1 is $N(H)$ invariant.

Lemma 15. Suppose $\rho_{1j} = s \oplus \bar{\rho}_{1j}$ where s denotes the trivial representation of dimension s and $\bar{\rho}_{1j}$ has no trivial subrepresentation. Then $\bigcup_j F(\rho_{1j}) = \bigcup_j F(\bar{\rho}_{1j}) \times B(O, K)_s$. Moreover $E_K \times_K \bigcup_j F(\rho_{1j}) = (E_K \times_K F(\bar{\rho}_{1j})) \times BO_s$.

Proof. For the first statement we must show that $F(s) = B(O, K)_s$ and then apply Lemma 14. $F(s)$ is the classifying space for $N(H)$ vector bundles E over $N(H)/H$ spaces with trivial fibre representation of H i.e. E is also a $K = N(H)/H$ space. But such bundles are classified by $B(O, K)_s$. For the second statement consider the diagram of fibrations

$$\begin{array}{ccccc}
E_K \times_K (\bigcup_j F(\bar{\rho}_{1j}) \times B(O, K)_s) & \longrightarrow & E_K \times_K B(O, K)_s & = & BK \times BO_s \\
\downarrow B(O, K)_s & & \downarrow B(O, K)_s & & \downarrow BO_s \\
E_K \times_K \bigcup_j F(\bar{\rho}_{1j}) & \longrightarrow & BK & = & BK
\end{array}$$

It was established in the proof of Theorem 2 that $E_K \times_K B(O, K)_s = BK \times BO_s$ hence the induced (vertical) fibration on the left is also a product. To complete step 3) we write each ρ_{ij} as $\bar{\rho}_{ij} \oplus s_i$ and then have $E_K \times_K F(B(O, G)_n, H) = \bigcup_i E_K \times_K (\bigcup_j F(\rho_{ij})) = \bigcup_i E_K \times_K (\bigcup_j F(\bar{\rho}_{ij} \oplus s_i)) = \bigcup_i E_K \times_K (\bigcup_j F(\bar{\rho}_{ij}) \times B(O, K)_{s_i}) = \bigcup_i (E_K \times_K \bigcup_j F(\bar{\rho}_{ij})) \times BO_{s_i} = \bigcup_i B_i' \times BO_{s_i}$.

Step 4). We must produce a map θ from $B = BP(H, \rho)$ to $\bigcup_i B_i'$ and show that is a monomorphism.

Lemma 16. B_i' is the classifying space for $P(H, \bar{\rho}_{ij})$ for any j.

Proof: $B_i' = E_K \times_K \bigcup_j F(\bar{\rho}_{ij})$. If $E \longrightarrow X$ is an $N(H)$ vector bundle with $G_x = H$ for each $x \in X$ and with fibre representations $\bar{\rho}_{ij}$ then there is a $h: X \longrightarrow \bigcup_j F(\bar{\rho}_{ij})$ classifying E by Lemma 13. Let $\ell : X \longrightarrow E_K$ be an equivariant map: ℓ and h are unique up to equivariant homotopy. We then have

$$\begin{array}{ccc} \ell \times h: \quad X & \longrightarrow & E_K \times \bigcup_j F(\bar{\rho}_{ij}) \\ \downarrow & & \\ X/K & \longrightarrow & E_K \times_K \bigcup_j F(\bar{\rho}_{ij}) = B_i' \end{array}$$

where $X/K \longrightarrow B_i'$ clearly classifies the fibre bundle $E \longrightarrow X/K$. Conversely, any such map induces a bundle of the required type.

If γ is a representation of $N(H)$ and $i : H \subset N(H)$ the inclusion then there is an equivariant map $\bigcup_j F(\rho_{ij}) \longrightarrow \bigcup_j F(\rho_{ij} \oplus \gamma \circ i)$ defined by Whitney sum of the universal bundle $\mu \longrightarrow \bigcup_j F(\rho_{ij})$ with the bundle $\bigcup_j F(\rho_{ij}) \times V(\gamma)$ where $N(H)$ operates as usual on the product. Note that $V(\gamma)$ must be an $N(H)$ representation space to have such a map. The map $\theta : B \longrightarrow B'$ that we are to investigate is defined by utilizing the above map with γ the representation of $N(H)$ on $V = T(G/N(H))_e$ and $\rho_{ij} = \rho$ as in step 1) yielding

$$\begin{array}{ccc} E_K \times_K F\{\rho\} & \longrightarrow & E_K \times_K F\{\rho \oplus \gamma \circ i\} \\ \| & & \\ \theta : B & \longrightarrow & B' \end{array}$$

where $F\{\rho\}$ indicates $\bigcup F(\rho \circ i_n)$ and the union is taken over all $n \in N(H)$. Recall that $B' = E_K \times_K \bigcup_j F(\bar{\rho}_{ij})$ where $\bar{\rho}_{ij}$ has no trivial subrepresentation. We must show that V has no points other than the origin fixed under H to conclude that B' is one of the B_i' as defined in step 3).

<u>Lemma</u> 17. $V = T(G/N(H))_{eN(H)}$ has no fixed points under H other than the origin.

Proof: H acts on $T(G)_e$ via the adjoint action and the fixed point set, F, is $T(Z(H))_e$ where $Z(H)$ is the centralizer of H in G. The H equivariant split epimorphism $T(G)_e \longrightarrow T(G/N(H))_{eN(H)}$ takes F to zero since $Z(H) \subset N(H)$ hence V has no fixed points.

Finally we have

Proposition 18. $\theta_* : \mathfrak{N}_*(B) \longrightarrow \mathfrak{N}_*(B')$ is a monomorphism.

Proof. We shall use the fact that B, B' are classifying space to restate the problem. $\mathfrak{N}_*(B)$ is the cobordism group of N(H) vector bundles $E \xrightarrow{\pi} M$ where K operates freely on M and the fibre representations of H are translates of ρ; $\mathfrak{N}_*(B')$ has a similar interpretation with fibre representations translates of $\rho + \gamma \circ i$. $\theta(E \longrightarrow M) = (E \times V \longrightarrow M)$ in this interpretation and we must prove:

> If there is a free K manifold W with N(H) bundle $E' \longrightarrow W$ such that $\partial W = M$, $E' | \partial W = E \times V$ then there exists a free K manifold Q and N(H) bundle $E'' \longrightarrow Q$ such that $\partial Q = M$ and $E'' | \partial Q = E$.

Note 1). The fibre representations ρ, $\rho \oplus \gamma \circ i$ have no trivial subrepresentations of H but the proof will not use that fact.

Note 2). If N(H) = H, Lemma 17 specializes to: $\mathfrak{N}_*(BO_k) \longrightarrow \mathfrak{N}_*(BO_{k+1})$, $\mathfrak{N}_*(U_k) \longrightarrow \mathfrak{N}_*(U_{K+1})$ and $\mathfrak{N}_*(Sp_k) \longrightarrow \mathfrak{N}_*(Sp_{k+1})$ are monomorphisms. To prove the theorem it would be sufficient to construct a N(H) equivariant bundle monomorphism $W \times V \longrightarrow E'$ which is compatible on $M = \partial W$ with the identification $E' | \partial M = E \times V$. Since N(H) equivariant bundle homomorphisms $W \times V \longrightarrow E'$ can be identified with N(H) equivariant sections of $\mathrm{Hom}_H(W \times V, E')$, we must try to construct a section s which is a monomorphism at each $w \in W$ and is the identity on ∂W. As a first step, we construct a section s_0 which has rank > 0 except on a submanifold $S_0 \subset$ interior of W. By constructing a new manifold W_1 by "blowing up along S_0" and modifying E' to get E_1 and a section s_1 which has rank > 0 for all $w \in W_1$. We then continue by "blowing up along S_1" = $\{x \in W | \text{rank } s_1 = 1\}$ etc. until we find $E_r \longrightarrow W_r$, $s_r : W_r \longrightarrow \mathrm{Hom}(W_r \times V, E_r)$ with rank $s_r = r$ where $r = \dim V$ i.e. until we have a monomorphism $W_r \times V \longrightarrow E_r$.

More formally, we shall construct, inductively, manifolds W_i and bundles $E_i \longrightarrow W_i$ such that $E_0 \longrightarrow W_0 = E' \longrightarrow W$ and $E_r \longrightarrow W_r = E'' \times V \longrightarrow Q$.

Given N(H) vector bundles A, B over a K space X let $\mathcal{H} = \mathrm{Hom}_H(A, B)$ and denote by $\mathcal{H}_j = \mathcal{H}_j(A, B) = \{T \in \mathcal{H} | \text{rank } T = j\}$. We then have $\mathcal{H} = \bigcup_{j=0}^{r} \mathcal{H}_j$; $\mathcal{H}_0$ may be identified with the zero section of $\mathcal{H}$; the $\mathcal{H}_j$'s are subfibrebundles of $\mathcal{H}$ with perhaps $\mathcal{H}_j = \phi$ for some values of j e.g. if all irreducible subrepresentations of H on A are even dimensional $\mathcal{H}_j = \phi$ for j odd.

If A, B are differentiable vector bundles, then the $\mathcal{H}_j$ are differentiable fibre bundles. $\mathcal{H}_i$ is a closed submanifold of $\mathcal{H} - \bigcup_{j<i} \mathcal{H}_j$. Of course $\mathcal{H}$, $\mathcal{H}_i$ are all N(H) spaces. Given a vector bundle A, D(A), S(A), P(A) will denote, respectively, the disk, sphere and projective bundle associated to A. L(A) will denote the line bundle associated to the Z_2 principal bundle $S(A) \longrightarrow P(A)$. If $X \subset Y$ are G manifolds, X compact, we shall identify $D(\nu(X, Y))$ with a neighborhood of X in Y.

We now proceed to construct W_i, E_i, s_i, S_i, $i = 0, 1, \ldots, r$ such that

i) K operates freely on W_i

ii) $\partial W_i = M$

iii) $E_i \longrightarrow W_i$ is an N(H) vector bundle

iv) $E_i | \partial W_i = E \times V$

v) $E_0 \longrightarrow W_0$ is just $E' \longrightarrow W$

vi) s_i is an N(H) equivariant section of $\mathrm{Hom}_H(W_i \times V, E_i)$

vii) $s_i | \partial W_i : \partial W_i \longrightarrow \mathrm{Hom}_H(\partial W_i \times V, E \times V)$ is the obvious map

viii) $s_i(W_i) \subset \mathcal{H} - \bigcup_{j<i} \mathcal{H}_i$ i.e. s_i has rank $\geq i$ at each point of W_i

ix) $s_i : W_i \longrightarrow \mathcal{H} - \bigcup_{j<i} \mathcal{H}_i$ is transverse to $\mathcal{H}_i$

x) $S_i = s_i^{-1}(\mathcal{H}_i) \subset$ interior W_i

xi) $W_{i+1} = W_i - D(\nu(S_i, W_i) \bigcup_{S(\nu(S_i, W_i)} D(L(\nu(S_i, W_i))$

xii) $E_{i+1} | W_i - D(\nu(S_i, W_i)) = E_i | W_i - D(\nu(S_i, W_i))$

Condition v) determines E_0, W_0 and vii) determines a section σ of $\mathcal{H}$ over ∂W. σ may be extended to all of W_0 (the fibre is contractible) and then averaged over N(H) to obtain an N(H) equivariant section $\bar{s}_0$. Conditions i) through viii) are then satisfied for $i = 0$ and we need only satisfy ix) since x), xi), xii) may be taken as definitions.

To make $\bar{s}_0$ transverse to $\mathcal{H}_0 = W_0 \subset \mathcal{H}$, one may pass to quotient spaces and apply the usual transversality theorem. Hence W_0, E_0, s_0, S_0 are defined. Assume now, inductively, that W_i, E_i, s_i, S_i have been constructed satisfying i) through xii). W_{i+1} is given by xi) or, alternatively, we may think of W_{i+1} as the quotient space of $W_i - D(\nu(S_i, W_i))$ under the relation $x \sim T(x)$ where $x \in S(\nu(S_i, W_i))$ and T is the antipodal involution of $S(\nu(S_i, W_i))$. Suppose that $T : S(\nu(S_i, W_i)) \longrightarrow S(\nu(S_i, W_i))$ can be covered by an N(H) bundle map $\tilde{T} : E_i | S(\nu(S_i, W_i)) \longrightarrow E_i | S(\nu(S_i, W_i))$ and such that $s_i(T(x))(T(x) \times v) = \tilde{T}(s_i(x)(x \times v))$.

Then the quotient space of $E_i|W_i - D(\nu(S_i, W_i))$ under the relation $y \sim \tilde{T}(y)$ $y \in E_i|S(\nu(S_i, W_i))$ defines a bundle $E_{i+1} \longrightarrow W_{i+1}$ and a section $\bar{s}_{i+1}$ of $\mathrm{Hom}_H(W_{i+1} \times V, E_{i+1})$ by the diagram

$$\begin{array}{ccccc} W_i - D(\nu(S_i, W_i)) & \longleftarrow & E_i|W_i - D & \xleftarrow{s_i} & (W_i - D) \times V \\ \downarrow & & \downarrow & & \downarrow \\ W_{i+1} & \longleftarrow & E_{i+1} & \xleftarrow{\bar{s}_{i+1}} & (W_{i+1}) \times V \end{array}$$

where the vertical maps are the projections onto the quotient spaces. The condition $s_i(T(x)) = \tilde{T} \circ s_i(x)$ imposed above guarantees that $\bar{s}_{i+1}$ is well defined. Clearly $\bar{s}_{i+1}(W_{i+1}) \subset \mathcal{H} - \bigcup_{j<i+1} \mathcal{H}_j$. Applying transversality as before yields s_{i+1} completing the construction (except for the construction of $\tilde{T}$). We shall construct $\tilde{T} : E_i|S \longrightarrow E_i|S$ where $S = S(\nu(S_i, W_i))$ as follows: $E_i|S = \pi_0^*(E_i|S_i)$, $\pi_0 : S \longrightarrow S_i$. But $s_i|S_i : S_i \times V \longrightarrow E_i|S_i$ has rank i at each point of S_i therefore $E_i|S_i = I \oplus I^\perp$ where I is the image of s_i and $I^\perp$ is the orthogonal complement to I with respect to some invariant riemannian metric. $S_i \times V$ also splits as $K \oplus K^\perp$ where $K = $ Kernel of s_i and $K^\perp$ is again the orthogonal complement. Let $T_1 : E_i|S_i \longrightarrow E_i|S_i$ be given by $T_1 = \mathrm{id}\, I \oplus -\mathrm{id}\, I^\perp$. Then $E_i|S_i$ is an $N(H) \times Z_2$ bundle and $\pi_0 : S(\nu(S_i, W_i)) \longrightarrow S_i$ is a $N(H) \times Z_2$ equivariant map where $T \in Z_2$ acts antipodally as before. Hence, $\pi_0^*(E_i|S_i)$ is an $N(H) \times Z_2$ vector bundle and $\tilde{T}$ is defined, and hence E_{i+1} is defined. It remains to check the condition $s_i(T(x)) = \tilde{T} \circ s_i(x)$.

Remark. An alternative description of E_{i+1} may be given as follows: regard W_{i+1} as constructed from W_i, S_i via condition xi). Then $E_{i+1}|W_i - D = E_i|W_i - D$ via xii). $E_{i+1}|D(L(\nu(S_i, W_i))) = (\pi_2 \circ \pi_1)^* I \oplus (\pi_2 \circ \pi_1)^* I^\perp \otimes \pi_1^* L(\nu(S_i, W_i))$ where $\pi_1 : D(L(\nu(S_i, W_i)) \longrightarrow P(\nu(S_i, W_i))$ and $\pi_2 : P(\nu(S_i, W_i)) \longrightarrow S_i$. $E_{i+1}|\partial(W_i - D) = E_i|S$ and $E_{i+1}|\partial D(L) = E_{i+1}|S = (\pi_2 \circ \pi_1)^* I \oplus (\pi_2 \circ \pi_1)^* I^\perp$ (since $\pi_1^* L(\nu(S_i, W_i))|S$ is trivial) $= \pi_0^* E_i|S_i = E_i|S$ therefore E_{i+1} is well defined. One easily checks that both definitions agree; we shall not use the second construction.

We now investigate $s_i|D : D \times V \longrightarrow E_i|D = \pi^*(E_i|S_i)$. For each $x \in D$, $s_i(x)$ is an H equivariant linear map $V \longrightarrow \pi^*(E_i|S_i)_x$ i.e. $s_i(x) : V \longrightarrow (E_i)_{\pi(x)}$. Hence, s_i determines an $N(H)$ equivariant fibre preserving map $\sigma : D \longrightarrow \mathrm{Hom}_H(S_i \times V, E_i|S_i) = \mathcal{H}$ and conversely, given such a σ, we can construct a section s_i of $\mathrm{Hom}_H(D \times V, E_i|D)$. Since $\sigma(D) \subset \mathcal{H} - \bigcup_{j<i} \mathcal{H}_j$ and σ is

transverse to $\mathcal{H}_i$, $d\sigma$ induces an isomorphism

$$\begin{array}{ccc} \widetilde{d\sigma} : \nu(S_i, D) & \longrightarrow & \nu(\mathcal{H}_i, \mathcal{H} - \bigcup_{j<i} \mathcal{H}_j) \\ \downarrow & & \downarrow \\ \sigma : S_i & \longrightarrow & \mathcal{H}_i \end{array}$$

Now let $p : \mathcal{H}_i \longrightarrow S_i$ be the projection. Then $p^*(S_i \times V) = \mathcal{H}_i \times V$ and $p^*(E_i | S_i)$ both split as $\mathcal{H}_i \times V = K \oplus K^{\perp}$, $p^*(E_i|S_i) = I \oplus I^{\perp}$ since $p^*\mathcal{H}$ has a canonical section, s, of constant rank i given by $s(h) = (h, h) \in p^*\mathcal{H}$. $\mathrm{Hom}_H(K, I^{\perp})$ can be embedded as an $N(H)$ subbundle of $p^*\mathcal{H}$ by sending $f : K \longrightarrow I^{\perp}$ to $\bar{f} : K \oplus K^{\perp} \longrightarrow K \longrightarrow I^{\perp} \longrightarrow I^{\perp} \oplus I$; furthermore $\nu(\mathcal{H}_i, \mathcal{H} - \bigcup_{j<i} \mathcal{H}_j) = \mathrm{Hom}_H(K, I^{\perp})$ a fact easily verified by dimension count. The addition map $+ : p^*\mathcal{H} \longrightarrow \mathcal{H}$ sending $(h, f) \longrightarrow h + f$, $h \in \mathcal{H}_i$, $f \in \mathcal{H}$ sends an open neighborhood, $\mathcal{O}$, of the zero section in $\mathrm{Hom}_H(K, I^{\perp})$ diffeomorphically onto a neighborhood $\mathcal{V}$ of $\mathcal{H}_i$ in $\mathcal{H} - \bigcup_{j<i} \mathcal{H}_j$. Let $\psi : \mathcal{V} \longrightarrow \mathcal{O}$ be the inverse and let D_1 be a disk bundle of smaller radius than D such that $\sigma(D_1) \subset \mathcal{V}$. Let σ_t be a homotopy of σ such that $\sigma_0 = \sigma$, $\widetilde{d\sigma_t} = \widetilde{d\sigma} : \nu(S_i, D) \longrightarrow \mathrm{Hom}(K, I^{\perp})$, $\sigma_t | S = \sigma | S$, $\sigma_t(D) \subset \mathcal{H} - \bigcup_{j<i} \mathcal{H}_j$, $\sigma_t^{-1}(\mathcal{H}_i) = S_i$ and σ_1 is such that the following diagram commutes

$$\begin{array}{ccc} \nu(S_i, D) & \xrightarrow{\widetilde{d\sigma}} & \mathrm{Hom}(K, I^{\perp}) \\ \cup & & \uparrow \psi \\ D_1 & \xrightarrow{\sigma} & \mathcal{V} \end{array}$$

σ_1 is just the linearization of σ ([13]). We have $\sigma_1(x) = \sigma_1(\pi(x)) + \widetilde{d\sigma}(x)$, $\sigma_1(T(x)) = \sigma_1(\pi(T(x))) + \widetilde{d\sigma}(T(x)) = \sigma_1(\pi(x)) + \widetilde{d\sigma}(-x) = \sigma_1(\pi(x)) - \widetilde{d\sigma}(x)$ and therefore $\widetilde{T} \circ \sigma_1(x) = \sigma_1(\pi(x)) - \widetilde{d\sigma}(x) = \sigma_1(T(x))$. Using D_1 and its associated sphere bundle to construct E_{i+1}, W_{i+1} we then have the necessary section condition fulfilled.

Step 5). $T(M)|M_H = T(M_H) \oplus \nu(M_H, M_{(H)}) \oplus \nu(M_{(H)}, M)|M_H$. By Lemma 17, H has no fixed points on V and therefore none on the bundle $\nu(M_H, M_{(H)}) = M_H \times V$. Also H has no fixed points on $\nu(M_{(H)}, M)|M_H$ since the fibre representations are all translates of ρ which has no trivial subrepresentations. On the other hand, H operates trivially on M_H and therefore on $T(M_H)$ also. Hence $\tau : M \longrightarrow B(O, G)_n$ restricts to $\tau/M_H : M_H \longrightarrow F(B(O, G)_n, H) = \bigcup_{ij} F(\rho_{ij})$ by Lemma 12 and $\tau(M_H) \subset F(\{\rho \oplus \gamma \circ i \oplus s\}) = F(\{\rho \oplus \gamma \circ i\}) \times B(O, K)_s$ by Lemma 15. Let the projections on the factors be denoted by ℓ, a respectively i.e. $T/M_H = \ell \times a : M_H \longrightarrow F(\{\rho \oplus \gamma \circ i\}) \times B(O, K)_s$. ℓ is the classifying map of

$\nu(M_{(H)}, M)|M_H \oplus \nu(M_H, M_{(H)})$ and α is the classifying map for $T(M_H)$. Crossing with $E_K \times_K$ yields $\ell \times \alpha$; $N \longrightarrow E_K \times_K F(\{\rho \oplus \gamma \circ i\}) \times BO_s$ by Lemma 15 where $\tilde{\ell}$ classifies $\nu(M_{(H)}, M)|M_H \oplus \nu(M_H, M_{(H)}) \longrightarrow N$ i.e. $\bar{\ell} = \theta \circ f$ and $\bar{\alpha}$ classifies $T(M_H)/K \longrightarrow N$ as in Theorem 2.

Step 6). The image of $\alpha^* : H^*(BO_s; Z_2) \longrightarrow H^*(N; Z_2)$ is the ring generated by the Stiefel-Whitney classes of $T(M_H)/K = T(M_H/K) \oplus T_F = T(N) \oplus T_F$ as in Theorem 2. The image of $H^*(BK; Z_2) \longrightarrow H^*(E_K \times_K F(B(O, G)_n, H); Z_2) \longrightarrow H^*(N; Z_2)$ contains the Stiefel-Whitney classes of T_F. Finally, $H^*(B'; Z_2) \xrightarrow{\cong} H^*(E_K \times_K F(\{\rho \oplus \gamma \circ i\}); Z_2) \longrightarrow H^*(E_K \times_K F(B(O, G)_n, H); Z_2) \longrightarrow H^*(N; Z_2)$ is just $(\theta \circ f)^*$. Thus image $\tilde{\tau}^* : h(B(O, G)_n) \longrightarrow H^*(N; Z_2)$ contains the Stiefel-Whitney classes of N and the pullback under $\theta \circ f$ of the cohomology of B' and thus by Theorem 4 $(N, \theta \circ f) = 0 \in \mathfrak{N}_*(B')$. But $(N, \theta \circ f) = \theta_*(N, f)$, $(N, f) \in \mathfrak{N}_*(B)$, and θ_* is a monomorphism by Proposition 18 hence $(N, f) = 0 \in \mathfrak{N}_*(B)$ and $M = 0 \in \mathfrak{N}(\mathcal{C}, \mathcal{C}')$ by step 1).

Note: $j : B(O, G)_n \longrightarrow B(O, G)_{n+1}$ induces $\bar{j} : B_i' \times BO_s \longrightarrow B_i' \times BO_{s+1}$; hence $j^* : h^*(B(O, G)_{n+1}) \longrightarrow h^*(B(O, G)_n)$ is an epimorphism and the characteristic numbers defined are cobordism invariants by Theorem 1.

2

We now consider actions of Z_2 on oriented manifolds. Following Connor [3], we denote by $\mathcal{O}_n(Z_2) = \mathcal{O}_n$ the cobordism group of triples (M, σ, T) where M is a closed n-dimensional orientable manifold, σ is an orientation of M and T is an involution on M. Note that we do not require that T preserve the orientation. Let $\mathcal{F}_n$ denote the cobordism group of triples (M, σ, T) as above where T acts freely on M. Finally let $A_n(k)$ denote the cobordism group of pairs $(E \longrightarrow M, \sigma)$ where $E \longrightarrow M$ is a vector bundle with fibre R^k over the closed n manifold M, E is orientable as a manifold and σ is an orientation of E or, equivalently, of the vector bundle $T(M) \oplus E$. Letting $\mathcal{A}_n = \sum_k A_{n-k}(k)$ one has a long exact sequence $\mathcal{F}_n \xrightarrow{i} \mathcal{O}_n \xrightarrow{j} \mathcal{A}_n \xrightarrow{\partial} \mathcal{F}_{n-1} \longrightarrow$ where $i(M, \sigma, T)$ forgets that T is free, $j(M, \sigma, T) = (\nu(F(M), M) \longrightarrow F(M), \tilde{\sigma})$ where $F(M)$ is the fixed point set of T, ν is the normal bundle of $F(M)$ in M, and $\tilde{\sigma}$ is the orientation induced from M. $\partial(E \longrightarrow M, \sigma) = (S(E), \bar{\sigma}, A)$ where $\bar{\sigma}$ is the induced orientation on the sphere bundle and A is the antipodal involution on each fibre of $S(E)$. The groups $\mathcal{O}_n$, $\mathcal{F}_n$, $\mathcal{A}_n$ decompose as $\mathcal{O}_n = \mathcal{O}_n^+ \oplus \mathcal{O}_n^-$,

$\mathcal{F}_n = \mathcal{F}_n^+ \oplus \mathcal{F}_n^-$, $\mathcal{A}_n = \mathcal{A}_n^+ \oplus \mathcal{A}_n^-$ where $(M, \sigma, T) \in \mathcal{O}_n^+$, (respectively $\mathcal{O}_n^-$) if T preserves the orientation (respectively, reverses orientation) and similarly for $\mathcal{F}_n^+$, $\mathcal{F}_n^-$. $\mathcal{A}_n^+ = \sum_k A_{n-2k}(2k)$, $\mathcal{A}_n^- = \sum_k A_{n-2k-1}(2k+1)$. The long exact sequence is easily seen to be a direct sum of the sequences $\mathcal{F}_n^+ \longrightarrow \mathcal{O}_n^+ \longrightarrow \mathcal{A}_n^+ \longrightarrow \mathcal{F}_{n-1}^+$ and $\mathcal{F}_n^- \longrightarrow \mathcal{O}_n^- \longrightarrow \mathcal{A}_n^- \longrightarrow \mathcal{F}_{n-1}^- \cdots$.

Lemma 19. $\mathcal{F}_n^- \approx \mathcal{N}_n$ and $i : \mathcal{F}_n^- \longrightarrow \mathcal{O}_n^-$ is the zero map. Hence $\mathcal{O}_n^- \longrightarrow \mathcal{A}_n^-$ is a monomorphism.

Proof: If Q is a manifold with boundary, $\hat{Q}$, the orientation cover of Q satisfies $\hat{Q}|\partial Q = \partial\hat{Q}$. Hence, $\theta : \mathcal{N}_n \longrightarrow \mathcal{F}_n^-$ defined by $\theta(N) = (\hat{N}, \sigma, T)$ where σ is the canonical orientation on $\hat{N}$ and T the bundle involution, is a well defined homomorphism. $\psi : \mathcal{F}_n^- \longrightarrow \mathcal{N}_n$ defined by $\psi(M, \sigma, T) = M/Z_2$ is similarly well defined and $\psi \circ \theta =$ identity. To see that $\theta \circ \psi$ is the identity we need only show that $\hat{N}$ is uniquely specified among 2-fold coverings by the conditions that $\hat{N}$ is orientable and T, the deck transformation, reverses orientation. Connected 2-fold coverings are classified by subgroups H of $\pi_1(N)$ of index 2. If N is not orientable, the subgroup H of orientation preserving loops is of index 2. A connected orientable 2-fold cover has an orientable quotient space iff the deck transformation is orientation preserving.

To see that $i : \mathcal{F}_n^- \longrightarrow \mathcal{O}_n^-$ is zero, let $(M, \sigma, T) \in \mathcal{F}_n^-$ and let $L \xrightarrow{\pi} M/Z_2$ be the line bundle associated to $M \longrightarrow M/Z_2$. Then $w_1(D(L)) = \pi^* w_1(T(L)|M/Z_2) = \pi^*(w_1(M/Z_2) \oplus w_1(L)) = 0$ since $w_1(L) = w_1(M/Z_2)$. Hence $D(L)$ is orientable and hence can be given an orientation such that $\partial D(L) = M$. The bundle involution on L and hence $D(L)$, restricts to T on M and is orientation reversing since the fixed point set M/Z_2 has codimension 1.

Lemma 20. $\rho : A_r(2k+1) \longrightarrow \mathcal{N}_r(BO_{2k+1})$ is an injection onto a direct summand.

Proof. Note first that $(E \longrightarrow M, \sigma) = (E \longrightarrow M, -\sigma) \in A_r(2k+1)$ since E admits an orientation reversing bundle map.

Define $\gamma : \mathcal{N}_r(BO_{2k+1}) \longrightarrow A_r(2k+1)$ as follows: given $E \longrightarrow W$ a $2k+1$ plane bundle over the manifold with boundary W, let $L = L(E)$ be the line bundle over W with $w_1(L) = w_1(E) + w_1(T(W))$. Then $L(E)|\partial W = L(E|\partial W)$ and $L \otimes E \longrightarrow W$ is $2k+1$ plane bundle which is orientable as a manifold. Let $\gamma(E \longrightarrow M) = (L(E) \otimes E \longrightarrow M, \sigma)$ where σ is any orientation for $L(E) \otimes E$.

If $M = \partial W$, $E' \longrightarrow W$, $E'|\partial W = E$ then $L(E')|\partial W = L(E)$ hence $(L(E') \otimes E' \longrightarrow W, \sigma)$ provides a cobordism of $\gamma(E \longrightarrow M)$ to zero. Both ρ and γ are maps of Ω_* modules and $\gamma \circ \rho$ is clearly the identity.

To obtain information about $\mathcal{O}_n^+$ we first note that $\mathcal{F}_n^+ = \Omega_n(BZ_2)$ [4] which by Burdick's theorem [7] is $\Omega_n \oplus \mathfrak{N}_{n-1}$. The map $\theta: \Omega_n \oplus \mathfrak{N}_{n-1} \longrightarrow \mathcal{F}_n^+$ is given by $\theta((M, \sigma), N) = (M \times Z_2, \sigma \times \bar{\sigma}, A) + (S(L\oplus 1), \sigma_1, T_1)$ where $\bar{\sigma}$ is the orientation on Z_2 given by both points having positive orientation. A acts on Z_2 nontrivially and trivially on M, L is the line bundle over N associated to $\hat{N} \longrightarrow N$, σ_1 is any orientation on the sphere bundle $S(L\oplus 1)$ and T is the antipodal involution. Clearly $i \circ \theta(\mathfrak{N}_{n-1})$ where $i: \mathcal{F}_n^+ \longrightarrow \mathcal{O}_n^+$ since $(S(L+1), \sigma_1, T_1) = \partial(D(L+1), \sigma, T_1)$. Letting $\delta: \mathcal{O}_n^+ \longrightarrow \Omega_n$ be given by $\delta(M, \sigma, T) = (M, T)$ we see that $\delta \circ i \circ \theta|\Omega_n : \Omega_n \longrightarrow \Omega_n$ is multiplication by 2 hence since Ω_n has only 2 torsion $\Omega_n \approx \Omega_n/\mathrm{Tor} \oplus \mathrm{Tor}$ where Tor is the torsion subgroup of Ω_*, $\delta \circ i \circ \theta$ injects Ω_n/Tor into $\mathcal{O}_n^+$. In fact, Rosensweig has shown that $i \circ \theta(\mathrm{Tor}) = 0$ [16]. Hence, the long exact sequences reduce to

$$0 \longrightarrow \Omega_*/\mathrm{Tor} \longrightarrow \mathcal{O}_n^+ \longrightarrow \mathcal{A}_n^+ \longrightarrow \mathrm{Tor} \oplus \mathfrak{N}_{n-1} \longrightarrow 0$$

$$0 \longrightarrow \mathcal{O}_n^- \longrightarrow \mathcal{A}_n^- \longrightarrow \mathfrak{N}_n \longrightarrow 0$$

Let $\mathcal{C} = \{(Z_2, a\rho \oplus b)(e, c)\}$, $\mathcal{C}' = \{(e, d)\}$ where ρ is the nontrivial representation of Z_2 and a, b, c, d are non negative integers. Then $\mathcal{G}(\mathcal{C})$ is the category of all Z_2 actions, $\mathcal{G}(\mathcal{C}')$ the category of free Z_2 actions. Following Connor, let $\mathfrak{N}_n(\mathcal{C}) = I_n$, $\mathfrak{N}_n(\mathcal{C}, \mathcal{C}') = \mathfrak{M}_n = \sum_k \mathfrak{N}_{n-k}(BO_k)$, $\mathfrak{N}_n(\mathcal{C}') = \mathfrak{N}_n(BZ_2)$. We have maps $r_1: \mathcal{F}_n \longrightarrow \mathfrak{N}_n(BZ_2)$, $r_2: \mathcal{O}_n \longrightarrow I_n$, $r_3: \mathcal{A}_n \longrightarrow \mathfrak{M}_n$ which forget the orientation.

<u>Proposition</u> 21 ([3] 5.8). If $x \in \mathcal{O}_n$, $r_2(x) = 0$ then $x = 2y \oplus i(z)$, $y \in \mathcal{O}_n$, $z \in \mathcal{F}_n^+$.

Proof. Let $x = x^+ + x^-$, $x^+ \in \mathcal{O}_n^+$, $x^- \in \mathcal{O}_n^-$. We have the maps of exact sequences

$$\begin{array}{ccccc} \mathcal{F}_n & \xrightarrow{i} & \mathcal{O}_n & \xrightarrow{j} & \mathcal{A}_n \\ \downarrow r_1 & & \downarrow r_2 & & \downarrow r_3 \\ \mathfrak{N}_n(BZ_2) & \xrightarrow{\tilde{i}} & I_n & \xrightarrow{\tilde{j}} & \mathfrak{M}_n \end{array}$$

Since $r_3|\mathcal{A}_n^-$ is a monomorphism by Lemma 20 and $j|\mathcal{O}_n^-$ is a monomorphism by Lemma 19, $r_3 \circ j = \tilde{j} \circ r_2$ is a monomorphism when restricted to $\mathcal{O}_n^-$.

But $r_2(x) = 0$ implies $\tilde{j} \circ r_2(x) = 0$ hence $\tilde{j} \circ r_2(x^+) + \tilde{j} \circ r_2(x^-) = 0$ and since $\tilde{j} \circ r_2(x^+) \in \sum_{k \text{ even}} \mathfrak{N}_{n-k}(BO_k)$, $\tilde{j} \circ r_2(x^-) \in \sum_{k \text{ odd}} \mathfrak{N}_{n-k}(BO_k)$ we have $\tilde{j} \circ r_2(x^-) = 0$ and hence $x^- = 0$. We may assume, therefore, that $x = x^+$.

Let x be represented by (M, σ, T). Then

$$(M, \sigma, T) = (M_1, \sigma_1, T_1) \cup (M_2, \sigma_2, T_2) \cup \ldots \cup (M_r, \sigma_r, T_r) \cup (N \times S^0, \sigma, A)$$

where M_i is connected and invariant and A interchanges the two copies of N. Since $N \times S^0 = i(z)$, $r_2 \circ i = \tilde{i} \circ r_1 = 0$, $r_2(x - i(z)) = 0$. We shall therefore establish the proposition $x - i(z) = 2y + i(z') \in \mathcal{O}_n$, i.e. assume $N = \emptyset$. Since $r_2(M, \sigma, T) = 0$, let $(W, T_0) \in I_n$, $\partial W = M$, $T_0 | \partial W = T$. Let $\hat{T}_0$ be the orientation preserving involution on $\hat{W}$, the orientation cover of W, which covers T_0 on W. The Z_2 line bundle $L \longrightarrow W$ associated to $\hat{W} \longrightarrow W$ is subordinate to the trivial representation of Z_2, i.e. for each $x \in W$ there is a $(Z_2)_x$ equivariant map $L_x \longrightarrow \mathbb{R}$, hence by [13], there is a Z_2 equivariant map $f: W \to RP^N$ classifying L. Altering f to make it transverse to $\mathbb{R}P^{N-1}$ produces an invariant closed submanifold $Q = f^{-1}(\mathbb{R}P^{N-1}) \subset$ interior of W such that $W - Q$ is orientable. Choosing an orientation for $W - D\nu(Q)$ produces a cobordism in $\mathcal{O}_n$ between $(M_1, \pm\sigma_1, T_1) \cup (M_2, \pm\sigma_2, T_2) \cup \ldots \cup (M_r, \pm\sigma_r, T_r)$ and $(S(\nu(Q)), \sigma, T_0 | S(\nu(Q)))$. Thus x differs from $(S(\nu(Q)), \sigma, T_0)$ by an element in $2\mathcal{O}_n$ namely 2 times those M_i which have the "wrong" orientation. Hence, it is sufficient to prove the theorem for $(\bar{Q}, \sigma, T_0) = (S(\nu(Q)), \sigma, T_0)$; we exploit the fact that $\bar{Q}$ has a free involution A which commutes with the Z_2 action defined by T_0. There are two cases to be considered: A preserves the orientation on $\bar{Q}$ or A reverses the orientation on $\bar{Q}$; we are assuming $Q | Z_2$ is connected, otherwise we argue with each component separately.

Case 1. A preserves the orientation of $\bar{Q}$. Then Q is oriented by the map $\bar{Q} \longrightarrow Q$. Since $\bar{Q} \longrightarrow Q$ is classified by $f|Q : Q \longrightarrow P^N$, there is an invariant codimension one submanifold of Q. Hence, one may perform the Dold construction as described in [7], equivariantly with respect to the Z_2 action to obtain a cobordism (X, σ_1, T_1, A) between $(\bar{Q}, \sigma, T_0)$ and $2(Q, \sigma, T)$. X is oriented, $\partial X = \bar{Q} - 2Q$, $T_1 | \bar{Q} = T_0$, $T_1 | Q = T$; ignoring the action of A we have shown $(\bar{Q}, \sigma, T_0) \in 2\mathcal{O}_n$ in this case.

Case 2. A reverses the orientation on $\bar{Q}$. Then $\bar{Q}$ is the orientation cover of Q. Let $c : I_n \longrightarrow \mathcal{O}_n$ be defined by $c(M, T) = (\hat{M}, \sigma, \hat{T})$ where $\hat{M}$ is the orientation cover of M, σ is the canonical orientation on $\hat{M}$ and $\hat{T}$ is the orientation preserving lift of T; clearly c is well defined. Then $(\bar{Q}, \sigma, T_0) = c(Q, T)$ in this case. We shall show that $c \equiv 0$. Up to cobordism, every element in I_n is a union of manifolds of the form $(P(E\oplus 1), T)$ where $E \longrightarrow F$ is a k plane bundle over the n-k dimensional manifold F, $k \neq 1$, and the action of T on P(E+1) is induced by the involution on $E \oplus 1$ given by $(e, t) \longrightarrow (e, -t)$. Choosing generators for $\mathfrak{N}_*(BO_k)$ as an $\mathfrak{N}_*$ module we may also express $P(E\oplus 1)$ as the quotient space of Z_2^{k+2} of $Y = \hat{X} \times S^{i_1} \times S^{i_2} \ldots \times S^{i_k} \times S^k$ where $i_1 \geq i_2 \ldots \geq i_k$; Z_2^{k+2} is generated by A_0, A_1, A_{k+1}; A_0 is the trivial extension to Y of an orientation reversing involution on the oriented manifold $\hat{X}$; A_j, $1 \leq j \leq k$, acts on $S^{i_j} \times S^k$ by $A_j(x, y_0, \ldots, y_k) = (-x, y_0, \ldots, -y_j, \ldots, y_k)$ $x \in S^{i_j}$, $y \in S^k$ and is extended trivially to Y; A_{k+1} is the extension to Y of the antipodal map on S^k. The involution on Y/Z_2^{k+2} is induced by the involution on S^k given by $(y_0, \ldots, y_k) \longrightarrow (y_0, \ldots, y_{k-1}, -y_k)$. We have then that $c(Y/Z_2^{k+2}, T) = (Y/G, \sigma, \hat{T})$ where $G \subset Z_2^{k+2}$ is the subgroup of orientation preserving elements. Let

$$A_i' = \begin{cases} A_i & \text{if } A_i \text{ preserves orientation} \\ A_iA_0 & \text{if } A_i \text{ reverses orientation} \end{cases}$$

$1 \leq i \leq k+1$. Then $G \approx Z_2^{k+1}$ is generated by $A_1', \ldots, A_{k+1}'$. If $A_j' \neq A_j$, for some $j \leq k$, then G acts freely on $W = \hat{X} \times S^{i_1} \times D^{i_j+1} \times S^{i_k} \times S^k$ and $\partial(W/G) = (Y/G)$. If $A_{k+1}' \neq A_{k+1}$, let $W = \hat{X} \times S^{i_1} \ldots \times S^{i_k} \times D^{k+1}$ then $\partial(W/G) = Y/G$ as before. $\hat{T}$ can be extended in an obvious way to W, W/G and hence $c(Y/Z_2^{k+2}, T) = 0$. If $A_i = A_i'$, $1 \leq i \leq k+1$, $c(Y/Z_2^{k+2}, T) = \partial(U \times S^{i_1} \ldots \times S^{i_k} \times S^k/G, \sigma, T)$ where U is an oriented manifold with boundary $\hat{X}$; note that $\hat{X}$ is in fact an oriented boundary since its Stiefel-Whitney numbers are all zero $(\hat{X} = \partial(\hat{X} \times_{Z_2} I)$ where $I = [-1, 1]$ and for $g \neq e \in Z_2$ we have $g(x, t) = (A_0x, -t))$ and the Pontryagin numbers of $\hat{X}$ are also zero since $\hat{X}$ admits an orientation reversing diffeomorphism A_0 and hence $2\hat{X} = 0 \in \Omega_*$. ∎ Let $\alpha : \mathfrak{N}_n(BO_{2k+1}) \longrightarrow \mathfrak{N}_n(BO_{2k})$, $\bar{\alpha} : A_n(2k-1) \longrightarrow A_n(2k)$ be given by taking the Whitney sum with a trivial bundle. Let $\beta : \mathfrak{N}_n(BO_{2k}) \longrightarrow \mathfrak{N}_{n-2k}(BO_{2k})$, $\bar{\beta} : A_n(2k) \longrightarrow \Omega_{n-2k}(BO_{2k})$ be the self intersection homomorphisms defined as follows: given $E \longrightarrow M$ is a 2k plane bundle over M, choose a section $s : M \longrightarrow E$ which is transverse regular to the zero section $M \subset E$; $\beta(E \longrightarrow M) = (E|s^{-1}(M) \longrightarrow s^{-1}(M))$. Note that $s^{-1}(M)$ has dimension n - 2k. If E is oriented as a manifold $\nu(s^{-1}(M), E) = \nu(s^{-1}(M), M) \oplus \nu(M, E)|s^{-1}(M) = E|s^{-1}(M) \oplus E|s^{-1}(M)$ hence $s^{-1}(M)$ has an oriented normal bundle and hence is oriented, thus $\bar{\beta}$ is defined.

Proposition 22. The following commutative diagram has exact rows.

$$\begin{array}{ccccccccc} 0 & \longrightarrow & A_n(2k-1) & \xrightarrow{\bar{\alpha}} & A_n(2k) & \xrightarrow{\bar{\beta}} & \Omega_{n-2k}(BO_{2k}) & \longrightarrow & 0 \\ & & \downarrow \rho & & \downarrow \rho_0 & & \downarrow r & & \\ 0 & \longrightarrow & \mathfrak{N}_n(BO_{2k-1}) & \xrightarrow{\alpha} & \mathfrak{N}_n(BO_{2k}) & \xrightarrow{\beta} & \mathfrak{N}_{n-2k}(BO_{2k}) & \longrightarrow & 0 \end{array}$$

Proof. α is a monomorphism by Lemma 18 and ρ is a monomorphism by Lemma 20; hence $\bar{\alpha}$ is a monomorphism.

If $\bar{\beta}(E \longrightarrow M, \sigma) = 0$ then $S = s^{-1}(M) = \partial W$ as oriented manifolds and $E|S$ extends to $E' \xrightarrow{\pi} W$. Since $\nu(S, M) = E|S$, we can construct a cobordism of $(E \longrightarrow M, \sigma)$ by forming $Y = M \times I \cup_f D(E')$ where $f : D(\nu(S, M)) \times 0 \longrightarrow D(E'|\partial W)$. Let $E'' \longrightarrow Y$ by $E''|M \times I = E \times I$, $E''|D(E') = \pi^* E'$, where $\pi : D(E') \longrightarrow W$. Note that $E''|M \times I$ is oriented as is $E''|D(E')$ and that f is orientation reversing hence E'' is oriented. $\partial(E'' \longrightarrow Y, \sigma) = (E \longrightarrow M, \sigma) + (E''|M' \longrightarrow M', \sigma)$ where $M' = M - D\nu(S, M) \times 0 \cup_g S(E')$, $g : S\nu(S, M) \longrightarrow S(E'|\partial W)$. We shall show that $E''|M'$ has a non zero section i.e. $(E''|M' \longrightarrow M', \sigma)$ is in the image of $\bar{\alpha}$. Let $\bar{s} : D(E') \longrightarrow \pi^* E'$ be given by $\bar{s}(x) = (x, x) \in \pi^* E'$; $\bar{s}|S(E')$ is non zero. Let $s_t : M \longrightarrow E$ be a homotopy of s_1 such that $s_t^{-1}(M) = S$, $s_0|D\nu(S, M) = \bar{s}|D(E'|S)$. Such a homotopy is just a linearization of s_1. Letting $s : M \times I \longrightarrow E \times I$ by $s(m, t) = s_t(m)$ defines a section of $E'' \longrightarrow Y$ which is non zero on M'.

If one omits the orientation questions in the above discussion, then we have also shown that $\beta(x) = 0$ implies $x = \alpha(y)$ for some $y \in \mathfrak{N}_n(BO_{2k-1})$.

To show that $\bar{\beta}$ is onto we shall construct a splitting $\gamma : \Omega_{n-2k}(BO_{2k}) \longrightarrow A_n(2k)$. Let $\pi : E \longrightarrow S$, S oriented, represent an element of $\Omega_{n-2k}(BO_{2k})$. We shall construct a bundle $E' \longrightarrow P(E \oplus 1)$ such that E' is oriented as a manifold and such that there exists a section $s : P(E \oplus 1) \longrightarrow E'$ transverse to the zero section $P(E \oplus 1)$ and such that $s^{-1}(P(E \oplus 1)) = S$ and $E'|S = E$. To that end let $\pi : D(E) \longrightarrow S$. Then $\pi^* E$ has a natural orientation induced from S. Let $\bar{s} : D(E) \longrightarrow \pi^* E$ by $\bar{s}(x) = (x, x)$; $\bar{s}$ is transverse to the zero section $D(E)$, $\bar{s}^{-1}(D(E)) = S$, and $\pi^* E|S = E|S$. We must "close up" $D(E)$ to complete the definition of $\bar{\gamma}$. Let $T : S(E) \longrightarrow S(E)$ be the antipodal involution; then the quotient space of $D(E)$ under the equivalence relation $x \sim T(x)$, $x \in S(E)$, is $P(E \oplus 1)$. To construct $E' \longrightarrow P(E \oplus 1)$ it will be sufficient to cover T by a bundle map $\tilde{T} : \pi^* E|S(E) \longrightarrow \pi^* E|S(E)$ such that $\tilde{T}(\bar{s}(x)) = \bar{s}(T(x))$ and $\tilde{T}$

is orientation reversing on the manifold $\pi^*E|S(E)$. Now $\pi^*E|S(E) = \pi_0^*E$ where $\pi_0 : S(E) \longrightarrow S$ and π_0 factors as $\pi_1 \circ \pi_2$ where $\pi_2 : S(E) \longrightarrow P(E)$ and $\pi_1 : P(E) \longrightarrow S$. But $\pi_1^*E = L(E) \oplus E_1$ where E_1 is a 2k-1 dimensional bundle. $\pi_0^*E = \pi_2^*\pi_1^*E = \pi_2^*(L(E) \oplus E_1) = 1 \oplus \pi_2^*E_1$ and $\bar{s}|S(E)$ can be expressed as $\bar{s}(x) = (1, 0) \in 1 \oplus \pi_2^*E_1$. The map $\pi_2 : S(E) \longrightarrow P(E)$ may be regarded as a Z_2 equivariant map if we let T act trivially on P(E) and the bundle E_1 may be regarded as a Z_2 bundle if we let $\tilde{T}_1 : E_1 \longrightarrow E_1$ by $\tilde{T}_1(e) = -e$. Then $\pi_2^*E_1$ is a Z_2 bundle over the Z_2 space S(E) as is $1 = S(E) \times R$ with the trivial action on R. Thus π_0^*E has the required orientation reversing involution and $\bar{\gamma}$ is defined. We may define γ in precisely the same way and the splittings will then commute.

We now have some immediate corollaries

Corollary 23. If $\bar{\beta}(x) = 0$, $r_2(x) = 0$ then $x = 0 \in A_n(2k)$.

Corollary 24. $A_n(2k) \approx A_n(2k-1) \oplus \Omega_{n-2k}(BO_{2k})$ as Ω_* modules.

Corollary 25. All torsion in $A_n(2k)$ is two torsion and r_2 maps the torsion subgroup of $A_n(2k)$ monomorphically into $\mathfrak{N}_n(BO_{2k})$.

Corollary 26. If $x \in \mathcal{O}_n$, $r_2(x) = 0$, $\delta(x) = 0$, $\bar{\beta} \circ j(x) = 0$ then $x = 0$.

Proof. $\delta : \mathcal{O}_n \longrightarrow \Omega_n$ is a monomorphism on the image of i.

To define characteristic numbers for oriented G manifolds requires a classifying space for oriented G vector bundles $B(SO, G)_n$. Since $B(O, G)_n$ has the ordinary homotopy type to BO_n by Lemma 3, we may form the universal cover $\widehat{B(O, G)}_n \approx BSO_n$ and, using the fact that $B(O, G)_n$ has fixed points, lift the action of G to $\widehat{B(O, G)}_n$. The liftings are in one-to-one correspondence with the homomorphisms of G into Z_2. Alternatively, one may construct the universal principal O_n bundle, P, over $B(O, G)_n$ which has a $G \times O_n$ action and note that P/SO_n is an $H = G \times Z_2$ space (namely $\widehat{B(O, G)}_n$). For each homomorphism $\omega : G \longrightarrow Z_2$, we get a distinct G space $B(SO, G)_n^{\omega}$ by mapping $G \longrightarrow H$ via $g \longrightarrow (g, \omega(g))$ and taking the induced action. If $G = Z_2$ there are only two homomorphisms and we let $B(SO, G)_n^+$ (resp. $B(SO, G)_n^-$) correspond to the trivial (resp. nontrivial) homomorphism. If $M \in \mathcal{O}_n^+$ (resp. $\mathcal{O}_n^-$) the orientation on M defines a lifting of $\tau_M : M \longrightarrow B(O, G)_n$ to $\tilde{\tau}_M : M \longrightarrow B(SO, G)_n^+$ (resp. $\tilde{\tau}_M : M \longrightarrow B(SO, G)_n^-$) which is equivariant. Characteristic numbers are defined in an entirely analogous way. In particular, the equivariant map $B(SO, G)_n \longrightarrow B(O, G)_n$ allows us to apply the numbers defined previously.

Let $h^*(X) = H^*(X; Q) \oplus H^*(E_{Z_2} \times_{Z_2} X; Z_2)$. Then if $M \in \mathcal{I}_n$ there is a top class defined in $h_*(X)$ and we have

Theorem 27. $M = 0 \in \mathcal{I}_n$ iff all h^* characteristic numbers vanish.

The proof is similar to the proof of Theorem 2 and is omitted.

To define characteristic numbers for $\mathcal{O}_n$, we must consider the category $\mathcal{G}_0$ of pairs (X, x) where X is a Z_2 space and $x \in H^1(F(X, Z_2); Z_2)$ and maps $f : (X, x) \longrightarrow (Y, y)$ where f is equivariant and $f^*y = x$. If (M, σ, T) represents an element of $\mathcal{O}_n$, we may regard M as being in $\mathcal{G}_0$ via $M \longrightarrow (M, w_1)$ where w_1 is the first Stiefel-Whitney class of $F(M, Z_2)$. Let $h^*((X, x)) = H^*(X; Q) \oplus H^*(F(X; Z_2); Z_2) \oplus H^*(F(X, Z_2); \{Q\})$ where $\{Q\}$ denotes the local coefficient system on $F(X, Z_2)$ defined by x. Then $h_*((M, w_1)) = H_*(M; Q) \oplus H_*(F(M; Z_2); Z_2) \oplus H_*(F(M, Z_2); \{Q\})$ has a "top class". We shall see that $\tau_M : M \longrightarrow B(SO, G)_n$ is a map in $\mathcal{G}_0$ and hence h^* characteristic numbers are defined.

Theorem 28. $(M, \sigma, T) = 0 \in \mathcal{O}_n$ iff all h^* characteristic numbers vanish.

Proof. By theorem 8, $r_2(M, \sigma, T) = 0$. Also $\delta(M, \sigma, T) = 0$ hence we need only show $\bar{\beta} \circ j(M, \sigma, T) = 0$ by cor. 26, Since ρ_0 maps the torsion of $\mathcal{A}_n$ monomorphically, it is sufficient to show that the Pontrjagin numbers of $\bar{\beta} \circ j(M, \sigma, T)$ are all zero. $\bar{\beta} \circ j(M, \sigma, T)$ is represented by $(E \longrightarrow S, \sigma)$, $S \subset F(M, Z_2) \subset M$ where $\nu(S, F) = \nu(F, M)|S = E$ and the orientation on S is induced by that on M. Since $H^*(BO_{2k}; Q) = Q[\bar{p}_1, \ldots, \bar{p}_k]$ [17] where $\bar{p}_i$ is the ith Pontrjagin class of E, we must show that all numbers of the form $\langle [S], p_1^{i_1} \cdots p_s^{i_s - j_1} \bar{p}_1 \cdots \bar{p}_k^{-j_k} \rangle$ are zero where p_i is the ith Pontrjagin class of S. Alternatively, we could use $\bar{p}_i$'s and $\tilde{p}_i$'s where the $\tilde{p}_i$'s are the Pontrjagin classes of $T(S) \oplus E$ since one can solve for p_i's, $\bar{p}_i$'s in terms of $\bar{p}_i$'s, $\tilde{p}_i$'s. But $\bar{p}_i = p_i(\nu(F, M)|S) = \alpha^* p_i(\nu(F, M))$ and $\tilde{p}_i = p_i(T(S) \oplus E) = p_i(T(S) \oplus \nu(S, F)) = p_i(T(F)|S) = \alpha^* p_i(F)$ where $\alpha : S^{n-4k} \longrightarrow F^{n-2k}$ is the inclusion. Let $\alpha^! : H_p(F; \{Q\}) \longrightarrow H_{p+2k}(S; Q)$ be defined by $H_p(F; \{Q\}) \overset{P}{\approx} H^{n-2k-p}(F; Q) \xrightarrow{\alpha^*} H^{n-2k-p}(S, Q) \overset{P}{\approx} H_{p+2k}(S; Q)$ where P stands for Poincare duality (with local coefficients). Then $\alpha^!(\{F\}) = [S]$ since $P(\{F\}) = 1$, $\alpha^*(1) = 1$, $P(1) = [S]$. We are considering the numbers

$\langle [S], a^*\bar{p}_I(\nu(F, M)) \cup a^* p_J(F)\rangle = \langle a^!\{F\}, a^*(\bar{p}_I \cup p_J)\rangle = \langle \{F\}, a_!(a^*(\bar{p}_I \cup p_J) \cup 1)\rangle =$ $\langle \{F\}, \bar{p}_I \cup p_J \cup a_!(1)\rangle$ where $a_!$ is defined in an analogous manner and $p_I = p_1^{i_1} \cdots p_s^{i_s}$ etc., and we have used the standard properties of the umkehr homomorphisms. To complete the proof we must investigate $h^*(B(SO, G)_n)$; in fact, it is sufficient to investigate $H^*(F(B(SO, G)_n^+); \{Q\})$. Since Z_2 preserves orientation on $B(SO, G)_n^+$, $F(B(SO, G)_n^+)$ is just the inverse image of the components of $F(B(O, G)_n)$ of the form $BO_{n-2k} \times BO_{2k}$ where the representation is trivial on the first factor and nontrivial on the second. Therefore $F(B(SO, G)_n^+) =$ $\bigcup_k BSO_{n-2k} \times_{Z_2} BSO_{2k}$ where Z_2 acts by reversing orientation on both factors. Let x be the nontrivial element of $H^1(F(B(SO, G)_n^+; Z_2)$.

Lemma 29. $(\tau_M|F)^* x = w_1(F)$.

Proof. The composite $\pi_1 \circ \tau_M|F : F \longrightarrow BSO_{n-2k} \times_{Z_2} BSO_{2k} \longrightarrow BO_{n-2k}$ classifies $T(F)$, $\pi_2 \circ \tau_M|F : F \longrightarrow BO_{2k}$ classifies $\nu(F, M)$. From the diagram below we see that $\pi_1^* w_1$ is the nontrivial element of $H^1(BSO_{n-2k} \times_{Z_2} BSO_{2k}; Z_2)$ since $BSO_{n-2k} \times BSO_{2k}$ is connected.

$$\begin{array}{ccccc} BSO_{n-2k} \times BSO_{2k} & \longrightarrow & BSO_{n-2k} & \longrightarrow & E_{Z_2} \\ \downarrow & & \downarrow & & \downarrow \\ BSO_{n-2k} \times_{Z_2} BSO_{2k} & \xrightarrow{\pi_1} & BO_{n-2k} & \xrightarrow{w_1} & BZ_2 \end{array}$$

$H^*(X; \{Q\})$ is a module over $H^*(X; Q)$ and the module structure is natural, i.e. if $f : X \longrightarrow Y$, $a \in H^*(Y; \{Q\})$, $b \in H^*(Y, Q)$ then $f^*(a \cdot b) = f^*(a) f^*(b)$. We are trying to show that there is a class $z \in H^*(F(B(SO, G)_n^+; \{Q\})$ with $(\tau_M|F)^* z =$ $\tau^* z = p_I(F) \cup p_J(\nu(F)) \cup a_!(1)$. By [17] we have $H^*(BSO_{n-2k} \times_{Z_2} BSO_{2k}; Q) =$ $Q[p_1, p_2, \ldots, p_{n-2k/2}, \bar{p}_1, \ldots, \bar{p}_k, X\bar{X}]$ when n is even and $Q[p_1, p_2 \cdots p_{n-2k/2}, \bar{p}_1, \ldots, \bar{p}_k]$ if n is odd where $\tau^* p_i = p_i(F)$, $\tau^* \bar{p}_i =$ $p_i(\nu(F))$. ($X\bar{X}$ is the product of the Euler classes and is unstable i.e. not in the image of j^* as required by Theorem 1.) Hence, to prove the theorem, it is sufficient to show that $a_!(1) = \tau^* e$ for some $e \in H^{2k}(BSO_{n-2k} \times_{Z_2} BSO_{2k}; \{Q\})$.

Let $\tilde{G}_p(\mathbb{R}^t)$ (resp. $G_p(\mathbb{R}^t)$) be the grassman manifold of oriented (resp. unoriented) p planes in $\mathbb{R}^t$ and let $\tilde{\mu}$ (resp. μ) be the canonical p-plane bundle over $\tilde{G}_p(R^t)$.

Lemma 30. The self intersection of $G_p(\mathbb{R}^t)$ in μ is $G_p(R^{t-1})$ and similarly for $\tilde{G}_p(\mathbb{R}^t)$.

Proof. Let $M(p, t)$ denote p by t matrices of rank p with the obvious action of GL_p. We then have the diagram

$$\begin{array}{ccccc} M(p,t)\times\mathbb{R}^p & \longrightarrow & M(p,t)\times_{GL_p}\mathbb{R}^p & = & \mu \\ {\scriptstyle s_1}\uparrow\downarrow & & {\scriptstyle s_2}\uparrow\downarrow & & \downarrow \\ M(p,t) & \longrightarrow & M(p,t)/GL_p & = & G_p(\mathbb{R}^t) \end{array}$$

where $s_1(x) = (x, t^{th}$ row of $x)$ and s_2 is induced by s_1 since s_1 is equivariant. s_1 and hence s_2 are transverse to the zero section and vanish on $M(p, t-1)$, $G_p(\mathbb{R}^{t-1})$ respectively.

Corollary 31. If $E \longrightarrow F$ is classified by $f: F \longrightarrow \tilde{G}_p(\mathbb{R}^t)$ and if f is transverse to $G_p(\mathbb{R}^{t-1}) \subset G_p(\mathbb{R}^t)$ then $f^{-1}(G_p(\mathbb{R}^{t-1})) = S \subset F$.

To complete the proof of the theorem approximate BSO_p by $\tilde{G}_p(\mathbb{R}^t)$ with $t > n$. We then have the diagram

$$\begin{array}{ccccc} F^{n-2k} & \longrightarrow & \tilde{G}_{n-2k}(\mathbb{R}^t)\times_{Z_2}\tilde{G}_{2k}(\mathbb{R}^t) & \longrightarrow & G_{2k}(\mathbb{R}^t) \\ a\cup & & \tilde{a}\cup & & \cup \\ S^{n-4k} & \longrightarrow & \tilde{G}_{n-2k}(\mathbb{R}^t)\times_{Z_2}\tilde{G}_{2k}(\mathbb{R}^{t-1}) & \longrightarrow & G_{2k}(\mathbb{R}^{t-1}) \end{array}$$

Since $S = \tau^{-1}(\tilde{G}_{n-2k}(\mathbb{R}^t) \times_{Z_2} (\mathbb{R}^{t-1}))$ we have $a_!(\tau|S)^*(x) = \tau^*(\tilde{a}_!(x))$ and hence for $x = 1$ $a_!(\tau|S)^*(1) = a_!(1) = \tau^*(\tilde{a}_!(1)) = \tau^*(e)$.

Remark. Reasoning as above one can show that $H^*(BSO_{n-2k} \times_{Z_2} BSO_{2k}; \{Q\})$ is a free module over $H^*(BSO_{n-2k} \times_{Z_2} BSO_{2k}; Q)$ on one generator in dimension $2k$ - the Euler class. Hence, one may identify $a_!(1) \in H^{2k}(F; \{Q\})$ as the Euler class, X, of $\nu(F, M)$ (with twisted coefficients if F is not orientable).

BIBLIOGRAPHY

1. Atiyah, M. F., K-theory, W. A. Benjamin, Inc., New York, 1967.

2. Bredon, G. E., Equivariant Cohomology Theories, Lecture Notes in Math. Vol. 34. Berlin-Heidelberg-New York: Springer, 1967.

3. Conner, P. E., Lectures on the Actions of a Finite Group, Lecture Notes in Math. Vol. 73. Springer, 1968.

4. Conner, P. E. and Floyd, E. E., Differentiable periodic maps, Berlin-Heidelberg: Springer, 1964.

5. ——————————, Maps of odd period, Ann. of Math., 84 (1966).

6. tom Dieck, T., Faserbündel mit Gruppen operation, Archiv der Math., 20, 1969.

7. Hirzebruch, F. and Jänick, K., Involutions and Singularities, Int. Colloquium on Algebraic Geometry, Bombay, 1968.

8. Lee, C. N., Equivariant Homology Theories, Proc. of the Conference on Transformation Groups at Tulane, pp. 237-244, 1967.

9. Pontrjagin, L. S., Characteristic cycles on differentiable manifolds, Math. Shor. (N.S.) 21 (63), 1947.

10. Stong, R. E., Equivariant bordism and $(Z_2)^k$ actions, Duke Math. Journal, Vol. 37, No. 4, 1970.

11. ——————————, Unoriented bordism and actions of finite groups, AMS Memoirs No. 103, 1970.

12. Thom, R., Quelques proprietés globales der varietés differentiable, Comm. Math. Helv., 28, 1954.

13. Wasserman, A. G., Equivariant Differential Topology, Vol. 8 (1968), pp. 127-150.

14. ——————————, A product theorem for equivariant cobordism, Proceedings of 1969 Georgia Topology Conference.

15. ——————————, Cobordism of group actions, BAMS, 72, 1966.

16. Rosenzweig, H., Bordism groups of all orientation preserving involutions, Dissertation, Univ. of Virginia, 1967.

17. Milnor, J., Lectures on characteristic classes, Princeton University, 1957.

COBORDISM OF DIFFEOMORPHISMS OF (k-1)-CONNECTED 2k-MANIFOLDS

SANTIAGO LOPEZ DE MEDRANO

Universidad Nacional Autónoma de México

Introduction

Consider pairs (M,f), where M is a compact smooth manifold and $f:M \longrightarrow M$ is a diffeomorphism. Such a pair corresponds with a smooth action of $\mathbb{Z}$ on M, where the action of a generator is given by f. The study and the classification of such pairs, even for a given manifold M, have proved to be very difficult problems. A vast literature exists dealing with the study of delicate properties of diffeomorphisms, such as the nature of the set of fixed points, the set of non-wandering points, etc., aiming at giving at least a generic classification of them. Many of the difficulties found in this subject are related to the fact that $\mathbb{Z}$ is not a compact group.

One can hope that a coarser classification of these pairs (M,f), such as a cobordism classification, might be useful as a partial step in the solution of these problems. We know that the methods and results of cobordism theory have been very useful in the study and classification of manifolds (Browder, Novikov, Sullivan, Wall, and many others), a development that has led to the solution of very delicate topological problems, such as the Annulus Conjecture and the Triangulation Problem (Kirby-Siebenmann). On the other hand, equivariant cobordism theory has been very useful in the study of smooth actions of finite and compact groups on manifolds (Conner-Floyd, Atiyah-Bott-Singer, Hirzebruch, etc.), and many beautiful results have been obtained relating cobordism invariants of such actions with properties of their fixed point sets. It is not clear what properties of the fixed point set, etc., of a diffeomorphism can be cobordism invariants, since they vary widely even within an isotopy class, but still it seems plausible that a cobordism classification of diffeomorphisms, or at least the study of their cobordism invariants, could produce useful information for the study of their more delicate properties.

Unfortunately, the cobordism classification of diffeomorphisms seems to be in itself a problem of a much higher degree of difficulty than any other cobordism problem that has been considered. In particular, it has been shown that the cobordism groups obtained are in many of the cases not finitely generated (Winkelnkemper [6]). Perhaps this is not so bad in itself, but it makes very unlikely the possibility of applying the standard homotopy-cobordism methods to this problem.

Winkelnkemper also remarks that "Another cobordism problem which leads to infinitely generated groups is the problem of cobordism of knots of Fox and Milnor". It is the purpose of this note to show that there is more substance in this analogy than the mere fact that in both cases we arrive at groups that are not finitely generated. The cobordism group of diffeomorphisms is related to a cobordism group of isometric structures, which is in turn formally very similar to Levine's algebraic description of knot cobordism groups ([3]). This enables us to apply many of the well-known methods of knot theory to our problem; in particular, the characteristic polynomial of an isometric structure can be used to extend Winkelnkemper's result to other dimensions, in a way that is completely analogous to the use of the Alexander polynomial by Fox, Milnor and Kervaire ([1],[2]) when they show that knot cobordism groups are not finitely generated. It seems clear that the invariants of Milnor and Levine ([4],[3]) can be used to prove many more results about the cobordism group of diffeomorphisms.

Originally we had hoped for a more definite result: that the cobordism class of its isometric structure completely determines the cobordism class of a diffeomorphism of a (2n-1)-connected 4n-manifold. But the proof ran into difficult obstructions, and we can only show this under certain conditions. The simplest cases where this obstruction is not zero are still unsolved, except for one case where R. Schultz proved that the conjecture is valid. This proof is included in an appendix.

Nevertheless, it doesn't seem unwise to conjecture that "the isometric structure is the ultimate discretization of a diffeomorphism".

Cobordism of Diffeomorphisms

We will assume that all manifolds are oriented and all diffeomorphisms are orientation preserving.

Two diffeomorphisms (M_1^n,f_1), (M_2^n,f_2) are called <u>cobordant</u> (in symbols $(M_1,f_1)\sim(M_2,f_2)$) if there is a diffeomorphism (Q,F), such that

$\partial Q = M_1 \cup (-M_2)$ and $F|M_i = f_i$, $i=1,2$. We will denote this situation by $\partial(Q,F) = (M_1,f_1) \cup (-M_2,f_2)$. It is clear that cobordism classes of diffeomorphisms form a group under disjoint sum, called the (oriented) cobordism group of diffeomorphisms and denoted by Δ^n.

The problem of computing Δ^n was first formulated, to our knowledge, by Browder, and was attacked unsuccesfully by several of his students for a number of years. The first to prove any results about this problem was Winkelnkemper, who showed the following ([6]):

(i). If M^n is a homotopy sphere, then any (M,f) is null-cobordant.
(ii). Δ^{4n+2} is not finitely generated.

One way of attacking this problem could be the following: Let $\Delta^n_{(k)}$ be the set of cobordism classes representable by diffeomorphisms of k-connected manifolds. Clearly $\Delta^n_{(k)} \subset \Delta^n_{(k-1)}$ and the fact that $\Delta^n_{(k)}$ is a subgroup of Δ^n follows easily from the connected sum construction: If (M,f) is a diffeomorphism of a connected manifold M, then f is isotopic to a diffeomorphism f' which is the identity on an open disc; then given two such (M_i,f_i), $i = 1,2$, we can find f_i', isotopic to f_i, which is the identity on a disc $D_i \subset M_i$, and they naturally define a diffeomorphism of the connected sum $M_1 \# M_2$, denoted by $(M_1,f_1') \# (M_2,f_2')$, which is clearly cobordant to the disjoint sum of the (M_i,f_i). (Remark: If the manifolds M_i are simply-connected, then the isotopy class of $(M_1,f_1') \# (M_2,f_2')$ depends only on those of the (M_i,f_i). This is not true in general, as can be seen by simple examples. Compare with the case of "connected" sum of disconnected manifolds.)

Then we can try to find the conditions under which an element of $\Delta^n_{(k-1)}$ is in $\Delta^n_{(k)}$, that is, to find the invariants of a diffeomorphism of a (k-1)-connected manifold which determine whether it is cobordant to a diffeomorphism of a k-connected manifold or not. Such a result can conceivably be proved using the methods of surgery, possibly combined with cobordism theory and the methods of Winkelnkemper. Starting from k=0, such procedure would produce a complete set of invariants of the diffeomorphism class of a diffeomorphism, since by Winkelnkemper's first result quoted above, $\Delta^n_{(k)} = 0$ if $k \geq [n/2]$.

We tried to carry out what would be the last step of this procedure: to compute $\Delta^{2n}_{(n-1)}$ by finding an invariant which determines when is a diffeomorphism of an (n-1)-connected 2n-manifold cobordant to a diffeomorphism of a homotopy sphere (and therefore null-cobordant). A first invariant is the cobordism class of the isometric structure of a diffeomorphism, but we haven't been able to show that this invariant completely determines its cobordism class. We restrict ourselves to the case when n is even.

Cobordism of isometric structures

The following definitions are analogous to those in [3], [4], where they are given for vector spaces.

An isometric structure (over $\mathbb{Z}$) is a triple (G,B,T), where G is a free abelian group of finite rank, $B:G\times G\longrightarrow\mathbb{Z}$ is a symmetric, unimodular bilinear form and $T:G\longrightarrow G$ is an endomorphism which preserves the bilinear form, i.e., $B(T(x),T(y)) = B(x,y)$ for all $x,y\in G$. An isometric structure (G,B,T) is said to be null-cobordant if it has an invariant subkernel, i.e., a subgroup $K\subset G$ such that rank $K = \frac{1}{2}$ (rank G), $B(x,y) = 0$ for all $x,y\in K$, and $T(K)\subset K$. With the natural definition of direct (orthogonal) sum of isometric structures, we say that two isometric structures (G_i,B_i,T_i), $i=1,2$, are cobordant if $(G_1,B_1,T_1)\oplus(G_2,-B_2,T_2)$ is null-cobordant. The group of cobordism classes of isometric structures will be denoted by I_+, and the subgroup formed by those which can be represented by isometric structures whose bilinear form is even will be denoted by I_+^e.

To every diffeomorphism (M^{4n},f) we can associate an isometric structure (G,B,T), where $G = H_{2n}(M)/\text{torsion}$, B is the intersection pairing and T is induced by $f_*:H_{2n}(M)\to H_{2n}(M)$. If $(M,f) = \partial(Q,F)$ then the kernel of $H_{2n}(M)\longrightarrow H_{2n}(Q)$ defines an invariant subkernel of G, as can be seen from the following diagram:

$$\begin{array}{ccc} H_{2n}(M) & \xrightarrow[\approx]{f_*} & H_{2n}(M) \\ \downarrow & & \downarrow \\ H_{2n}(Q) & \xrightarrow[\approx]{F_*} & H_{2n}(Q) \end{array}$$

This defines a homomorphism

$$I:\Delta^{4n}\to I_+$$

which restricts, for $n\neq 1,2,4$, to:

$$I: \Delta^{4n}_{(2n-1)} \longrightarrow I^e_+$$

One can similarly define the cobordism group of skew-symmetric isometric structures I_- and a homomorphism

$$I: \Delta^{4n+2} \longrightarrow I_-$$

An application

We study here the cokernel of $I: \Delta^{4n}_{(2n-1)} \longrightarrow I^e_+$. Given an isometric structure with even bilinear form, it can be realized as the isometric structure of a diffeomorphism (M', f'), where M' is $(2n-1)$-connected, $4n$-manifold with $\partial M'$ a homotopy sphere, for $n > 1$ (See [5]). To complete it to one of a closed manifold we would have to extend the diffeomorphism $f'|\partial M'$ to one of a disc with boundary $\partial M'$, and therefore the obstruction to doing this is an element of the group of isotopy classes of diffeomorphisms of $(4n-1)$-homotopy spheres, under connected sum (as defined above), which is clearly isomorphic to $\theta^{4n} \oplus \theta^{4n-1}$. This obstruction defines a homomorphism $I^e_+ \longrightarrow F$, where F is the quotient of $\theta^{4n} \oplus \theta^{4n-1}$ by the subgroup formed by those diffeomorphisms of homotopy spheres which bound diffeomorphisms of $(2n-1)$-connected $4n$-manifolds whose isometric structures are null-cobordant. (One can conjecture that this subgroup is 0, and in that case F would actually be $\theta^{4n} \oplus \theta^{4n-1}$. This conjecture follows from one similar to the one we state below.) We have an exact sequence $(n \neq 1,2,4)$

$$\Delta^{4n}_{(2n-1)} \longrightarrow I^e_+ \longrightarrow F$$

and F is in any case a finite group.

We will now show that I^e_+ is not finitely generated, and therefore, by the above exact sequence, that $\Delta^{4n}_{(2n-1)}$ is not finitely generated either.

The characteristic polynomial of an isometric structure (G,B,T) is, by definition, the characteristic polynomial $\Delta_T(t)$ of $T:G \longrightarrow G$. If A is the matrix of T and B the matrix of B with respect to some basis, we have $ABA^t = B$, and therefore, if $k = \operatorname{rank} G$,

$$\Delta_T(t) = \det(A-tI) = \det(BA^{-1}B^{-1}-tI) = \det(A^{-1}-tI) = t^k \det(t^{-1}I-A) = \pm\, t^k \Delta_T(t^{-1})$$

Clearly the characteristic polynomial of the direct sum of two isometric structures is the product of their characteristic polynomials. Now, if (G,B,T) is null-cobordant, there is a basis $e_1,\dots,e_s$, $f_1,\dots,f_s$ of G, such that $e_1,\dots,e_s$ is a basis of K and $B(e_i,f_j) = \delta_{ij}$. With respect to this basis T has a matrix of the form

$$\begin{pmatrix} A & 0 \\ B & C \end{pmatrix}$$

where $C^t = A^{-1}$. Therefore,

$$\Delta_T(t) = \det(A-tI)\det(C-tI) = \det(A-tI)\det(A^{-1}-tI)$$

$$= \pm t^s \det(A-tI)\det(A-t^{-1}I) = \pm t^s f(t)f(t^{-1}),$$

where $f(t) = \det(A-tI)$. Let P be the multiplicative group of equivalence classes of polynomials $F(t)$ satisfying $F(1) = \pm 1$ and $F(t) = \pm t^d F(t^{-1})$, under the equivalence relation $F_1(t) \sim F_2(t)$ if $F_1(t)F_2(t)$ is of the form $\pm t^k f(t)f(t^{-1})$. The above computation shows that we have a homomorphism $\Delta : I_+^e \longrightarrow P$, and since every element of P is of order 2, to show that I_+^e is not finitely generated, it is enough to exhibit infinitely many different elements in the image of Δ. To do this consider the isometric structures (G,B,T_r), $r \in \mathbb{Z}$, where G has a basis $e_1,e_2,f_1 f_2$, $B(e_i,e_j) = B(f_i,f_j) = 0$ and $B(e_i,f_j) = \delta_{ij}$, and T_r is given by the matrix

$$\begin{pmatrix} 0 & 0 & 1 & r \\ 0 & 0 & 0 & 1 \\ 1 & 0 & 0 & 0 \\ -r & 1 & 0 & 0 \end{pmatrix}$$

Then the characteristic polynomial of (G,B,T_r) is $\Delta_r(t) = t^4 + (r^2-2)t^2 + 1$, and a direct computation shows that $\Delta_r(t) \sim \Delta_s(t)$ if, and only if, $r = \pm s$. We have proved

<u>Theorem 1</u>. $\Delta^{4n}_{(2n-1)}$ <u>is not finitely generated for</u> $n > 1$.

It is clear that only formal modifications are needed to include the cases $n = 2,4$.

By a similar, but even simpler computation, we can give another proof of Winkelnkemper's result: $\Delta^{4n+2}_{(2n)}$ is not finitely generated.

By using more subtle invariants of isometric structures, such as those given in [3],[4], one could try to decide other questions; for example, whether $\Delta^{2n} \otimes \mathbb{Q}$ is finitely generated or not.

A Conjecture

Regarding the kernel of $I: \Delta^{4n}_{(2n-1)} \longrightarrow I_+$ we conjecture that it is 0. That is, that a diffeomorphism of a (2n-1)-connected 4k-manifold is null-cobordant if, and only if, its isometric structure is null-cobordant.

We have some evidence supporting this conjecture: Theorem 2 below, where it is proved under certain conditions, and the result of R. Schultz given in the appendix, which solves a typical case not included in the Theorem, but which also shows that the general proof would be rather difficult.

Our attempt to prove this conjecture is based on the following idea: If we have a diffeomorphism (M^{4n},f) such that $f_*:H_{2n}(M) \longrightarrow H_{2n}(M)$ is the identity, and if $x \in H_{2n}(M)$ is a primitive element that can be represented by an embedding $\varphi:S^{2n} \longrightarrow M$ with trivial normal bundle, then φ and $f\varphi$ are homotopic (under our (2n-1)-connectivity assumption) and therefore isotopic if $n>1$, and therefore we can assume that f is the identity on the image of φ, by changing it through an isotopy. Then f extends to $M \cup_\varphi D^{2n+1}$ and now the problem arises of extending f to the trace of the surgery on φ, $M \cup D^{2n+1} \times D^{2n}$ which is possible if, and only if, f doesn't twist the normal bundle of φ. Assuming this can be done, we would have shown that (M,f) is cobordant to (M',f'), where rank $H_{2n}(M')$ = rank $H_{2n}(M) - 2$. If this procedure can be carried out to the end, we arrive at a diffeomorphism of a homotopy sphere, which, by Winkelnkemper's theorem is null-cobordant, so f itself is null-cobordant.

With this guiding idea, assume now that (M,f) is such that its isometric structure is null-cobordant, with invariant subkernel $K \subset H_{2n}(M)$. To remove the condition $f_* = 1$, we will use results of Wall ([5]) on diffeomorphisms of handlebodies. To represent a basis of K by embedded spheres we will assume that M is almost parallelizable. The most difficult part to handle is the obstruction that arises when f twists the normal bundles of embedded spheres, because the relevant obstruction groups are almost never 0. This obstruction can be interpreted as an obstruction to isotopy ([5]).

Assuming, then, that M is almost parallelizable, we can find a cobordism W between M and a homotopy sphere Σ by doing surgery on a basis of K. Then it follows from Wall [5] that we can find a diffeomorphism $F:W \longrightarrow W$ which is the identity on a neighborhood of Σ and such that $(F|M)_* = f_*: H_{2n}(M) \to H_{2n}(M)$, so $F|M$ is homotopic to f. The obstruction to constructing an isotopy between $F|M$ and f in the complement of a disc is a homomorphism $\beta: H_{2n}(M) \to \pi_{2n}(SO_{2n})$ ([5]). If $\beta = 0$, then (M,f) is null-cobordant: First of all, f is cobordant to $F|M$, because there exists a diffeomorphism $(G,(M\times I)_0)$, where $(M\times I)_0$ is $M\times I$ minus an open disc, such that $G|M\times 0 = f$ and $G|M\times 1 = F|M$, so the cobordism classes of f and $F|M$ differ by a diffeomorphism of a sphere, which, by Winkelnkemper's result, is null-cobordant. And $F|M$ is null-cobordant, since the homotopy sphere Σ bounds a manifold Q and we can extend F to $W \cup Q$ as the identity on Q.

So our obstruction to making (M,f) null-cobordant in the way described above is the homomorphism $\beta: H_{2n}(M) \to \pi_{2n}(SO_{2n})$ which measures the difference between f and $F|M$ with respect to the way they twist the normal bundle of embedded spheres: If $x \in H_{2n}(M)$ is represented by an embedded S^{2n} with normal bundle ν, we can assume that f and $F|M$ coincide on S^{2n} and that they differ by a bundle automorphism of ν on a neighborhood of S^{2n}. This automorphism is given by the element $\beta(x) \in \pi_{2n}(SO_{2n})$.

The construction described above depends on the choice of W and of F. For a fixed W, by varying F we can assume that $\beta = 0$ on a subgroup of $H_{2n}(M)$ complementary to K, and we can change $\beta(x)$ for $x \in K$ by an element in the image of $\partial: \pi_{2n+1}(S^{2n}) \to \pi_{2n}(SO_{2n})$, i.e., in the kernel of the suspension $S: \pi_{2n}(SO_{2n}) \to \pi_{2n}(SO_{2n+1})$. We are left then with a homomorphism $\beta' = S\beta : K \to S\pi_{2n}(SO_{2n})$, and this is all we can achieve by varying F ([5]). So β' is the obstruction to carrying out the above construction with a given W.

We still have the freedom to change W, and it amounts precisely to changing the trivilization of the embedded spheres on which we performed surgery to construct W. So, if W_1 and W_2 are two such cobordisms, their difference can be measured by a homomorphism $\alpha: K \to \pi_{2n}(SO_{2n})$, and the corresponding obstructions β_1' and β_2' are related by

$$\beta_2' = \beta_1' + (S\alpha)(1 - f_*)$$

Therefore, if the induced homomorphism

$$(1 - f_*)^* : \mathrm{Hom}(K, S\pi_{2n}(SO_{2n})) \to \mathrm{Hom}(K, S\pi_{2n}(SO_{2n}))$$

is onto we can always find α such that $\beta_2' = 0$. Since $S\pi_{2n}(SO_{2n})$

is always a 2-torsion group, this amounts to the condition that $\det(1-f_*)$ is odd. We have proved:

Theorem 2. Let (M^{4n},f), $n>1$, be a diffeomorphism satisfying
(i) M is $(2n-1)$-connected and almost parallelizable,
(ii) $I(M,f) = 0$,
(iii) $\det(1-f_*)$ is odd,
then (M,f) is null-cobordant.

(So instead of our original idea of proving the conjecture first when $f_* = 1$, we have proved it only when f_* is far away from the identity.)

For n=3, only, the obstruction group for almost parallelizability, $\pi_5(SO)$, and the obstruction group for isotopy, $S\pi_6(SO_6) \subset \pi_6(SO_7)$ are both 0, so we can prove the full conjecture:

Corollary. $I: \Delta^{12}_{(5)} \longrightarrow I_+$ is a monomorphism.

In general, the obstruction to carrying out the above construction is an element

$$\bar{\beta} \in \mathrm{Hom}(K, S\pi_{2n}(SO_{2n}))/\mathrm{Im}(1-f_*)^*$$

This is really a secondary obstruction, for it is defined only when the isometric structure is null-cobordant and depends on the choice of an invariant subkernel. It would be interesting to define it as a primary obstruction. The difficulty consists in defining, a priori, how much f twists the normal bundle of a sphere representing $x \in H_{2n}(M)$, when $f_*(x) \neq x$.

In any case, $\bar{\beta} = 0$ is not a necessary condition for (M,f) being null-cobordant. The simplest type of examples where our proof breaks down can be constructed as follows:

Let $M = S^{2n} \times S^{2n}$, and $\alpha_1, \alpha_2 \in S\pi_{2n}(SO_{2n})$. Represent α_i by a map $\alpha_i : S^{2n} \longrightarrow SO_{2n} \subset SO_{2n+1}$ and define diffeomorphisms (M,f_i) by $f_1(x,y) = (x, \alpha_1(x)y)$ and $f_2(x,y) = (\alpha_2(y)x, y)$. Let $f = f_1 f_2$. Then $f_* = 1$ so $I(M,f) = 0$, but if $\alpha_1, \alpha_2 \neq 0$, then our obstruction $\bar{\beta}$ is different from 0 for any invariant subkernel.

Question. Is (M,f) null-cobordant when $\alpha_1, \alpha_2 \neq 0$?

R. Schultz has given a positive answer to some cases of this question, thus showing that there are null-cobordant diffeomorphisms such that

$\bar{\beta} \neq 0$ for any invariant subkernel. His result is given in the appendix. The above description of the examples is also due to him.

Another thing we have shown is that under conditions (i) and (ii) of Theorem 2, f is homotopic to a null-cobordant diffeomorphism, the second condition being again necessary for this. So our conjecture is equivalent, for manifolds satisfying condition (i), to the assertion that homotopic diffeomorphisms are cobordant.

Other cobordism problems to which our ideas would apply more naturally can be formulated by varying the conditions on the manifolds and the cobordisms. For example, consider the group of cobordism of diffeomorphisms, where the manifolds are (2n-1)-connected parallelizable 4n-manifolds with a homotopy sphere as boundary, and the cobordisms are (2n-1)-connected and parallelizable and restrict to h-cobordisms between the boundary spheres. For this case our method shows that a diffeomorphism is null-cobordant if, and only if, its isometric structure admits an invariant subkernel for which $\bar{\beta} = 0$. If $\bar{\beta}$ could be defined as a primary obstruction, we could incorporate it in the definition of the isometric structure and thus obtain a completely algebraic description of this cobordism group. Similarly, we could deal with manifolds which are not almost-parallelizable, by including the tangential information in the definition of the isometric structure.

The case of diffeomorphisms of 2n-connected (4n+2)-manifolds is further complicated by the presence of Kervaire invariant problems, but it seems that these can be solved satisfactorily by applying some ideas of Browder, related to the generalized Kervaire invariant, so here, again, the main difficulty comes from the β obstructions.

References

[1]. Fox, R., and Milnor J., Singularities of 2-spheres on 4-space and cobordism of knots. Osaka J. Math., 3 (1966), 257-67
[2]. Kervaire, M., Les noeuds de dimensions supérieures. Bull. Soc. Math. France, 93 (1965), 225-71.
[3]. Levine, J., Invariants of knot cobordism, Inventiones math. 8 (1969), 98-110.
[4]. Milnor, J. On isometries of inner product spaces, Inventiones math. 8 (1969).
[5]. Wall, C.T.C., Classification problems in differential topology-

III. Applications to special cases. Topology, 3 (1965), 291-304.
[6]. Winkelnkemper, H.E., On equators of manifolds and the action of θ^n. Princeton, Ph. D. Thesis, 1970.

Appendix

SPECIAL CASES OF LOPEZ DE MEDRANO'S PROBLEM

REINHARD SCHULTZ
Purdue University

Let $\sigma_k \in \pi_{8k-1}(SO_{8k+1}) = \mathbb{Z}$ be a generator, and let $\eta : S^{h+1} \to S^h$ be homotopically nontrivial. It is well-known that $\pi_{8k}(SO_{8k+1})$ and $\pi_{8k+1}(SO_{8k+2})$ are generated by $\sigma_k\eta$ and $S\sigma_k\eta^2$ (resp.) along with the class of the tangent bundle. In the problem's notation, if $\alpha_1 = \alpha_2 = \sigma_k\eta$ or $\sigma_k\eta^2$ and $k \geqslant 2$, then the diffeomorphism under consideration extends to a diffeomorphism of a manifold with boundary. The proofs for $\sigma_k\eta$ and $\sigma_k\eta^2$ are parallel, so only the proof for $\sigma_k\eta$ is sketched. Since $k \geqslant 2$, σ_k pulls back to $\pi_{8k-1}(SO_{8k-2})$. Embed S^{8k-3} in $R^{8k-2} \times \{0\} \subset \mathbb{R}^{8k+1}$ as the sphere of radius $\frac{1}{2}$, and take a nice tubular neighborhood $S^{8k-3} \times D^4 \subset \mathbb{R}^{8k+1}$ contained in the unit disk D^{8k+1}. Let $W = D^{8k+1} - \text{Int}\, S^{8k-3} \times D^4$. Then SO_{8k+2} acts on W, the action on the $S^{8k-3} \times S^3$ component of ∂W being standard on the first coordinate and fixed on the second. Since $\sigma_k\eta$ pulls back to SO_{8k+2}, it is easy to see that f_2 extends to a diffeomorphism of $W \times S^{8k}$. On the other hand, S^{8k} is a retract of W, and hence α_1 extends to a homotopy class in $[W, SO_{8k+1}]$, which implies that f_1 also extends to a diffeomorphism of $W \times S^{8k}$. Let $f_1^! f_2^!$ be the induced diffeomorphism of the other component of $\partial(W \times S^{8k})$, i.e., $S^{8k-3} \times S^3 \times S^{8k}$. Then $f_1^!$ is defined by a homotopy class $\alpha_1^!$ in $[S^{8k-3} \times S^3, SO_{8k+1}]$ and $f_2^! = d \times id_{S^3}$, where d is a diffeomorphism of $S^{8k-3} \times S^{8k}$. Let ξ be the canonical complex line bundle over S^2, $E(\xi)$ its disk bundle, $S^3 \subset E(\xi)$ the unit sphere bundle. Then $\sigma_k\eta = \alpha_1$ implies that the homotopy class $\alpha_1^!$ extends to $[S^{8k-3} \times E(\xi), SO_{8k+1}]$, and hence $f_1^!$ extends to a diffeomorphism of $S^{8k-3} \times E(\xi) \times S^{8k}$. But $f_2^! = d \times id_{S^3}$ implies $f_2^!$ extends to a diffeomorphism of $S^{8k-3} \times E(\xi) \times S^{8k}$ trivially. Thus $f_1^! f_2^!$ extends to a diffeomorphism of $S^{8k-3} \times E(\xi) \times S^{8k}$, which implies that $f_1 f_2$ extends to a diffeomorphism of $W \times S^{8k} \cup S^{8k-3} \times E(\xi) \times S^{8k}$.

THE INDEX OF MANIFOLDS WITH TORAL ACTIONS AND GEOMETRIC INTERPRETATIONS OF THE $\sigma(\infty, (S^1, M^n))$ INVARIANT OF ATIYAH AND SINGER

Katsuo Kawakubo*
Osaka University and Institute for Advanced Study

and

Frank Raymond*
University of Michigan and Institute for Advanced Study

§1. Introduction

In the theory of transformation groups the smooth theory is certainly the most tractable. However, not all operations that one would like to perform remain in the smooth category. For example, analysis of the orbit space usually must go outside the smooth category. Furthermore, one would also like to study symmetries of interesting geometric spaces which often fail to be locally Euclidean such as spaces having manifolds as ramified coverings and analytic spaces.

It has long been recognized that cohomology manifolds or generalized manifolds encompass all manifolds as well as many typical analytic and ramified spaces. The most important feature of generalized manifolds, from the point of view of transformation groups, is that one is often able to work with orbit spaces (cf. [8]). In addition, one does have characteristic classes and with care one can often define workable invariant tubular neighborhoods of fixed point sets.

In this paper we develop and exploit some of these ideas to prove results which even when specialized to the smooth category seem to be unobtainable from standard smooth methods. We introduce the notion of nicely embedded fixed points. This enables us to find invariant closed tubular neighborhoods which are cohomologyfiber spaces [2] over the fixed point sets. The mapping cylinder of the orbit map is an important construction. For circle actions, without fixed points on generalized manifolds, this mapping cylinder is a generalized manifold with boundary with a natural circle action having nicely embedded fixed points identical with the orbit space. We establish by cohomological methods alone formulae for

* Both authors are partially supported by the National Science Foundation.

the Atiyah-Singer invariant $\sigma(\infty, (S^1, M))$, Theorems 2, 3, 4 and 5. This enables us to conclude that its value is an integer and also to give explicit geometric interpretations of it in terms of the orbit space and the mapping cylinder of the orbit map. This makes it much easier to compute than by using smooth techniques alone, e.g. §5. A very interesting geometric interpretation of $\sigma(S^1, M^{4k-1}) = 0$ is also given in terms of fibering M^{4k-1} over a circle, Theorems 4 and 5. The detailed version of these results will appear in [4].

§2. The index of manifolds with toral actions

Let us recall that a (closed) mapping cylinder neighborhood of a closed subset A of a space X is the closure of an open subset $U \supset A$ of X, a map f of the frontier of U onto the frontier of A, and a homeomorphism h of the closure of U minus the interior of A onto the mapping cylinder of f such that h restricted to the frontiers of U and A is the identity. Note that closure U minus A is homeomorphic to (frontier of U) $\times [0, 1)$, with (frontier of U) $\times (0, 1)$ being an open subset of X. (In [6] it is shown that any two such open mapping cylinder neighborhoods (MCN) are essentially unique.)

We shall say that an S^1-action on an orientable closed $\mathbb{R}$-generalized n-manifold M^n has nicely embedded fixed points if each component $F_{k,j}$ of the fixed point set F has an invariant closed mapping cylinder neighborhood $T(F_{k,j})$ satisfying the conditions (i) - (iv) listed below.

Without loss of generality we may assume that S^1 is not acting trivially upon M and $F_{k,j}$ is an $(n-2k)$-dimensional orientable generalized manifold over $\mathbb{R}$ with $k > 0$. If we let $r_t : T \longrightarrow T$ denote the retraction along the "fibers," then $r_1|_{\partial T} = f$. We shall postulate that (i) $(r_1|_{\partial T})^{-1}(x) = S_x^{2k-1}$ is a $(2k-1)$-generalized manifold over $\mathbb{R}$ having the real cohomology of the $(2k-1)$-sphere, for each $x \in F_{k,j}$. This means that $r_1^{-1}(x) = D_x^{2k}$ is the cone over S_x^{2k-1} and is consequently a generalized $2k$-cell over $\mathbb{R}$. We shall assume that (ii) $r_1^{-1}(x)$ is invariant and the S^1-action on $S_x^{2k-1} \times [0, t)$ is independent of t, $0 \leqq t < 1$. To handle how the "fibers" fit together homologically we assume that (iii) the Leray sheaf of the map f is simple, coefficients in $\mathbb{R}$. Finally, to eliminate "exotic" pathology we assume that (iv) (S^1, M^n) has only a finite number of distinct orbit types (always holds if M^n is a Z-generalized manifold).

Finally, if (T^s, M^n) is an action of an s-dimensional torus on an orientable $\mathbb{R}$-generalized manifold, we shall say that it has (very) nicely embedded fixed points if (every) some circle subgroup $S^1 \subset T^s$ for which $F(S^1, M^n) = F(T^s, M^n)$ has nicely embedded fixed points. Then we have

Theorem 1. *If* (T^s, M^n) *denotes an action with nicely embedded fixed points on an oriented closed* $\mathbb{R}$ *generalized manifold, then each component of the fixed point set can be oriented so that*

$$I(M^n) = I(F) ,$$

where I *dentoes the Thom-Hirzebruch index.*

The proof is a mixture of the techniques of [3] and [5].

§3. An invariant for actions without fixed points

Let (T^s, M^n) be a smooth action without fixed points. Suppose (T^s, M^n) is a smooth equivariant boundary of (T^s, B^{n+1}) where B^{n+1} is a smooth oriented compact manifold with oriented boundary M^n. The action (T^s, B^{n+1}) may have fixed points. Define

$$'\sigma(T^s, M^n) = I(F(T^s, B^{n+1})) - I(B^{n+1}).$$

Theorem 2. $'\sigma(T^s, M^n)$ *depends only upon* (T^s, M^n).

When $s = 1$, Atiyah and Singer arrive by application of their G-signature theorem to the same formula for their invariant $'\sigma(t, (S^1, M^n))$ when $t \longrightarrow \infty$ [1]. It is known [7] that for every smooth action (S^1, M^n) without fixed points there is a multiple rM^n (r is actually of the form 2^s), which is an equivariant smooth boundary. Thus one can define:

$$\sigma(\infty, (S^1, M^n)) = \frac{1}{r}\, '\sigma(\infty, (S^1, rM)) .$$

Of course, from the Atiyah-Singer theorem it is not apparent that $\sigma(\infty, (S^1, M^n))$ is an integer. In the next section we shall see that this rational number is actually an integer. In order to have this, we extend Theorem 2 to actions on $\mathbb{R}$-generalized manifolds. We shall assume that all our actions considered on compact $\mathbb{R}$-generalized manifolds have a finite number of distinct orbit types.

Theorem 3. (i) *Let* (S^1, M^n) *be an action without fixed points on a closed oriented* $\mathbb{R}$*-generalized manifold. Then, the induced* S^1*-action on the* $\mathbb{R}$*-generalized* $(n+1)$*-manifold with boundary* $(S^1, \mathrm{Map}(\pi))$*, where* $\pi : M \longrightarrow M/S^1$ *is the orbit map, has very nicely embedded fixed point sets.*

(ii) *Let* (T^s, M^n) *be an action without fixed points on an oriented* $\mathbb{R}$*-*

generalized manifold. Suppose (T^s, M^n) is an equivariant boundary of the oriented compact (T^s, B^{n+1}) with very nicely embedded fixed points. Then,

$$'\sigma(T^s, M^n) = I(F(T^s, B^{n+1})) - I(B^{n+1})$$

depends only upon (T^s, M^n).

§4. Geometric interpretation of the $\sigma(\infty, (S^1, M))$ invariant of Atiyah and Singer.

Suppose (S^1, M^n) is an action without fixed points on an oriented closed $\mathbb{R}$-generalized n-manifold. Define:

$$\sigma(S^1, M^n) = \begin{cases} I(M^n/S^1), & \text{if } n = 4k+1 \\ -I(\mathrm{Map}(\pi)), & \text{if } n = 4k-1 \\ 0\ , & n \text{ even} \end{cases}$$

Then we have

Corollary 1. $'\sigma(S^1, M^n) = \sigma(S^1, M^n)$. In particular, if (S^1, M^n) is smooth then the Atiyah-Singer invariant $\sigma(\infty, (S^1, M^n))$ is an integer and equals $\sigma(S^1, M^n)$.

Remark. To obtain the analogue of Corollary 1 for toral actions we need to further strengthen the notion of very nicely embedded fixed points. Let (T^s, M^n) be an action without fixed points. For $S^1 \subset T^s$, such that $F(S^1, M^n) = F(T^s, M^n) = \phi$, we may write $T^s = S^1 \times T^{s-1}$. We form $(S^1 \times T^s, D^2 \times M^n)$, where S^1-acts diagonally and T^s on the second factor. We obtain a toral action, $(T^s, D^2 \times_{S^1} M^n) = (T^s, \mathrm{Map}(\pi))$, where $\pi : M \longrightarrow M/S^1$. We require that $(T^s, D^2 \times_{S^1} M^n)$ has very nicely embedded fixed points. In particular, if (T^s, M^n) is a smooth action, then $(T^s, D^2 \times_{S^1} M^n)$ has very nicely embedded fixed points. We may then define

$$\sigma(T^s, M^n) = \begin{cases} I(F(T^{s-1}, M^n/S^1)), & \text{if } n = 4k+1 \\ -I(\mathrm{Map}(\pi)), & \text{if } n = 4k-1 \\ 0, & n \text{ even} \end{cases}$$

Of course, $\partial(T^s, D^2 \times_{S^1} M^n) = (T^s, M^n)$ and we apply Theorem 3, (ii) and identify $'\sigma(T^s, M^n) = \sigma(T^s, M^n)$ for the purpose of computation. (Observe that $I(F(T^s, \mathrm{Map}(\pi))) = I(F(T^{s-1}, M^n/S^1)) = I(M^n/S^1)$.)

It is not known whether a smooth fixed point free (T^s, M^n) has a multiple

(T^s, rM^n) which is a smooth equivariant boundary. If this is true, then $'\sigma(T^s, rM^n)$ is defined as a smooth invariant and hence is clearly divisible by r and the quotient is $\sigma(T^s, M^n)$.

It should also be observed that the smooth $\sigma(T^s, M^n)$ invariant really is defined up to topological equivalence.

§5. Geometric interpretation of $\sigma(S^1, M^n) = 0$.

Let (S^1, M^3) be closed oriented and without fixed points. Then we have

Theorem 4. $\sigma(S^1, M^3) = 0$ or ± 1. Furthermore, $\sigma(S^1, M^3) = 0$, if and only if (S^1, M^3) fibers (equivariantly) over S^1 with finite structure group.

We generalize a part of Theorem 4, which seems to suggest an interesting geometric meaning for the vanishing of $\sigma(S^1, M^{4k-1})$.

Theorem 5. Let M^{4k-1} be an orientable closed $\mathbb{R}$-generalized manifold. If M^{4k-1} fibers over the circle with finite structure group then there exists an S^1-action (S^1, M^{4k-1}) which fibers equivariantly over $(S^1, S^1/Z_p)$ and $\sigma(S^1, M^{4k-1}) = 0$. If M^{4k-1} and the fibering are smooth then (S^1, M^{4k-1}) is smooth.

References

[1] Atiyah, M. F. and Singer, I. M., The index of elliptic operators: III, Ann. of Math. 87 (1968), 546-604.

[2] Bredon, G. E., Cohomology fiber spaces, the Smith-Gysin sequence, and orientation in generalized manifolds, Mich. Math. J. 10 (1963), 321-333.

[3] Conner, P. E. and Raymond, F., Injective operations of the toral groups, to appear in Topology.

[4] Kawakubo, K. and Raymond, F., The index of manifolds with toral actions and geometric interpretations of $\sigma(\infty, (S^1, M^n))$ invariant of Atiyah and Singer, to appear in Inventiones Math.

[5] Kawakubo, K. and Uchida, F., On the index of a semi-free S^1-action, Proc. Japan Acad. 46 (1970), 620-622 and J. Math. Soc. Japan 23 (1971), 351-355.

[6] Kwun, K. W. and Raymond, F., Mapping cylinder neighborhoods, Mich. Math. J. 10 (1963), 353-357.

[7] Ossa, E., Fixpunktfreie S^1-Aktionen, Math. Ann. 186 (1970), 45-52.

[8] Raymond, F., The orbit spaces of totally disconnected groups of transformations on manifolds, Proc. A.M.S. 12 (1961), 1-7.

INVOLUTIONS ON HOMOTOPY COMPLEX PROJECTIVE SPACES AND RELATED TOPICS

Ted Petrie
Department of Mathematics
Rutgers University, The State University of New Jersey
New Brunswick, New Jersey

I. NONEXISTENCE THEOREMS

§0. Introduction

The motivation for this paper is the question of Hsiang [6] and Sullivan (unpublished): Does every closed C^∞ manifold admit a non-trivial involution? The idea here is that a good class of manifolds to study for counterexamples are the smooth manifolds homotopy equivalent to CP^n complex projective n space. The reason is that these manifolds are well understood in terms of

a) their Pontriagin classes which serve as data for applying the Atiyah-Singer G index theorem $G = Z_2$.

b) the mod 2 cohomology ring structure of the fixed point set of an involution on one of these manifolds. (Bredon)

These two pieces of information provide the key material for our results. We prove two technical lemmas of a general nature. One gives a condition under which a complex line bundle over a manifold X with a smooth involution can be given an involution making it a Z_2 line bundle over X i.e. an element of $K_{Z_2}(X)$, Corollary 1.2. The other shows that a $\text{spin}^c(2n)$ manifold X always has a class $\delta_{Z_2} \in K_{Z_2}(TX)$ (TX = tangent bundle of X) coming from an equivariant Dirac operator, Corollary 2.5.

Before stating the application of these tools, let's examine some examples of involutions on CP^n.

Let $(z_0:z_1:\dots z_n)$ be homogenous coordinates for CP^n. There are two kinds of "linear" involutions on CP^n:

a) Conjugation, $(z_0:\dots z_n) \to (\bar{z}_0:\dots\bar{z}_n)$ where $\bar{z}_i$ denotes the complex conjugate of z_i. The fixed point set consists of the single component RP^n real projective n space.

b) Involutions of type k denoted τ_k, $\tau_k(z_0:z_1:\dots z_n) = (-z_0:-z_1:\dots -z_k:z_{k+1}\dots z_n)$. The fixed point set consists of two components CP^k, CP^{n-k-1}.

These two examples in a sense typify the general case of involutions on manifolds homotopy equivalent to CP^n. In fact a theorem of Bredon states [3]: if X is a manifold having the same mod 2 cohomology ring as CP^n, then the fixed point set has either 0 one or two components and the mod 2 cohomology ring of the fixed point set is that of RP^n in the second case and that of $CP^k \cup CP^{n-k-1}$ in the third. Moreover, in the third case the restriction map from $H^2(X,Z_2)$ to the second cohomology of either component of the fixed point set is onto and each component has the mod 2 cohomology ring structure of a CP^j, $j = k$ or $n-k-1$.

In this paper we restrict ourselves to studying the third case. We make this:

Definition: Let τ be a smooth involution on a closed smooth manifold homotopy equivalent to CP^n. Suppose that the fixed point set consists of two components X_0 and X_1. We say τ is of type k if 2k is the dimension of the component of least dimension.

Remark: CP^n has involutions of type k for every k $0 \leq k \leq [n/2]$ namely τ_k.

Question: What types of involutions exist on manifolds homotopy equivalent to CP^n?

If τ is an involution on a manifold X with tangent bundle TX, it defines an action of Z_2 on X; $K_{Z_2}(X)$ and $K_{Z_2}(TX)$ denote the equivariant complex K theory of X and TX and $Id^X_{Z_2} : K_{Z_2}(TX) \to R(Z_2)$ denotes the Atiyah-Singer index homomorphism to $R(Z_2)$ the complex representation ring of Z_2. For $V \in K_{Z_2}(TX)$ $Id^X_{Z_2}(V)(g)$ is the value of the character of $Id^X_{Z_2}(V)$ at $g \in Z_2$.

The observation we exploit is that the condition that $Id^X_{Z_2}(V)(g)$ be an integer for every $V \in K_{Z_2}(TX)$ and $g \in Z_2 = (1,\tau)$ impose stringent restrictions on the Pontryagin classes of X.

In order to make use of this observation we need elements of $K_{Z_2}(TX)$. In the spirit of [8] we show that the line bundle $h^*(H) = \eta$ where H is the Hopf bundle over CP^n and $h: X \to CP^n$ is a homotopy equivalence can be given an involution making it an element $\hat{\eta}$ of $K_{Z_2}(X)$. Then we construct an element $\delta_{Z_2} \in K_{Z_2}(TX)$ from an equivarient $spin^c(2n)$ structure on X [8]. Of course the powers $\hat{\eta}^k$ are elements of $K_{Z_2}(X)$ and the products $\delta_{Z_2}\hat{\eta}^k \in K_{Z_2}(TX)$ are the elements we use.

Using these ideas we obtain a condition on the Pontryagin classes of X imposed by the existence of a type 0 involution in Theorem 3.1. We specialize this to the case of homotopy CP^3's and CP^4's showing that many cannot support involutions of type 0. In particular if X is a homotopy CP^3 with an orientation preserving involution τ, then τ is either of type 0 or of type 1. If $x \in H^2(X,Z)$ is a generator then the first Pontryagin class of X is $(24j+4)x^2$ for some integer j. If $j \not\equiv 0(2)$ then X doesn't support an involution of type 0, Theorem 3.2 ii). I believe that all such X with first Pontryagin class of the form $(48j+4)x^2$ admit involutions of type 1. It remains to deal with the case $(24j+4)x^2$ where j is odd.

I contemplate a second part to this paper discussing existence of involutions on homotopy complex projective spaces. The appendix gives some information in this direction.

§1. Equivariant Complex Line Bundles

The main technical tool utilized here is a condition which insures that a complex line bunlde over a Z_2 manifold X can be given the structure of a Z_2 bundle.

Let X^1 be the Z_2 manifold whose underlying space is CP^{2N+1} where N is large. The involution on X^1 is τ_N. The fixed point set F^1 consists of two components F_0^1 and F_1^1 both diffeomorphic to CP^N. We denote by $\sigma \in H^2(X^1)$ the canonical generator and $\sigma_\varepsilon \in H^2(F_\varepsilon^1)$ its restriction to $F_\varepsilon^1, \varepsilon = 0,1$.

Let $G = Z_2$ and E be a contractible C.W. complex on which G acts freely on the left. Then the orbit space is B_G. For any right G space Y, $Y \times_G E$ denotes the orbit space of the right G action on $Y \times E$ defined by $(y,e)g = (yg, g^{-1}e)$. If Z is a subspace of a space W and if α is a cohomology class of W its restriction to Z is denoted by $\alpha|Z$.

Lemma 1.1. Let X^0 be a manifold with involution with fixed point set F^0. If $\alpha \in H^2(X^0 \times_G E)$ and $\eta \in H^2(X^0)$ is its restriction to X^0, there is an equivariant map $\Gamma\colon X^0 \to X^1$ such that $\Gamma^*(\sigma)|F^0 = \eta|F^0$.

Proof: The following diagram is provided to facilitate the proof. Definitions follow. Let $G = Z_2$.

I recommend that the reader proceed with the rest of the paper and then return to the proof of Lemma 1.1 as it is quite complicated.

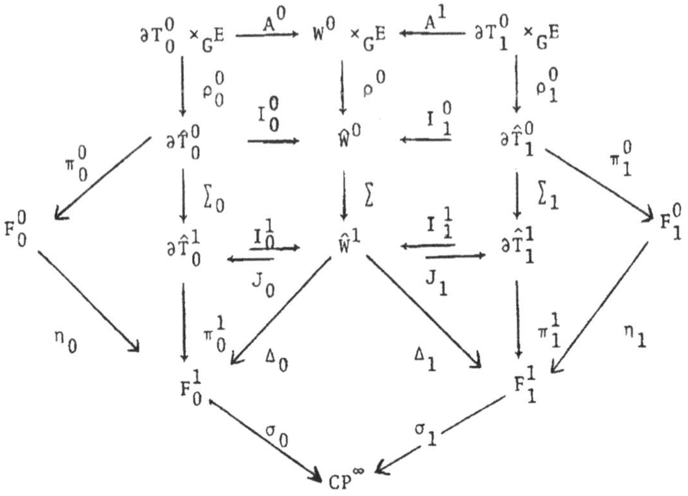

(i) The fixed point set of the involution on X^i is denoted by F^i. It is partitioned into two sets F^i_ε $\varepsilon = 0,1$ by this rule: F^i_ε is the union of those components of F^i such that for every $y \in F^i_\varepsilon$ the cohomology class $\alpha^i \in H^2(X^i \times_G E)$ restricts to $\varepsilon\mu \in H^2(y \times_G E) = H^2(B_G) = Z_2$. Here $\mu \in H^2(B_G)$ is the generator and $\alpha^0 = \alpha$ while α^1 is a class in $H^2(X^1 \times_G E)$ which maps to the first Chern class of the Hopf bundle over X^1 under the restriction

$$H^2(X^1 \times_G E) \to H^2(X^1).$$

(ii) Note that F^1_ε is CP^N for $\varepsilon = 0,1$. Let $\sigma_\varepsilon \in H^2(F^1_\varepsilon)$ be the canonical generator. Since N is large we can choose a map $\eta_\varepsilon \colon F^0_\varepsilon \to F^1_\varepsilon$ such that $\eta^*_\varepsilon(\sigma_\varepsilon) = \eta | F^0_\varepsilon$.

(iii) Denote by T^i_ε an equivariant tubular neighborhood of F^i_ε and ∂T^i_ε its boundary. Let W^i be the closure of the complement of $T^i_0 \cup T^i_1$. Then W^i is an invariant submanifold of X^i whose orbit space we designate as $\hat{W}^i$ and $\rho^i \colon W^i \times_G E \to \hat{W}^i$ the map which sends the equivalence class $[w,e]$ to the orbit of w in $\hat{W}^i$.

(iv) Let k_ε be the map from $T^0_\varepsilon \times_G E$ to $F^0_\varepsilon \times B_G$ defined for the equivalence class $[t,e] \in T^0_\varepsilon \times_G E$ by $k_\varepsilon[t,e] = (\pi(t),\bar{\pi}(e))$ where π respectively $\bar{\pi}$ is the projection of T^0_ε on F^0_ε respectively of E on B_G. The map k_ε is a homotopy equivalence.

(v) Let $\partial\hat{T}^i_\varepsilon$ denote the orbit space of ∂T^i_ε with its involution. Then $\partial\hat{T}^i_\varepsilon$ is a projective space bundle over F^i_ε. Let ρ^0_ε be the projection of $\partial T^0_\varepsilon \times_G E$ on $\partial\hat{T}^0_\varepsilon$ which sends $[t,e]$ to the orbit of t in $\partial\hat{T}^0_\varepsilon$ and let π^i_ε denote the projection of $\partial\hat{T}^i_\varepsilon$ on F^i_ε. Then we have a commutative diagram:

$$\begin{array}{ccccc}
\partial T^0_\varepsilon \times_G E & \xrightarrow{i_\varepsilon} & T^0_\varepsilon \times_G E & \xrightarrow{k_\varepsilon} & F^0_\varepsilon \times B_G \\
\downarrow \rho^0_\varepsilon & & & \swarrow \phi_\varepsilon & \searrow \tau_\varepsilon \\
\partial\hat{T}^0_\varepsilon & \xrightarrow{\pi^0_\varepsilon} & F^0_\varepsilon & & B_G
\end{array}$$

Since $\alpha|y \times B_G = \varepsilon\mu$ when $y \in F^0_\varepsilon$ and since $\alpha|F^0_\varepsilon = \eta|F^0_\varepsilon = \eta^*_\varepsilon(\sigma_\varepsilon)$ we have $\alpha|T^0_\varepsilon \times_G E = k^*_\varepsilon(\phi^*_\varepsilon\eta^*_\varepsilon(\sigma_\varepsilon)+\varepsilon\tau^*_\varepsilon(\mu))$ so $\alpha|\partial T^0_\varepsilon \times_G E = \rho^{0*}_\varepsilon\pi^{0*}_\varepsilon\eta^*_\varepsilon(\sigma_\varepsilon) + i^*_\varepsilon k^*_\varepsilon(\varepsilon\tau^*_\varepsilon(\mu))$. The first uses the fact that k_ε is a homotopy equivalence.

(vi) Let $\hat{\lambda}^i: W^i \times_G E \to B_G$ be defined by

$$\hat{\lambda}^i[w,e] = \bar{\pi}(e) \qquad w \in W^i,\ e \in E \text{ and}$$

set $\lambda^i = \hat{\lambda}^i h^i$ where h^i is a homotopy inverse of ρ^i. Note that $\hat{W}^1$ is an h cobordism between $\partial\hat{T}^1_0$ and $\partial\hat{T}^1_1$. Let J_0 and J_1 respectively be retractions on $\partial\hat{T}^1_0$ and $\partial\hat{T}^1_1$. Set $\Delta_\varepsilon = \pi^1_\varepsilon J_\varepsilon$. Observe also that the map $G^1_0: \hat{W}^1 \to F^1_0 \times B_G$ defined as $G^1_0(W) = (\Delta_0(W),\lambda^1(W))$ induces an isomorphism in cohomology through a large range of dimensions. This means that there is a unique map

$$\sum : \hat{W}^0 \to \hat{W}^1$$

with these properties:

1) $\rho^{0*}\Sigma^*\Delta_0^*(\sigma_0) = \alpha|W^0 \times_G E$

2) $\lambda^1 \Sigma$ is homotopic to λ^0

We can also require

3) Σ restricted to $\partial\hat{T}^0_\varepsilon$ gives a map $\Sigma_\varepsilon\colon \partial\hat{T}^0_\varepsilon \to \partial\hat{T}^1_\varepsilon$ i.e. $\Sigma I^0_\varepsilon = I^1_\varepsilon\Sigma_\varepsilon$ where I^i_ε are inclusions

(vii) Let $i \in H^1(B_{G_1} Z_2) \cong Z_2$ be the generator and $\beta^1\colon H^1(B_G,Z_2) \to H^2(B_G,Z)$ the Bockstein homomorphism. Then $\beta^1(i) = \mu$. Let A^ε denote the inclusion of $\partial T^0_\varepsilon \times_G E$ in $W^0\times_G E$, $\varepsilon = 0,1$. One easily checks that $\lambda_0\rho^0 A^1 = \tau_1 k_1 i_1$, so

$$A^{1*}\rho^{0*}\lambda_0^*(i) = i_1^* k_1^* \tau_1^*(i).$$

(viii) By calculation one finds that $\Delta_1^*(\sigma_1) = \Delta_0^*(\sigma_0) + \beta^1\lambda^{1*}(i)$.

We can now draw these inclusions from our data: From (v) we have

4) $$\alpha|\partial T^0_\varepsilon \times_G E = \rho^{0*}_\varepsilon \pi^{0*}_\varepsilon \eta^*_\varepsilon(\sigma_\varepsilon) + i^*_\varepsilon k^*_\varepsilon(\varepsilon\tau^*_\varepsilon(\mu))$$

so $\rho^{0*}_0 \Sigma^*_0 \pi^{1*}_0(\sigma_0) = A^{0*}\rho^{0*}\Sigma^*\Delta_0^*(\sigma_0) \underset{(1)}{=} A^{0*}\alpha|W \times_G E = \alpha|\partial T^0_0 \times_G E \underset{(4)}{=} \rho^{0*}_0 \pi^{0*}_0 \eta^*_0(\sigma_0)$ i.e.

5) $$\Sigma^*_0\pi^{1*}_0(\sigma_0) = \pi^{0*}_0 \eta^*_0(\sigma_0)$$

Also we have

$$\rho^{0*}_1\Sigma^*_1\pi^{1*}_1(\sigma_1) = A^{1*}\rho^{0*}\Sigma^*\Delta_1^*(\sigma_1) \underset{(viii)}{=}$$

$$A^{1*}\rho^{0*}\Sigma^*(\Delta_0^*(\sigma_0)+\beta^1\lambda^{1*}(i)) \underset{(1)}{=}$$

$$A^{1*}\alpha|W \times_G E + A^{1*}\rho^{0*}\Sigma^*\beta^1\lambda^{1*}(i) \underset{(4),\ (2)}{=}$$

$$\rho^{0*}_1 \pi^{0*}_1 \eta^*_1(\sigma_1) + i^*_1 k^*_1 \tau^*_1(\mu) + A^{1*}\rho^{0*}\beta^1\lambda^{0*}(i)$$

and by (vii)
$= \rho_1^{0*}\pi_1^{0*}\eta_1^*(\sigma_1)$

Thus we have established

(6) $$\sum_1^* \pi_1^{1*}(\sigma_1) = \pi_1^{0*}\eta_1^*(\sigma_1)$$

Note that (5) and (6) imply that $\sum_\varepsilon$ is homotopic to a fiber map (of projective space bundles) from $\partial\hat{T}_\varepsilon^0$ to $\partial\hat{T}_\varepsilon^1$. Let $\tilde{\sum}: W^0 \to W^1$ be an equivariant map covering $\sum$. Then $\tilde{\sum}_\varepsilon = \tilde{\sum}|\partial T_\varepsilon^0$ is a fiber map (of sphere bundles) which commutes with the involution in the fibers and it may be extended to a map $\tilde{\Lambda}_\varepsilon: T_\varepsilon^0 \to T_\varepsilon^1$ commuting with the involution in the disk fibers.

Now define $\Gamma: X^0 \to X^1$ by $\Gamma|T_\varepsilon^0 = \tilde{\Lambda}_\varepsilon$, $\Gamma|W^0 = \tilde{\sum}$. Then Γ is an equivariant map and $\Gamma|F^0 = \eta|F^0$.

<u>Corollary 1.2</u>. Suppose the involution on X^0 induces the trivial map in $H^*(X^0,Z)$, $H^1(X^0,Z) = 0$ and $H^2(X^0,Z)$ maps monomorphically to $H^2(F^0,Z)$. Then any complex line bundle η over X^0 can be given the structure of a Z_2 bundle over X^0 i.e. $\eta \in K^*(X^0)$ comes from an element $\hat{\eta}$ of $K^*_{Z_2}(X^0)$.

<u>Proof</u>: The line bundle η is defined by a map from X^0 to $X^1 = CP^\infty$ i.e. $\eta \in H^2(X^0,Z)$. By the lemma we can find an equivariant map Γ from X^0 to X^1 such that

$$\Gamma^*(\sigma)|F^0 = \eta|F^0.$$

Since the restriction map from $H^2(X^0,Z)$ to $H^2(F^0,Z)$ is a monomorphism, $\Gamma^*(\sigma) = \eta$.

The line bundle defined by σ over X^1 comes from an element $\hat{\sigma}$ in $K_{Z_2}(X^1)$. Since Γ is equivariant, $\Gamma^*(\hat{\sigma}) = \hat{\eta} \in K_{Z_2}(X^0)$ and

$\hat{\eta}$ maps to η via the forgetful map $K_{Z_2}(X^0) \to K(X)$.

Corollary 1.3. If τ is an involution of type k on a manifold homotopy equivalent to CP^n, then any line bundle η over X comes from an element $\hat{\eta} \in K_{Z_2}(X)$.

Proof: τ^* is the trivial map in cohomology. It follows from Bredon's theorem that $H^2(X) \to H^2(F)$ is a monomorphism; thus the preceding corollary applies.

§2. The Class $\delta_{Z_2} \in K_{Z_2}(TX)$

As observed in [8], if X is a homotopy CP^n, its tangent bundle has a $spin^c(2n)$ structure i.e. there is a principle $spin^c(2n)$ bundle over X with total space P such that

$$P \times_{spin^c(2n)} R^{2n} = TX.$$

Suppose there is an involution on P commuting with the principle $spin^c(2n)$ action on P which covers the canonical involution on the frame bundle Q of TX defined by sending a frame $[V_1, V_2, \ldots, V_{2n}] \in Q$ to the frame $[d\tau_* V_1, d\tau_* V_2 \ldots d\tau_* V_{2n}]$ where $v_i \in TX$ and $d\tau_*$ is the differential of the involution τ on X. Then the half $spin^c(2n)$ modules Δ_+ and Δ_- give Z_2 vector bundles ω_+ and ω_- over TX

$$\omega_{\pm} = P \times_{spin^c(2n)} (R^n \times \Delta_{\pm})$$

and there is a Z_2 complex E [2] over TX; $E_0 \to E_1$, $E_0 = \omega_+$, $E_1 = \omega_-$ which defines an element

$$\delta_{Z_2} \in K_{Z_2}(TX) \qquad [8].$$

In order to have this element we need to know when an involution on X lifts to P with the specified properties. To follow are conditions under which the lifting can be achieved. We do <u>not</u> assume X is a homotopy complex projective space.

<u>Lemma 2.1</u>. Let G_1 and G_2 be two compact Lie groups. We suppose that $G_1 \times G_2$ acts on the left on Q where $H^1(Q,Z) = 0$. We also suppose P is the total space of a principle right S^1 bundle over Q and there are actions of G_i on P commuting with the principle S^1 action and covering the actions of G_i on Q for $i = 4,2$. Then the action of G_1 on P may be modified so that $G_1 \times G_2$ acts on P covering the action of $G_1 \times G_2$ on Q.

<u>Proof</u>: Define a function $\psi_1\colon G_1 \times G_2 \times P \to S^1$ by $g_2 g_1 x = g_1 g_2 x \psi_1(g_1,g_2,x)$ where $g_i \in G_i$ $i = 1,2$ and $x \in P$. Since

i) $\psi_1(g_1,g_2;xt) = \psi_1(g_1,g_2,x)$ for $t \in S^1$, there is a function $\psi\colon G_1 \times G_2 \times Q \to S^1$ such that $\psi_1(g_1,g_2,x) = \psi(g_1,g_2,z(x))$ where $z\colon P \to Q$ is the projection. The function ψ has these properties

ii) $\psi(1,g_2,z) = \psi(g_1,1,z) = 1$, $z \in Q$ and 1 is the identity of the appropriate group.

iii) $\psi(g_1 g_1',g_2,z) = \psi(g_1,g_2,g_1' z)\psi(g_1',g_2,z)$

iv) $\psi(g_1,g_2 g_2',z) = \psi(g_1,g_2,g_2' z)\psi(g_1,g_2',z)$.

Because of ii) and $H^1(Q,Z) = 0$, ψ is null homotopic and there is a unique lifting $\tilde{\psi}$ of ψ to R^1 which satisfies

$$\tilde{\psi}(1,z_0,1) = 1 \text{ for some fixed } z_0 \in Q \text{ and}$$

$$\pi\tilde{\psi} = \psi \text{ where } \pi\colon R^1 \to S^1 \text{ is the covering}$$

homomorphism. Moreover $\tilde{\psi}$ will satisfy ii), iii) and iv) except we change from a multiplicative to an additive notion. It follows from iv) that

v) $\tilde{\psi}(g_1, g_2 g_2'^{-1}, g_2'z) - \tilde{\psi}(g_1, g_2'^{-1}, g_2'z) = \tilde{\psi}(g_1, g_2, z)$

Define

$$\gamma(g_1, z) = \int_{G_2} \tilde{\psi}(g_1, g_2'^{-1}, g_2'z)dg_2'$$

where dg_2 denotes the normalized Haar measure on G_2. Then

vi) $\gamma(g_1, g_2 z) - \gamma(g_1, z) =$

$$\int_{G_1} \tilde{\psi}(g_1, g_2'^{-1}, g_2'g_2 z)dg_2' - \int_{G_1} \tilde{\psi}(g_1, g_2'^{-1}, g_2'z)dg_2'$$

$$= \int_{G_1} [\tilde{\psi}(g_1, g_2 g_2''^{-1}, g_2''z) - \tilde{\psi}(g_1, g_2''^{-1}, g_2''z)]dg_2''$$

$$= \int_{G_1} \tilde{\psi}(g_1, g_2, z)dg_2'' = \tilde{\psi}(g_1, g_2, z)$$

by v). Also

vii) $\gamma(g_1 g_2, z) = \gamma(g_1, g_2 z) + \gamma(g_2, z)$ by iii).

Set $\bar{\gamma}(g_1, z) = \pi\gamma(g_1, z)$ and define a new left action of S^1 on P by

$$g_1 \circ x = g_1 x \bar{\gamma}(g_1, z(x)). \text{ Then}$$

$$g_2(g_1 \circ x) = g_2 g_1 x \bar{\gamma}(g_1, z(x)) = g_1 g_2 x \psi(g_1, g_2, x)\bar{\gamma}(g_1, z(x))$$

$g_1 \circ (g_2 x) = g_1 g_2 x \bar{\gamma}(g_1, g_2 x)$. By vi) we have

viii) $g_2(g_1 \circ x) = g_1 \circ (g_2 x)$ for $g_i \in G_i$ and $x \in P$.

Also

ix) $g_1 \circ (g_1' \circ x) = g_1 g_1' \circ x$ by vii).

Property ix) shows that our definition of o gives an action of G_1. Property viii) shows that the actions of G_1 and G_2 commute. It follows from the definition that the new action of G_1 on R covers the action of G_1 on Q.

Let $Z_2^n = Z_2 \times Z_2 \times \ldots Z_2$ be the product of n factors of Z_2.

Corollary 2.2. Suppose Z_2^n acts on the left of Q where $H^1(Q,Z) = 0$, that P is the total space of a principle S^1 bundle over Q and that each of the n actions defined by the n factors of Z_2 in Z_2^n lifts to P commuting with the principle S^1 action and covering the Z_2 action on Q. Then the actions of the factors Z_2 on P may be modified so that Z_2^n acts on P covering the Z_2^n action on Q.

Lemma 2.3. Suppose Z_2 acts on Q freely and $H^1(Q,Z) = 0$. Let P be the total space of a principle right S^1 bundle over Q. Then the action of Z_2 lifts to P commuting with the action of S^1.

Proof: Let $\overline{Q}$ be the orbit space of the Z_2 action on Q and let $f: Q \to CP^\infty$ be a map inducing the given principle S^1 bundle. Since $H^2(\overline{Q},Z) \to H^2(Q,Z)$ is onto, there is a map $\overline{f}: \overline{Q} \to CP^\infty$ such that $\overline{f}\pi \sim f$ where $\pi: Q \to \overline{Q}$ is the orbit map. So we may assume $f(gx) = f(x)$ where $g \in Z_2$ is the generator and $x \in Q$.

Since $P \subset Q \times S^\infty$ is the subset of points (q,s) with $f(q) = \pi(s)$ and $\pi: S^\infty \to CP^\infty$ the S^1 orbit map, we may define an action of Z_2 on P by

$$g(x,s) = (gx,s).$$

Then this action commutes with the right action of S^1 on P defined by

$$(x,s)t = (x,st) \quad s \in S^\infty,\ t \in S^1,\ x \in Q$$

and covers the given action on Q.

Corollary 2.4. Suppose Z_2^n acts on the left of Q where $H^1(Q,Z) = 0$ and that P is the total space of a principle S^1 bundle over Q. Then the Z_2^n action on Q lifts to a Z_2^n action on P commuting with the principle S^1 action on P.

Corollary 2.5. Suppose that X is a smooth n manifold with $H^1(X,Z) = 0$ which supports an action of Z_2^k. Let P be the total space of a principle $\mathrm{spin}^c(n)$ bundle associated to TX. Then there is a lifting of the canonical action of Z_2^k on Q, the total space of the principle $SO(n)$ bundle associated to TX, to P. Moreover, the action of Z_2^k on P commutes with the principle $\mathrm{spin}^c(n)$ action on P.

Proof: Let $G_1 = Z_2^k$ and $G_2 = \mathrm{spin}^c(n)$. We have an action of $G_1 \times G_2$ on Q by setting $(g_1,g_2)oq = g_1 q\rho(g_2^{-1})$ where $\rho: G_2 \to SO(n)$ is the canonical epimorphism of groups. By Corollary 2.5 the action of G_1 lifts to P. Since G_2 automatically acts on the left of P by

$$g_2 oq = qq_2^{-1},$$

the hypothesis of Lemma 2.1 are satisfied, hence the conclusion.

Corollary 2.6. Suppose that X is a $\mathrm{spin}^c(2n)$ manifold supporting an action of Z_2^k and $H^1(X,Z) = 0$. Let $R^{2n} \times \Delta_+ \to \Delta_-$ be the bilinear pairing defined by clifford multiplication in the Clifford algebra of R^{2n}; denoted by $(v,x) \to v\cdot x$ $v \in R^n$, $x \in \Delta_+$. Then the bundles $\omega_\pm = P \times_{\mathrm{spin}^c(2n)} (R^{2n} \times \Delta_\pm)$ over $TX = P \times_{\mathrm{spin}^c(2n)} R^{2n}$ together with the mapping $\omega_+ \xrightarrow{\sigma} \omega_-$ defined by

$$\sigma[p,(v,x)] = [p,(v,v\cdot x)]$$

for $p \in P$, $v \in R^{2n}$, $x \in \Delta_+$ define a complex of vector bundles over TX [2] and hence an element $\delta_{Z_2^k} \in K_{Z_2^k}(TX)$. The action of Z_2^k on ω_+ is defined by

$$g[p,(v,x)] = [gp,(v,x)] \quad \text{when} \quad g \in Z_2^k.$$

§3. Applications to Involutions on Homotopy Complex Projective Spaces

Henceforth X is a smooth 2n dimensional manifold homotopy equivalent to CP^n. Suppose that X supports an involution of type k. Let $h: X \to CP^n$ be a homotopy equivalence, $\eta = h^*(H)$ where H is the Hopf bundle over CP^n and $\hat{\eta} \in K_{Z_2}(X)$ an equivariant line bundle whose underlying line bundle is η (Corollary 1.3). Let $\delta_{Z_2} \in K_{Z_2}(TX)$ be the element provided by Corollary 2.6. The cohomology ring of X is the truncated polynomial ring

$$Z[x]/(x^{n+1})$$

where $x = c_1(\eta)$ is the first Chern class of η.

Theorem 3.1. Suppose that X supports an involution of type 0; X^τ the fixed point set of τ is $X_0 \cup X_1$ and X_0 is a point; (So X_1 has the same mod 2 cohomology ring as CP^{n-1} by Bredon's theorem). Then there is an odd integer k defined by

$$i_*[X_1] = k\, x \cap [X] \quad \text{where} \quad i: X_1 \to X \quad \text{is the inclusion},$$

$[X_1]$ and $[X]$ are the orientation classes of X_1 and X respectively and $\cap$ denotes the cap product. Moreover

$$\text{i)} \quad \langle e^{(\frac{n+1}{2}+\ell)x} \tanh {}^{kx}/_2 \, \hat{A}(X), [X]\rangle \equiv \pm {}^1/_{2^n} \mod 1$$

$$\text{ii)} \quad \langle e^{(\frac{n+1}{2}+\ell)x} \hat{A}(X), [X]\rangle \equiv 0 \mod 1$$

Here $\langle y,[X]\rangle$ denotes the number obtained by evaluation of the cohomology class y on the orientation class $[X]$.

Proof: By Bredon's Theorem,

$$i_*: H_*(X_1, Z_2) \to H_*(X, Z_2)$$

is an isomorphism for $* \leq 2(n-1)$. Since the mod 2 reduction of $[X_1]$ generates $H_{2n}(X_1, Z_2)$ and since $i_*[X_1] = kx \cap [X]$ for some integer k, we see that k must be odd.

Since the values of characters of the group Z_2 are integers we have $Id^X_{Z_2}(\delta_{Z_2} n^\ell)(\tau) \in Z$ for every integer ℓ. It follows from the proof of Proposition 5.2 of [8] that

$$Id^X_{Z_2}(\delta_{Z_2} \hat{n}^\ell)(\tau) =$$

$$\sum_{j=0}^{1} \varepsilon_j \left\langle ch_\tau(n^\ell | X_j^\tau)\, e^{\frac{(n+1)x_j}{2}} \frac{\hat{A}(X_j^\tau)}{ch\Delta(NX_j^\tau)}, [X_j^\tau] \right\rangle$$

where

a) ch_τ: $K_{Z_2}(X^\tau) = R(Z_2) \otimes K(X^\tau) \to H^*(X^\tau, Q)$ is defined by $ch_\tau(X \otimes \alpha) = X(\tau) \cdot ch\alpha$; $X(\tau)$ is the value of the character X on τ ; $X \in R(Z_2)$ and $\alpha \in K(X^\tau)$

b) $n^\ell | X_j^\tau$ denotes the restriction of $n^\ell \in K_{Z_2}(X)$ to $K_{Z_2}(X_j^\tau)$ and x_j denotes the restriction of x to $H^*(X_j, Z)$

c) $\varepsilon_j = \pm 1$

d) $ch\Delta(NX_j^\tau)$ is the unit of $H^*(X^\tau, Q)$ defined by the formal power series

$$2^m \, \Pi_{i=1}^m \cosh \omega_i / 2$$

where the elementary symmetric functions of the ω_i^2 give the Pontriagin classes of the normal bundle NX^τ of X^τ in X and $2m$ is the real dimension of NX^τ. Of course for the case at hand X^τ has two components as does NX^τ, $\dim_R NX_1^\tau = 2$ and

$$ch\Delta(NX_1^\tau) = \cosh kx_{1/2} \, .$$

e) $\hat{A}(X_j^\tau)$ is the cohomology class defined by the power series

$$\Pi z_i/2(\sinh z_1/2$$

where the elementary symmetric functions of the z_1^2 give the Pontriagin classes of X_j^τ. Actually the class $\hat{A}(\xi)$ is defined for every oriented bundle ξ and by definition $\hat{A}(M) = \hat{A}(TM)$ where TM is the tangent bundle of the smooth manifold M. If i denotes the inclusion of X^τ in X, then

$$i*TX = TX^\tau \oplus NX^\tau \quad \text{so}$$

$$\hat{A}(X^\tau) = i*\hat{A}(X)A(NX^\tau)^{-1}$$

We find that for the component X_1^τ,

$$\hat{A}(NX_1^\tau)^{-1} = i* 2k^{-1}x^{-1} \sinh kx/2$$

$$ch_\Delta(NX_1^\tau) = i* 2 \cosh kx/2$$

$$ch_\tau(\eta^\ell | X_1^g) = \pm i* e^{\ell x} .$$

Thus

$$\left\langle e^{\frac{n+1}{2}x} ch_\tau(\eta^\ell|X_1^\tau)\hat{A}(X_1^\tau) \middle/ ch_\Delta(NX_1^\tau) \; , \; [X_1^\tau] \right\rangle =$$

$$\pm \left\langle i*\{e^{(\frac{n+1}{2}+\ell)x} \cdot \hat{A}(X) \cdot \frac{\sin k \, kx/2}{kx \cosh kx/2}\} \; , \; [X_1^\tau] \right\rangle =$$

$$\pm \left\langle e^{(\frac{n+1}{2}+\ell)x} \hat{A}(X) \frac{\tan k \, kx/2}{kx} \; , \; kx \cap [X] \right\rangle$$

$$= \pm \left\langle e^{(\frac{n+1}{2}+\ell)x} \hat{A}(X) \tan k \, kx/2 \; , \; [X] \right\rangle .$$

The contribution to $Id_{Z_2}^X(\delta_{Z_2}\hat{\eta}^\ell)(\tau)$ of the component X_0 = point is easily seen to be $\pm 1/2n$. The condition that $Id_{Z_2}^X(\delta_{Z_2}\hat{\eta}^\ell)(\tau)$ be an integer gives condition i) in view of the above computation.

The condition ii) is well known and follows from the fact that $Id^X_{Z_2}(\delta_{Z_2}\hat{\eta}^{\ell})(1)$ is an integer. Here $1 \subset Z_2$ is the identity.

<u>Theorem 3.2</u>: Suppose in addition to the hypothesis of the preceding theorem that $n=3$. Then

i) k is an odd square

ii) $P_1(X) = (48j+4)x^2$ for some integer j. Here $P_1(X)$ is the first Pontriagin class of X.

<u>Proof</u>: X_1 is an oriented 4 dimensional manifold having the same mod 2 cohomology ring as $\mathbb{C}P^2$; thus if $\hat{H}^1(X_1)$ denotes $H^1(X)$ modulo torsion we have

$$\hat{H}^*(X_1) \cong Z[x_1]/(x_1^2)$$

$x_1 \in \hat{H}^2(X_1)$. This is a consequence of Poincare duality.

Let $i^*(x) = dx_1$ where d is an integer. Then

$$k = \langle x^2 \cdot kx, [X] \rangle = \langle x^2, kx \cap [X] \rangle$$

$$= \langle x^2, i_*[X_1] \rangle = \langle (i^*X)^2, [X_1] \rangle =$$

$$\langle d^2 x_1{}^2, [X_1] \rangle = d^2 .$$

Since

$$\tanh {}^{kx}/_2 = {}^{kx}/_2 (1 - {}^{k^2x^2}/_{12}) \text{ and}$$

$$\hat{A}(X) = (1 - p_1\ x_2/\ 24)$$

where p_1 is the integer defined by $P_1(X) = p_1 \cdot x^2$, we obtain

a) $kp_1 \equiv 6\varepsilon - 2k^3 \mod 48$

by applying condition i) of the preceding theorem with $\ell = -2$, $n=3$. Here $\varepsilon = \pm 1$. In addition it follows from condition ii) of the preceding theorem that

b) $p_1 = 24j+4$ for some integer j.

Thus

c) $k(24j+4) \equiv 6\varepsilon - 2k^3 \mod 48$ so

d) $k(12j+2) \equiv 3\varepsilon - k^3 \mod 8$

But k is an odd square and hence $k \equiv 1 \mod 8$ and we obtain

e) $12j+2 \equiv 3\varepsilon - 1 \mod 8$

Since $\varepsilon = \pm 1$, e) requires $j \equiv 0(2)$ if $\varepsilon=1$. There is no solution for $\varepsilon = -1$.

Corollary 3.3: There are infinitely many smooth manifolds homotopy equivalent to $\mathbb{C}P^3$ which support no involution of type 0.

Proof: Montgomery-Yang [7] have shown that the oriented smooth manifold homotopy equivalent to $\mathbb{C}P^3$ are in one-one correspondence with the integers. Their correspondence $j \to X_j$ is characterized by $P_1(X_j) = (24j+4)x^2$ where x generates $H^2(X,Z)$.

From the preceding two corollaries we see that at most half of the manifolds homotopy equivalent to $\mathbb{C}P^3$ admit involutions of type 0. We shall now see that at most a finite number of manifolds homotopy equivalent to $\mathbb{C}P^4$ admit involutions of type 0.

Theorem 3.4 If X is homotopy equivalent to $\mathbb{C}P^4$ and admits an involution of type 0, then X is P. L. homeomorphic to CP^4.

Proof: It follows from the work of Sullivan [9] that to show that X is P. L. homeomorphic to $\mathbb{C}P^3$ we must show that the surgery obstructions

$$\sigma_1(f) \in Z_2 \text{ and } \sigma_2(f) \in Z$$

associated to the normal maps [4]

$$f|f^{-1}(\mathbb{C}P^3): \quad f^{-1}(\mathbb{C}P^3) \to \mathbb{C}P^3$$

$$f|f^{-1}(\mathbb{C}P^2): \quad f^{-1}(\mathbb{C}P^2) \to \mathbb{C}P^2$$

are zero. Here $f: X \to \mathbb{C}P^4$ is a homotopy equivalence which is transverse regular to $\mathbb{C}P^3$ and $\mathbb{C}P^2$ as submanifolds of $\mathbb{C}P^4$.

Remark: To prove that $\sigma_1(f)$ is zero, it suffices to show that X contains a submanifold X^1 of dimension 6 such that the homology class of X defined by X^1 is the dual of a generator of $H^2(X,Z)$ and such that $H_*(X^1,Z^2) \to H_*(X,Z_2)$ is an isomorphism for $* \leq 6$. This will imply that f can be deformed so that $f^{-1}(\mathbb{C}P^3) = X^1$ and the normal bundle of X^1 in X is the pull back of the normal bundle of $\mathbb{C}P^3 \subset \mathbb{C}P^4$. Moreover the condition that $H^*(X,Z_2) \to H^*(X^1,Z_2)$ be an isomorphism for $* \leq 6$ will imply that $f^{1*}: H^*(\mathbb{C}P^3,Z_2) \to H^*(X^1,Z_2)$ is an isomorphism when $f^1 = f|X^1$. This in turn shows that $\sigma_1(f) = 0$.

The Atiyah-Singer G Index Theorem for $G=Z_2$ ([2] page 583) gives

$$\text{Sign}\ (\tau,X) = \text{Index}\ (X^\tau \cdot X^\tau)$$

where $X^\tau \cdot X^\tau$ denotes the self intersection of X^τ in X. Since τ^* acts trivially on cohomology, Sign (τ,X) = Sign $(1,X)$ = Index $(X) = 1$. By hypothesis

$$X^\tau = X_o \cup X_1 \text{ where } X_o \text{ is a point}$$

and X_1 is the six dimensional component.

If the normal bundle of X_1 in X is ν, then the normal bundle of $X_1 \cdot X_1$ in X_1 is the restriction ν^1 of ν to $X_1 \cdot X_1$ and the normal buhdle of $X_1 \cdot X_1$ in X is $2\nu^1$. Let η be the complex line bundle over X which restricts to ν and c_1 its first chern class. Then if j and i are the inclusions of X_1 and $X_1 \cdot X_1$ in X

$$j_*[X_1] = c_1 \cap [X]$$

$$i_*[X_1 \cdot X_1] = c_1^2 \cap [X]$$

are the homology classes of X determined by X_1 and $X_1 \cdot X_1$.

The first Pontriagin class of X, $P_1(X)$ is p_1x^2 where p_1 is an integer and $x \in H^2(X,Z)$ is a generator. By the Hirzebruch Index Theorem [5].

$$\text{Index } (X_1 \cdot X_1) = \langle L(\tau_{X_1} \cdot X_1), [X_1 \cdot X_1] \rangle$$

where $L(\tau_{X_1} \cdot X_1)$ is a rational cohomology class depending on the Pontriagin classes of the tangent bundle of $X_1 \cdot X_1$ i.e., $\tau_{X_1 \cdot X_1}$. But $\tau_{X_1 \cdot X_1} + i^* 2\eta = i^* \tau_X$ and $L(\tau_{X_1 \cdot X_1}) = i^* L(\tau_X) \cdot L(\eta)^{-2}$ so

$$\text{Index } (X_1 \cdot X_1) = \langle i^* L(\tau_X) L(\eta)^{-2}, [X_1 \cdot X_1] \rangle =$$

$$\langle L(\tau_X) \cdot L(\eta)^{-2}, c_1{}^2 \cap [X] \rangle .$$

But $L(\tau_X)L(\eta)^{-2} = 1 + {}^1/_3\ (p_1(\tau_X) - 2c_1{}^2) +$ higher terms. Let c be the integer defined by $c_1 = cx$. Then

$$\text{Index } (X_1 \cdot X_1) = {}^1/_3\ (p_1 - 2c^2)c^2$$

Since $X_1 \cdot X_1 = X^\tau \cdot X^\tau$ we have from above $1 = \text{Index } (X^\tau \cdot X^\tau) = \text{Index } (X_1 \cdot X_1) = {}^1/_3(p_1 - 2c^2)c^2$ and $3 = (p_1 - 2c^2)c^2$. Clearly $c = \pm 1$ and $p_1 = 5$. We suppose X oriented so that $c=1$.

We are now in position to show $\sigma_1(f) = 0$. Note

a) $H^*(X,Z_2) \to H^*(X_1,Z_2)$ is an isomorphism for $* \leq 6$ by Bredon's theorem.

b) $i_*[X_1] = x \cap [X]$ where x generates $H^2(X,Z)$ because $c=1$.

By the remark above applied to $X^1 = X_1$ we conclude $\sigma_1(f) = 0$.

To see that $\sigma_2(f)$ is zero, we observe that

$$f^* P_1(\mathbb{C}P^4) = P_1(X)$$

because $P_1(X) = 5x^2$. Let $V = f^{-1}\ \mathbb{C}P^2$ and γ the line bundle over $\mathbb{C}P^4$ whose first chern class $c_1(\gamma)$ is dual to $\mathbb{C}P^3$. Then the dual of

$\mathbb{C}P^2$ in $\mathbb{C}P^4$ is $c_1(\gamma)^2$. By transversality the dual of V in X is x^2 and the normal bundle of V in X is $2j^*f^*(\gamma)$ where $j = V \to X$ is the inclusion. Thus

$$\text{Index }(V) = \langle L(j^*(\tau_X - 2f^*\gamma)), [V]\rangle =$$

$$\langle L(\tau_X)L(f^*\gamma)^{-2}, j_*[V]\rangle$$

$$= \langle L(\tau_X)L(f^*(\gamma))^{-2}, x^2 \cap [X]\rangle$$

$$= \langle \tfrac{1}{3}(5-2)x^2, x^2 \cap [X]\rangle = 1$$

But $\sigma_2(f) = \frac{1}{8}(\text{Index }(V) - 1) = 0$.

<u>Corollary 3.5</u> At most a finite number of manifolds homotopy equivalent to $\mathbb{C}P^4$ admit an involution of type zero.

<u>Proof</u>: If X admits an involution of type zero, X is P. L. homeomorphic to $\mathbb{C}P^3$. By [9], there are only a finite number of distinct smooth manifolds P. L. homeomorphic to $\mathbb{C}P^4$.

APPENDIX

Here we show the existence of orientation reversing fixed point free involutions on homotopy $\mathbb{C}P^3$'s. The question of the existence of an orientation preserving involution on every homotopy $\mathbb{C}P^3$ is still open.

(A.1) Every manifold X homotopy equivalent to CP^3 admits an orientation reversing fixed point free involution.

Proof: CP^3 fibers S^4 with the two sphere as fiber. The anti-podal map in the fibers gives rise to a fixed point free orientation reversing orientation on CP^3. The orbit space of the induced action of Z_2 on CP^3 is denoted by Y and we have a commutative diagram

$$\begin{array}{ccccc} S^2 & \longrightarrow & CP^3 & \xrightarrow{f} & S^4 \\ \downarrow{\scriptstyle p} & & \downarrow{\scriptstyle \Pi} & & \downarrow{\scriptstyle Id} \\ RP^2 & \longrightarrow & Y & \xrightarrow{g} & S^4 \end{array}$$

where the rows are fibrations and the maps p, Π are orbit maps (induced by the actions of Z_2) and Id is the identity map of S^4.

The set of unoriented diffeomorphism classes of manifolds homotopy equivalent to CP^3, written $hS(CP^3)$, is in one-one correspondence with the integers [7]. In fact, the correspondence is characterized by this property: Let $x \in H^2(CP^3,Z)$ be a generator and let X_i be the homotopy CP^3 attached to the integer i by the Montgomery-Yang correspondence. Then there is a homotopy equivalence $h_i\colon X_i \longrightarrow CP^3$ such that the first Pontryagin class of X_i, $P_1(X_i)$, is $h_i^*(24i+4)x^2$. More precisely: if $G \in \Pi_4(B_0) = Z$ is a generator, then $24iG$ is fiber homotopically trivial and the stable tangent bundle of X_i τ_{X_i} is $h_i^*(\tau_{CP^3} \oplus f^* 24iG)$.

Our aim is to produce smooth manifolds Y_i with $\Pi_1(Y_i) = Z_2$ and homotopy equivalent to Y with

$$\tau_{Y_i} = \bar{h}_i^* (\tau_Y \oplus g^* 24iG)$$

if $\bar{h}_i\colon Y_i \longrightarrow Y$ is the homotopy equivalence. If this is achieved, the universal cover of Y_i, X_i will be a homotopy CP^3 with

(A.2) $$\tau_{X_i} = \Pi^* \tau_{Y_i} = h_i^*(\tau_{CP^3} \oplus f^* 24iG)$$

where h_i is the map to Y covering $\bar{h}_i$. This shows that every homotopy CP^3 arises as the universal cover of such a Y_i.

To produce the manifolds Y_i we study the surgery sequence of [9], [10]: $hS(Y) \xrightarrow{i} [Y, G/_0] \xrightarrow{\sigma_0} L_6(Z_2,-)$. Here $[Y, G/_0]$ denotes the homotopy classes of base point preserving maps of Y to $G/_0$ and σ_0 is the smooth surgery obstruction with values in the abelian group $L_6(Z_2,-)$ [10]. The set $hS(Y)$ consists of equivalence classes of smooth manifolds homotopy equivalent to Y. See [19] for additional details.

We note these facts: $\dim Y=6$, $\Pi_i(PL/_0) = 0 \quad i \leq 7$ so $[Y, G/PL] \longrightarrow [Y, G/_0]$ is an isomorphism and we can deal with the P. L. situation:

$$hT(Y) \longrightarrow [Y, G/_{PL}] \xrightarrow{\sigma_{PL}} L_6(Z_2,-).$$

In this case the Kervaire-Arf invariant defines an isomorphism $c: L_6(Z_2,-) \longrightarrow Z_2$ [10] and if $c\,\sigma_{PL} = \Delta$, we have the Sullivan formula for Δ,

$$\Delta(f) = \langle W(Y) \cdot f^*(k) \, , \, [Y] \rangle$$

where $[Y] \in H_6(Y,Z_2)$ is the fundamental class, $f: Y \longrightarrow G/_{PL}$ represents an element of $[Y, G/_{PL}]$, $W(Y)$ is the total Stiefel Whitney class of Y and $k = (1+S_q{}^2+S_q{}^2+S_q{}^2)(x_2+x_6)$ where x_2, x_6 are elements of $H^*(G/_{PL},Z_2)$ of dimensions 2 and 6 respectively.

Using this formula we see that the composition

$$[S^4, G/_0] \xrightarrow{g^*} [Y, G/_0] \xrightarrow{\Delta} Z_2$$

is zero (S^4 has no cohomology in dimensions 2 or 6). Thus there is an element

$$[Y_\ell, \bar{h}_\ell] \in hS(Y)$$

such that

$$i[Y_\ell, \bar{h}_\ell] = g^*(z_\ell) \in [Y, G/_0]$$

where $z_\ell = \ell z$ and $z \in [S^4, G/_0] = Z$ is the generator of this infinite cyclic group. It follows from surgery theory that

(A.3) $\tau_{Y_\ell} = \bar{h}_\ell^*(\tau_Y \oplus j\ g^*(z_\ell)) \in [Y, BO]$. Here $j = [Y, G/_0] \longrightarrow [Y, BO]$ is induced by the natural map of $G/_0$ into the classifying space BO of the infinite orthogonal group.

By computation $j(z) = 24 \cdot G$ and the first Pontryagin class of G is a generator of $H^4(S^4)$. It follows from (A.3) that

(A.4) $\tau_{Y_\ell} = \bar{h}_\ell^*(\tau_Y + 24\ell g^*(G))$

(A.5) $\tau_{X_\ell} = h_\ell^*(\tau_{CP^3} + 24\ell f^*(G))$ where $h_\ell\colon X_\ell \longrightarrow CP^3$ is a homotopy equivalence covering the homotopy equivalence $h_\ell\colon Y_\ell \longrightarrow Y$. Thus

(A.6) $P_1(\tau_{X_\ell}) = h_\ell^*(4x^2 + 24\ell\ x^2) = h_\ell^*(24\ell + 4)x^2$.

We have thus shown that each smooth manifold X_ℓ is the universal cover of a manifold Y_ℓ with $\Pi_1(Y_\ell) = Z$ and hence admits a free orientation reversing involution

BIBLIOGRAPHY

[1] Atiyah, M. F., Vector Fields on Manifold, Arbitsgemeinschaft Fur Forshung Des Landes Nordrhein-Westfalen, Heft 200 (1969).

[2] __________, Singer I., The Index of Elliptic Operators I and III, Annals of Math, Vol. 87, No. 3, (1968), pp. 484-530. and 546-604.

[3] Bredon, G., The Cohomology Ring Structure of a Fixed Point Set, Annals of Math, Vol. 80, No. 3, (1964) pp. 524-537.

[4] Browder, Wm., Surgery on Simply Connected Manifolds, Mimeographed notes, Princeton University.

[5] Hirzebruch, F., Topological Methods in Algebraic Geometry, Springer-Verlag (1966).

[6] Hsiang, W. C., Manifolds-Amsterdam 1970 Springer-Verlag Lecture Series 197 (1971).

[7] Montgomery D. and Yang C. T., Free Differentiable Actions on Homotopy Seven Spheres, Proceedings of the Conference on Transformation Groups, Springer-Verlag, (1968).

[8] Petrie, T., Smooth S^1 Actions on Homotopy Complex Projective Spaces and Related Topics, to appear in Bull. A. M. S.

[9] Sullivan, D., Geometric Topology Seminar Notes, Princeton (1967).

[10] Wall, C. T. C., Free Piecewise Linear Involutions on Spheres, Bull. A. M. S. Vol. 74, No. 3, (1968) pp. 554-558.

ON THE HOMOLOGY OF WEIGHTED HOMOGENEOUS MANIFOLDS

Peter Orlik*

University of Wisconsin and University of Oslo

The object of this talk is to describe the characteristic map of an isolated singularity of a hypersurface defined by a weighted homogeneous polynomial and the homology information obtained in [4] for its neighborhood boundary and to give an explicit conjecture for the torsion in homology, not yet known in general.

1. Introduction

Let $\mathbb{C}^m$ denote complex m-space and consider the algebraic variety (hypersurface) defined by the zeros of a polynomial function $f(\underline{z}) = f(z_1,\ldots,z_m)$,

$$V = \{\underline{z} \in \mathbb{C}^m \mid f(\underline{z}) = 0\}.$$

A point $\underline{x} \in V$ is called <u>singular</u> if all partial derivatives $\partial f/\partial z_i$ vanish at $\underline{x}$, otherwise it is <u>simple</u>. A singular point is <u>isolated</u> if in some neighborhood of $\underline{x}$ all other points are simple. To study the topology of V at $\underline{x}$ we follow Milnor [3] and intersect V with a small sphere $S_\varepsilon = S_\varepsilon^{2n-1}$ centered at $\underline{x}$ and let $K = V \cap S_\varepsilon$. The mapping $\Phi(\underline{z}) = f(\underline{z})/|f(\underline{z})|$ from $S_\varepsilon - K$ to the unit circle is the projection map of a smooth fiber bundle. Each fiber $F_\theta = \Phi^{-1}(e^{i\theta}) \in S_\varepsilon - K$ is a smooth, parallelizable $(2m-2)$-manifold.

If $\underline{w}$ is an isolated singularity on V then K is a closed $(m-3)$-connected $(2m-3)$-manifold and F_θ has the homotopy type

* Partially supported by NSF.

of a bouquet of $(m - 1)$-spheres. Their number, μ, is strictly positive. The Wang sequence [3, 8.4] corresponding to the above fibration reduces for $m \geq 3$ to the following short exact sequence:

$$0 \longrightarrow H_{m-1}K \longrightarrow H_{m-1}F \xrightarrow{I_*-h_*} H_{m-1}F \longrightarrow H_{m-2}K \longrightarrow 0$$

Here I is the identity map of a typical fiber F and $h : F \longrightarrow F$ is the characteristic map of the fibration. It is the homeomorphism by which $F \times 0$ and $F \times 2\pi$ are identified in $F \times [0,2\pi]$ to obtain $S_\epsilon - K$.

Thus the matrix, H of the map $h_* : H_{m-1}F \longrightarrow H_{m-1}F$ contains all information about the homology of the $(m - 3)$-connected manifold K. Denote its characteristic polynomial by $\Delta(t) = \det(tI - H)$. Clearly

$$H_{m-1}K = \ker(I - H) \quad \text{and} \quad H_{m-2}K = \operatorname{coker}(I - H).$$

By Poincaré duality $H_{m-1}K$ is isomorphic to the free part of $H_{m-2}K$, so we only need to compute the latter group. Over the principal ideal domain $\mathbb{Z}$ the matrix $(I - H)$ is equivalent to a diagonal matrix, i.e. there are unimodular matrices U and V so that

$$U(I - H)V = \operatorname{diag}(m_1, m_2, \ldots, m_\mu)$$

where m_i divides m_{i+1} for all i. Then

$$H_{m-2}K = \mathbb{Z}_{m_1} \oplus \ldots \oplus \mathbb{Z}_{m_\mu}$$

where $\mathbb{Z}_1$ is the trivial group and $\mathbb{Z}_0$ is the infinite cyclic group. We shall investigate this group for a special class of manifolds defined below.

2. Weighted homogeneous manifolds

Let $w_1, \ldots, w_m$ be positive rational numbers. The polynomial $f(z_1, \ldots, z_m)$ is called weighted homogeneous of type $(w_1, \ldots, w_m)$ if it can be expressed as a linear combination of monomials $z_1^{i_1} \dots z_m^{i_m}$ for which

$$i_1/w_1 + \dots + i_m/w_m = 1.$$

The fact that $f(\underline{z})$ is weighted homogeneous implies that there is an action of the multiplicative group of complex numbers, $\mathbb{C}^*$, on $\mathbb{C}^m$ leaving the variety V invariant. Let $w_i = u_i/v_i$, $i = 1,\dots,m$ be the irreducible representations of the positive rational numbers w_i as fractions of positive integers. Let d be the least common multiple of the numerators and define positive integers $q_i = d/w_i$. Let $t \in \mathbb{C}^*$ act on $\mathbb{C}^m$ by

$$t(z_1,\dots,z_m) = (t^{q_1}z_1,\dots,t^{q_m}z_m).$$

It follows that $f(t\underline{z}) = t^d f(\underline{z})$ and hence V is invariant under the action.

The following is essentially proved in [5] and [6].

Proposition 2.1. If V is defined by a weighted homogeneous polynomial, then the origin, $\underline{0}$ is contained in V. For any two positive numbers ε, ε' the neighborhood boundaries of $\underline{0}$, K and K' are equivariantly diffeomorphic and V is the open cone over K with cone point $\underline{0}$.

From now on we shall study weighted homogeneous polynomials whose varieties have isolated singularities. Call the corresponding neighborhood boundaries weighted homogeneous manifolds. First consider the following four types of polynomials:

$$f(z) = z^a$$

$$f(z_1,z_2) = z_1z_2$$

$$f(z_1,\dots,z_k) = z_1^{a_1} + z_1z_2^{a_2} + z_2z_3^{a_3} + \dots + z_{k-1}z_k^{a_k}, \qquad k > 1,$$

$$f(z_1,\dots,z_k) = z_1^{a_1}z_2 + z_2^{a_2}z_3 + \dots + z_{k-1}^{a_{k-1}}z_k + z_k^{a_k}z_1, \qquad k > 1.$$

The next result is obtained along the lines of [5,3.1].

Proposition 2.2. Let $h(z_1,\ldots,z_m)$ be a weighted homogeneous polynomial with weights $(w_1,\ldots,w_m)$ whose variety V_h has an isolated singularity at 0. Then there exists a (not necessarily unique) weighted homogeneous polynomial $g(z_1,\ldots,z_m)$ with the same weights so that

(i) $g(z_1,\ldots,z_m)$ is a sum of the above four types of polynomials in disjoint variables,

(ii) V_g has an isolated singularity at 0,

(iii) there is an analytic deformation of V_f into V_g giving an equivariant diffeomorphism of some neighborhood boundary K_f into some K_g.

Now let us return to the fibration mentioned in the introduction. Milnor [3, 9.4] proves that the fiber F is diffeomorphic to the non-singular variety

$$F' = \{\underline{z} \in \mathbb{C}^m \mid f(\underline{z}) = 1\}$$

and the characteristic map may be chosen as

$$h(z_1,\ldots,z_m) = (\xi^{q_1} z_1,\ \xi^{q_2} z_2,\ldots,\ \xi^{q_m} z_m)$$

where $\xi = \exp(2\pi i/d)$. In particular h is a map of order d. Recall that F has the homotopy type of $\vee_\mu S^{m-1}$. According to [4]:

Proposition 2.3. The group $H_{m-1}F$ is free abelian of rank $\mu = (w_1 - 1)(w_2 - 1) \ldots (w_m - 1)$.

Next the characteristic polynomial $\Delta(t)$ can be computed [4]. It is crucial to note that in order to find $\Delta(t)$ we may consider the map

$$h_*^{\mathbb{C}} : H_{m-1}(F;\ \mathbb{C}) \longrightarrow H_{m-1}(F;\ \mathbb{C})$$

since $\Delta(t) = \det(tI - H) = \det(tI - H^{\mathbb{C}})$.

To each monic polynomial $(t-\alpha_1)\dots(t-\alpha_k)$ with $\alpha_1 \dots \alpha_k \in \mathbb{C}^*$ assign the divisor

$$\text{divisor}\left[(t-\alpha_1)\dots(t-\alpha_k)\right] = <\alpha_1> + \dots + <\alpha_k>$$

thought of as an element of the integral group ring $\mathbb{Z}\mathbb{C}^*$. In particular let

$$\Lambda_n = \text{divisor}(t^n - 1) = <1> + <\xi> + \dots + <\xi^{n-1}>$$

where $\xi = \exp(2\pi i/n)$.

For integers $a_1,\dots,a_k$ denote by $[a_1,\dots a_k]$ their least common multiple and $(a_1,\dots,a_k)$ their greatest common divisor. Note the multiplication rule

$$\Lambda_a\Lambda_b = (a,b)\,\Lambda_{[a,b]}.$$

With the above notation we have in $Q\mathbb{C}^*$ [4]:

Proposition 2.4.

$$\text{divisor}\,\Delta(t) = \left(\frac{1}{v_1}\Lambda_{u_1} - 1\right)\left(\frac{1}{v_2}\Lambda_{u_2} - 1\right)\dots\left(\frac{1}{v_m}\Lambda_{u_m} - 1\right)$$

$$= \sum_I (-1)^{m-s} \frac{w_{i_1}\dots w_{i_s}}{[u_{i_1},\dots,u_{i_s}]}\Lambda_{[u_{i_1},\dots,u_{i_s}]}$$

where I denotes the 2^m subsets $\{i_1,\dots,i_s\}$ of $\{1,\dots,m\}$.

Thus $\Delta(t)$ may be computed explicitly in terms of the weights $(w_1,\dots,w_m)$. The ring $\mathbb{C}[t]$ is a principal ideal domain, hence the matrix $(tI - H)$ is equivalent to a diagonal matrix, i.e. there exist unimodular matrices $U(t)$ and $V(t)$ with entries in $\mathbb{C}[t]$ so that

$$U(t)[tI - H]\,V(t) = \text{diag}[m_1(t),\dots,m_\mu(t)] \qquad (2.5)$$

where $m_i(t)$ divides $m_{i+1}(t)$ for all i.

The minimal polynomial, $m_\mu(t)$ contains each irreducible factor of $\Delta(t)$. On the other hand $H^d - I = 0$ so $m_\mu(t)$ divides $t^d - 1$ and therefore it is square free. Thus $\Delta(t)$ determines the right hand side of (2.5) uniquely.

Since $\mathrm{rank}(\ker h_*) = \mathrm{rank}(\ker h_*^{\mathbb{C}})$ it follows from above that the rank of $H_{m-1}K$ (and $H_{m-2}K$) equals $\varkappa$, the exponent of $(t-1)$ in $\Delta(t)$. The latter is easily computed [4]:

Proposition 2.6. _The rank of_ $H_{m-2}K$ _equals_

$$\varkappa = \sum_I (-1)^{m-s} \frac{w_{i_1} \cdots w_{i_s}}{[u_{i_1}, \ldots, u_{i_s}]} .$$

3. Torsion conjectures

Unfortunately, the proof of (2.4) depends on working with $h_*^{\mathbb{C}}$. Thus no information about the torsion subgroup of $H_{m-2}K$ is obtained. The following conjecture would correct this.

Conjecture 3.1. _The matrices_ $U(t)$ _and_ $V(t)$ _of_ (2.5) _may be chosen with entries in the subring_ $\mathbb{Z}[t]$ _so that they are unimodular in this subring_.

This immediately implies:

Conjecture 3.2.

$$H_{m-2}K = \mathbb{Z}_{m_1(1)} \oplus \mathbb{Z}_{m_2(1)} \oplus \cdots \oplus \mathbb{Z}_{m_\mu(1)}$$

where $\mathbb{Z}_1$ _is the trivial group and_ $\mathbb{Z}_0$ _is the infinite cyclic group._

J. Milnor and D. McQuillan pointed out to me that there are examples of matrices with properties similar to those of H not satisfying (3.1); e.g. let

$$A = \begin{bmatrix} 1 & 0 & 0 \\ 0 & 0 & 1 \\ 0 & -1 & -1 \end{bmatrix} .$$

Then $A^3 = I$ and its characteristic polynomial is $\Delta(t) = t^3 - 1$. Over $\mathbb{C}[t]$ the matrix $tI - A$ is equivalent to $\mathrm{diag}(1,1, t^3-1)$ but over $\mathbb{Z}[t]$ only to $\mathrm{diag}(1, t^2+t+1, t-1)$. The difficulty is, of course, that $\mathbb{Z}[t]$ is not a principal ideal domain and (3.1) holds only if every ideal generated by the determinants of $k \times k$ minors is principal.

We shall now make (3.2) more specific in terms of the weights $(w_1,\dots,w_m)$. Given an index set $\{i_1,\dots,i_s\}$ let I denote all its subsets and J all its proper subsets. For all subsets $\{i_1,\dots,i_s\}$ of $\{1,\dots,m\}$ with $i_1 < i_2 < \dots < i_s$ define inductively

$$c_{i_1,\dots,i_s} = (u_1,\dots,\hat{u}_{i_1},\dots,\hat{u}_{i_s},\dots,u_m)/\prod_J c_{j_1,\dots,j_t}$$

where "^" means "delete". Note that $c_{i_1,\dots,i_s}$ is the part of the greatest common divisor of $\{u_1,\dots,\hat{u}_{i_1},\dots,\hat{u}_{i_s},\dots,u_m\}$ which is not a common divisor of any larger set of u_i - s properly containing the above set. Similarly define

$$\varkappa(w_{i_1},\dots,w_{i_s}) = \sum_I (-1)^{s-t} \frac{w_{j_1}\cdots w_{j_t}}{[u_{j_1},\dots,u_{j_t}]} .$$

Let

$$\epsilon_{m-s} = \begin{cases} 0 & \text{if } m-s \text{ is even} \\ 1 & \text{if } m-s \text{ is odd} \end{cases}$$

and

$$k_{i_1,\dots,i_s} = \epsilon_{m-s}\varkappa(w_{i_1},\dots,w_{i_s}).$$

Finally, define

$$d_j = k_{i_1,\dots,i_s}\prod_{\geq j} c_{i_1,\dots,i_s}$$

and let $\quad r = \max\{k_{i_1,\dots,i_s}\}$.

Conjecture 3.3. *The torsion subgroup of* $H_{m-2}K$ *is isomorphic to*

$$\mathbb{Z}_{d_1} \oplus \mathbb{Z}_{d_2} \oplus \dots \oplus \mathbb{Z}_{d_r}.$$

To see that (3.3) follows from (3.1) define

$$g_{i_1,\dots,i_s} = \prod_I c_{j_1,\dots,j_t}$$

and

$$f_{i_1,\dots,i_s}(t) = \frac{t^{g_{i_1,\dots,i_s}} - 1}{(t-1)\prod_J f_{j_1,\dots,j_t}(t)} .$$

Then $f_{i_1,\dots,i_s}(1) = c_{i_1,\dots,i_s}$ and the product of irreducible polynomials, $f_{i_1,\dots,i_s}(t)$ occurs exactly with exponent $- (-1)^{m-s}\,\varkappa(w_{i_1},\dots,w_{i_s})$ in $\Delta(t)$. Thus when $(m-s)$ is even this will not contribute to the torsion and when $(m-s)$ is odd the contribution will be as stated, according to conjecture (3.1).

Proposition 3.4. *For* $m = 3$ *conjecture* (3.3) *holds*. This follows from the computations of [5, § 3].

In the special case when the weighted homogeneous polynomial is of the form

$$f(\underline{z}) = z_1^{u_1} + z_2^{u_2} + \dots + z_m^{u_m}$$

the associated weighted homogeneous manifolds have been studied by Brieskorn [1] and are usually called Brieskorn manifolds. In this case the matrix H of the characteristic map is computed as follows [1], [2], [3].

Let H_u denote the $(u-1) \times (u-1)$ matrix

$$H_u = \begin{bmatrix} 0 & 1 & 0 & \dots & & 0 \\ 0 & 0 & 1 & 0 & \dots & 0 \\ & & & & & \\ & & & & & 1 \\ -1 & -1 & \dots & & & -1 \end{bmatrix}$$

It is the companion matrix of the polynomial $t^{u-1} + t^{u-2} + \dots + t + 1$. Then

$$H = H_{u_1} \otimes H_{u_2} \otimes \dots \otimes H_{u_m}$$

where $\otimes$ denotes tensor (Kronecker) product of matrices. It is therefore a matter of straightforward calculation to check the conjectures for each Brieskorn manifold.

Example 3.5. Consider the polynomial

$$f(\underline{z}) = z_1^2 + z_2^3 + z_3^3 + z_4^4 + z_5^6.$$

Here $c_{2,3} = 2$, $c_{1,4} = 3$, $c_{1,2,3,5} = 2$ and all other c's equal 1. $k_{2,3} = \varkappa(3,3) = 2$, $k_{1,4} = \varkappa(2,4) = 1$, $k_{1,2,3,5} = \varkappa(2,3,3,6) = 4$ so $d_1 = 12$, $d_2 = 4$, $d_3 = d_4 = 2$. Note also that $\varkappa = \varkappa(2,3,3,4,6) = 2$. Thus according to (2.6) and (3.3) we have

$$H_3K^7 = \mathbf{Z} \oplus \mathbf{Z} \oplus \mathbf{Z}_{12} \oplus \mathbf{Z}_4 \oplus \mathbf{Z}_2 \oplus \mathbf{Z}_2.$$

Diagonalizing the 60×60 matrix $(I - H_2 \otimes H_3 \otimes H_3 \otimes H_4 \otimes H_6)$ gives the same result. Performing the diagonalization of $(tI - H)$ shows that also (3.1) holds in this case.

References

1. E. Brieskorn: Beispiele zur Differentialtopologie von Singularitäten, Invent. Math. 2. 1-14 (1966).

2. F. Hirzebruch and K. Meyer: O(n) - Mannigfaltigkeiten, exotische Sphären und Singularitäten, Lecture Notes 57, Springer Verlag (1968).

3. J. Milnor: Singular points of complex hypersurfaces. Ann. Math. Stud. 61, (1968). Princeton U. Press.

4. J. Milnor and P. Orlik: Isolated singularities defined by weighted homogeneous polynomials, Topology 9. 385-393 (1970).

5. P. Orlik and P. Wagreich: Isolated singularities of algebraic surfaces with $\mathbb{C}^*$ action, Ann. of Math. 93, 205-228 (1971).

6. P. Orlik and P. Wagreich: Singularities of algebraic surfaces with $\mathbb{C}^*$ action, Math. Annalen (to appear).

EQUIVARIANT RESOLUTION OF SINGULARITIES WITH C* ACTION

Peter Orlik*
University of Wisconsin and University of Oslo
Philip Wagreich*
University of Pennsylvania

1. Introduction

The connection between the algebraic properties of a singular point of an algebraic variety and the topological properties of its neighborhood was first investigated by Brauner [4] in the case of complex curves. For isolated singularities of complex surfaces Mumford [19] proved that if a neighborhood boundary is simply connected then the point is simple (non-singular). The connection between the singularities of certain algebraic surfaces and Seifert manifolds was first observed by Hirzebruch [9 , 10, 11, 12] and further studied by von Randow [27], Brieskorn [5] and Neumann [20].

There are, of course, the general results of Brieskorn [6] and Milnor [16] for higher dimensional hypersurfaces but these have a different point of view. While the latter derive information about the singularity from the embedding of the neighborhood boundary in the ambient space, the approach of Mumford and Hirzebruch adopted in our papers [22], [23] relies on detailed knowledge of the neighborhood boundary itself and therefore it is intrinsic and independent of the embedding dimension. In this paper we shall give an updated survey of our results on complex surfaces with C* action [22,23] and prove, in sections 4 and 6, some new theorems which are the first steps toward generalizing our work to higher dimensions.

Suppose $V \subset \mathbb{C}^{n+1}$ is a complex algebraic variety of complex dimension m with $\underline{0} \in V$ and let $f_1,\ldots,f_r$ be polynomials so that $V = \{\underline{z} \mid f_1(\underline{z}) = \ldots = f_r(\underline{z}) = 0\}$. A. Wallace [27] proved that there is a topological space K of dimension $2m-1$ and $\epsilon_0 > 0$ so that

$$V \cap S_\epsilon^{2n+1} \approx K \quad \text{and} \quad V \cap D_\epsilon^{2n+2} \approx \text{cone}(K) , \quad \epsilon \le \epsilon_0$$

* Partially supported by NSF.

where D_ϵ^{2n+2} is the ball of radius ϵ centered at $\underline{0}$, $S_\epsilon^{2n+1} = \partial D_\epsilon^{2n+2}$ and $\approx$ denotes homeomorphism. If $\underline{0}$ is an isolated singular point of V then Milnor [16] showed that K^{2m-1} is a C^∞ manifold and the above statements hold with $\approx$ denoting diffeomorphism in the first expression. Assume for the time being that $\underline{0}$ is an isolated singular point of V. We may pose two general problems.

Problem I. Determine K (up to homeomorphism or diffeomorphism) from the polynomials $f_1,\dots,f_r$.

Before stating the second problem recall that K bounds. This follows from Hironaka's proof of the resolution of singularities [9] which we state below in a special case.

Theorem 1.1: If V is an algebraic variety with one singular point $\underline{0} \in V$ then there is a non-singular algebraic variety $\tilde{V}$ and a proper morphism $\pi : \tilde{V} \to V$ so that for $X = \pi^{-1}(\underline{0})$

(i) $X = \bigcup_{i=1}^{r} X_i$, where each X_i is compact non-singular, $\operatorname{codim}_{\mathbb{C}} X_i = 1$ and the X_i meet with normal crossings,

(ii) $\pi : \tilde{V} - X \to V - \{\underline{0}\}$ is an isomorphism.

(A map is proper if the inverse image of any compact set is compact.) This theorem gives a manifold M^{2m} so that $\partial M^{2m} = K$ by letting $M = \pi^{-1}(D_\epsilon \cap V)$.

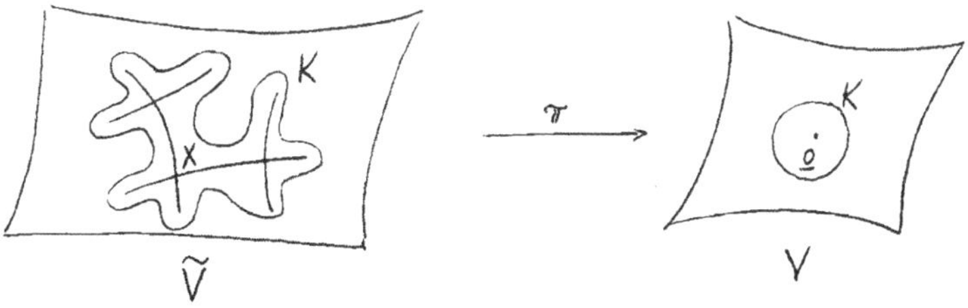

The theorem says that any isolated algebraic singularity is obtained by collapsing to a point some submanifolds of a manifold $\tilde{V}$. Hironaka's method gives an algorithm for actually finding π but the construction cannot always be carried out in a reasonable amount of time. This technique also has the defect of not being canonical (there are choices along the way) and of giving an M which is "too big". For example M may be a connected sum $M = M' \# \mathbb{CP}^m \# \ldots \# \mathbb{CP}^m$.

Problem II. Find an effective algorithm for obtaining a suitable M. Our aim is to show how to solve these two problems for algebraic surfaces $(m = 2)$ under the assumption that there is an action of the multiplicative group of complex numbers, $\mathbb{C}^*$ on V.

Suppose $q_0, \ldots, q_n$ are integers and $(q_0, \ldots, q_n) = 1$. Then we can define an effective action σ of $\mathbb{C}^*$ on $\mathbb{C}^{n+1}$ by

$$(1.2) \qquad \sigma(t,(z_0,\ldots,z_n)) = (t^{q_0}z_0,\ldots,t^{q_n}z_n)$$

If $q_i > 0$, $\forall_i$ we say the action is good. Suppose V is invariant under σ, then there is an induced S^1 action on K.

Examples 1.3: (i) Suppose f is a Brieskorn polynomial, i.e.

$$f(z_0,\ldots,z_n) = z_0^{a_0} + \ldots + z_n^{a_n}$$

and $d = \operatorname{lcm}(a_0,\ldots,a_n)$, $q_i = d/a_i$. Then $V = \{\underline{z} \mid f(\underline{z}) = 0\}$ is invariant under the action (1.2).

(ii) More generally, we say f is weighted homogeneous if there are positive integers $q_0,\ldots,q_n$ and d so that

$$f(t^{q_0}z_0,\ldots,t^{q_n}z_n) = t^d f(z_0,\ldots,z_n)$$

for all $t \in \mathbb{C}^*$ and $(z_0,\ldots,z_n) \in \mathbb{C}^{n+1}$. Then V is invariant under the action (1.2). For example,

$$f(\underline{z}) = z_0 z_1^{a_1} + z_1 z_2^{a_2} + \ldots + z_n z_0^{a_0}$$

is weighted homogeneous but not Brieskorn in general.

(iii) Suppose $V = \{\underline{z} \mid f_1(\underline{z}) = \ldots = f_r(\underline{z}) = 0\}$ where

$$f_i(t^{q_0}z_0,\ldots,t^{q_n}z_n) = t^{d_i} f(z_0,\ldots,z_n)$$

then V is invariant under (1.2). In fact it can be shown that if V is an algebraic subvariety of $\mathbb{C}^{n+1}$ invariant under a good $\mathbb{C}^*$-action (1.2) then V can be described as above [22, Prop.1.1.2].

Now we can ask for equivariant solutions to problems I and II.

I'. Determine K as a manifold with S^1 action.

II'. Find a $2m$-manifold M with S^1 action so that $K = \partial M$ equivariantly. We allow M to have fixed points even when the action on K is fixed point free.

The first non-trivial case is when $m = 2$. If σ is a good $\mathbb{C}^*$-action then K is an oriented 3-manifold with fixed point free circle action. The equivariant classification of such manifolds is given in terms of certain Seifert invariants [21]. Let $0_1, \ldots, 0_r$ be the orbits of K with non-trivial isotropy groups $\mathbb{Z}_{\alpha_1}, \ldots, \mathbb{Z}_{\alpha_r}$ respectively and let β_i be the integer $0 < \beta_i < \alpha_i$ determining the slice representation $\mathbb{Z}_{\alpha_i} \to SO(2)$. Then $(\alpha_i, \beta_i) = 1$. Finally let g equal the genus of $X = K/S^1$ and denote by b the "Chern class" of K over X. Then up to a permutation of $\{1, \ldots, r\}$ the invariants $\{-b; g; (\alpha_1, \beta_1), \ldots, (\alpha_r, \beta_r)\}$ determine K up to orientation preserving equivariant homeomorphism.

2. The relation between K and the resolution

Recall that we are assuming that V has complex dimension 2. We shall now indicate how to describe, for any resolution $\pi : \tilde{V} \to V$, the topology of $M = \pi^{-1}(D_\epsilon \cap V)$. In the case when there is a $\mathbb{C}^*$-action we shall show the relation between M and the Seifert invariants of K.

Suppose $\pi : \tilde{V} \to V$ is a resolution satisfying the conditions of (1.1). Then the X_i are Riemann surfaces of genus g_i. It is easy to see that M is determined by the g_i and the intersection matrix $[(X_i \cdot X_j)]$. We can tabulate this information by associating to π a weighted graph Γ in the following way. To each curve X_i assign a vertex e_i of Γ, $i = 1, \ldots, r$. An edge joins e_i and e_j for each point of $X_i \cap X_j$. The vertices are weighted by

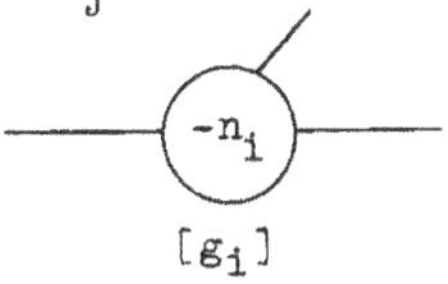

where $-n_i = (X_i \cdot X_i)$ and $g_i = \text{genus} X_i$. Clearly Γ determines M up to diffeomorphism.

Remark: It is not hard to show that the intersection matrix is negative definite [19]. Conversely, one can show, using theorems of

Grauert [7] that any negative definite weighted graph as above can be realized from the resolution of some complex analytic isolated singularity. M. Artin [1,3.8] proves that any isolated analytic singularity is in fact algebraic.

Theorem 2.1 [22]: If V is a complex surface with a good $\mathbb{C}^*$ action, $\underline{0} \in V$ is an isolated singular point and K has Seifert invariants $\{-b; g; (\alpha_1, \beta_1), \ldots, (\alpha_s, \beta_s)\}$ then there is a canonical equivariant resolution so that Γ is star shaped and equal to

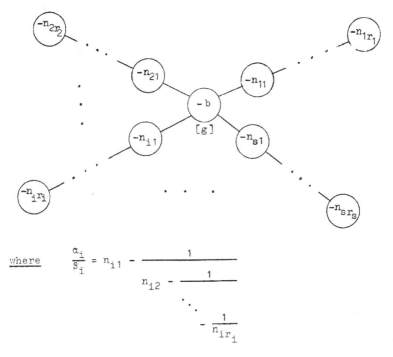

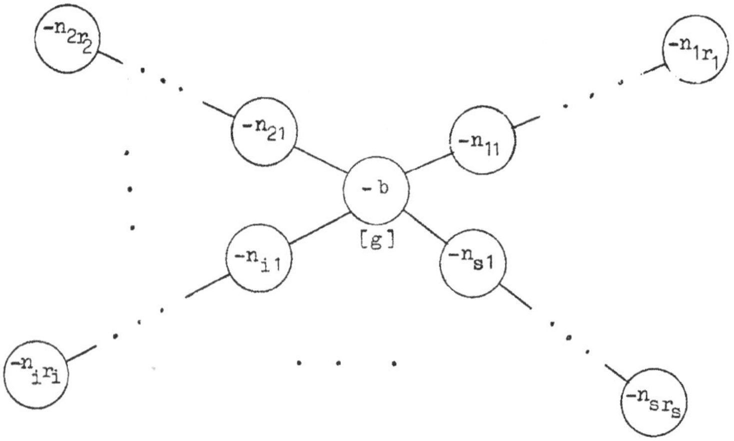

where
$$\frac{\alpha_i}{\beta_i} = n_{i1} - \cfrac{1}{n_{i2} - \cfrac{1}{\ddots - \cfrac{1}{n_{ir_i}}}}$$

for $i = 1, \ldots, s$ and $n_{ij} \geq 2$. All vertices other than the center have $g_j = 0$.

Idea of proof: The following construction is analogous to the constructions (called monoidal transforms) which resolve arbitrary singularities. Let $\pi: V - \{\underline{0}\} \to X$ be the quotient map. Then the graph of π, G_π is contained in $(V - \{\underline{0}\}) \times X$. Let F be the closure of G_π in $V \times X$. If $p_1\colon F \to V$ is the map induced by projection on the first factor then $p_1^{-1}(\underline{0}) = \underline{0} \times X$. Identifying $p_1^{-1}(\underline{0})$ with X we note that $p_1\colon F - X \to V - \{\underline{0}\}$ is an isomorphism. The other projection has the property that $p_2^{-1}(x) = \mathbb{C}$ for all $x \in X$. In fact $p_2\colon F \to X$ is precisely the Seifert (singular) fibration with fiber $\mathbb{C}$ associated to the Seifert (singular) fibration $\eta\colon V - \{\underline{0}\} \to X$ with fiber $\mathbb{C}^*$. For any point $x \in X$ there is a neighborhood of x in F which is isomorphic to $\mathbb{C}^2/\mathbb{Z}_\alpha$ where $\mathbb{Z}_\alpha$ is the isotropy subgroup of the orbit corresponding to x. Moreover the action of $\mathbb{Z}_\alpha$ on $\mathbb{C}^2$ is given by $\xi(z_0, z_1) = (\xi z_0, \xi^\nu z_1)$ $\forall \xi \in \mathbb{Z}_\alpha$, where $0 < \nu < \alpha$, $(\alpha, \nu) = 1$ and where $\xi(z) = \xi^\nu z$ is the action of $\mathbb{Z}_\alpha$ in the slice. We conclude that F has a finite number of singular points, precisely at the points of X corresponding to exceptional orbits and each is of the above type. Now it remains to resolve this finite number of singularities of F. This is carried out in our paper [22]. We will sketch the reason why the graph is star-shaped and all vertices except perhaps the center have genus zero. Suppose $\tilde{\rho}\colon \tilde{V} \to F$ is an equivariant resolution of the singularities of F. There is a unique irreducible complex curve $X_0 \subset \tilde{V}$ so that $\tilde{\rho}(X_0) = X$ (in fact $\tilde{\rho}$ restricted to X_0 is an isomorphism). Let $\rho = p_1 \circ \tilde{\rho}$ and $\rho^{-1}(\underline{0}) = \bigcup_{i=0}^{r} X_i$ where X_i is a non-singular curve. The $\mathbb{C}^*$ action leaves X_0 fixed, hence each of the X_i meeting X_0 must have a non-trivial $\mathbb{C}^*$ action. But the only complex curve with a non-trivial $\mathbb{C}^*$ action is $\mathbb{CP}^1$. Moreover any action on $\mathbb{CP}^1$ has exactly two fixed points. One proceeds inductively to show that every X_i, $i \neq 0$, has a non-trivial action and that two of these curves meet only at a fixed point. Therefore we conclude that Γ is star-shaped.

Remark: The construction of F can be carried out when $\dim V > 2$. This reduces the resolution of singularities with $\mathbb{C}^*$ action to the resolution of (not necessarily isolated) singularities locally of the form $\mathbb{C}^m/\mathbb{Z}_\alpha$. Note that the set of singular points on F has complex codimension ≥ 2.

3. Solution of I' and II' when $m = 2$

We will indicate how to find the invariants (α_i, β_i), b and g.

The (α_i, β_i) are found using the following facts.

Proposition 3.1: Suppose $V \subset \mathbb{C}^{n+1}$ is a variety invariant under a $\mathbb{C}^*$ action (1.2).

(i) Suppose that the only fixed point of the action is $\underline{0}$. If $w \in K$ then the slice through w in K is equivariantly homeomorphic to the slice through $\underline{w}$ in V.

(ii) Let $\underline{w} = (w_0, \ldots, w_n)$ and $w_{i_1} = \ldots = w_{i_s} = 0$ and all other $w_j \neq 0$. Choose $j_0 \neq i_k$ for $k = 1, \ldots, s$. Then

$$V \cap \{\underline{z} \mid z_{j_0} - w_{j_0} = 0\}$$

is a slice through $\underline{w}$.

Proof: The first follows from the fact that $K = V - \{0\}/\mathbb{R}^*$. The second is true when $V = \mathbb{C}^{n+1}$ and hence is true in general by [3, VIII.3.7]. Explicit computations for all hypersurfaces in $\mathbb{C}^3$ with isolated singularities are given in [22].

To find the invariant b we first consider the case that $V = \{\underline{z} \mid f(\underline{z}) = 0\}$ and f is homogeneous of degree d i.e. $q_0 = q_1 = q_i = 1$. In this case the action is free and $-b$ is the Chern class of the bundle $K \to X$. The quotient of $\mathbb{C}^3 - \{\underline{0}\}$ by this $\mathbb{C}^*$ action is $\mathbb{CP}^2$, therefore there is an embedding of X in $\mathbb{CP}^2$. Now the bundle map $K \to X$ is induced by the $\mathbb{C}^*$ bundle $\mathbb{C}^3 - \{\underline{0}\} \to \mathbb{CP}^2$ and the latter has Chern class -1. Since the map $H^2(\mathbb{CP}^2, \mathbb{Z}) \to H^2(X, \mathbb{Z})$ is multiplication by d we have $b = d$.

The general case can be reduced to the homogeneous case by letting $\varphi(z_0, z_1, x_2) = (z_1^{q_0}, z_1^{q_1}, z_2^{q_2})$ and $W = \varphi^{-1}(V)$. Now W is defined by a homogeneous polynomial and hence we can apply the above remarks to W (carefully, since W need not have an isolated singularity !). To find the invariant b one must analyse how b changes under the finite cover φ. This is carried out in [22,3.6] for arbitrary embedding dimension.

Theorem 3.2: If V is a complex surface in $\mathbb{C}^{n+1}$ with an isolated singularity at 0 and V is invariant under a good $\mathbb{C}^*$ action (1.2) then

$$b = \frac{d}{q_o \cdots q_n} + \sum_i \frac{\beta_i}{\alpha_i}$$

where d is the degree of the curve $Y = V - \{0\}/\mathbb{C}^*$ in $\mathbb{CP}^n$. This is the same as the d of example (1.3.ii). This theorem makes sense and holds unchanged when the singularity of V is not isolated [23,4.3].

Finally we consider the problem of computing g. If V is a hypersurface with an isolated singularity in $\mathbb{C}^{n+1}$ then Milnor and Orlik [17] computed the rank of the free part of $H_{n-1}(K)$. One can check that for $n = 2$ this rank equals $2g$ hence we get an explicit formula for g in terms of d and q_i. In [22] we did not have the Milnor-Orlik result hence we resorted to analysing the covering φ starting from the fact that a non-singular curve of degree d in $\mathbb{CP}^2$ has genus $(d-1)(d-2)/2$. This method does not generalize, however, either for higher embedding dimension or for non-isolated singularities. Here is a different approach to finding g. First let us state the results and then give the motivation.

Theorem 3.3: Suppose $V \subset \mathbb{C}^{n+1}$ is a variety with $\mathbb{C}^*$ action and $I = \{f \in \mathbb{C}[Z_o,\ldots,Z_n] \mid f$ vanishes on $V\}$. Suppose $r = \operatorname{codim}_{\mathbb{C}} V$ and we can find polynomials $f_1,\ldots,f_r$ so that the f_i generate I and $f_i(t^{q_o}z_o,\ldots,t^{q_n}z_n) = d^i f_i(z_o,\ldots,z_n)$. The ring $R = \mathbb{C}[Z_o,\ldots,Z_n]/I$ is graded by letting $\deg Z_i = q_i$. Then $R = \bigoplus_{i \geq o} R_i$ where R_i is the subspace generated by polynomials of "degree" i. Define

$$\mathcal{P}(t) = \sum_{i=o}^{\infty} \dim R_i \, t^i .$$

Then

$$\mathcal{P}(t) = \frac{(1-t^{d_1}) \ldots (1-t^{d_r})}{(1-t^{q_o}) \ldots (1-t^{q_n})} .$$

Recall $X = V - \{0\}/C^*$. Let $\underline{O}_X$ denote the sheaf of holomorphic functions on X and for any sheaf F on X define $\chi(X,F) = \sum_{i=0}^{\infty} (-1)^i \dim H^i(X;F)$.

Theorem 3.4: *There is a polynomial* p *so that* $p(i) = \dim R_{q_0\ldots q_n\, i}$. *Moreover* $p(0) = \chi(X,\underline{O}_X)$. *If* X *is non-singular then* $\chi(X,\underline{O}_X) = T(X)$, *the Todd genus of* X *[10,20.2.2]. Thus when* $\dim_C V = 2$ *we have* $g(X) = 1 - p(0)$.

When $n = 2$ we used (3.3) to compute $p(0)$ in (3.4) and obtained the following.

Corollary 3.5: *If* V *is a surface in* $\mathbb{C}^3$ *with an isolated singularity then*

$$g = \tfrac{1}{2}\left[\frac{d^2}{q_0q_1q_2} - \frac{d(q_0,q_1)}{q_0q_1} - \frac{d(q_1,q_2)}{q_1q_2} - \frac{d(q_2,q_0)}{q_2q_0} + \frac{(d,q_0)}{q_0} + \frac{(d,q_1)}{q_1} + \frac{(d,q_2)}{q_2} - 1\right].$$

We shall show later how this generalizes when the singularity is not isolated.

The theorems above were motivated by the following semi-classical theorem of algebraic geometry.

Let X be an algebraic subvariety of $\mathbb{CP}^n$, $I = \{f \in \mathbb{C}[Z_0,\ldots,Z_n] \mid f \text{ vanishes on } X\}$. Then the homogeneous coordinate ring of X, $S = \mathbb{C}[Z_0,\ldots,Z_n]/I$ is a graded ring i.e. $S = \bigoplus_{i\geq 0} S_i$ where S_i equals the subspace of S generated by homogeneous forms of degree i.

Let $O_{\mathbb{P}}(1)$ be the normal bundle of $\mathbb{CP}^n$ in $\mathbb{CP}^{n+1}$ and $O_X(1)$ the restriction of $O_{\mathbb{P}}(1)$ to X. Define $O_X(i) = O_X(1) \otimes \ldots \otimes O_X(1)$ i times.

Theorem 3.6: [26,§6]. *There is an integer* N *so that*

$$\dim S_i = \chi(X,O_X(i)) = \sum_{k\geq 0}^{\dim X} (-1)^k \dim H^k(X,O_X(i))$$

for all $i \geq N$. *Moreover* $p(i) = \chi(X,O_X(i))$ *is a polynomial function of* i, *for all* i. *The degree of* p *is* $\dim_{\mathbb{C}} X$.

If we can compute $p(i) = \dim S_i$ for all large i then we can find the constant term of p.

Theorem 3.4 is proven by showing that $S = \bigoplus_{i\geq 0} R_{q_0\dots q_n i}$ is the homogeneous coordinate ring of a projective embedding of X.

4. Application

Here is an example that was not included in [22].

Theorem 4.1: Suppose $V^{n-t+1} \subset \mathbb{C}^{n+1}$ is a complex variety invariant under the $\mathbb{C}^*$ action (1.2) and the ideal I of polynomials vanishing on V is generated by weighted homogeneous polynomials $f_1, \dots, f_t$. Let $f_i(t\underline{z}) = t^{d_i} f(\underline{z})$ and assume that for all i and j q_i divides d_j. (For example the f_i are Brieskorn.) If $X = V - \{0\}/\mathbb{C}^*$ then

$$\chi(X, \underline{O}_X) = \frac{d_1 \dots d_t}{2q_0 \dots q_n} \left[\sum_{i=1}^{t} d_i - \sum_{i=0}^{n} q_i p_i \right]$$

where $p_i = (q_0, \dots, \hat{q}_i, \dots, q_n)$. If X is non-singular then $\chi(X, \underline{O}_X) = T(X)$, the Todd genus of X.

Proof: Let

$$\mathcal{P}(t) = \frac{(1 - t^{d_1}) \dots (1 - t^{d_t})}{(1 - t^{q_0}) \dots (1 - t^{q_n})} = \sum_{i=0}^{\infty} a_i t^i .$$

By theorems (3.3), (3.4) $T(X)$ is the constant coefficient of the polynomial $p(i) = a_{qi}$, $q = q_0 \dots q_n$. If we decompose $\mathcal{P}(t)$ as a partial fraction we get

$$\mathcal{P}(t) = \frac{A_{-2}}{(1 - t)^2} + \frac{A_{-1}}{(1 - t)} + \sum_{k=0}^{n} \sum_{j=1}^{p_k - 1} \frac{B_{kj}}{(1 - \rho_k^j t)} + \sum_{i=1}^{s} A_i t^i$$

where $\rho_k = \exp(2\pi i/q_k)$, A_i and $B_{kj} \in \mathbb{C}$ and $s \geq 0$. The constant term of $p(t)$ is $A_{-2} + A_{-1} + \sum_{k=0}^{n} \sum_{j=1}^{p_k - 1} B_{kj}$. The B_{kj} can be found by observing that $B_{kj} = \lim_{t \to \rho_k^{-j}} \mathcal{P}(t)(1 - \rho_k^j t)$. This yields

$$B_{kj} = \frac{d_1 \dots d_t q_k}{q_0 \dots q_n (1 - \rho_k^{-j\, q_k})} \quad \text{and} \quad \sum_{j=1}^{p_k - 1} B_{kj} = \frac{d_1 \dots d_t q_k}{q_0 \dots q_n} \left(\frac{p_k - 1}{2} \right) .$$

Similarly one obtains $A_{-1} + A_{-2} = \dfrac{(q_0+\dots+q_n - d_1 - \dots - d_t)\ \ d_1\dots d_t}{2q_0\dots q_n}$

giving the desired result.

This can be applied to find surfaces where K is a homology 3-sphere. Seifert [24] showed that K is a homology 3-sphere different from S^3 with an S^1 action if and only if

$$K = \{-b;\ 0;\ (\alpha_1,\beta_1),\dots,(\alpha_r,\beta_r)\}$$

where $r \geq 3$, the α_j are pairwise relatively prime and

$$b\alpha_1\dots\alpha_r - \beta_1\alpha_2\dots\alpha_r - \dots - \alpha_1\dots\alpha_{r-1}\beta_r = \pm 1 \ .$$

<u>Corollary</u> 4.2: <u>Suppose</u> $\alpha_1,\dots,\alpha_r$ <u>are pairwise relatively prime positive integers</u>. <u>Choose</u> $\mu_{ij} \in \mathbb{C}$ <u>so that</u>

$$V = \{z_1,\dots,z_r \in \mathbb{C}^r \mid \sum_{i=1}^{r} \mu_{ij} z_i^{\alpha_i} = 0\ ,\ j = 1,\dots,r-2\}$$

<u>is a surface with an isolated singularity</u>. <u>Then</u> K <u>is the homology sphere described above</u>.

It is not hard to see that such a set of μ_{ij} exists.

<u>Proof</u>: We have $d = \prod_{i=1}^{r} \alpha_i$, $q_j = d/\alpha_j$. One can easily see that there are r exceptional orbits on K with respective stability groups α_i . Now

$$g = 1 + \frac{d^{r-2}\alpha_1\dots\alpha_r}{2d^r}\left[(r-2)d - \sum_{i=1}^{r} d\right] = 1 + \frac{1}{2d}[-2d] = 0$$

Finally, by (3.2)

$$b = \frac{d^{r-2}}{q_1\dots q_r} + \sum_{i=1}^{r}\frac{\beta_i}{\alpha_i} = \frac{d^{r-2}\alpha_1\dots\alpha_r}{d^r} + \sum_{i=1}^{r}\frac{\beta_i}{\alpha_i} = \frac{1}{\alpha_1\dots\alpha_r} + \sum_{i=1}^{r}\frac{\beta_i}{\alpha_i}$$

hence

$$b\alpha_1\dots\alpha_r - \beta_1\alpha_2\dots\alpha_r - \dots - \alpha_1\dots\alpha_{r-1}\beta_r = 1 \ .$$

This equation determines b and β_i uniquely [24].

Remark 4.3: The map $\varphi(z_1,\ldots,z_r) = (z_1^{\alpha_1},\ldots,z_r^{\alpha_r})$ shows that K is a d-fold branched cover of S^3. This action of $\mathbb{Z}_d$ commutes with the S^1 action on K.

5. Non-isolated singularities of surfaces

Most isolated singularities of surfaces cannot be analytically embedded in $\mathbb{C}^3$. Brieskorn [5] proved that if K is homeomorphic to the oriented lens space $L(n,q)$ then the smallest embedding dimension equals $3+\sum_{j=1}^{r}(n_j-2)$ where the n_j are the entries of the continued fraction expansion of n/q as in (2.1) with $n_j \geq 2$. As the embedding dimension gets large the number of equations needed to describe V increases even faster since V need not be a complete intersection of hypersurfaces. On the other hand, given an algebraic variety $U \subset \mathbb{C}^m$ we can choose a generic linear map $\pi : \mathbb{C}^m \to \mathbb{C}^3$ and study the image $V = \pi(U)$. Now V no longer has an isolated singularity in general but it is a hypersurface and hence the locus of one polynomial, $f(\underline{z}) = 0$. Moreover, there is a functorial way of recovering U.

Definition 5.1: Suppose $V \subset \mathbb{C}^{n+1}$ is an algebraic variety and R is the coordinate ring of V i.e. the ring of polynomial functions $g(Z_0,\ldots,Z_n)$ on V. Let $\bar{R}$ be the integral closure of R in its field of fractions. Then $\bar{R}$ is an algebra of finite type [32, V.§4, Thm 9], $\bar{R} = \mathbb{C}[Y_0,\ldots,Y_s]/I$. Let $\bar{V} = \{\underline{y} \in \mathbb{C}^{s+1} \mid f(\underline{y}) = 0 \quad \forall f \in I\}$. Then $\bar{V}$ is called the normalization of V and there is an induced map

$$\Theta : \bar{V} \to V .$$

If S is the set of singular points of V then $\Theta|\bar{V} - \Theta^{-1}(S)$ is an isomorphism onto an open dense subset of V. The set of singular points of a normal surface has codimension ≥ 2 hence if V is a surface then $\bar{V}$ has isolated singularities. If U is as above and U is analytically irreducible near $\underline{0}$ (i.e. for any disk D_ε about $\underline{0}$,

$U \cap D_\epsilon$ is not the union of two closed proper analytic subvarieties) then U is homeomorphic to $\bar{V}$. This discussion motivated us to consider singularities which are not necessarily isolated [23].

Let $V \subset \mathbb{C}^{n+1}$ be a complex surface invariant under a good $\mathbb{C}^*$ action (1.2). Let $\Theta : \bar{V} \to V$ denote the normalization. Then using the universal mapping property of normalization one can show that $\bar{V}$ has a unique $\mathbb{C}^*$ action making Θ equivariant. We shall describe the resolution of the isolated singularity of $\bar{V}$ and the topology of $K = V \cap S^{2n+1}$. First note that K need not be a manifold but it is what we call a singufold.

Definition 5.2: Let $B^3 = \{(x,y,z) \in \mathbb{R}^3 \mid x^2+y^2+z^2 < 1\}$ and $A = \{(x,0,0) \mid x^2 < 1\}$. We call U a standard neighborhood if it has closed subsets $X, U_1, \ldots, U_n$ so that $U = \bigcup_{i=1}^{n} U_i$, $U_i \cap U_j = X$, for all $i \neq j$, and there are homeomorphisms $\varphi_i : U_i \to B^3$ with $\varphi_i(X) = A$ for all i. If $x \in X$ and $n > 1$ we call x a singular point of U and U_i a branch through x. We say K is a 3-dimensional singufold if K is a compact separable Hausdorff space and every point of K has a standard neighborhood.

Let $\bar{K} = \bar{V} \cap S^{2m+1}$. Then $\bar{K}$ is an S^1-manifold and there is an induced equivariant map $\theta : \bar{K} \to K$. This map may identify several orbits of $\bar{K}$, some by maps of different degrees. In addition $\theta : \bar{K} \to \mathbb{C}^n$ may not be differentiable along some orbits. Let S be the set of singular points of K. Then $S = S_1 \cup \ldots \cup S_r$ where S_i is an orbit and $\theta^{-1}(S_i) = \bigcup_{j=1}^{t_i} \bar{S}_{ij}$ where $\bar{S}_{ij}$ is an orbit in $\bar{K}$. Let α_i be the order of the isotropy group of S_i and n_{ij} the degree of the map $\theta \mid \bar{S}_{ij}$. Clearly $\alpha_i = n_{ij}\, \alpha_{ij}$, where α_{ij} is the order of the isotropy group of $\bar{S}_{ij}$.

Definition 5.3: The Seifert invariants of K are

$$\{-b;\ g;\ (\alpha_{11},\beta_{11}) \approx (\alpha_{12},\beta_{12}) \approx \ldots \approx (\alpha_{1t_1},\beta_{1t_1}) \sim \alpha_1,$$
$$\ldots,\ (\alpha_{r1},\beta_{r1}) \approx \ldots \approx (\alpha_{rt_r},\beta_{rt_r}) \sim \alpha_r\}$$

where $b, g\ (\alpha_{ij}, \beta_{ij})$ are the corresponding Seifert invariants of $\bar{K}$, $\approx$ indicates that corresponding orbits are identified by θ and $\sim$ indicates the order of the isotropy group of the image orbit in K. The Seifert invariants are well defined up to permutation in the obvious way and determine K up to orientation preserving equivariant homeomorphism [23, Proposition 3.6]. Clearly the Seifert invariants of K determine those of $\bar{K}$ and hence we can obtain the resolution of $\bar{V}$ in the way described in Theorem 2.1. Moreover, the analogue of Theorem 3.2 is true in this case.

Theorem 5.4: Suppose V is an irreducible algebraic surface invariant under a good $\mathbb{C}^*$ action and V is not contained in any coordinate hyperplane. Let $(\alpha_1, \beta_1), \ldots, (\alpha_r, \beta_r)$ be the orbit invariants of $\bar{K}$. Then

$$b = \frac{d}{q_o \cdots q_n} + \sum_{i=1}^{r} \frac{\beta_i}{\alpha_i}$$

where the q_i are as in (1.2) and d is the degree of the cone V' over V (i.e. let $X' = V' - \{0\}/\mathbb{C}^* \subset \mathbb{CP}^m$, then d is the number of points of intersection of X' and a general hyperplane in $\mathbb{CP}^m$).

An analogue of (3.5) is also true when the singularities of V are not isolated. In this case X may have singularities. The theorem we want should tell us the genus of the normalization $\tilde{X}$ of X, $g(\tilde{X})$. Theorems (3.3) and (3.4) enable us to compute the arithmetic genus of X, $p_a(X)$.

Theorem 5.5: Suppose $V \subset \mathbb{C}^3$ is as above. Then V is defined by a weighted homogeneous polynomial f so that $f(t^{q_o}Z_o, t^{q_1}Z_1, t^{q_2}Z_2) = t^d f(Z_o, Z_1, Z_2)$ and

$$p_a(X) = 1 + \frac{d^2}{2q_oq_1q_2} - \frac{d(q_o,q_1)}{2q_oq_1} - \frac{d(q_1,q_2)}{2q_1q_2} - \frac{d(q_2,q_o)}{2q_2q_o}$$

$$- \rho(q_o;q_1,q_2;d) - \rho(q_1;q_2,q_o;d) - \rho(q_2;q_o,q_1;d) .$$

Here $\rho(k;n_1,n_2;d) = \frac{1}{k}\sum_j \frac{1-\xi^{jd}}{(1-\xi^{jn_1})(1-\xi^{jn_2})}$, $\xi = \exp(2\pi i/k)$ and j ranges over all integers so that $0 < j < k$, $jn_1 \not\equiv 0 \pmod{k}$ and $jn_2 \not\equiv 0 \pmod{k}$.

Remark: It is not too hard to see that $\rho(k;n_1,n_2;n_1) = \frac{1}{2k}[k-(k,n_1) - (k,n_2)+1]$. It would be helpful to have an algorithm for computing ρ in general.

Once we know $p_a(X)$ we can find g by the relation

$$g = p_a(X) - \sum_{x\in X} \delta_x$$

where δ_x is an invariant of the singular point x . An algebraic definition of δ_x is given by $\delta_x = \dim(\bar{\mathcal{O}}_x/\mathcal{O}_x)$ where $\mathcal{O}_x$ is the local ring of germs of holomorphic functions at x and $\bar{\mathcal{O}}_x$ is its integral closure [25,p.68]. For a topological description when X is a plane curve see [16,p.85].

Finally we shall describe how to find the (α_{ij},β_{ij}) . With the notation of (3.1) we have the slice $V \cap H$ through z which is a complex curve. A neighborhood $\mathcal{D}$ of z in $V \cap H$ is a union of 2-disks $\mathcal{D}_1,\ldots,\mathcal{D}_p$ meeting at z . Suppose that the set $\mathcal{D}_1,\ldots,\mathcal{D}_{n_1}$ is the smallest subset of $\mathcal{D}$ containing $\mathcal{D}_1$ and invariant under the action of $\mathbb{Z}_\alpha$. Then there is an orbit, say $\bar{S}_1$, over S so that the map $\theta|\bar{S}_1 : \bar{S}_1 \to S$ has degree n_1 . Thus the order of the isotropy group of $\bar{S}_1$ is $\alpha_1 = \alpha/n_1$. In order to determine the action of α_1 in a slice through $\bar{S}_1$ it is sufficient to find the action of $\mathbb{Z}_{\alpha_1}$ on $\mathcal{D}_1$ (since Θ induces a homeomorphism of a slice onto $\mathcal{D}_1$). Let $D_\varepsilon = \{ z \mid |z| < \varepsilon \}$. There is an analytic homeomorphism $\varphi: D_\varepsilon \to \mathcal{D}_1$ so that the induced $\mathbb{Z}_{\alpha_1}$ action on D_ε is linear of the form $(\zeta,x) \to \zeta^\nu x$ where $\zeta = \exp(2\pi i/\alpha_1)$. Now $\mathcal{D}_1 \subset \mathbb{C}^n$ and if we write down the coordinate functions $\varphi_i(u) = \sum_{j=1}^{\infty} a_{ij}\, u^j$ one can easily see that

$$\text{(5.6)} \qquad j\nu \equiv q_i \pmod{\alpha_1}$$

for all i,j with $a_{ij} \neq 0$. These equations determine ν uniquely.

Now ν and β_1 are related by $\nu\beta_1 \equiv -1 \pmod{\alpha_1}$. Hence we can determine β_1 from the φ_i. In the case that $n = 2$ there is an algorithm due to Puiseux for finding φ_1 and φ_2 [28, IV §2]. It can be shown that one of the φ_i is of the form $\varphi_i(u) = u^e$ where e is the multiplicity of the singularity of $V \cap H$ at z (the degree of the leading form of the defining polynomial).

Example [23]: $f(Z_0,Z_1,Z_2) = Z_0^2Z_1^3 + Z_1^2Z_2^5 + Z_0^4Z_2^3$.

This f is a weighted homogeneous polynomal with $q_0 = 3$, $q_1 = 4$, $q_2 = 2$, $d = 18$. There are three singular orbits in K, $S_1 = \{Z_0 = Z_1 = 0\}$ with isotropy group $\mathbb{Z}_2$, $S_2 = \{Z_0 = Z_2 = 0\}$ with isotropy group $\mathbb{Z}_4$ and $S_3 = \{Z_1 = Z_2 = 0\}$ with isotropy group $\mathbb{Z}_3$. The slice through S_1 at $(0,0,1)$ is the curve $f(Z_0,Z_1) = Z_0^2Z_1^3 + Z_1^2 + Z_0^4 = 0$. This has two branches through $(0,0)$ approximated by $\mathcal{D}_1 = \{Z_1 + iZ_0^2 = 0\}$ and $\mathcal{D}_2 = \{Z_1 - iZ_0^2 = 0\}$, [23, 5.5]. The action of the isotropy group $\mathbb{Z}_2$ sends (Z_0,Z_1) into $(-Z_0,Z_1)$, hence each $\mathcal{D}_i$ is invariant under $\mathbb{Z}_2$. We conclude that S_1 is covered by two orbits of type $(2,1)$ in $\bar{K}$. The slice $\mathcal{D}$ through S_2 at $(0,1,0)$ is the curve $Z_0^2 + Z_2^5 + Z_0^4Z_2^3 = 0$ which has only one branch. Thus S_2 is covered by one orbit of type $(4,\beta_2)$ in $\bar{K}$ and K is a topological (but not C^∞) manifold along S_2. It is easy to check (the Puiseux theorem is not needed here) that if $\varphi : D_\epsilon \to \mathcal{D}$ is as above then

$$\varphi(t) = (t^5,\ t^2 + \lambda_3 t^3 + \dots) \quad \text{where} \quad \lambda_3 \in \mathbb{C}.$$

By (5.6) we have $q_0\beta_2 \equiv -5 \pmod 4$ and hence $\beta_2 = 1$. The slice $\mathcal{D}$ through S_3 at $(1,0,0)$ is the curve $Z_1^3 + Z_1^2Z_2^5 + Z_2^3 = 0$. This has three branches approximated by $\mathcal{D}_1 = \{Z_1 + Z_2 = 0\}$, $\mathcal{D}_2 = \{Z_1 + \rho Z_2\} = 0$, $\mathcal{D}_3 = \{Z_1 + \rho^2 Z_2 = 0\}$ where $\rho = \exp(2\pi i/3)$. The action of $\mathbb{Z}_3$ on $\mathcal{D}$ sends (Z_1,Z_2) into $(\rho Z_1, \rho^2 Z_2)$ (since $q_1 = 4$, $q_2 = 2$) and hence permutes the $\mathcal{D}_i$. Therefore S_3 is covered by a single principal orbit in $\bar{K}$.

If one carries out the computations of g and b it can be seen that $g = 2$, $b = 2$ and hence the graph associated to the resolution of $\bar{V}$ is

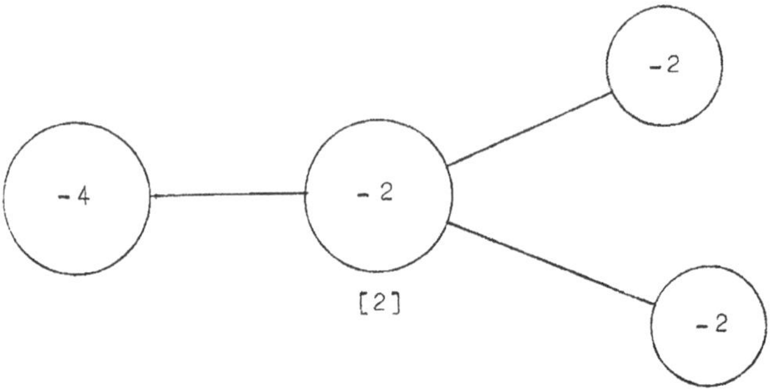

6. Remarks on higher dimensions

It is natural to try to extend the previous results to higher dimensions. In particular one can ask for a classification of all S^1 actions on the 5-sphere which can arise from singularities. It seems to be very difficult to extend the work of Montgomery and Yang [18] on 7-spheres to the 5-sphere since one runs into problems with the 4-dimensional Poincaré conjecture. We will indicate one result which gives us hope that these difficulties do not occur if the action on the sphere comes from a singularity.

Theorem 6.1: *Suppose* $V \subset \mathbb{C}^{n+1}$ *is a complex 3-dimensional algebraic variety with an isolated singularity at* 0 , *invariant under a good* $\mathbb{C}^*$ *action. Suppose* $K = V \cap S^{2n+1}$ *is the 5-sphere and the quotient space* $X = K/S^1 = V - \{0\}/\mathbb{C}^*$ *is a manifold. If the action on* K *is not free then* X *is algebraically isomorphic to* $\mathbb{CP}^2$.

Proof: Borel [3,IV,5.9] has shown that X has the same cohomology groups as $\mathbb{CP}^2$. But X is a manifold hence by Poincaré duality the square of a generator of $H^2(X,\mathbb{Z})$ is the generator of $H^4(X,\mathbb{Z})$ so X has the same cohomology ring as $\mathbb{CP}^2$. Now X is a projective algebraic variety [22] hence a Kähler variety [10,15.6]. The argument which follows uses several facts about Kähler varieties which are summarized in [10]. Let H be a <u>positive</u> generator of $H^2(X,\mathbb{Z})$ [10,18.1]. Then the first Chern class of X is an integer multiple of H, $c_1 = nH$. The Riemann-Roch theorem for algebraic surfaces [10,20.7] states that

$$\chi(X) = \frac{1}{12}(c_1^2 + c_2)[X]$$

Now $c_2[X]$ is just the topological Euler characteristic so $c_2[X] = 3$. On the other hand $\chi(X) = \sum_{q=0}^{2}(-1)^q h^{0,q}$ where $h^{p,q}$ is the dimension of the space of harmonic q-forms of type p,q [10, 15.4.2 and 15.5(13)]. Each Betti number of X can be expressed in the form $b_i = \sum_{p+q=i} h^{p,q}$ and $h^{j,j} \geq 1$ for $0 \leq j \leq 2$. Thus we see that $\chi(X) = 1$ and therefore $c_1^2[X] = 9$ and $c_1 = \pm 3H$. We claim that $c_1 = 3H$. Suppose $\mathbb{Z}_\alpha$ is the isotropy group of a non-principal orbit. Then the fixed point set of $\mathbb{Z}_\alpha$, $W \subset V$ is a complex surface with an isolated singularity at $\underline{0}$ (since the action in the slice leaves an <u>analytic</u> submanifold invariant [15]). Now $Y = W - \{0\}|C^*$ is a non-singular complex curve on X. Borel [3,IV.6.3] has shown that $K' = W \cap S^{2n+1}$ is a rational cohomology 3-sphere and hence by Orlik and Raymond [21] $Y = K'/S^1$ is a 2-sphere i.e. Y is a curve of genus 0. The Riemann-Roch theorem for a curve on a surface [30] states that

$$\text{genus}(Y) = \tfrac{1}{2}(Y.Y - Y.c_1) + 1 .$$

Now Y determines a cohomology class mH in $H^2(X,\mathbb{Z})$ where $m \geq 1$, [2]. Thus we must have $m = 1$ or 2 and $c_1 = 3H$. The Kodaira vanishing theorem [10,18.2.1] implies that $H^j(X,\underline{O}_X(iH)) = 0$ for $i \geq 0$ and $j \geq 1$. Therefore by the Riemann-Roch theorem

$$\dim H^0(X,\underline{O}_X(iH)) = \tfrac{1}{2}(i^2 + 3i) + 1 = \frac{(i+1)(i+2)}{2} .$$

Therefore by Hirzebruch and Kodaira [14,Thm.6] we have that X is analytically (and therefore algebraically) isomorphic to $\mathbb{CP}^2$.

References

1. M. Artin, Algebraic approximations of structures over complete local rings, Publ.Math. No 36. IHES, Paris (1969) 23-58.

2. M. Berger and A. Lascoux, Variétés Kählerienne Compactes, Lecture Notes no. 154, Springer Verlag, Berlin. 1970.

3. A. Borel et al., Seminar on transformation groups, Princeton University Press, 1960.

4. K. Brauner, Zur Geometrie der Funktionen zweier komplexen Veränderlichen III, IV, Abh.Math.Sem. Hamburg, 6.(1928), 8-54.

5. E. Brieskorn, Rationale Singularitäten komplexer Flächen, Invent.Math. 4 (1968), 336-358.

6. E. Brieskorn, Beispiele zur Differentialtopologie von Singularitäten, Invent.Math. 2 (1966), 1-14.

7. H. Grauert, Über modifikationen und exzeptionelle analytische Mengen, Math.Ann. 146 (1962), 331-368.

8. H. Hironaka, Resolution of singularities of an algebraic variety over a field of characteristic zero, Ann. of Math. 79 (1964), 109-326.

9. F. Hirzebruch, Über vierdimensionale Riemannsche Flächen mehrdeutiger analytisher Funktionen von zwei komplexen Veränderlichen, Math.Ann. 126 (1953), 1-22.

10. F. Hirzebruch, Topological methods in algebraic geometry, Springer Verlag, Berlin, 1966.

11. F. Hirzebruch, Differentiable manifolds and quadratic forms, revised by W.D. Neumann, to be published by Marcel Dekker Inc.,New York.

12. F. Hirzebruch, Involutionen auf Mannigfaltigkeiten, Proceedings of the Conference on Transformation Groups, Springer Verlag, Berlin (1968), 146-166.

13. F. Hirzebruch and K. Jänich, Involutions and singularities, Symposium on Algebraic Geometry, Tata Institute, Oxford University Press (1969), 219-240.

14. F. Hirzebruch and K. Kodaira, On the complex projective spaces, J.Math.Pures Appl. 36 (1957), 201-216.

15. H. Holmann, Quotienten komplexer Räume, Math.Ann. 142 (1961), 407-440.

16. J. Milnor, Singular points of complex hypersurfaces, Princeton University Press, 1968.

17. J. Milnor and P. Orlik, Isolated singularities defined by weighted homogeneous polynomials, Topology 9. (1970), 385-393.

18. D. Montgomery and C.T. Yang, Differentiable pseudo-free circle actions, these Proceedings.

19. D. Mumford, The topology of normal singularities of an algebraic surface and a criterion for simplicity, Publ.math. No. 9. IHES, Paris (1961).

20. W.D. Neumann, S^1-actions and the α-invariant of their invosutions, Bonner Math.Schriften 44, 1970.

21. P. Orlik and F. Raymond, Actions of SO(2) on 3-manifolds, Proceedings of the Conference on Transformation Groups, Springer Verlag, Berlin(1968), 297-318.

22. P. Orlik and P. Wagreich, Isolated singularities of algebraic surfaces with $\mathbb{C}^*$ action, Ann. of Math. 93 (1971), 205-228.

23. P. Orlik and P. Wagreich, Singularities of algebraic surfaces with $\mathbb{C}^*$ action, Math.Ann. 193 (1971), 121-135.

24. H. Seifert, Topologie dreidimensionaler gefaserter Räume, Acta math. 60 (1933), 147-238.

25. J.P. Serre, Groupes algébrique et corps de classes, Hermann, Paris, 1959.

26. J.P. Serre, Faisceaux algébriques coherents, Ann. of Math.61 (1955), 197-278.

27. R. von Randow, Zur Topologie von dreidimensionalen Baummannigfaltigkeiten, Bonner Math. Schriften 14, 1962.

28. R.J. Walker, Algebraic Curves, Dover Publications, New York, 1950.

29. A. Wallace, Linear sections of algebraic varieties, Indiana Univ. Math.J. 20 (1971), 1153-1162.

30. O. Zariski, Algebraic Surfaces, Springer Verlag, Berlin, 1971.

31. O. Zariski, An introduction to the theory of algebraic surfaces, Lecture Notes No. 83, Springer Verlag, Berlin, 1969.

32. O. Zariski and P. Samuel, Commutative Algebra I, II, Van Nostrand, Princeton, 1960.

STRANGE CIRCLE ACTIONS ON PRODUCTS OF ODD DIMENSIONAL SPHERES, AND RATIONAL HOMOTOPY

Glen E. Bredon, Rutgers University

Let X be a finite dimensional CW-complex and let $G = S^1$ act on X. Suppose that the fixed point set F is a simply connected CW-complex. We are interested in connections between $\pi_*(X) \otimes Q$ and $\pi_*(F) \otimes Q$, and we shall give two examples in this note with a bearing on this question.

In this direction, there is the following general theorem. (The proof is difficult and has not yet been published.) We let S_r^{2n+r} denote the join $S^{2n-1} * S^r$ with the standard free S^1-action on S^{2n-1} and the trivial action on S^r.

THEOREM A. Let $0 \neq \tau \in \pi_r(F) \otimes Q$. Then there is an integer n_0 depending only on τ and on the dimension of X such that if $f\colon S_r^{2n+r} \longrightarrow X$ is an equivariant map with $[f^G] = q\tau$, for $0 \neq q \in Q$, then $n \leq n_0$.

For $f\colon S_r^{2n+r} \longrightarrow X$ equivariant, note that if $0 = [f] \in \pi_{2n+r}(X)$ then f can be extended to the disk $(\text{point}) * S^{2n+r}$ and then by equivariance to $f_1\colon S_r^{2n+2+r} = S^1 * S_r^{2n+r} \longrightarrow X$. If we only have that $0 = [f] \otimes 1 \in \pi_{2n+r}(X) \otimes Q$, then $0 = k[f] \in \pi_{2n+r}(X)$ for some integer $k \neq 0$, and the usual definition of addition in homotopy groups provides an equivariant map $kf\colon S_r^{2n+r} \longrightarrow X$ with $[kf] = k[f] = 0$. As before, kf extends to $f_1\colon S_r^{2n+2+r} \longrightarrow X$ and $[f_1{}^G] = k[f^G] \in \pi_r(F)$.

Thus, starting with $0 \neq \tau \in \pi_r(F) \otimes Q$, we can perform the above constructions, by induction on n, of equivariant maps $f\colon S_r^{2n+r} \longrightarrow X$ with $[f^G] = k\tau$ for some integer $k \neq 0$, and this process can be continued until we reach such an f with $0 \neq [f] \in \pi_{2n+r}(X) \otimes Q$. Theorem A shows that this construction must be so obstructed for some $n \geq 0$. (As usual in obstruction theory, the obstruction need not be unique but may depend on choices involved in preceding constructions.)

As a corollary of these facts one can prove the following inequality (see [1])

$$(1) \qquad \dim(\pi_r(F) \otimes Q) \leq \sum_{n \geq 0} \dim(\pi_{2n+r}(X) \otimes Q).$$

One of the main test cases for results in this direction is the case in which X is a product $X = S^{n_1} \times S^{n_2} \times \ldots \times S^{n_k}$ for n_i odd. (More generally, spaces of this rational homotopy type are of interest, the main examples being the connected Lie groups.) For such spaces there are two main known theorems. The first of these follows from (1) and states that

$$\pi_i(F) \otimes Q = 0 \quad \text{for} \quad i \quad \text{even.}$$

(This assumes that $\pi_1(F) = 0$.) The second main result for actions on $X = S^{n_1} \times \ldots \times S^{n_k}$, n_i odd, states that if X is totally non-homologous to zero in X_G, then F has the rational cohomology ring of $S^{r_1} \times \ldots \times S^{r_k}$ for r_i odd. The easy proof of this may be found in [2]. If $n_1 \leq n_2 \leq \ldots \leq n_k$ and $n_k < n_1 + n_2$ (e.g., if $k = 2$), then X is automatically totally non-homologous to zero in X_G as $H^*(X)$ is then transgressively generated. (We are assuming that $F \neq \emptyset$.) It was thought for some time that this may always be true without the restriction $n_k < n_1 + n_2$. Our first example (which will also appear in [2]) shows that such is not the case.

EXAMPLE 1. (X homologous to zero in X_G.). Let τ be the tangent bundle of S^8 with the trivial S^1-action, and let ε^2 be the trivial 2-plane bundle on S^8 with the standard $S^1 = SO(2)$-action in the fibers. Let $f\colon S^3 \times S^5 \longrightarrow S^8$ have degree one and let X be the unit sphere bundle $S(f^*(\tau \oplus \varepsilon^2))$. Since $\tau \oplus \varepsilon^2$ is trivial, $X = S^3 \times S^5 \times S^9$. Clearly $F = S(f^*(\tau))$ which is an S^7-bundle over $S^3 \times S^5$. The transgression $Z \approx H^7(S^7) = E_2^{0,7} \longrightarrow E_2^{8,0} = H^8(S^3 \times S^5) \approx Z$ is multiplication by ± 2 since this is the case for $S(\tau) \longrightarrow S^8$. Thus we have

$$H^i(F;Q) \approx \begin{cases} Q & \text{for } i = 0, 3, 5, 10, 12, 15 \\ 0 & \text{otherwise,} \end{cases}$$

which shows that F does not have the rational cohomology of a product of odd dimensional spheres. Also note that

$$\pi_i(F) \otimes Q \approx \begin{cases} Q & \text{for } i = 3, 5, 7 \\ 0 & \text{otherwise} \end{cases}$$

and that the Whitehead product $\pi_3 \quad \pi_5 \longrightarrow \pi_7$ is rationally non-trivial. (Otherwise, F would have the rational homotopy type of $S^3 \times S^5 \times S^7$.)

From this example it would appear that the best, presently known, general result about S^1-actions on $S^{n_1} \times \ldots \times S^{n_k}$, n_i odd, is that $\pi_i(F) \otimes Q = 0$ for i even (when $\pi_1(F) = 0$).

We now turn to the case in which X is totally non-homologous to zero in X_G. In fact we shall now restrict our attention to the case

$$X = S^n \times S^m \quad \text{with } n,m \text{ odd, } n \leq m.$$

Then F has the rational cohomology ring of $S^q \times S^r$ with q,r odd, and we shall, in fact, assume that $F \approx S^q \times S^r$.

Let $\alpha \in H^n(X_G)$ and $\beta \in H^m(X_G)$ restrict to an exterior basis in $H^*(X)$. Then $H^*(X_G)$ is the exterior algebra over $H^*(B_G) = Z[t]$ with α and β as a basis. (Note that $\deg t = 2$.) Let $u,v \in H^*(F)$ be an exterior basis. Let

$$j^*\colon \ H^*(X_G) \longrightarrow H^*(F_G) = H^*(B_G) \otimes H^*(F)$$

be the restriction. There is a theorem stated by W. Y. Hsiang in [3] which contains, as a special case, the statement that α, β, u, v can be so chosen that

$$j^*(\alpha) = t^{(n-q)/2} \otimes u$$
$$j^*(\beta) = t^{(m-r)/2} \otimes v.$$

This result has, as an easy consequence, the following interesting implication for the rational homotopy of these S^1-spaces: u and v have dual classes $0 \neq [f] \in \pi_q(F) \otimes Q$ and $0 \neq [g] \in \pi_r(F) \otimes Q$ such that the construction indicated below Theorem A is obstructed by distinct classes in $\pi_*(X) \otimes Q$. If one could prove such a result in more generality, then perhaps one could prove the stronger inequality (for $\pi_1(F) = 0$)

$$\sum_{n \geq 0} \dim(\pi_{i+2n}(F) \otimes Q) \leq \sum_{n \geq 0} \dim(\pi_{i+2n}(X) \otimes Q)$$

which would be a very substantial improvement of (1). This inequality may well be true, but the following example shows that this simple-minded approach fails.

EXAMPLE 2. (Counterexample to the theorem of W. Y. Hsiang.)

We wish to construct an S^1-action on $S^n \times S^m$. One family of such actions is obtained from the orthogonal S^1-vector bundle structures on $S^n \times R^{m+1}$ over a given linear S^1-action on S^n. Our example will be one of these. Let us first describe some general principles on the construction of such G-vector bundles. The following proposition from [2] is an easy computation:

PROPOSITION. *Let* G *be a compact group and make the space* Map(G,e;O(k),I), *with the compact-open topology, into a* G-*space by defining* $(gT)(h) = T(hg)T(g)^{-1}$. *Then, for a given* G-space X, *there is a one-one correspondence between the set of orthogonal* G-*vector bundle structures on* $X \times R^k$ *and the set of equivariant maps*

$X \longrightarrow Map(G,e;O(k),I)$. If $x \longmapsto T_x$ is such an equivariant map, then the corresponding G-vector bundle structure on $X \times R^k$ is given by the action $g(x,v) = (gx, T_x(g)\cdot v)$.

Note that if X has the trivial G-action then the equivariant maps $X \longrightarrow Map(G,e;O(k),I)$ are just the maps into the fixed set, which is clearly just the space $Homo(G, O(k))$ of homomorphisms.

Our program is to find a suitable map $S^n \longrightarrow Homo(S^1, O(k))$, with n odd, k even, and the trivial action on S^n, such that the composition into $Map(S^1,1;O(k),I) = \Omega O(k)$ is homotopically trivial. Then, just as in the argument below Theorem A, this will extend to an equivariant map $S_n^{n+2} \longrightarrow \Omega O(k)$ and, by the proposition, this will give an orthogonal S^1-bundle structure on $S^{n+2} \times R^k$ over S_n^{n+2}, and hence an S^1-action on $S^{n+2} \times S^{k-1}$. If $S^{n+2} \longrightarrow \Omega O(k)$ is homotopically trivial we can continue the construction to obtain an action on $S^{n+4} \times S^{k-1}$, and so on.

If one examines the structure of the space $Homo(S^1,O(k))$, one sees that the main class of maps of S^n into it are given by the following procedure: Let φ: $S^1 \longrightarrow SO(k)$ be a given representation, let θ: $S^n \longrightarrow O(k)$ be a map taking x to $\theta_x \in O(k)$, and define $S^n \longrightarrow Homo(S^1,O(k))$ by $x \longrightarrow \theta_x\varphi(\cdot)\theta_x^{-1}$. If, moreover, φ is homotopically trivial as a map, then so is the composition $S^n \longrightarrow Homo(S^1,O(k)) \longrightarrow \Omega O(k)$, and we can apply our construction to get $S_n^{n+2} \longrightarrow \Omega O(k)$.

There is an obvious choice to try for θ, namely the characteristic map for the tangent bundle of S^n. Also there are obvious choices for the representation φ. Thus we shall now replace n by $2m-1$ and k by $2m$ and define

$$\theta:\ S^{2m-1} \longrightarrow O(2m)$$

to be the map taking x to the reflection θ_x through the hyperplane $x^\perp$. For any $1 \leq r < m$ let φ: $S^1 \longrightarrow SO(2m)$ be $m-r$ times the standard representation plus a trivial 2r-dimensional representation for m-r even. If m-r is odd we let S^1 act twice as fast as usual in one of the standard representations. (This is so that we will have $0 = [\varphi] \in \pi_1(SO(2m))$.)

Let $\varphi(z,t)$ define a null-homotopy of φ ; that is, $\varphi(z,0) = I$ and $\varphi(z,1) = \varphi(z)$.

Applying the above remarks to this situation then gives an equivariant map $S^{2m+1} = S^1 * S^{2m-1} \longrightarrow Map(S^1,1;O(2m),I) = \Omega O(2m)$ (where S^1 acts as usual on S^1 and trivially on S^{2m-1}) and this gives, in turn, an S^1-action on $X = S^{2m+1} \times S^{2m-1}$. Using the join coordinates $<w,x,t>$ for $S^1 * S^{2m-1} = S^{2m+1}$, where $w \in S^1$, $x \in S^{2m-1}$, and $t = 0$ on S^1, $t = 1$ on S^{2m-1}, we can write down this action on $S^{2m+1} \times S^{2m-1}$ explicitly as follows:

$$z(<w,x,t>,y) = (<zw,x,t>,\ \theta_x\varphi(zw,t)\varphi(w,t)^{-1}\theta_x \cdot y).$$

It is a good exercise to see that this formula does indeed define an S^1-action. Note that $\theta_x = \theta_x^{-1}$.

We now turn to the analysis of the fixed set F of this action. Note that F is contained in the subspace $S^{2m-1} \times S^{2m-1}$ and the action on this subspace is given by

$$z(x,y) = (x,\ \theta_x\varphi(z)\theta_x \cdot y).$$

We see immediately that

$$F = \{(x,\ \theta_x(y)) \mid x \in S^{2m-1},\ y \in S^{2r-1}\} \approx S^{2m-1} \times S^{2r-1}$$

where S^{2r-1} denotes the fixed set of the representation φ on S^{2m-1}.

For this action on $X = S^{2m+1} \times S^{2m-1}$ let $\alpha \in H^{2m-1}(X_G, x_G)$ be a generator, where $x \in F$. This is unique up to sign, and its choice in the relative group is a normalization which will have no effect on our conclusions. Let $u \in H^{2r-1}(F)$ and $v \in H^{2m-1}(F)$ be generators.

Note that the composition

$$S^{2m-1} \longrightarrow F \subset X = S^{2m+1} \times S^{2m-1} \xrightarrow{\text{proj.}} S^{2m-1}$$

is $x \longrightarrow \theta_x(y_0)$, where the first map is $x \longmapsto (x, \theta_x(y_0))$ which is dual to v. But this map has degree two. From the diagram

$$\begin{array}{ccc} H^{2m-1}(X_G) & \xrightarrow{j^*} & H^{2m-1}(F_G) \supset H^0(B_G) \otimes H^{2m-1}(F) \\ \downarrow & & \downarrow \\ H^{2m-1}(X) & \xrightarrow{\text{restriction}} & H^{2m-1}(F) \end{array}$$

one sees that $j^*(\alpha)$ must contain the term $1 \otimes 2v$ (up to sign).

On the other hand, note that if x_0 is perpendicular to the fixed set R^{2r} of φ, then $\{x_0\} \times S^{2m-1} \approx S^{2m-1}$ is invariant and intersects F in $\{x_0\} \times S^{2r-1} \approx S^{2r-1}$. In fact, S^1 acts via the representation φ on this S^{2m-1}. The diagram

$$\begin{array}{ccccc} H^{2m-1}(X_G) & \xrightarrow{j^*} & H^{2m-1}(F_G) \supset H^{2m-2r}(B_G) & \otimes & H^{2r-1}(F) \\ \downarrow & & \downarrow & & \downarrow \\ H^{2m-1}(S_G^{2m-1}) & \xrightarrow{j^*} & H^{2m-1}(S_G^{2r-1}) \supset H^{2m-2r}(B_G) & \otimes & H^{2r-1}(S^{2r-1}) \end{array}$$

and the known value of j^* for this linear action φ, shows that $j^*(\alpha)$ must contain the term (up to sign) $t^{m-r} \otimes u$ (or $t^{m-r} \otimes 2u$ when $m-r$ is odd). Thus, with the proper choice of signs, we have

$$j^*(\alpha) = \begin{cases} 1 \otimes 2v + t^{m-r} \otimes u & \text{for } m-r \text{ even} \\ 1 \otimes 2v + t^{m-r} \otimes 2u & \text{for } m-r \text{ odd.} \end{cases}$$

This finishes the example. As a consequence, one sees easily that the duals in $\pi_*(F) \otimes Q$ of u and v are obstructed by the <u>same</u> class in $\pi_{2m-1}(S^{2m+1} \times S^{2m-1}) \otimes Q$, and the obstructions are unambiguous in this case.

We can continue the construction as follows to provide a wide class of such examples. Recall that the action corresponds, by the proposition (and by construction), to an equivariant map

$$S^{2m+1}_{2m-1} \longrightarrow \mathrm{Map}(S^1,1;\ SO(2m),I) = \Omega SO(2m),$$

and an integer multiple of this map produces examples of the same sort. Since $\pi_i(SO(2m))$ is finite for all even $i \geq 2m$, we can extend multiples of this map to equivariant maps

$$S^{2n-1}_{2m-1} \longrightarrow \Omega SO(2m)$$

for any $n > m$, producing actions on $S^{2n-1} \times S^{2m-1}$ fixing $F \approx S^{2m-1} \times S^{2r-1}$. By adding a non-trivial representation (i.e., by including $SO(2m) \longrightarrow SO(2m+2) \longrightarrow \ldots \longrightarrow SO(2k)$ with non-trivial representation of S^1 on the extra factors) we produce actions on $S^{2n-1} \times S^{2k-1}$ fixing $F \approx S^{2m-1} \times S^{2r-1}$.

In this way, for $n > k \geq m > r$, we can find an S^1-action on $X = S^{2n-1} \times S^{2k-1}$ fixing $F \approx S^{2m-1} \times S^{2r-1}$ and such that

$$j^*(\alpha) = A\ t^{k-m} \otimes v + B\ t^{k-r} \otimes u$$

for <u>non-zero</u> integers A and B, where $\alpha \in H^{2k-1}(X_G, x_G) \approx Z$, $u \in H^{2r-1}(F) \approx Z$ and $v \in H^{2m-1}(F) \approx Z$ are generators (unique up to sign).

Now $|A|$ and $|B|$ are clearly invariants for such actions. We shall call

$$\sigma = |AB|$$

the <u>strangeness invariant</u> for the action and we say that the action is <u>strange</u> if $\sigma \neq 0$. Clearly $\sigma = 0$ for products of S^1-actions on S^{2n-1} and S^{2k-1}. Thus, for any $n > k \geq m > r$, there are strange actions on $S^{2n-1} \times S^{2k-1}$ with fixed set $F \approx S^{2m-1} \times S^{2r-1}$.

Perhaps it is of interest to ask what strangeness invariants σ are realizable. One can achieve $\sigma = 1$ for $(n,k,m,r) = (5,4,4,2)$. The case $\sigma = 2$ can be achieved for $(n,k,m,r) = (3,2,2,1)$, $(5,4,4,1)$, $(5,4,4,3)$ and $(m+1,m,m,r)$ for $m-r$ even. If one keeps k, m and r

fixed, then it can be shown that $\sigma \longrightarrow \infty$ as $n \longrightarrow \infty$, and, in fact, σ is eventually divisible by any given prime for n sufficiently large. Probably $\sigma \longrightarrow 0$ in the "adic-topology" as $n \longrightarrow \infty$.

REFERENCES

[1] Bredon, G.E., Homotopical properties of fixed point sets of circle group actions, Amer. J. M., 91 (1969) pp. 874-888.

[2] ______, Introduction to Compact Transformation Groups, Academic Press, New York (to appear).

[3] Hsiang, W.-Y., On generalizations of a theorem of A. Borel and their applications in the study of topological actions, Topology of Manifolds, pp. 274-290, Markham Publ. Co., Chicago, 1970.

EXAMPLES OF ACTIONS ON MANIFOLDS ALMOST DIFFEOMORPHIC TO $V_{n+1,2}$

Michael Davis
Yale University and
Princeton University

In [7] Hirzebruch discusses a relationship between transformation groups, knot theory, and the study of Brieskorn varieties. This interplay originally represented the convergence of the work of K. Jänich [9] and of W. C. and W. Y. Hsiang [8] on classifying the type of $O(n)$-manifolds called "knot manifolds" with the work of Brieskorn, Milnor, and others [1], [7], [12] on the behaviour of certain complex varieties near isolated singularities. Hirzebruch pointed out that the Brieskorn spheres provide examples of knot manifolds. These examples have since been used in work on smooth actions of other compact Lie groups, notably S^1 and $\mathbb{Z}_p$, on homotopy spheres (e.g. [2]). In this paper, we exhibit analogous examples which differ from Hirzebruch's in three ways. First of all, rather than being concerned with actions on homotopy spheres, in our examples the ambient manifold is almost diffeomorphic to $V_{n+1,2}$, the Stiefel manifold of 2-frames in $\mathbb{R}^{n+1}$. Secondly, it will be necessary to use manifolds defined by weighted homogeneous polynomials (see [12; p. 75] for definition of these) rather than the Brieskorn manifolds. Finally, in our examples the action will be associated with a link in S^3 rather than a knot.

---

[1] This paper represents part of the author's senior thesis at Princeton University.

The author is an NSF graduate fellow.

I would like to thank Dieter Erle, Lou Kauffman, John Morgan, Robert Szczarba, and Steve Weintraub for many valuable conversations. I am particularly indebted to my advisor William Browder for guiding my work and for suggesting the topic of this paper.

Throughout this paper "manifold" will mean "smooth, compact, orientable manifold" (with or without boundary), all group actions will be smooth, and "$\cong$" will mean diffeomorphic. Also, "Σ" will be used to denote a homotopy sphere and bP_{2m} will denote the subgroup of homotopy spheres which bound parallelizable manifolds.

1. The Examples, $K_m^{p,q}$

Consider the weighted homogeneous polynomial $g: \mathbb{C}^{m+1} \to \mathbb{C}$ defined by

$$g(z) = (z_1)^p + (z_1)(z_2)^q + (z_3)^2 + \dots\dots + (z_{m+1})^2$$

where p and q are odd and $\gcd(p-1,q) = 1$. Let

$$K_m^{p,q} = g^{-1}(0) \cap S^{2m+1} ,$$

where $S^{2m+1} \subset \mathbb{C}^{m+1}$ is the unit sphere. We will be interested in the examples $K_m^{p,q}$ when $m = 2n$, although similar results also hold if m is odd. We will show that $K_{2n}^{p,q} = V_{2n+1,2} \# \Sigma$, for some $\Sigma \in bP_{4n}$. Then, examining the natural action of $O(2n-1)$ on $K_{2n}^{p,q}$, we will show that $K_{2n}^{p,q}$ is an example of what we shall call a "prime link manifold."

First, we must recall some facts proved in [12]. Let $f : \mathbb{C}^{m+1} \to \mathbb{C}$ be a polynomial such that $f(0) = 0$, the origin is a critical point, and $f^{-1}(0) \cap (D^{2m+2} - 0)$ contains no critical points. (g satisfies these conditions.) If $V_m^f = f^{-1}(0) \cap S^{2m+1}$, then

1) $\varphi : S^{2m+1} - V_m^f \to S^1$ defined by $\varphi(z) = f(z)/\|f(z)\|$, is the projection map of a smooth fibre bundle.

2) V_m^f is an $(m-2)$-connected, compact, $(2m-1)$-dimensional Π - manifold.

3) V_m^f bounds the $(m-1)$-connected, parallelizable $2m$ - manifold $\bar{F}_\theta$, where $\bar{F}_\theta$ is the closure of a typical fibre. (If $f = g$, we denote this fibre by $F_{2n}^{p,q}$.)

Associated with any such fibre bundle over the circle is a characteristic polynomial $\Delta(t)$. (See [12; p. 67] for the definition of $\Delta(t)$.) A trivial modification of the proof of Theorem 8.5 in [12] shows,

(1.1) Lemma: V_{2n}^f is a homology $V_{2n+1,2}$ (that is $H_{2n-1}(V_{2n}^f) = \mathbb{Z}_2$) if and only if $\Delta(1) = \pm 2$.

Remark: If

$$f(z) = (z_1)^{a_1} + \dots\dots + (z_{m+1})^{a_{m+1}}$$

V_m^f is called a Brieskorn manifold and often denoted by $V_m(a_1, \dots a_{m+1})$. It is not difficult to show that if the characteristic polynomial associated with $V_{2n}(a_1, \dots a_{2n+1})$ satisfies $\Delta(1) = \pm 2$ each of the $a_i = 2$ [3; Prop. 2.3]. Since $V_m(2, 2 \dots 2)$ can be identified with $V_{m+1,2}$ in a natural way, it follows that the Brieskorn manifolds do not provide non-trivial examples of manifolds homeomorphic to $V_{2n+1,2}$.

Using [12; Theorem 9.6] it is possible to compute $\Delta(t)$

for any manifold defined by a weighted homogeneous polynomial.

(1.2) <u>Lemma</u>: The characteristic polynomial of $K_{2n}^{p,q}$ is

$$\Delta(t) = \frac{(t+1)(t^{pq}+1)}{(t^p+1)}$$

Hence $\Delta(1) = 2$ and so $H_{2n-1}(K_{2n}^{p,q}) = \mathbb{Z}_2$.

We will say that a manifold M^{4n-1} <u>satisfies</u> (<u>A</u>) if and only if

(A) M is a 1-connected π - manifold with the integral homology of $V_{2n+1,2}$.

Summarizing the above results, we have:

(1.3) <u>Corollary</u>: $K_{2n}^{p,q}$ satisfies (A). Furthermore, it bounds the parallelizable manifold $F_{2n}^{p,q}$.

The following proposition shows that this corollary is all that is needed to prove $K_{2n}^{p,q}$ is homeomorphic to $V_{2n+1,2}$.

(1.4) <u>Proposition</u>: If M^{4n-1} satisfies (A), $n > 2$, then

$$M \equiv V_{2n+1,2} \# \Sigma .$$

If M also bounds a parallelizable manifold, then $\Sigma \in bP_{4n}$.

This proposition is an analog of the fact that 1-connected homology spheres are homotopy spheres. Undoubtedly, it is a special case of a more general theorem (for example, a theorem of Wall's) but we give a direct proof based on the next two lemmas.

Let $E(\gamma)$ denote the total space of the closed 2n-disc bundle over S^{2n} classified by $\gamma \in \pi_{2n-1}(SO(2n))$. Let $E_o(\gamma) = \partial E(\gamma)$ be the associated sphere bundle. Let $\tau \in \pi_{2n-1}(SO(2n))$ classify the tangent bundle (so that $E_o(\pm \tau) = \pm V_{2n+1,2}$) and let σ generate the stable part of $\pi_{2n-1}(SO(2n))$. The proof of the following lemma can essentially be found in [11].

(1.5) <u>Lemma</u> (Kosinski): $E_o(\gamma)$ satisfies (A) if and only if

$$\gamma = \begin{cases} \pm\tau & ; \text{ if } n \text{ is odd} \\ \pm\tau + 2m\sigma ; & \text{if } n \text{ is even} \end{cases}$$

where $m \in \mathbb{Z}$.

The substance of the proof of the next lemma is contained in [10].

(1.6) <u>Lemma</u>: If M satisfies (A), then there exists a framed surgery on M so that the resulting manifold is a homotopy sphere.

<u>Proof</u>: Let $L(\lambda, \mu) \in \mathbb{Q}/\mathbb{Z}$ be the rational linking number of two homology classes. (See [10; p. 524].) If λ is the non-zero element of $H_{2n-1}(M) = \mathbb{Z}_2$, then it follows from Poincaré duality for torsion groups that $L(\lambda, \lambda) = \frac{1}{2}$. By Lemmas 6.3 and 6.4 in [10], a framed surgery can be chosen so that H_{2n-1} of the new manifold is definitely smaller than $\mathbb{Z}_2$; hence 0.

<u>Proof of Proposition 1.4</u>: Since M can be obtained by one surgery on a homotopy sphere Σ, we may assume (subtracting and adding Σ) that we get it by a single surgery on S^{4n-1}. The

result of the surgery is completely determined by the isotopy class of an embedding $S^{2n-1} \times D^{2n} \subset S^{4n-1}$. Applying the results of [6], the isotopy class is unique on $S^{2n-1} \times 0$, $(n > 2)$, so by the tubular neighborhood theorem the isotopy class of a bundle map

$$\hat{\gamma}: S^{2n-1} \times D^{2n} \longrightarrow S^{2n-1} \times D^{2n} .$$

It follows that $M \equiv E_0(\gamma)$, where $\hat{\gamma}$ is a characteristic map for $\gamma \in \pi_{2n-1}(SO(2n))$. If n is odd, Lemma 1.5 completes the proof. If n is even, $\gamma = \pm \tau + 2m\sigma$. According to [11, 5.7.1],

$$E_0(\pm \tau + 2m\sigma) \equiv E_0(\pm \tau) \# m^2 \Sigma(\sigma, \sigma)$$

where $\Sigma(\sigma, \sigma)$ is a homotopy sphere. Noting that M bounds a π-manifold if and only if Σ does, completes the proof.

(1.5) <u>Remark</u>: If M^{4m+1} is a 1-connected π-manifold with the integral homology of $V_{2m+2,2}$ (i.e., $H_{2m+1}(M) = H_{2m+2}(M) = \mathbb{Z}$), then the same argument shows that either

$$M \equiv V_{2m+2,2} \# \Sigma \quad \text{or}$$
$$M \equiv (S^{2m+1} \times S^{2m}) \# \Sigma$$

This is a special case of a theorem of DeSapio [4].

2. Link Manifolds

Since $K_m^{p,q}$ is defined by the polynomial

$$(z_1)^p + (z_1)(z_2)^q + (z_3)^2 + \dots + (z_{m+1})^2$$

it admits an action of $O(m-1)$ in the usual fashion by operating on the last $m-1$ coordinates. This is an example of an $O(m-1)$-action with exactly three orbit types. When dealing with such actions we shall use the term <u>knot manifold</u> as originally defined. (See [7; 314-10].) In particular, we shall mean that the orbit space of a knot manifold is D^4 and that the fixed point set is a circle. The term <u>link manifold</u> will be used to distinguish the case where the fixed point set consists of more than one circle. (For our purposes, the fixed point set will be precisely two circles.) It is easily checked that $K_m^{p,q}$ is a link manifold. Since the fixed point set of a link manifold is an embedded submanifold of S^3, the boundary of the orbit space, we can think of the fixed point set as a link in S^3.

A well known theorem of K. Jänich [9] shows that for each $m \geq 3$ there is a one-to-one correspondence between smooth unoriented knot classes (S^3, F) and equivariant diffeomorphism classes of $(2m-1)$-dimensional knot manifolds $M^{2m-1}(F)$. Jänich's results also give a corresponding theorem for link manifolds (as communicated to me by Dieter Erle), which is only slightly more complicated.

Two oriented links (S^3, L_i), $i = 1,2$, will be called <u>equivalent</u> if and only if there exists a diffeomorphism of pairs $(S^3, L_1) \longrightarrow (S^3, L_2)$ which preserves the link orientations (but not necessarily the orientation of S^3).

(2.1) <u>Theorem</u> (Classification of link manifolds): If $m \geq 3$, then for each link equivalence class (S^3, L) there corresponds a unique (up to equivariant diffeomorphism) $(2m-1)$-dimensional link manifold $M^{2m-1}(L)$.

So if (S^3, L) is a link with 2 components, then depending on whether or not L is amphicheiral there are either one or two link manifolds with L (unoriented) as fixed point set.

<u>Remark:</u> For no knot (S^3, F) is $M^{4n-1}(F)$ homeomorphic to $V_{2n+1,2}$. For if $H_{2n-1}(M^{4n-1}(F)) = \mathbb{Z}_2$, then it follows from Hirzebruch [7; 314-16] that $\det F = \pm 2$ and that $M^{4n+1}(F)$ is therefore a homotopy sphere. But $M^{4n+1}(F)$ has a $\mathbb{Z}_2$- action with $M^{4n-1}(F)$ as fixed point set, which, by P. A. Smith theory, is a contradiction.

Using Theorem 2.1, we can reduce questions of equivariant diffeomorphism to knot theoretic questions. We have two applications of this technique.

<u>Application 1</u> (Equivalent link manifolds): Let $M_o(E_7)$ denote the plumbing of 7 copies of the unit tangent S^{m-1}-bundle to S^m along the graph

(the Dynkin diagram of E_7). $O(m-1)$ operates on each copy of the unit tangent bundle, $V_{m+1,2}$, and the plumbing can be taken equivariantly. This gives $M_o(E_7)$ the structure of a link manifold.

On the other hand, $K_m^{p,q}$ is the link manifold corresponding to the oriented link $(S^3, K_1^{p,q})$. $K_1^{p,q}$ is a torus link of two components consisting of an unknotted circle linked q times with the torus knot of type $(p-1,q)$, denoted by $t(p-1,q)$. (See

[12, chapter 10].) The author has verified that the fixed point link of $M_0(E_7)$ is equivalent to $K_1^{3,3}$. (These links are pictured on the next page.) Thus, as a consequence of Theorem 2.1:

(2.2) Proposition: If $m \geq 3$, $M_0(E_7)$ is $O(m-1)$-diffeomorphic to $K_m^{3,3}$. (Compare [7, 314-14].)

Application 2 (Prime actions): One can construct link manifolds almost diffeomorphic to $V_{m+1,2}$ as equivariant connected sums. For example, the inclusion $O(m-1) \subset O(m+1)$ gives $V_{m+1,2}$ the structure of the link manifold, where the fixed point link consists of two unknotted circles linked once. If the knot manifold $M^{2m-1}(F)$ is a homotopy sphere, the connected sum $V_{m+1,2} \# M^{2m-1}(F)$ can be performed equivariantly at fixed points. Moreover, the link which classifies the connected sum is just the connected sum of the fixed point sets, that is, a trivial knot linked once with F. This leads us to the following definition.

Definition: A link manifold $O(m-1) \times M \longrightarrow M$ is decomposable if the action is the connected sum of two non-trivial actions, i.e., M is equivariantly diffeomorphic to $M_1 \# M_2$ where M_i is a link manifold or a knot manifold, the connected sum is equivariant, and M_i is not equivalent to S^{2m-1} with the standard diagonal action (that is, the knot manifold classified by the trivial knot). If the action is not decomposable, the link manifold is prime. Clearly, this definition can be generalized to G-actions with fixed points, where G is a Lie group which acts in some standard fashion on S^n.

Note, for example, that a knot manifold is prime if and only if the corresponding knot is prime.

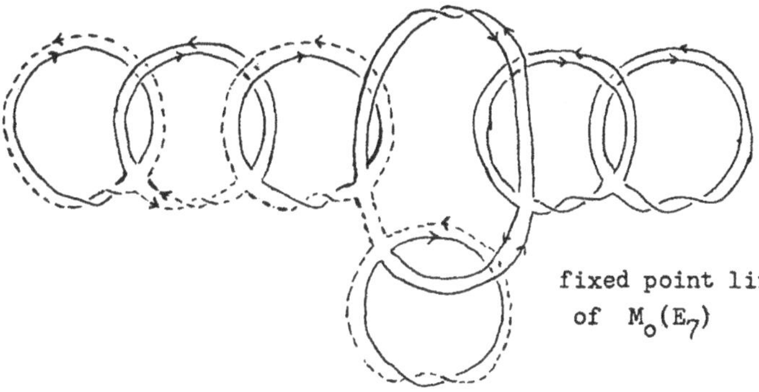

fixed point link of $M_o(E_7)$

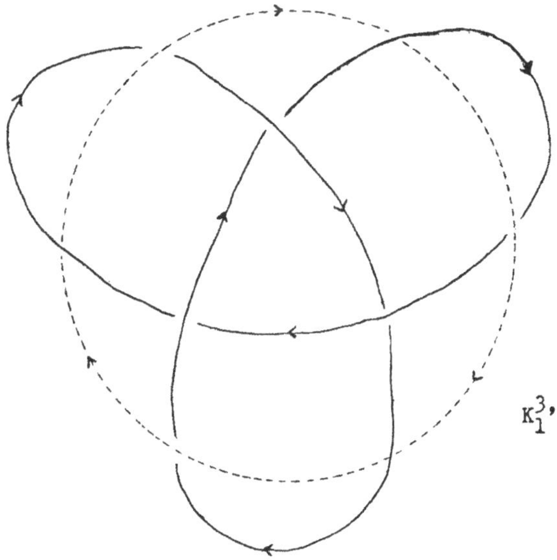

$K_1^{3,3}$

If a link manifold M is homeomorphic to $V_{2n+1,2}$ and $M \equiv M_1 \# M_2$ is decomposable, then

$$\mathbb{Z}_2 = H_{2n-1}(M) = H_{2n-1}(M_1) \oplus H_{2n-1}(M_2) .$$

So using Proposition 1.4, one of the M_i must be homeomorphic to $V_{2n+1,2}$ while the other is a homotopy sphere. (Link manifolds are, of course, Π-manifolds and bound Π-manifolds [5], [7; 314-16].)

(2.3) <u>Proposition:</u> $K_{2n}^{p,q}$ is prime.

<u>Proof:</u> Let L denote the link corresponding to $K_{2n}^{p,q}$. If $K_{2n}^{p,q}$ were decomposable, then we would have $L = F \# L'$, where F is a knot and L' is a link of two components, one of which is trivial and the other of which is a knot F' satisfying $t(p-1,q) = F \# F'$. Since torus knots are prime [13; p. 250], F' must be trivial and $F = t(p-1,q)$. But the knot manifold $M^{4n-1}(F)$ corresponding to $t(p-1,q)$ is the Brieskorn manifold $V_{2n}(p-1,q,2_1,\ldots,2)$ which satisfies $H_{2n-1}(M^{4n-1}(F)) = \mathbb{Z}_q$. Since $H_{2n-1}(M^{4n-1}(F))$ is a direct summand of $H_{2n-1}(K_{2n}^{p,q}) = \mathbb{Z}_2$, this is a contradiction.

3. Further Questions and Remarks

Semifree S^1-actions: If $k \leq n-1$, let ρ_k be the representation of S^1 in $O(2n-1)$ which takes $g \in SO(2)$ to the matrix

$$\rho_k(g) = \begin{bmatrix} I & & & \\ & g & & \\ & & \ddots & \\ & & & g \end{bmatrix}$$

where there are k copies of g. This defines a semifree

S^1-action on $K_{2n}^{p,q}$. The proof of Theorem 1.5 in [2] also shows,

(3.1) Proposition (Browder-Petrie): If $k = n/2$, $k \leq n-1$ and if the semifree S^1- action defined by ρ_k on $K_{2n}^{p,q}$ and $K_{2n}^{p',q'}$ are equivalent, then $I(\bar{F}_{2n}^{p,q}) = I(\bar{F}_{2n}^{p',q'})$, where I means index.

Link manifolds diffeomorphic to $V_{m+1,2}$, m odd: As pointed out in Remark 1.5, in these dimensions the homology of the ambient manifold does not distinguish between link manifolds diffeomorphic to $V_{m+1,2}$ and those almost diffeomorphic to $S^m \times S^{m-1}$. Suppose that a link manifold M^{2m-1} is a homology $V_{m+1,2}$. Suppose further that the fixed point set has two components linked with linking number ℓ. There is some evidence for the following conjecture.

(3.2) Conjecture: If m is odd,

$$M^{2m-1} \equiv V_{m+1,2} \quad ; \quad \text{if } \ell \text{ is odd}$$

while

$$M^{2m-1} \equiv S^m \times S^{m-1} \# \Sigma \; ; \quad \text{if } \ell \text{ is even}$$

$\Sigma \in bP_{2m}$.

Restricting to $O(m-k) \subset O(m-1)$: $O(m-k)$ acts on $K_m^{p,q}$ $(1 \leq k \leq m-4)$ via the inclusion. This again is an action with 3 orbit types, the orbit space is D^{2k+4} and the embedding of the fixed point manifold is $K_k^{p,q} \subset S^{2k+1}$ = the boundary of the orbit space. The question naturally arises - for which values of k is this action on $K_m^{p,q}$ prime? Not much is known about this question, although some observations are made in [3].

We have discussed prime actions in this paper, because this is a particularly interesting question in the study of non-standard G-actions on manifolds almost diffeomorphic to $V_{2n+1,2}$. Since a G-action on such a manifold is decomposable only if one of the manifolds in the connected sum is a homotopy sphere, the non-standard prime G-actions are, in some sense, precisely those actions whose "exoticness" depends on the topological structure rather than the specific differential structure of an exotic $V_{2n+1,2}$.

References

1. Brieskorn, E., Beispiele zur differential topologie von singularitaten, Inventiones Math., 2 (1966), 1-14.

2. Browder, W. and T. Petrie, Semifree and quasi-free S^1-actions on homotopy spheres, Essays on Topology and Related Topics Memoires dedies a Georges de Rham, Berlin-Heidelberg-New York, Springer-Verlag, 1970, pp. 136-146.

3. Davis, M., Group actions on exotic Stiefel manifolds, senior thesis, Princeton, 1971.

4. DeSapio, R., Action of θ_{2k+1}, Mich. Math. J., 14 (1967), 97-100.

5. Erle, D., Die quadratische form eines knotens und ein satz uber Knotenmannigfaltig-Keiten, J. Reine Angew. Math., 236 (1969), 174-218.

6. Haefliger, A., Differentiable imbeddings, Bull. Amer. Math. Soc., 67 (1961) 109-111.

7. Hirzebruch, F., Singularities and exotic spheres, Seminaire Bourbaki, 19e annee, No. 314, (1966-67).

8. Hsiang, W. C. and W. Y. Hsiang, Differentiable actions of compact connected classical groups I, Amer. J. Math., 89 (1967), 705-786.

9. Jänich, K., Differenzierbare mannigfaltigkeiten mit rand als orbitravme differenzierbarer G-mannigfaltigkeiten ahne rand, Topology, 5 (1966), 301-320.

10. Kervaire, M. and J. Milnor, Groups of homotopy spheres I, Annals of Math., 77 (1963) 504-537.

11. Kosinski, A., On the inertia group of π-manifolds, Amer. J. Math. 89 (1967), 227-248.

12. Milnor, J., Singular points on complex hypersurfaces, Annals of Math. Studies, No. 61, Princeton Univ. Press, 1969.

13. Schubert, H., Knoten und vollringe, Acta Math., 90 (1953), 131-286.

ON UNITARY AND SYMPLETIC KNOT MANIFOLDS

Dieter Erle

University of Dortmund

Abstract of lecture given at the Second Conference on Compact Transformation Groups, June 1971, at Amherst, Mass.

This is a report on joint work with Wu-chung Hsiang.

A knot manifold is a compact connected smooth manifold M without boundary, together with a smooth action of a compact Lie group $\Lambda(n)$ on M such that the following conditions are fulfilled:

1) $\Lambda(n)$ is the orthogonal group $O(n)$, the unitary group $U(n)$, or the symplectic group $Sp(n)$.
2) $n \geq 3$
3) The action has three orbit types, and the isotropy group of a point is conjugate to $\Lambda(n)$, $\Lambda(n-1)$, or $\Lambda(n-2)$.

Depending on what $\Lambda(n)$ is, the knot manifold is called orthogonal, unitary, or symplectic. The orbit space M' of M is a smooth manifold with boundary in a natural way, and the image F of the fixed point set in M' is a smooth submanifold of the boundary $\partial M'$ of M'. F has codimension 2, 3, or 5 in $\partial M'$ in the orthogonal, unitary, or symplectic case, respectively. The triple $(M',\partial M',F)$ is called the orbit triple, the pair $(\partial M',F)$ the orbit knot of the knot manifold.

Questions: Which triples are orbit triples of knot manifolds?
How many knot manifolds do have a given orbit triple?

The following theorems partially answer these questions.

Theorem: A triple $(M',\partial M',F)$ is the orbit triple of some knot manifold with trivial principal orbit bundle if and only if F has the correct codimension in $\partial M'$ and F bounds a framed manifold in

$\partial M'$.

If M' is a disk and F a homotopy sphere, i.e. $(\partial M',F)$ an ordinary (possibly higher dimensional) knot, then the above theorem means that, in the orthogonal case, all knots do occur. (This result is due to Wu-chung Hsiang, Wu-yi Hsiang, and K. Jänich.) In the unitary and symplectic cases, however, many knots are ruled out as orbit knots, because in codimensions 3 and 5, there are knots that do not bound a framed manifold.

Theorem (Wu-chung Hsiang, Wu-yi Hsiang, K. Jänich): If M' is a disk and F a homotopy sphere in $\partial M'$, then for each $n \geq 3$, there is at most one $O(n)$ - knot manifold with orbit triple $(M',\partial M',F)$.

Theorem: In the unitary and in the symplectic case, there are non-diffeomorphic knot manifolds (with the same group acting) having the same orbit triple $(M',\partial M',F)$ where M' is a disk and F a 1-sphere.

So in the unitary and symplectic cases, the orbit triple is not in general sufficient to classify the action.

A CLASSIFICATION OF 6-MANIFOLDS WITH FREE S^1 ACTIONS

Richard Z. Goldstein and Lloyd Lininger
State University of New York at Albany

In this paper we will define an operation which assigns to each pair of simply connected manifolds M^n and L^n an n-manifold K^n. Using this construction we are able to take two manifolds with non-trivial differentiable actions of S^1, (S^1_1, M) and (S^1_2, L), and define a circle action on the new manifold K. This allows us to explicitly classify all 6-manifolds with torsion free homology on which S^1 acts freely and differentiably. Wall classified such 6-manifolds under the additional hypothesis that the second Stiefel-Whitney class is zero [3]. We do not need this assumption.

To classify the simply connected 6-manifolds with torsion free homology, two actions of S^1 are defined on $S^3 \times S^3$, say G_1 and G_2. These actions were studied in [1]. Then using the construction defined above we prove that any such 6-manifold with a free circle action is obtained by applying the construction m times to $S^3 \times S^3$ with the G_1 action and n times to $S^3 \times S^3$ with the G_2 action for suitable non-negative integers m and n. In particular this completes the classification in [2].

1. CANONICAL AND TWISTED FRAMINGS

Let S^1 be a differentiably embedded circle in M^n, a simply connected manifold of dimension greater than 4. A framing $f_1, \ldots, f_{n-1}$ of the normal bundle of S^1 is called "canonical" if and only if there exists a 2-disk D^2 differentiably embedded in M^n with boundary S^1 and the frames $f_1, \ldots, f_{n-2}$ can be extended to a framing of the normal bundle of D^2. A framing of the normal bundle of S^1 is called twisted if and only if it is not canonical. An embedding of S^1 and a framing of its normal bundle will be denoted by (S^1, F). A given framing $F = (f_1, \ldots, f_{n-1})$ together with a map $h: S^1 \to SO(n-1)$ determines a new framing hF. If h is homotopic to g then hF is canonical if and only if gF is canonical. Throughout this paper we will denote a map representing a generator of $\pi_1(SO(n-1))$ by g. Since $\pi_1(SO(n-1)) = Z_2$, we only have to consider the action of g on a given framing.

<u>Theorem 1</u>. a) If (S^1, F) is twisted then (S^1, gF) is canonical. b) If the

second Stiefel-Whitney class of M^n is zero and (S^1, F) is canonical then (S^1, gF) is twisted. c) If $w_2(M^n) \neq 0$ and (S^1, F) is canonical then (S^1, gF) is canonical.

Corollary 2. If $w_2(M^n) \neq 0$ then any framing of the normal bundle of an embedded S^1 is canonical.

Proof of Theorem 1. a) Embed a disk D^2 with boundary S^1 such that f_{n-1} points inward. Choose a product structure for the tubular neighborhood of D^2. The framing $f_1,\dots,f_{n-2}$ determines then a map h of S^1 into $SO(n-2)$. Since F was twisted h is not null-homotopic. We can choose a map $g': S^1 \to SO(n-1)$ homotopic to g such that g' leaves f_{n-1} fixed. It follows that g'h is null-homotopic and hence (S^1, gF) is canonical.

To prove parts b and c, we will use the following observation. Let η be the non-trivial vector bundle over S^2 of a given dimension greater than 2. The framings of the normal bundle of the equator induced by the trivializations of the northern and southern hemispheres differ by the mapping g. To prove part b), suppose F extends over some D^2_+ and gF extends over some D^2_-. We may assume the union of D^2_+ and D^2_- is a differentiably embedded S^2. Using the above observation it follows that this S^2 has a non-trivial normal bundle in M which contradicts the fact that $w_2(M) = 0$.

For the proof of part c), we can assume our embedded S^1 is the equator of an S^2 embedded in M^n whose normal bundle is non-trivial. Hence F extends over the normal bundle of one of the hemispheres and gF extends over the normal bundle of the other hemisphere.

If $w_2(M)$ is zero, F a framing of the normal bundle of an embedded S^1, and any embedded D^2 with boundary S^1 such that f_{n-1} points into D^2, then $f_1,\dots,f_{n-2}$ extends over the normal bundle of D^2 if and only if (S^1, F) is canonical.

2. THE CONSTRUCTION ON MANIFOLDS

Let M and L be simply connected n-manifolds with orientation, n greater than 4. Let S^1 be embedded in M and L and let F and F' be framings of the normal bundle of S^1 in M and L respectively. Choosing an orientation for S^1, we assume moreover that $f_1,\dots,f_{n-1},t$, where t is the tangent direction of S^1, agrees with the orientation of M, and $f'_1,\dots,f'_{n-1},-t$ agrees with the orientation of L. Now identify M and L along the tubular neighborhoods of S^1 by the linear map such that f_j is identified with f'_j, and S^1 is identified by the identity. Now remove the interior of the tubular neighborhoods and "round the corners" to obtain a new oriented manifold. It is easily seen that this manifold depends only on the framings F and F', so we will denote it by (M, F) o (L, F'). Observe that if h: $S^1 \to SO(n-1)$ then (M, hF) o (L, hF') = (M, F) o (L, F'). This "sum" is associative and commutative.

Theorem 3. Suppose M and L are simply connected manifolds with orientation and (S^1, F) and (S^1, F') are framed normal bundles of S^1 in M and L respectively. Then a) If (S^1, F) and (S^1, F') are canonical then (M, F) o (L, F') is diffeomorphic to $M \# L \# S^2 \times S^{n-2}$. Here # denotes the usual connected sum of manifolds. b) If (S^1, F) is canonical and (S^1, F') is twisted or (S^1, F) is twisted and (S^1, F') is canonical, then (M, F) o (L, F') is diffeomorphic to $M \# L \# \eta$ where η is the non-trivial (n-2)-sphere bundle over S^2. And c) If (S^1, F) and (S^1, F') are twisted then (M, F) o (L, F') is diffeomorphic to $M \# L \# S^2 \times S^{n-2}$.

Proof. Part c) follows from part a) by composing F and F' with the map g, g representing a generator of $\pi_1(SO(n-1))$, and using Theorem 1.

To prove part a), embed D^2 in M and L with boundary S^1 and take a tubular neighborhood of each embedding. Now we observe that we can assume the identification is taking place in the interior of these neighborhoods. Thus, it follows that (M, F) o (L, F') is diffeomorphic to $M \# L \# X$, where X is formed by taking (S^n, T) o (S^n, T), T a canonically framed S^1 in S^n.

To prove that (S^n, T) o (S^n, T) is diffeomorphic to $S^2 \times S^{n-2}$ we consider the following. Regard S^n as the boundary of $D^2 \times D^{n-1}$ and we may regard (S^1, T) as bdy $D^2 \times D^{n-1}$. Now take two copies of $D^2 \times D^{n-1}$ and identifying the normal bundles in S^n of S^1 as subsets of D^{n+1} gives $S^2 \times D^{n-1}$. Now (S^n, T) o (S^n, T) is just the boundary of $S^2 \times D^{n-1}$. See Figure 1.

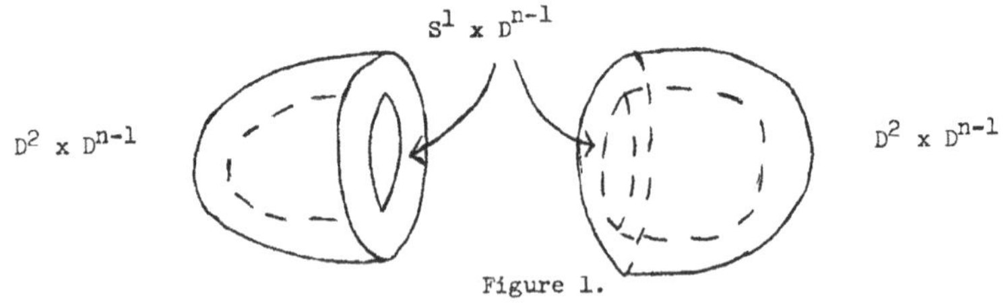

Figure 1.

Proof of part b). Using the ideas of the proof of part a), we only have to determine $(S^n, T) \circ (S^n, T')$ to prove the first part of b), where T is the canonical framing of S^1 in S^n and T' is gT. The second statement of part b) follows from the first statement of part b) and the commutativity of "o".

The construction in the proof of part a) can now be used, and we observe that when the identification is made on the two copies of $D^2 \times D^{n-1}$, we have the non-trivial D^{n-1} bundle over S^2. Hence $(S^n, T) \circ (S^n, T')$ is the non-trivial (n-2)-sphere bundle over S^2. This completes the proof of Theorem 3.

In classical surface theory one has the following result.

Theorem 4. If M^2 is a non-orientable manifold then $M^2 \# S^1 \times S^1$ is diffeomorphic to $M^2 \# K$ where K is the Klein bottle.

This result easily generalizes to the following.

Theorem 5. If $w_1(M^n) \neq 0$ then $M^n \# S^1 \times S^{n-1}$ is diffeomorphic to $M^n \# K$ where K is the non-trivial (n-1)-sphere bundle over S^1.

As a corollary to Theorem 3 we have the following.

Theorem 6. If $w_2(M^n) \neq 0$, $n \geq 5$ then $M^n \# S \times S^{n-2}$ is diffeomorphic to $M^n \# \eta$, η the non-trivial (n-2)-sphere bundle over S^2.

Proof. Let (S^n, T) be such that T is the canonical framing of S^1 in S^n. Let D^n be an n-cell in M^n and S^1 an embedded circle in the interior of D^n with F a framing of its normal bundle. Then from Theorem 1 above, F and gF are canonical. By Theorem 3, $(M, F) \circ (S^n, T)$ is diffeomorphic to $M^n \# S^2 \times S^{n-2}$, $(M, gF) \circ (S^n, gT)$ is diffeomorphic to $M^n \# S^2 \times S^{n-2}$, and since gF is also canonical, $(M, gF) \circ (S^n, gT)$ is diffeomorphic to $M^n \# \eta$.

3. FREE CIRCLE ACTIONS

Suppose (G, M) denotes a free circle action on M, M a simply connected manifold of dimension greater than four. Then G will induce a framing of an invariant normal bundle of an orbit C. Whether or not this framing is canonical is independent of the orbit.

Recall that a simply connected manifold M is a spin manifold if and only if $w_2(M)$ is zero. For this section it will be convenient to use the fact that for a simply connected manifold M, one can regard w_2 as a homomorphism from $H_2(M; Z)$

into Z_2.

Since (G, M) is a free circle action, as a circle bundle it will be classified by a homotopy class of maps from M/G into CP^∞, the base space of the universal S^1 bundle. We will let f denote a classifying map for G. Since CP^∞ is a K(Z, 2), the homotopy classes of maps from M/G into CP^∞ are classified by the image under the induced map of the generators of $H_2(M/G; Z)$. Now by looking at the Gysin sequences of the fibrations associated with G and of the universal bundle, it is easy to see that for a suitably chosen basis for $H_2(M/G; Z)$, the map f_* will map one generator α of $H_2(M/G; Z)$ to a generator of $H_2(CP^\infty; Z)$ and all other generators will be mapped to zero. This uses the fact that M is simply connected.

Theorem 7. If M/G is a spin manifold then M is a spin manifold.

Proof. Since M and M/G are orientable, and the tangent bundle of M is the Whitney sum of the tangent bundle of M/G and the bundle along the fiber S^1, the tangent bundle of M is stably the bundle induced by pulling back the tangent bundle of M/G. It follows that the induced map in cohomology maps the Stiefel-Whitney classes of M/G onto the Stiefel-Whitney classes of M. Hence $w_2(M/G) = 0$ implies $w_2(M) = 0$.

Theorem 8. Suppose M^n is simply connected, $n > 5$, and G is a free differentiable S^1 action on M. Then the orbit space M/G is a spin manifold if and only if the framing of an orbit C induced by G is twisted.

Proof. Case 1. Assume M/G is not a spin manifold and there exists a generator of $H_2(M/G; Z)$ such that $w_2(\gamma) \neq 0$ and $\gamma \neq \alpha$.

Now γ can be represented by an embedded 2-sphere S in M/G and we can assume that f(S) is a point in CP^∞. Since $w_2(\gamma) \neq 0$, the normal bundle of S in M/G is non-trivial. Since f(S) is a point, S will lift to a 2-sphere in M with normal bundle isomorphic to the non-trivial bundle plus a trivial line bundle. Hence M is not a spin manifold and by Corollary 2 any framing of a normal bundle of an orbit C will be canonical.

Case 2. Assume $w_2(\alpha) \neq 0$. Let S be an embedded 2-sphere in M/G which represents α . Let π denote the projection map from M to M/G. Then $\pi^{-1}(S)$ is a 3-sphere K in M and G restricted to K is the Hopf circle action. There is a 2-disk D in K such that a) the boundary of D is an orbit, and b) D intersects each orbit distinct from its boundary in exactly one point. The existence of D is seen by regarding S^3 as the join of two circles. The Hopf action is defined by the standard action on each S^1 and D can be taken as the join of the first circle and a point on the second circle. Let C denote the boundary of D, and $f_1,\ldots,f_{n-1}$ be a framing of the normal bundle of a point p on C. Then we can assume the image of f_1 and f_2 under G is contained in K, and the images of $f_3,\ldots,f_{n-1}$ give a framing of the normal bundle of K restricted to C.

Assume the framing induced by $f_1,\ldots,f_{n-1}$ is twisted. Then the framing of C induced by $f_1,\ldots,f_{n-1}$ does not extend over D. Since the two frames induced by f_1 and f_2 do not extend over D, the frames induced by $f_3,\ldots,f_{n-1}$ will extend over D. Now since the frames induced by $f_3,\ldots,f_{n-1}$ extend over D, their extended image under π will give a framing of the normal bundle of $\pi(K) = S$. Hence S has a trivial normal bundle in M/G. This is a contradiction.

Case 3. Suppose M/G is a spin manifold. Then it follows from Theorem 7 that M is a spin manifold. As in the proof of case 2, we embed a 2-sphere S in M/G which represents α. $\pi^{-1}(S)$ is a 3-sphere K and G restricted to K is the Hopf action. Since M is a spin manifold we can assume we have picked C and a disk D as in case 2. The orbit space M/G is a simply connected spin manifold, hence the normal bundle of S is trivial. Let $e_3,\ldots,e_{n-1}$ be a framing of the normal bundle of S in M/G. This framing will lift by π^{-1} to a framing $f_3,\ldots,f_{n-1}$ of the normal bundle of K in M restricted to D. At a point p of C, choose points f_1 and f_2 in the normal bundle of C in K, such that $f_1,\ldots,f_{n-1}$ gives a framing of the normal bundle of C at p in M. We will let $f_1,\ldots,f_{n-1}$ denote also the framing of C induced by G.

Now if g' acts on the framing $f_1,\ldots,f_{n-1}$ by changing only f_1 and f_2, g' homotopic to g, then the new framing can be extended over D. Therefore, the new framing is canonical and the original framing was twisted.

We can now classify the simply connected 6-manifolds with torsion free homology which admit free differentiable circle actions. The actions on these manifolds were classified in [2]. The case where the homology has torsion is only slightly improved over the results in [2].

Let G_1 be the free circle action on $S^3 \times S^3$ defined by the Hopf action on the first factor and the identity on the second factor. Let G_2 be the free circle action on $S^3 \times S^3$ defined as follows. Consider $S^3 \times S^3$ as $\{(z_0, z_1, z_2, z_3) | z_i$ is complex, $z_0\bar{z}_0 + z_1\bar{z}_1 = 1$, $z_2\bar{z}_2 + z_3\bar{z}_3 = 1\}$ and S^1 as $\{z | z$ is complex and $z\bar{z} = 1\}$. Define G_2: $S^1 \times S^3 \times S^3 \to S^3 \times S^3$ by $G_2(z, z_0, z_1, z_2, z_3) = (zz_0, zz_1, zz_2, z_3)$. See [1] for further details. Now it is easily checked that the framing of the normal bundle of an orbit by G_1 is twisted, while the framing of a normal bundle of an orbit by G_2 is canonical. This also follows from Theorem 8 and [1].

Suppose $mG_1 \circ nG_2$ denotes the action of S^1 on the manifold obtained by applying the "o" construction m times to $(G_1, S^3 \times S^3)$, n times to $(G_2, S^3 \times S^3)$ and then to the resulting manifolds.

Theorem 9. Suppose M^6 is a simply connected manifold, the homology of M^6 has no torsion, the rank of $H_2(M; Z)$ is k-1, $k \geq 1$, and G is a free action on M^6. Then (G, M) is either kG_1 or kG_2 if M is a spin manifold, if M is not a spin manifold then (G, M) is $G_1 \circ (k-1)G_2$.

Proof. The case k = 1 was proved in [1]. Suppose k > 1. It follows from Theorem 3 that each of the total spaces of kG_1, kG_2 and $G_1 \circ (k-1)G_2$ is simply connected, has torsion free homology, and the rank of the second homology group is k-1. Also from Theorem 3 we know that kG_1 and kG_2 have diffeomorphic total spaces while we know from Theorem 8 that their orbit spaces are distinct. Theorem 3 also implies that the total space of $G_1 \circ (k-1)G_2$ is not diffeomorphic to the total space of kG_1. Now the main classification theorem of [2] proved that there are exactly two manifolds satisfying the hypothesis of Theorem 9, one admits two actions of S^1 and one admits exactly one free action. This concludes the proof of Theorem 9.

We observe from Theorem 3 that the manifolds satisfying the hypothesis of Theorem 9 are $S^2 \times S^4 \# \dots \# S^2 \times S^4 \# S^3 \times S^3 \# \dots \# S^3 \times S^3$ and n $\# S^2 \times S^4 \# \dots \# S^2 \times S^4 \# S^3 \# \dots \# S^3 \times S^3$, k copies of $S^3 \times S^3$ and (k-1) other factors.

BIBLIOGRAPHY

1. R. Goldstein and L. Lininger, "Actions on 2-connected 6-manifolds" Amer. J. of Math., 91 (1969), 499-504.

2. L. Lininger, "S^1 actions on 6-manifolds", to appear in Topology.

3. C. T. C. Wall, "Classification problems in differential topology V", Invent. Math., 1(1966), 355-375.

Partially supported by National Science Foundation Grant GU 3171

SU(n) ACTIONS ON MANIFOLDS WITH VANISHING FIRST AND SECOND INTEGRAL PONTRJAGIN CLASSES

Edward Andrew Grove*
University of Rhode Island

INTRODUCTION

Let G be a compact classical group acting differentiably on a manifold M. For $x \in M$, let $G_x = G \cap \{g : g \cdot x = g\}$, and let G_x^o be the connected component of the identity of G_x. One of the problems studied in [2] was the following: Let $P_j(M)$ be the j'th rational Pontrjagin class of M. Assume $P_1(M)=0$, and try to determine all G_x^o for $x \in M$. By assuming $\dim M \leq \frac{(n-1)^2}{2}$, they were able to determine all possible G_x^o of the given action.

In this paper, we let G be $SU(n)$, and we assume $P_1(M)=0=P_2(M)$. For our methods to work, it is necessary to assume all isotropy subgroups have positive dimension, and so the requirement $\dim M \leq n^2-2$ is a natural one. Using this restriction, one can actually find all possible G_x^o for x a regular element. If we strengthen the restriction slightly further to $\dim M \leq n^2 - \frac{8}{3}n - 2$, however, the results are extremely simplified, and so this is the dimension requirement we shall adopt.

Our main results are the following.

(1) We determine all possible G_x^o for x a regular element of the above action. The list is quite short.

(2) For a given G_x^o in the above list, we show that for any other $y \in M$, G_y^o is of the same type as G_x^o; i.e. if $G_x^o \hookrightarrow SU(n) = SU(k) \xrightarrow{\mu_k \oplus \mu_k^* \oplus \theta} SU(n)$, where μ_k is the standard representation of $SU(k)$ and θ is some trivial representation, then after conjugating if necessary,

$$(G_x^o \hookrightarrow G_y^o \hookrightarrow SU(n)) = (SU(k) \xrightarrow{\mu_k \oplus \theta} SU(\ell) \xrightarrow{\mu_\ell \oplus \mu_\ell^* \oplus \theta} SU(n))$$

§1 Let G be a compact connected Lie group acting differentiably on a manifold M. Let $P_1(M)=0=P_2(M)$.

We first wish to compute G_x^o, where $x \in M$ is a regular element of M.

* Research supported in part by National Science Foundation Grant GP-6371. These results are included in the author's Ph.D. thesis.

Let $e: G/G_x \approx G(x) \hookrightarrow M$ be the natural map. Let ξ_1 be the principal bundle $(G_x \to G \to G/G_x)$.

Then $$e^{!}\tau(M) = \tau(G/G_x) \oplus \nu(G/G_x)$$

$$= \alpha_{\xi_1}(\iota_{G/G_x}) \oplus \alpha_{\xi_1}(\phi_x)$$

where ι_{G/G_x} is the isotropy representation of G_x in G and ϕ_x is the slice representation of G_x at x. As x is regular, ϕ_x is trivial [3]. Hence $e^{!}\tau(M) = \alpha_{\xi_1}(\iota_{G/G_x}) \oplus \theta$ where θ is some trivial vector bundle.

<u>Remark</u> Let $\xi = (H \to E \to B)$ and $\eta = (G \to E' \to B')$ be principal bundles, and let

$$\begin{array}{ccc} H & \xrightarrow{\Lambda} & G \\ \downarrow & & \downarrow \\ E & \longrightarrow & E' \\ \downarrow & & \downarrow \\ B & \xrightarrow{x} & B' \\ \vdots & & \vdots \\ \xi & & \eta \end{array}$$

be a homomorphism of principal bundles. Then it follows trivially that

$$\begin{array}{ccc} RO(H) & \xleftarrow{\Lambda^{!}} & RO(G) \\ \downarrow{\scriptstyle \alpha_\xi} & & \downarrow{\scriptstyle \alpha_\eta} \\ KO(B) & \xleftarrow{x^{!}} & KO(B') \end{array}$$

is a commutative diagram.

So if $P: G/G_x^{\circ} \to G/G_x$ is the natural map, and $\xi_2 = (G_x^{\circ} \to G \to G/G_x^{\circ})$, then

$$P^{!}e^{!}\tau(M) = \alpha_{\xi_2}(\iota_{G/G_x^{\circ}}) \oplus \theta$$

$$= \tau(G/G_x^{\circ}) \oplus \theta$$

and so $P^* e^* P_j(M) = P_j(G/G_x^{\circ})$.

So we must first find all those connected subgroups H of G with $P_1(G/H) = 0 = P_2(G/H)$.

Let ξ be the principal bundle $(H \to G \to G/H)$.

Now $\tau(G/H) = \alpha_\xi(\iota_{G/H}) = \alpha_\xi(\mathrm{Ad}_G|H - \mathrm{Ad}_H)$

or

$$\tau(G/H) \oplus \alpha_\xi(\mathrm{Ad}_H) = \alpha_\xi(\mathrm{Ad}_G|H).$$

By the remark, $\alpha_\xi(\mathrm{Ad}_G|H)$ is a trivial bundle. Hence $\tau(G/H) \oplus \alpha_\xi(\mathrm{Ad}_H) = \theta$.

Corollary 1. [1] Let T be a torus in G. Then G/T is stably parallelizable.

Corollary 2. $\sum_{\alpha+\beta=j} P_\alpha(\tau(G/H))\cdot P_\beta(\alpha_\xi(Ad_H))=0$ mod 2 torsion.

Let T be a maximal torus in H, and let $\pi:G/T \to G/H$ be the natural map. Then π is a fibre map with fibre H/T, and hence the kernel of π^*: $H^*(G/H;Z) \to H^*(G/T;Z)$ consists just of torsion elements.

Let $\xi'=(T \to G \to G/T)$ and let ϕ be a real representation of H. Then again by the remark,

$$\pi^! \alpha_\xi(\phi)=\alpha_{\xi'}(\phi|T) \quad \text{and hence}$$

$$\pi^* P_j(\alpha_\xi(\phi))=P_j(\alpha_{\xi'}(\phi|T)).$$

Let $\eta(G)=(G \to E_G \to B_G)$ be a classifying bundle for G. Then we may take

$\eta(T)=(T \to E_G \to B_T)$ as a classifying space for T.

Given a fibre bundle γ, we also let $E(\gamma)$ be the Serre cohomology spectral sequence associated with γ.

Let $\lambda:H \to G$ be the embedding of H as a subgroup of G. Let $\eta(T)$ be the universal T bundle, and let $i:G/\lambda[T]\to B_T$ be the classifying map of ξ'. Then as before,

$$\begin{aligned}\pi^*P_j(\alpha_\xi(\phi))&=P_j(\alpha_{\xi'}(\phi|T))\\ &=i^*(P_j(\alpha_{\eta(T)}(\phi|T)))\\ &=(-1)^j i^*\sigma_{2j}(\tau_{\eta(T)}(W_1),\cdots,\tau_{\eta(T)}(W_N),-\tau_{\eta(T)}(W_1),\cdots,-\tau_{\eta(T)}(W_N))\end{aligned}$$

where σ_m is the m'th symmetric function, $\tau_{\eta(T)}$ is the transgression of $E(\eta(T))$ and $\{W_1,\cdots,W_N\}$ is the set of positive weights of [1]. So we need to compute the universal Pontrjagin classes of $\alpha_{\eta(T)}(\phi|T)$ (mod ker ι^*). Recall that as E_G is contractible, $E_G \times_{\lambda|T} G \to G/\lambda[T]$ is a homotopy equivalence. So we may take $\gamma=(G \to G/\lambda[T] \xrightarrow{i} B_T)$ to be a principal G bundle.

So $\iota^*:H^*(B_T;Z) \to H^*(G/\lambda[T];Z)$ is just the edge homomorphism

$$H^*(B_T)=E_2^{*,0}(\gamma) \twoheadrightarrow E_\infty^{*,0}(\gamma)\subseteq H^*(G/\lambda[T];Z).$$

Hence ker i^* = $\langle \text{im } d(\gamma)^+\rangle$, the ideal in $H^*(B_T;Z)$ generated by elements of positive degree in the image of the differential of $E(\gamma)$.

§2. We now take $G=SU(n)$. Then by examining $E(\gamma)$, we have (at least in low dimensions) that ker $\iota^* = \langle(\text{im } \tau_\gamma)^+\rangle$ where τ_γ is the transgression of $E(\gamma)$. Now $P(\lambda[T], SU(n))$ classifies γ, and so at least in low dimensions,

$$\ker \iota^* = \langle \operatorname{im} P^*(\lambda[T], SU(n))^+ \rangle$$
$$= \langle S^+(\tau_{\eta(T)}(\mu_1), \cdots, \tau_{\eta(T)}(\mu_n)) \rangle$$

where $\mu_1, \cdots, \mu_n$ are the weight vectors of λ.
Note $\mu_n = -(\mu_1 + \cdots + \mu_{n-1})$ as $\lambda : H \to SU(n)$.

Now $2\sigma_2(\tau_{\eta(T)}(\mu_1), \cdots, \tau_{\eta(T)}(\mu_n)) = -\sum_{i=1}^{n}(\tau_{\eta(T)}(\mu_i))^2$

as $\mu_n = -(\mu_1 + \cdots + \mu_{n-1})$. Recall by Corollary 2,

$$P_1(\tau(G/\lambda[H])) + P_1(\alpha_\xi(Ad_H)) = 0.$$

So $\pi^* P_1(\tau(G/\lambda[H])) = -\pi^* P_1(\alpha_\xi(Ad_H))$

$$= i^*\sigma_2(\tau_{\eta(T)}(W_1), \cdots, \tau_{\eta(T)}(W_N), -\tau_{\eta(T)}(W_1), \cdots, -\tau_{\eta(T)}(W_N))$$
$$= -i^* \sum_{j=1}^{N} (\tau_{\eta(T)}(W_j))^2$$

So $\pi^* P_1(\tau(G/\lambda[H])) = 0$

if and only if

for some integer K, $K \sum_{j=1}^{n}(\tau_{\eta(T)}(\mu_j))^2 = 2\sum_{j=1}^{N}(\tau_{\eta(T)}(W_j))^2$.
Let U be a normal subgroup of H. Then by using our remark together with a standard argument,

$$P_j(\tau(G/H)) = 0 \quad \text{implies} \quad P_j(\tau(G/U)) = 0.$$

So we now determine those compact subgroups (U, λ) of $SU(n)$ with U simple, simply connected, and $\ker \lambda$ finite satisfying

$$P_1(SU(n)/\lambda[U]) = 0 = P_2(SU(n)/\lambda[U]).$$

Recall $\pi^* P_1(G/\lambda[U]) = 0$

if and only if

there exists an integer K with

$$K \sum_{j=1}^{n}(\tau_{\eta(T)}(\mu_j)) = 2\sum_{j=1}^{N}(\tau_{\eta(T)}(W_j))^2.$$

It follows that $\pi^* P_1(G/\lambda[U]) = 0$

if and only if

there exists an integer K with

$$K \sum_{j=1}^{n}|\mu_j|^2 = 2\sum_{j=1}^{N}|W_j|^2$$

where $|\ |$ is induced by the Cartan-Killing form (all others are proportional as H is simple.)

Theorem 1. Let U be a compact, connected, simply connected, non-albelian, simple Lie group. Let $n \geq 28$. Let $\dim U > \frac{8}{3}n$. Then the following is the list of those almost faithful embeddings $\lambda : U \to SU(n)$ such that

$$P_1(SU(n)/\lambda[U]) = 0 = P_2(SU(n)/\lambda[U]).$$

G		λ			
SU(k)	$k \geq 9$	μ_k			
		$a\mu_k \oplus b\mu_k^*$		$a+b = 2$	
				$a+b = 3$	$3\|k$
				$a+b = 6$	$3\|k$
$S_p(k)$	$k \geq 6$	ν_{2k}			
		$3\nu_{2k}$	$3\|k+1$		
SO(2k+1)	$k \geq 6$	ρ_{2k+1}			
		$3\rho_{2k+1}$	$3\|2k-1$		
SO(2k)	$k \geq 7$	ρ_{2k}			
		$2\rho_{2k}$			
		$3\rho_{2k}$	$3\|k-1$		
		$6\rho_{2k}$	$3\|k-1$		

Proof:

The following is an example of the method.

Let $U=SU(k)$. If ϕ is a complex representation of U and $\{\mu_1,\cdots,\mu_t\}$ is the set of weight vectors of ϕ, let $n(\phi)=\sum_{j=1}^{t}|\mu_j|^2$. We have $n(Ad_{SU(k)})=2k(k-1)$. One can show easily that if ϕ is irreducible and $n(\phi)\leq 2k(k-1)$, then after conjugating if necessary, ϕ occurs on the following list:

$$\Lambda^3\mu_k \qquad 6\leq k\leq 8$$

$$\mu_k, S^2\mu_k, \Lambda^2\mu_k \qquad 2\leq k\ .$$

So we now have the restriction that after conjugating if necessary, all irreducible components of λ occur on the above list. We now determine by direct computation all those $\lambda: SU(k) \to SU(n)$ where $n(\lambda)$ divides $2k(k-1)$. This gives all subgroups $(SU(k),\lambda)$ of $(SU(k)$ with $\pi^*P_1(SU(n)/\lambda[SU(k)])=0$.

One handles all other simple Lie groups similarly. Recall

$$P_2(G/\lambda[H])+P_1(G/\lambda[H])P_1(\alpha_\xi(Ad_H))+P_2(\alpha_\xi(Ad_H))=0.$$

So if $P_1(SU(n)/\lambda[U])=0$, then $P_2(SU(n)/\lambda[U])=0$ if and only if $P_2(\alpha_\xi(Ad_U))=0$.

One now uses the condition that at least in low dimensions, $\ker \iota^* = \langle S^+(\tau_{\eta(T)}(\mu_1),\cdots,\tau_{\eta(T)}(\mu_n))$ to determine for which (U,λ) in our previous list we also have $P_2(\alpha_\xi(Ad_U))=0$. This concludes the proof.

<u>Theorem 2</u>. Let $SU(n)$ act differentiably on a manifold M. Let $P_1(M)=0=P_2(M)$. Let $\dim M \leq n^2-\frac{8}{3}n-2$. Let $n \geq 28$. Let $x \in M$ be a regular element. Then G_x^o occurs on the list of Theorem 1.

Proof:

As $n^2-\frac{8}{3}n-2 \geq \dim M \geq \dim SU(n)-\dim G_x^o = n^2-1-\dim G_x^o$, we see $\dim G_x^o > \frac{8}{3}n$. Hence

1) G_x^o is not a torus as $\dim G_x^o > n$ and $rk\, SU(n)=n-1$.

2) λ does not contain a copy of AdG_x^o, as if it did, we would have $\dim G_x^o \leq n$.

So let H be a compact connected Lie group. Let $\lambda: H \to SU(n)$ be an almost faithful homorphism. Let $\dim H > \frac{8}{3}n$.

We first consider $H=U_1\times U_2 \xrightarrow{\lambda} SU(n)$ with $P_1(SU(n)/\lambda[h])=0$. Let T_1 be a maximal for U_1 and T_2 for U_2. Let $\{v_1,\cdots,v_{rkU_1}\}$ be a base for $H^2(B_{T_2};Z)$, and $\{\bar v_1,\cdots,\bar v_{rkU_2}\}$ for $H^2(B_{T_2};Z)$. Now $\lambda=\sum_{j=1}^t \phi_j \otimes \psi_j$ where each ϕ_j and ψ_2 is either irreducible or trivial. For a representation ϕ with weight vectors $\{\mu_1,\cdots,\mu_s\}$ let $w(\phi)=\sum_{j=1}^s (\tau_{\eta(T)}(\mu_j))^2$. Then recall $P_1(SU(n)/\lambda[H]=0$ if and only if there exists an integer K with

$w(Ad_H)=Kw(\lambda)$. So $w(Ad_{U_1})+w(Ad_{U_2})=w(Ad_{U_1\times U_2})=Kw(\lambda)$.

$$=K\sum_{j=1}^t \dim \psi_j\cdot w(\phi_j)+K\sum_{j=1}^t \dim \phi_j\cdot w(\psi_j).$$

So as $w(Ad_{U_1})$ and the $w(\phi_j)$'s are polynomials in $\{v_1,\cdots,v_{rkU_1}\}$ and $w(Ad_{U_2})$ and the $w(\psi_j)$'s are polynomials in $\{\bar v_;\cdots,\bar v_{rkU_2}\}$, we have

$$w(Ad_{U_1}) = K\sum_{j=1}^t \dim \psi_j\cdot w(\phi_j), \quad \text{and}$$

$$w(Ad_{U_2}) = K\sum_{j=1}^t (\dim \phi_j)\cdot w(\psi_j).$$

So as $w(Ad_H)=0$ if and only if H is a torus, we have the following lemma.

<u>Lemma</u>. Let H be a compact, connected subgroup of $SU(n)$ such that $P_1(SU(n)/H)=0$. Then either H is a torus or H is semi-simple.

So we may assume U_1 and U_2 are simple. Suppose $U_1=SU(k)$ and $U_2=SU(\ell)$. Then we first find all those $\phi:SU(k)\to SU(n_1)$ and $\psi:SU(\ell)\to SU(n_2)$ satisfying

1) $\dim \psi\cdot n(\phi)\leq 2k(k-1)$

2) $\dim \phi\cdot n(\psi)\leq 2\ell(\ell-1)$.

These will be our only candidates to make up λ with.

We then use $P_1(SU(n)/\lambda[H])=0=P_2(SU(n)/\lambda[H])$ $n\geq 28$, and $\dim M\leq n^2-\frac{8}{3}n-2$ to derive a contradiction. We similarly handle the other cases. Hence H cannot contain precisely two normal simple subgroups. By modifying the above proof slightly, one can show that H must be simple.

Theorem 3. Let $SU(n)$ act differentiably on a manifold M. Let $P_1(M)=0=P_2(M)$. Let $\dim M\leq n^2-\frac{8}{3}n-2$. Let $n\geq 28$. Suppose that $y\in M$ is a regular element, and that $x\in M$. Then $(G_y^o\subseteq G_x^o\subseteq SU(n))$ occurs on the following list.

1) if $(G_y^o)=(SU(k)\xrightarrow{a\mu_k\oplus b\mu_k^*\oplus(n-(a+b)k)\theta} SU(n))$

where $a+b=1,2,3,6$ $k\geq 9$, and $3|k$ if $a+b=3,6$,

then

$$(G_y^o\subseteq G_x^o\subseteq SU(n))=(SU(k)\xrightarrow{\mu_k^*\oplus(\ell-k)\theta} SU(\ell)\xrightarrow{a\mu_\ell\oplus b\mu_\ell^*\oplus(n-(a+b)\ell)\theta} SU(n))$$

2) if $(G_y^o)=(S_p(k)\xrightarrow{a\gamma_{2k}\oplus(n-2ak)\theta} SU(n))$ where

$a=1,3$ $k\geq 6$, and $3|k+1$ if $a=3$, then

$$(G_y^o\subseteq G_x^o\subseteq SU(n))=(S_p(k)\xrightarrow{\gamma_{2k}\oplus 2(\ell-k)\theta} S_p(\ell)\xrightarrow{a\gamma_{2\ell}\oplus(n-2a\ell)\theta} SU(n))$$

3) if $(G_y^o)=(SO(2k+1)\xrightarrow{a\rho_{2k+1}\oplus(n-a(2k+1))\theta} SU(n))$ where

$a=1,3$ $k\geq 6$, and $3|2k-1$ if $a=3$, then

$$(G_y^o\subseteq G_x^o\subseteq SU(n))=(SO(2k+1)\xrightarrow{\rho_{2k+1}\oplus(\ell-2k-1)\theta} SO(\ell)\xrightarrow{a\rho_\ell\oplus(n-a\ell)\theta} SU(n))$$

4) if $(G_y^o)=(SO(2k)\xrightarrow{a\rho_{2k}\oplus(n-2ak)\theta} SU(n))$ where

$a=1,2,3,6$ $k\geq 7$, and $3|k-1$ if $a=3,6$, then

$$(G_y^o\subseteq G_x^o\subseteq SU(n))=(SO(2k)\xrightarrow{\rho_{2k}\oplus(\ell-2k)\theta} SO(\ell)\xrightarrow{a\rho_\ell\oplus(n-a\ell)\theta} SU(n)).$$

Proof:

As an example of the method of the proof, suppose

$(G_y^o \overset{\lambda}{\hookrightarrow} SU(n)) = (SU(k) \xrightarrow{a\mu_k \oplus b\mu_k^* \oplus (n-2k)\theta} SU(n))$ where $a+b=2$ and $k \geq 9$.

We shall first determine those subgroups H of $SU(n)$ containing G_y^o as a subgroup.

Let $G_y^o \xrightarrow{\phi} H \xrightarrow{\psi} SU(n)$ be the respective almost faithful homomorphisms, with $\psi \circ \phi = \lambda$. We may take $H = T^c \times U_1 \times \cdots \times U_m$ where T^c is a c dimensional torus, and U_j is a compact, simple Lie group for $1 \leq j \leq m$. We wish to determine the smallest normal subgroup N of H which contains G_y^o as a subgroup. Now $\Phi = \phi_T \oplus \phi_1 \oplus \cdots \oplus \phi_m$ and $\Psi = \sum_{j=1}^{t} \psi_{j,0} \otimes \cdots \otimes \psi_{j,m}$. As G_y^o is simple and T^c is abelian, ϕ_T is trivial. Note $n^2 - \frac{8}{3}n - 2 \geq \dim M \geq \dim G(y) = \dim SU(n) - \dim G_y^o$

$$= n^2 - 1 - \dim G_y^o$$

and so $\dim G_y^o > \frac{8}{3}n$. Using this fact together with $\psi \circ \phi = a\mu_k \oplus b\mu_k^* \oplus (n-2k)\theta$, we are led to such severe restrictions on the ϕ_j's and $\psi_{j,t}$'s, that the following lemma follows easily.

<u>Lemma</u>. H contains a normal subgroup N which contains G_y^o, and the embeddings $G_y^o \hookrightarrow N \hookrightarrow SU(n)$ are either

1) $(G_y^o \overset{\phi}{\hookrightarrow} N \overset{\psi}{\hookrightarrow} SU(n)) = (SU(k) \xrightarrow{\mu_k \oplus (\ell - k)\theta} SU(\ell) \xrightarrow{a\mu_\ell \oplus b\mu_\ell^* \oplus (n-2\ell)\theta} SU(n))$

or

2) $(G_y^o \overset{\phi}{\hookrightarrow} N \overset{\psi}{\hookrightarrow} SU(n)) = SU(k) \xrightarrow{a\mu_k \oplus b\mu_k^* \oplus (\ell-2k)\theta} SU(\ell) \xrightarrow{\mu_\ell \oplus (n-\ell)\theta} SU(n))$.

We shall now determine these subgroups H of $SU(n)$ for which it is possible to have $H = G_x^o$. Let $\xi_1 = (G_x \longrightarrow G \longrightarrow G/G_x)$. Then

$$0 = P_j(\tau(G/G_x) \oplus \nu(G/G_x)) = P_j(\alpha_{\xi_1}(\iota_{G/G_x}) \oplus \alpha_\xi(\phi_x)).$$

Let N be a connected normal subgroup of G_x such that

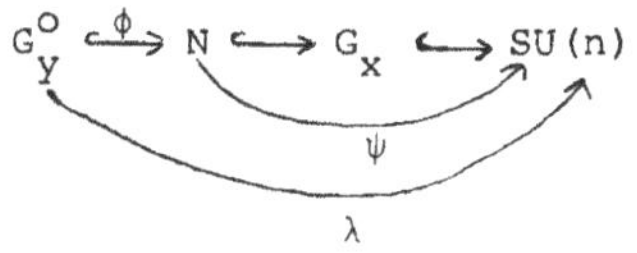

Let $\pi: G/N \longrightarrow G/G_x$, and $\xi_2 = (N \xrightarrow{\Psi} G \longrightarrow G/N)$.

Then $\pi^!(\alpha_{\xi_1}(\iota_{G/G_x}) \oplus \alpha_{\xi_1}(\phi_x))$

$$= \alpha_{\xi_2}(Ad_G|N - Ad_{G_x}|N) \oplus \alpha_{\xi_2}(\phi_x|N)$$

$=\alpha_{\xi_2}(Ad_G|N-Ad_N-\theta)\oplus\alpha_{\xi_2}(\phi_x|N)$

$=\tau(G/N)\oplus\alpha_{\xi_2}(\phi_x|N)\in\widetilde{K}\theta(G|N)$.

So possibly mod torsion,

$$* \quad \begin{cases} P_1(G/N)+P_1(\alpha_{\xi_2}(\phi_x|N))=0 \\ P_2(G/N)+P_1(G/N)P_1(\alpha_{\xi_2}(\phi_x|N))+P_2(\alpha_{\xi_2}(\phi_x|N))=0 \end{cases}$$

Now as G_y is the principal isotropy subgroup of ϕ_x,

$\phi_x|G_y=Ad_{G_x}|G_y-Ad_{G_y}\oplus\theta$.

So $\phi_x|G_y^o=((Ad_{G_x}|G_x^o)|N)|G_y^o-Ad_{G_y}|G_x^o\oplus\theta$

$=Ad_N|G_y^o-Ad_{G_x}\oplus\theta$

or

$$** \quad \phi_x|G_y^o=Ad_N\circ\phi-Ad_{G_y^o}\oplus\theta\ .$$

We use ** to compute $\phi_x|G_y^o$. Note that $\dim(\iota_{G/G_x}\oplus\phi_x)=\dim M\leq n-\frac{8}{3}n-2$. So $\dim\phi_x<n^2-\frac{8}{3}n-2$. We use this fact together with our knowledge of $\phi_x|G_y^o$ to determine what $\phi_x|N$ must be. We then use * to show that the inclusion of G_x^o into N in the lemma is the standard one, i.e.,

$$(G_x^o \hookrightarrow N \hookrightarrow SU(n))=(SU(k)\xrightarrow{\mu_k\oplus(\ell-k)\theta}SU(\ell)\xrightarrow{a\mu_\ell\oplus b\mu_\ell^*\oplus(n-2\ell)\theta}SU(n))\ .$$

So G_x^o contains $N=(SU(\ell)\xrightarrow{a\mu_\ell\oplus b\mu_\ell^*\oplus(n-2\ell)\theta}SU(n))$ as a normal subgroup. We shall show $G_x^o=N$.

Suppose not. We give here a slight modification of an argument found in [2]. We may assume x was chosen so that $\dim G_x$ is minimal among those G_t with G_t^o not simple.

We may write $\phi_x|G_x^o=\phi_1\oplus\beta\theta$ where ϕ_1 is a non-trivial representation containing no trivial copies, and $\beta\theta$ is the β dimensional trivial representation. We may assume $y\in S_x$, where S_x is the slice at x. If ν_x is the representation space of ϕ_x, we can write $\nu_x=V_{\phi_1}\oplus V_{\beta\theta}$. Let B be the unit ball of ν_x. We may consider

$B \subset S_x \subset M$. Let $Z \varepsilon B$. It follows by minimality that $(G_x^o(\phi_1)_z)^o \subseteq N$. So we may apply

Proposition [2]

Let K be a compact Lie group. Let N be a closed, connected, normal subgroup of K, with $N \subseteq K^o$. Let $\psi: K \to O(m)$ be a representation such that if $0 \neq x \varepsilon V_\psi$, $K_x^o \subseteq N$. Then

1) $N \subseteq \ker \psi$

2) rank $K/N=1$.

So $N \subseteq \ker \phi_1 = \ker \phi_x | G_x^o$, and rank $G_x^o/N=1$. So $\phi_x | G_y^o$ is trivial. But $\phi_x | G_x^o = \phi_x \circ \phi = (\ell-k)\ (\mu_k \oplus \mu_k^*) \oplus \theta$.

So $\ell=k$ and $N=G_y^o$. Hence G_x^o is locally isomorphic to $G_y^o \times L$, where $L=SU(2)$ or $L=T'$.

One then checks the possibilities to derive a contradiction.

Remark

It follows easily from examining the slice representation that no G_x^o are missing. For example, if $x, y \varepsilon M$ and

$$(G_x^o \subset G_y^o \subset SU(n)) = (SU(k) \xrightarrow{\mu_k} SU(\ell) \xrightarrow{a\mu_\ell \oplus b\mu_\ell^*} SU(n))$$

where $a+b=2$ and $k<\ell-1$, then if $k<t<\ell$, there exists $Z \varepsilon M$ so that

$$(G_x^o \subset G_z^o \subset G_y^o \subset SU(n)) = (SU(k) \xrightarrow{\mu_k} SU(t) \xrightarrow{\mu_t} SU(\ell) \xrightarrow{a\mu_\ell \oplus b\mu_\ell^*} SU(n)).$$

References

1. A. Borel and F. Hirzebruch, "Characteristic classes and homogeneous spaces I," American Journal of Mathematics, Vol. 80(1958), pp. 485-538; II, Vol. 81(1959), pp. 351-382; III, Vol. 82(1960), pp. 491-504.

2. W. C. Hsiang and W. Y. Hsiang, "Differentiable actions of compact connected classical groups I," American Journal of Mathematics, Vol. 89(1967), pp. 705-786; II, Annals of Math, Vol. 92(1970), pp. 189-223.

3. W. Y. Hsiang, "On the principal orbit type and P. A. Smith Theory of SU(p) actions," Topology, Vol. 6(1967), pp. 125-135.

ON THE SPLITTING PRINCIPLE AND THE GEOMETRIC WEIGHT SYSTEM OF TOPOLOGICAL TRANSFORMATION GROUPS I

Wu-yi Hsiang*

University of California at Berkeley

Introduction: In the study of the geometric behavior of transformation groups in the framework of modern topology, there are the following two natural settings:

(1) Topological actions of Lie groups (or more generally, topological groups) on topological manifolds (resp. topological spaces). (2) Differentiable actions of Lie groups on differentiable manifolds.

As usual, there are the local theory and the global theory in both the topological as well as the differentiable settings. In the local theory, one studies the geometric behavior of a topological (or differentiable) action in an invariant neighborhood of a given orbit. In the case of compact differentiable transformation groups, the so-called differentiable slice theorem [21,22] shows that the equivariant normal bundle of an orbit $G(x)$ is completely determined by the slice representation φ_x of G_x on normal vectors. The above theorem completely reduces the local theory of compact differentiable transformation groups to that of linear representations which are rather well understood and comparatively much simpler. It also readily introduces the theory of vector bundles and linear representations as powerful tools in the study of differentiable actions [Cf. 13, 14, 15, 16, 19]. This local linearity is exactly the major reason that makes compact differentiable transformation groups much more regular and technically more accessible than either the non-compact case or the topological case.

In the case of topological actions of compact Lie groups, there

---

*The author is an Alfred Sloan Fellow and is also partially supported by NSF

is a topological slice theorem which also proves the existence of a "slice" [22]. However, the local linearity fails miserably in the topological case. In §1, we shall include some examples of topological actions to show how perilous the failure of local linearity may occur. This failure makes the local theory of topological compact transformation groups much more complicated and challenging. Based on a deep theorem of A. Borel and the topological slice theorem, we shall introduce a fundamental local invariant for topological actions that we shall call it geometric weight system (Cf.§2). It seems to be a workable substitute of local linearity for topological compact transformation groups. It is well known that the global theory for actions on cohomology spheres and the (general) local theory are intimately related. Moreoever, one may construct the cone over a sphere and the one point compactification of an acyclic manifold in the category of topological G-spaces. Hence, we shall first study the global theory for topological actions on acyclic manifolds in §3,4 and 5 and then state the corresponding results in local theory in §8 without proof.

Historically, our approach is simply a continuation of the classical cohomology method of P.A. Smith and A. Borel [5]. However, it is the reformulation of classical cohomology theory of topological transformation groups into a type of characteristic class theory that opens up the present new horizon. In §2, we shall introduce a concept of F-varieties which is an analog of Zariski closure for algebraic varieties. It seems to me that this is exactly the geometric setting that one needs in order to fully unfold the actual significance of Borel's Theorem (Cf. §2) as a splitting principle of "characteristic classes". One may also regard the above mentioned splitting principle as a substitute of Schur lemma (in the theory of linear representations) for topological torus actions on acyclic manifolds. In fact, it gives us such a strong grip of the geometric situation of topological torus actions on acyclic manifolds that one can actually extend the central ideas of weight system of Cartan and Weyl to set up a corresponding theory of topological compact transformation groups as they did so beautifully for the theory of linear compact transformation groups. Of course, the geometric weight system for topological actions is simply a generalization of the geometric weight system for differentiable actions defined in [18] which is again a direct generalization of the ordinary weight system for linear actions via local linearity. In retrospect, this stage by stage extension of weight system, in fact, reveals a deeper insight of the simple-minded idea of "via maximal torus".

§1. Examples of compact topological transformation groups

In this section, we shall exhibit some types of compact topological transformation groups which will explain some of the basic difficulties in topological transformation groups.

Example of type I: Let G be a compact Lie group which has an irreducible complex (or real) representation of odd dimension > 1. (In the case G is connected, it is equivalent to noncommutativity.) Then, by the construction of [14,18], there exists a differentiable G-action on a suitable euclidean space R^m without fixed point. Hence, the one point compactification of the G-space R^m gives a topological G-action on S^m with exactly one fixed point. This example is non-differentiable at the point of infinity. (See [18] for a proof of this fact by means of geometric weight system.) However, it is still an open problem whether there exists an acyclic invariant neighborhood of $\infty \in S^m$.

Example of type II: In [12], Floyd and Richardson gave an example of differentiable actions of the alternating group of five letters $\mathfrak{a}_5$ on a disc without fixed point. So far, this is the only known example of compact transformation group on discs without fixed point. Based on the above example, we shall exhibit some constructions to show that almost all conceivable topological spaces can be realized as the fixed point set of a suitable $\mathfrak{a}_5$-action on a high dimensional disc.

Example II_a: Let K be an arbitrary finite complex and D_0 be the disc with an $\mathfrak{a}_5$-action without fixed point. Let $K \circ D_0$ be the join of K and D_0. There is a natural $\mathfrak{a}_5$-action on $K \circ D_0$ with K as the fixed point set. If we suitably imbed the contractible G-space $K \circ D_0$ equivariantly in a high dimensional linear G-space R^m and take a small equivariant regular neighborhood of $K \circ D_0$ in R^m, then we get a differentiable $\mathfrak{a}_5$-action on D^m with the fixed point set F of the same homotopy type as K, which is an arbitrary finite complex. Furthermore, let $C(D^m)$ be the cone of D^m with the induced G-action. Then the fixed point set is the cone of F, $C(F)$, and the local cohomology of $C(F)$ at the vertex is isomorphic to $H^*(C(K),K)$.

Example II_b: Let D_0 be an m-disc with a topological G-action without fixed point, D_1^k be a k-disc with trivial G-action. Then $D_1^k \circ D_0$ is a $(k+m+1)$-disc with a natural G-action. It is clear that

$$S_+ = \left\{(x,t,y);\ t \leq \tfrac{1}{2} \text{ and either } x \in \partial D_1^k \text{ or } y \in \partial D_0\right\}$$

and

$$S_- = \left(\overline{\partial(D_1^k \circ D_0) - S_+}\right)$$

are invariant hemi-spheres of $\partial(D_1^k \circ D_0)$. Hence

$$(D_1^k \circ D\ ) \cup_{S_+} C(S_+) \quad \text{and} \quad (D_1^k \circ D\ ) \cup_{S_-} C(S_-)$$

are naturally G-spaces with their fixed point set

$$C(D_1^k) \quad \text{or} \quad D_1^k + \text{point}$$

respectively. Furthermore, if one takes finite copies of the above examples and identifies them along suitable partial boundaries, it is not difficult to construct G-actions on D^{m+k+1} with arbitrary finite collection of disjoint (k+1)-discs, k-discs and isolated points as the fixed point set. For example, let D, D', D'' be three copies of $D_1^k \circ D_0$; and

$$S_+ \cup S_- = \partial D, \quad S_+' \cup S_-' = \partial D', \quad S_+'' \cup S_-'' = \partial D''$$

be the respective invariant hemispheres of ∂D, $\partial D'$, $\partial D''$. Then, the following is a G-disc with $D^k + D^{k+1} + 2$ points as $F(G)$,

$$C(S_-) \cup_{S_-} D \cup_{f_+} D' \cup_{S_-'} C(S_-') \cup_{g_-} C(S_-'') \cup_{S_-''} D'' \cup_{S_+''} C(S_+'')$$

where f_+ is the identification map between S_+ and S_+' and g_- is the identification map between $C(\partial S_-')$ and $C(\partial S_-'')$.

Example II_c: If one takes several copies of G-discs of type II_a or II_b, then their equivariant join is again a G-disc. It is easy to see that the fixed point set is the join of their respective fixed point sets which is clearly very complicated topologically.

Discussion: Let G be a compact Lie group which has at least one irreducible complex (or real) representation of odd dimension > 1. If there exists an open acyclic invariant neighborhood of ∞ in the example of topological G-action on S^m with $F(G) = \infty$, then the complement will be a compact acyclic G-space without fixed point. Hence, it is not difficult to use the method of equivariant thickening (Cf. [20]) to construct a G-action on a high dimensional disc without fixed point. Then, the method of example II applies to get all kinds of irregular behavior for topological G-actions. Hence, in either way, the nice property of "local linearity" for the differentiable actions of compact Lie groups breaks down miserably for topological actions.

Example of type III: Let G be a compact Lie group which has at least one irreducible complex (or real) representation of odd dimension > 1. Then, by the construction of [14, p. 715-718], there exists an orthogonal G-action on a suitable sphere S^m without fixed point such that S^m admits an equivariant map into itself with degree zero,

$$f: S^m \to S^m.$$

Let X be the inverse limit of

$$\cdots \to S^m \xrightarrow{f} S^m \xrightarrow{f} S^m \xrightarrow{f} S^m \cdots .$$

Then X is a compact acyclic topological G-space without fixed point. Let K be any finite complex with trivial G-action and $X \circ K$ be the join of X and K with the induced action. Then $X \circ K$ is a compact acyclic topological G-space with K as its fixed point set. The above examples show that theorems of P.A. Smith type are false for all such compact Lie groups which have some odd dimensional complex (or real) representations of dimension > 1.

For further examples, see a survey article of G. Bredon [8]. To all examples of exotic actions on sphere, the cone construction gives examples of topological actions on disc, $C(S^n)$, with bad singularity at the vertex of the cone.

2. F-varieties and a theorem of A. Borel

(A) The concept of F-varieties: Let X be a topological space (resp. smooth manifold) and G be a compact Lie group, Ψ be a given topological (resp. differentiable) G-action on X. We introduce the following basic concepts:

Definition: For a given point $x \in X$. Let G_x be the isotropy subgroup of x and G_x^0 be the connected isotropy subgroup of x, i.e., the identity component of G_x. We shall denote the set of orbit types and the set of connected orbit types by $\theta(\Psi)$ and $\theta^0(\Psi)$

respectively. Namely

$$\theta(\Psi) = \text{the conjugacy classes of subgroups in } \{G_x ; x \in X\}$$

$$\theta^0(\Psi) = \text{the conjugacy classes of subgroups in } \{G_x^0 ; x \in X\}.$$

Definition: The fixed point set of G_x, $F(G_x, X)$, is called the F-variety spanned by x and denoted by $F(x)$. For a subtle technical reason, it is usually more convenient to define the connected F-variety spanned by x to be the following subset:

$$F^0(x) = \text{the connected component of } x \text{ of the subset } F(G_x^0, X).$$

Remarks: (i) In the study of topological (resp. differentiable, linear) actions of compact connected Lie groups, those actions of tori play an outstanding role. It seems to be that one of the basic reasons behind this is the following nice property uniquely enjoyed by torus actions. Namely, the set of all F-varieties as well as the set of all connected F-varieties then constitute a network of natural invariant subspaces of the given G-space.

(ii) The situation of G-space (especially when G is a torus) is quite analogous to that of algebraic varieties. The concept of F-varieties spanned by x is an analog of Zariski closure in the case of algebraic varieties. As one may expect, a great deal of important information of a given G-space is contained in the topological invariants of the network of F-varieties.

(B) The Borel setting and characteristic class theory of G-spaces: Following A Borel [5], we shall denote the twisted product of a G-space X and the total space of universal G-bundle E_G by X_G. Namely X_G is the total space of the universal bundle

$$X \to X_G \xrightarrow{\pi_1} B_G$$

with the given G-space X as fibre. Notice that there is another natural mapping $\pi_2: X_G \to X/G$ with $\pi_2^{-1}(G(x)) = B_{G_x}$. One observes that this construction of X_G together with the two projections π_1, π_2 is clearly functorial. Hence, in the case G is a torus, the collection of F-varieties will give us a network of spaces with natural maps. One may then analyze this network of maps from the traditional algebraic topology. For example, one may apply the ordinary cohomology theory to get various algebraic invariants which can be viewed as the characteristic classes of the fibration $X_G \to B_B$. Here, the Serre spectral sequence of $X_G \to B_G$ and the Larrey spectral sequence of $\pi_2: X_G \to X/G$ offer a powerful tool in analyzing the algebraic relationships among the corresponding network of cohomology algebras.

(C) A splitting principle and a theorem of A. Borel [3]:

Let X be an Z-cohomology n-sphere with a topological T-action Ψ. Let $Y = F(T,X)$ be the fixed point set of T. It is well-known that Y is again a Z-cohomology sphere of dimension $r \equiv n \pmod 2$, $r = -1$ if Y is empty. Hence

$$H^i(X-Y;Z) \simeq \begin{cases} Z, & i = n, \quad r+1 \\ 0, & \text{otherwise.} \end{cases}$$

We shall denote the generators of $H^n(X-Y;Z)$ and $H^{r+1}(X-Y;Z)$ by ζ and $\partial\eta$ respectively. Then, the Serre spectral sequence of the fibration

$$(X-Y) \to (X-Y)_T \to B_T$$

consists of only two lines. Namely,

E_2:

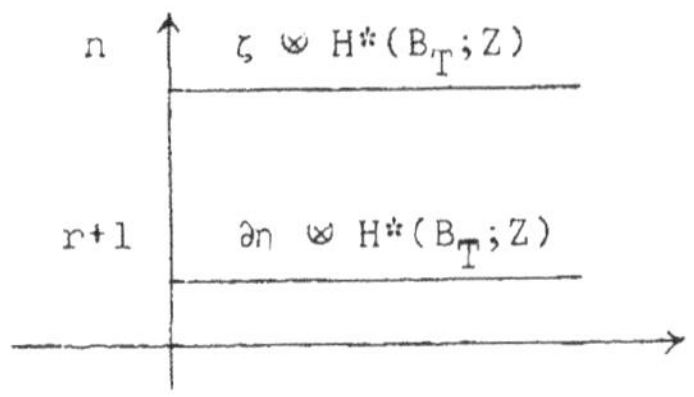

Hence, the transgression of ζ, $d\zeta = \partial\eta \otimes a$, $a \in H^{n-r}(B_T;Z)$ is the only possible non-vanishing differential. On the other hand, a must be non-zero, for otherwise, it follows from an easy theorem of Borel [5, p.164] that $(X-Y)$ will have non-empty fixed point of T.

A splitting theorem: The above non-zero element $a \in H^{n-r}(B_T;Z)$ splits into the product of degree two elements. Namely,

$$a = \ell\cdot\omega^{k_1}\cdot\omega_2^{k_2}\cdots\omega_4^{k_s};\quad 2(k_1 + k_2+\cdots+k_s) = (n-r) \tag{1}$$

for a suitable integer ℓ and suitable non-proportional elements $\omega_j \in H^2(B_T;Z)$.

Remark: Notice that in the case Ψ is a linear T-action on S^n with $m = \frac{1}{2}(n-r)$ pairs of non-zero weights $\{\pm\alpha_j\}$, then $a = \alpha_1\cdot\alpha_2\cdots\alpha_m$. If one collects all the integral factors among α_j into ℓ and groups those proprotionate weights into powers, one gets the above expression of the theorem. Once correctly formulated, the proof of the above theorem is, in fact, rather straightforward. We refer to [17] for such a proof of the above theorem. Instead, we shall explain the geometric significance of the above splitting theorem in the following:

Corollary 1: Let $T' \subseteq T$ be a subtorus of T and i^* be the induced homomorphism $i^* : H^*(B_T;Z) \longrightarrow H^*(B_{T'};Z)$. If $\omega_1,\ldots,\omega_j$ are those factors of a with $i^*(\omega_1) = \cdots = i^*(\omega_j) = 0$, then $Y' = F(T',X)$ is a Z-cohomology sphere of dimension $2(k_1+\cdots+k_j) + r$. Furthermore, for the restricted T'-action on X, one has

$$d\zeta = \partial\eta' \otimes a'$$

$$a' = \ell \cdot i^*(\omega_{j+1})^{k_{j+1}} \cdot \ldots \ldots i^*(\omega_s)^{k_s}$$

for the fibration $(X-Y') \to (X-Y')_{T'} \to B_{T'}$.

Proof: It follows readily from the naturality of Borel's construction and the naturality of transgression.

We shall identify elements of $H^2(B_T;Z)$ with those of $H^1(T;Z)$ via the transgression $H^1(T;Z) \approx H^2(B_T;Z)$, and then consider them as a homomorphism of T to S^1, or equivalently, a weight vector. We shall denote the connected component of the kernel $\omega_j: T \to S^1$ by $\omega_j^\perp$, which is a corank one subtorus with the perpendicular hyperplane of ω_j as its Lie algebra.

Corollary 2: (Borel's Theorem [5, p. 175]). Let $H_j = \omega_j^\perp$ be the kernel of those ω_j in the splitting theorem. Then, H_j are exactly those corank one subtori of T with

$$\dim F(H,X) - r = n(H) - r > 0$$

and $n(H_j) - r = 2 \cdot k_j$. Hence, we have the Borel formula

$$(n - r) = \Sigma 2k_j = \sum_H (n(H) - r). \tag{2}$$

Corollary 2 follows obviously from corollary 1.

Corollary 3: Let $x \in X$ be a point of X. Then

$$G_x^0 = \bigcap \left\{ \omega_j^\perp ; \omega_j^\perp \supseteq G_x^0 \right\}, \tag{3}$$

and the connected F-variety spanned by x, $F^0(x)$, is a Z-cohomology sphere with its dimension

$$n' = r + \Sigma 2k_j = r + \Sigma(n(\omega_j^\perp) - r) \tag{4}$$

where j run through those subset that $\omega_j^{\perp} \supseteq G_x^0$.

Proof: It follows from corollary 1 that

$$\mathrm{Ker}^0(\Psi) = \omega_1^{\perp} \cap \omega_2^{\perp} \cap \ldots \cap \omega_s^{\perp}.$$

Since $X' = F^0(x)$ is an <u>invariant</u> cohomology sphere, we may apply the above results to the restriction of Ψ to $F^0(x)$. It is clear that $d\zeta' = \partial\eta \otimes b$ with

$$b = \ell \cdot \omega_{j_1}^{k_{j_1}} \cdot \ldots \cdot \omega_{j_t}^{k_{j_t}}$$

where $\omega_{j_1}, \ldots, \omega_{j_t}$ are those weight with $\omega_j^{\perp} \supseteq G_x^0$. Hence, we have

$$G_x^0 = \mathrm{Ker}^0(\Psi | X') = \omega_{j_1}^{\perp} \cap \ldots \cap \omega_{j_t}^{\perp}$$

and

$$n' = \dim X' = r + 2(k_{j_1} + \ldots + k_{j_t}).$$

(D) <u>Localization</u>: It is well known that one may localize a theorem of <u>cohomology</u> natural for torus actions on spheres to obtain its local version. The procedure is roughly as follows. Let M be a cohomology manifold with a topological T-action Ψ and $p \in M$ be a fixed point. Let $U \subseteq V$ be two invariant neighborhoods of p and $r^*_{V-p,U-p}$ be the restriction map

$$H^*(V-p;Z) \to H^*(U-p;Z)$$

(cohomology with closed support). By the equivariant embedding theorem of Mostow [24], it is easy to see that invariant neighborhoods are cofinal in the projective system of all neighborhoods or p. Hence, the direct limit

$$I_p^*(M;Z) = \mathrm{dir\,lim}(H^*(V-p;Z), r^*_{V-p,U-p})$$

of the sub-projective system of invariant neighborhoods is isomorphic to $H^*(S^{n-1};Z)$, $n = \dim M$. Since the construction of A. Borel and the spectral sequences we used are obviously functorial, we may consider their limit as the local spectral sequences at $p \in M$. Hence, the local splitting theorem follows readily from an almost identical proof as that of [17].

Local splitting theorem: Let M be a Z-cohomology n-manifold with a topological T-action Ψ and $p \in M$ be a fixed point. Then the connected component of $F(T,M)$ at p is an r-dimensional cohomology manifold, $r \equiv n \pmod 2$; and there exist a set of non-zero local weights $\{\pm\omega_j\}$ with respective multiplicities k_j, $\Sigma k_j = (n-r)$. This system of local weights multiplicities $\{\pm\omega_j; k_j\}$ gives a complete discription of the local behavior of those F-varieties passing through p.

§3. Geometric weight system for topological actions and a fundamental fixed point theorem

In the study of differentiable actions of compact Lie groups, an invariant called geometric weight system was introduced in [18] for actions on acyclic manifolds. It plays an important role in recent development of differentiable compact transformation groups. Let ϕ be a differentiable action of a compact connected Lie group G on an acyclic manifold M and T be an arbitrarily chosen maximal torus of G. In [18], we exploit the connectedness of $F(T,M)$ and the local linearity of differentiable actions, and simply define the geometric weight system of Φ to be the ordinary weight system of the local representation of T at an arbitrary point of $F(T,M)$. Suppose now we consider a topological action Ψ of G on an acyclic cohomology manifold X. Then, the fixed point set $F(T,X)$ of T is also an acyclic cohomology manifold. However, we no longer have local linearity for topological actions. As we point out in §1, this failure of local linearity is exactly one of the major difficulties in the study of topological actions. In this section, we shall use the splitting theorem of §2 to set up a slightly weaker version of weight system for topological actions.

Definition: Let Ψ be a topological G-action on an acyclic manifold X and T be a maximal torus of G. Since the fixed point set, F(T,X), of T is also acyclic (and hence, connected), the local weights $\{\pm\omega_j, k_j\}$ given by the local splitting theorem at a point $p \in F(T,X)$ is clearly independent of the choice of p. Hence we shall define $\{\pm\omega_j, k_j\}$ as the system of non-zero weights of the topological action Ψ, denoted by $\Omega'(\Psi)$. The weight system of Ψ, $\Omega(\Psi)$, is then defined to be

$$\Omega(\Psi) = \Omega'(\Psi) \cup \{r \text{ zero-weights}\}, \quad r = \dim F(T,X).$$

Remark: (i) Notice that one has little control on the length of those weights in the splitting theorem, since proportionate weights are reduced to the shortest weight of the same direction with all the integral factors combined into one integer ℓ. (ii) Since F(T,X) is invariant under the Weyl group W(G), it is obvious that $\Omega(\psi)$ is also invariant under W(G). (iii) Let $G' \subseteq G$ be a subgroup of G and $\Psi' = \Psi|G'$ be the restriction of Ψ to G'. Let T', T be maximal tori of G' and G respectively and $T' \subseteq T$. Then it is clear that

$$\Omega(\Psi') = \Omega(\Psi)|T' \quad . \tag{5}$$

(A) The weight system of a slice: The topological slice theorem of Montgomery and Yang proved the existence of a slice [22]. In case the G-space X is a cohomology manifold, the slice at $x \in X$ is clearly also a cohomology manifold invariant under G_x. However, in order to make the above slice theorem into a useful tool for the study of local theory, one needs some kind of substitute of the missing local linearity of the topological G_x-action on a slice. For this purpose, we introduce the following invariant.

Definition: Let X be a cohomology manifold with a given topological G-action Ψ and x be a point of X. Let S_x be a slice at x and $T_1 \subseteq G_x$ be a maximal torus of G_x. We shall call the weight system of the T_1-action on S_x the weight system of the slice at x and denoted by $\Omega(S_x)$.

Remark: Let $\Omega_x(\Psi|T_1)$ be the system of local weights at x. Then, it is clear that

$$\Omega_x(\Psi|T_1) = \Omega(S_x) + \Omega\left(Ad_G|T_1 - Ad_{G_x}|T_1\right).$$

Hence $\Omega(S_x)$ is independent of the choice of the slice at x.

Proposition 1: Let Ψ be a topological G-action on an acyclic manifold X, $\Omega(\Psi)$ be the weight system of Ψ. For a given point $x \in X$, we may assume that the maximal torus T_1 of G_x is contained in the maximal torus T of G. Then there exist a suitable subcollection of weights in $\Omega(\Psi)$ such that

$$T_1 = (G_x \cap T)^0 = T_x = \omega_{j_1}^{\perp} \cap \dots \cap \omega_{j_t}^{\perp}, \tag{6}$$

and the weight system of slice at x is given by

$$\Omega(\Psi)|T_1 = \Omega\left(Ad_G|T_1 - Ad_{G_x}|T_1\right) + \Omega(S_x). \tag{7}$$

Proof: Since $T_1 = T_x^0$, it follows directly from (3) of Corollary 3, §2, that there exist suitable weights $\{\omega_{j_1}, \dots, \omega_{j_t}\}$ in $\Omega(\Psi)$ such that $T_1 = \omega_{j_1}^{\perp} \cap \dots \cap \omega_{j_t}^{\perp}$. The second equation follows easily from the connectedness of $F(T_1, X)$.

Corollary 1: Let (H_Ψ) be the principal orbit type of the topological G-action Ψ. Then the above equation reduces to the following powerful equation

$$\Omega(\Psi|H_\psi) \equiv \Omega\left(Ad_G|H_\psi - Ad_{H_\psi}\right), \text{ (mod zero weight).} \tag{8}$$

Corollary 2: Let $\Delta(G)$ and $\Delta(G_x)$ be the root system of G and G_x respectively, $x \in F(T,X)$. Then

$$\Delta(G_x) \supseteq \Delta(G) \setminus \Omega(\Psi) \text{ (the difference set).} \tag{9}$$

Definition: A subgroup $H \subseteq G_1 \times G_2 \times \dots \times G_k$ is called a splitting subgroup (with respect to the given decomposition) if

$$H = (H \cap G_1) \times (H \cap G_2) \times \dots \times (H \cap G_k).$$

A weight system $\Omega(\Psi)$ of $G_1 \times G_2 \times \dots \times G_k$ is called splitting if

$$\Omega'(\Psi) = \Omega'(\Psi|G_1) \cup \dots \cup \Omega'(\Psi|G_k).$$

Proposition 2: Let $G = G_1 \times \dots \times G_k$ be a given decomposition of G and Ψ be a topological G-action on an acyclic manifold X. If $\Omega'(\Psi)$ is splitting, then the identity component of every isotropy subgroups, G_x^0, is a splitting subgroup.

Proof: Let $T, T_1, \dots, T_k$ be maximal tori of $G, G_1, \dots, G_k$ respectively, and $T = T_1 \times \dots \times T_k$. Consider the T-action $\Psi|T$. Since $\Omega'(\Psi)$ is splitting, it follows from (6) that T_x^0 are splitting subgroups for all $x \in X$. On the other hand, it is an easy consequence of the maximal torus theorem that a connected subgroup $H \subseteq G$ is a splitting subgroup if and only if its maximal tori $T_1 \subseteq H \subseteq G$ are all splitting subgroups (Cf. [18] for a proof of this simple fact). Hence G_x^0 are splitting subgroups for all $x \in X$.

Conjecture: In case G is connected, the isotropy subgroups,

G_x, themselves should also be splitting subgroups.

Proposition 3: Let X be a cohomology Z-acyclic n-manifold and Ψ be a topological SO(3)-action on X. If $\Omega'(\Psi)$ consists of only one pair of non-zero weights, then

$$F(SO(3),X) = F(Z_2,F(SO(2),X)) \tag{10}$$

and is a cohomology Z-acyclic manifold of dimension (n-3).

Proof: Let $Y = F(SO(2),X)$. Since $\Omega'(\Psi)$ only consists of a single pair of weights, it is clear that Y is a Z-acyclic cohomology manifold of dimension (n-2). We claim that SO(2) acts freely on (X-Y). Since SO(3) contains no normal subgroup (except the identity group), we see that $\Psi|SO(2)$ must be effective. On the other hand, it follows easily from the fact dim Y = (n-2) that SO(2) acts freely in the neighborhood of the fixed point set Y. Now, suppose the contrary that there exists a point $x \in (X-Y)$ with $SO(2)_x \approx Z_m$, $m > 1$. Let p be a prime factor of m. Then $F(Z_p,X) \supseteq Y \cup \{x\}$ consists at least two connected components which contradicts to a theorem of P.A. Smith that $F(Z_p,X) \sim_{Z_p} pt$. Hence, SO(2) acts freely on (X-Y). As a consequence of the above fact, the SO(3)-action consists of only two types of orbits, namely

$$\theta(\Psi) = \left\{SO(3),(SO(2))\right\} .$$

Now, let $H = Z_2 + Z_2$ be a maximal Z_2-torus of SO(3), say

$$H = \left\{\begin{pmatrix} \varepsilon_1 & & 0 \\ & \varepsilon_2 & \\ 0 & & \varepsilon_3 \end{pmatrix};\ \varepsilon_j = \pm 1,\ \varepsilon_1 \cdot \varepsilon_2 \cdot \varepsilon_3 = 1\right\}.$$

Let $\Omega_2'(\Psi)$ be the system of non-zero 2-weights of Ψ. It is easy to see that $\Omega_2'(\Psi) = \{\varepsilon_1,\varepsilon_2,\varepsilon_3\}$. Hence

$$F(SO(3),X) = F(Z_2,F(SO(2),X)) = F(H,X)$$

is of dimension (n-3). Since Y is Z-acyclic and $F(Z_2,Y)$ is of codimension one in Y, it is not difficult to show that $F(Z_2,Y)$ is in fact Z-acyclic rather than Z_2-acyclic.

Q.E.D.

Following the idea of [16], one also has a slightly weaker fixed point theorem for topological an acyclic manifolds than the case of differentiable actions.

Theorem 1: Let X be an acyclic cohomology manifold and Ψ be a topological G-action on X. Let $\Omega(\Psi)$ be the weight system of Ψ and

$$\Sigma_0(\Psi) = \Delta(G) \setminus \Omega(\Psi)$$

$$\Sigma_1(\Psi) = \left\{\alpha \in \Delta(G) \text{ and with multi 1 in } \Omega(\Psi)\right\} .$$

Then, there exist a maximal rank subgroup $K \subseteq G$, such that $\Delta(K) \supseteq \Sigma_0 \cup \Sigma_1$ and $F(K,K)$ is also an acyclic cohomology submanifold.

Proof: Let $T \subseteq G$ be a maximal torus of G and $Y = F(T,X)$ be the fixed point set of T. Clearly, the Weyl group $W(G) = N(T)/T$ acts naturally on Y. For a given root $\alpha \in \Delta(G)$, let $T_\alpha = \alpha^\perp$ be the corank one subtorus perpendicular to α, $G_\alpha = N(T_\alpha)$ be the normalizor of T_α and $\tilde{G}_\alpha = G_\alpha/T_\alpha$. Let $Y_\alpha = F(T_\alpha,X)$ and Ψ_α be the induced action of $\tilde{G}_\alpha$ on Y_α. It is clear that $\Omega'(\Psi_\alpha) = \{\pm\alpha$, with the same multiplicity as in $\Omega'(\Psi)\}$. Hence, in the case $\alpha \in \Sigma_1(\Psi)$, the $\tilde{G}_\alpha$-action Ψ_α has only one pair of non-zero weights and Proposition 3 applies to show that

$$F(G_\alpha,X) = F(\tilde{G}_\alpha,Y_\alpha) = F(W(G_\alpha),Y) = H_\alpha \tag{11}$$

where $W(G_\alpha) \simeq Z_2$ and with $F(W(G_\alpha),Y)$ of codimension one in Y, i.e., a topological reflection. Hence, the subgroup W' of $W(G)$ generated by $\left\{W(G_\alpha);\alpha \in \Sigma_1(\Psi)\right\}$ acts as a group generated by topological reflections. On the other hand, it is not difficult to modify the proof of [16] for groups generated by differentiable reflections so that the fixed point theorem for groups generated by reflections also valid for the topological case. Hence

$$F(W',Y) = \cap \left\{H_\alpha ; \alpha \in \Sigma_1 \ (\Psi)\right\}$$

is also Z-acyclic. Let $z \in F(W',Y)$. Then, it follows from (9) and (11) that G_z is a maximal rank subgroup of G with

$$\Delta(G_z) \subseteq \left\{\Sigma_0 \cup \Sigma_1\right\}.$$

Hence, we may simply put $K = \cap\left\{G_z, z \in F(W',Y)\right\}$.

Q.E.D.

If one considers topological G-actions as representations of G via topological transformations, then it is rather natural to look for the lowest dimensional representations. For this purpose, we introduce the following notations:

Definition: For a given compact Lie algebra $\mathfrak{g}$, let

$$L_{top}(\mathfrak{g}) = \text{Min}\ \{\dim\ \Psi\}$$

where Ψ runs through all effective topological transformation groups on cohomology acyclic manifolds with $\mathfrak{g}$ as the Lie algebra. For a given compact Lie group G, let

$$L_{top}(G) = \text{Min}\ \{\dim\ \Psi\}$$

where Ψ runs through all possible topological effective G-action on cohomology acyclic manifolds.

Remark: Clearly, one may similarly define $L_{diff}(\mathfrak{g})$, $L_{diff}(G)$ and $L_{lin}(\mathfrak{g})$, $L_{lin}(G)$ in terms of differentiable actions and linear actions respectively. It is then obvious that

$$L_{top}(\mathfrak{g}) \leq L_{diff}(\mathfrak{g}) \leq L_{lin}(\mathfrak{g})$$

$$L_{top}(G) \leq L_{diff}(G) \leq L_{lin}(G).$$

Hence, the interesting problem here is to estimate $L_{top}(\mathfrak{g})$ and $L_{top}(G)$ from below. As an example, we shall generalize the estimate of $L_{diff}(Spin(k))$ in [18] to the topological case as follows

Theorem 2: Let $G = Spin\ (k)$, and $r = \left[\frac{k}{2}\right]$. Then

$$L_{top}(G) \geq \begin{cases} 2^r & \text{if } k \not\equiv 0 \pmod 4 \\ 2^{r-1}+2r & \text{if } k \equiv 0 \pmod 4). \end{cases}$$

Proof: Let Z_2 be the kernel of the covering homomorphism

$$Spin(k) \to SO(k)$$

and $T \subseteq Spin(k)$ be a maximal torus. Let $(\theta_1, \theta_2, \ldots, \theta_r)$ be the usual coordinate for the Cartan subalgebra of Spin (k), $r = \left[\frac{k}{2}\right]$. It is well-known that all weights of Spin(k) are of the form

$$\omega = k_1\theta_1 + \ldots + k_r\theta_r \quad \text{with} \quad k_1 \equiv \ldots \equiv k_r \equiv \begin{cases} 0 \\ \frac{1}{2} \end{cases} \pmod 1).$$

Let Ψ be an effective action of Spin(k). Then $\Psi|T$ is also effective. Let $\Omega'(\Psi)$ be the system of non-zero weights of Ψ. We claim that $\Omega'(\Psi)$ must contain some weights of the form $\omega = k_1 \cdot \theta_1 + \ldots + k_r \cdot \theta_r$ with $k_j \equiv \frac{1}{2} \pmod 1$, i.e. half integers. For otherwise one can show that $Y' = F(Z_2, X)$ will be a Q-acyclic manifold of the same dimension as X itself and hence $F(Z_2, X) = X$, Ψ is not effective (Cf. Corollary 1 of the Splitting Theorem, §2). Since $\Omega'(\Psi)$ is clearly invariant under the conjugation of the Weyl group, $\Omega'(\Psi)$ must contain the whole "orbit" of such a weight, $W(\omega)$, which consists of at least 2^r elements if $k = 2r + 1$ and at least 2^{r-1} elements if $k = 2r$. In the case $k \equiv 2 \pmod 4$, $-\omega$ is not a conjugate weight of ω and hence $\Omega'(\Psi)$ must also contain the "orbit" of $W(-\omega)$. Therefore,

$$\dim(\Psi) \geqslant \text{the number of weights in } \Omega'(\Psi) \geqslant 2^r$$

for the case $k \not\equiv 0 \pmod 4$. In the case $k \equiv 0$ mod (4), the center of Spin(k) is $Z_2 + Z_2$ which has three Z_2 subgroups. Suppose $\Omega'(\Psi)$ only consists of a single orbit $W(\omega)$ of the above type with exactly 2^{r-1} elements. Then the same reason will show that $\mathrm{Ker}(\Psi)$ is another Z_2 subgroup which contradicts the assumption that Ψ is effective. Hence, $\Omega'(\Psi)$ either contains one more orbit or the orbit $W(\omega)$ consists of more than 2^{r-1} elements. It is then easy to see that

$$\dim(\Psi) \geqslant \text{number of elements in } \Omega'(\Psi) \geqslant 2^{r-1} + 2r$$

for the case $k \equiv 0 \pmod 4$.

Q.E.D.

Remark: It is not difficult to see that

$$L_{lin}(\mathrm{Spin}(k)) = \begin{cases} 2^r, & k \equiv \pm 1, \pm 2, 4 \pmod 8 \\ 2^{r+1}, & k \equiv \pm 3 \pmod 8 \\ 2^{r-1}+2r, & k \equiv 0 \pmod 8) . \end{cases}$$

Hence, the above estimates are best possible for the case $k \equiv 0, \pm 1, \pm 2$ (mod 0). Namely

$$L_{top}(\mathrm{Spin}(k)) = L_{lin}(\mathrm{Spin}(k))$$

for $k \equiv 0, \pm 1, \pm 2 \pmod 8$.

As a preliminary step to the complete determination of $L_{top}(\mathfrak{g})$, we shall first compute the case of simple Lie algebras. For convenience, we shall introduce the following notations:

$$\dim \Omega(\Psi) = \dim (\Psi)$$
= the number of weights counting with multiplicities.

$\dim \Omega'(\Psi)$ = the number of non-zero weights counting with multiplicities.

Theorem 3: Let $\mathfrak{g}$ be a simple compact Lie algebra and Ψ be a topological G-action on an acyclic cohomology n-manifold X with $\mathfrak{g}$ as the Lie algebra of G. If

$$\dim \Omega'(\Psi) \leqslant L_{lin}(\mathfrak{g}),$$

then the fixed point set F(G,X) is also an acyclic cohomology manifold. Moreover, with the only exceptional case of $\mathfrak{g} = A_2$, $\Omega'(\Psi) = \Delta(A_2)$, codim F(G,X) = 8, one has

$$\operatorname{codim} F(G,X) = L_{lin}(\mathfrak{g})$$

$$\Omega'(\Psi) = \Omega'(\varphi)$$

where φ is the unique lowest dimensional real linear representation of $\mathfrak{g}$ given by the following table:

TABLE

$\mathfrak{g}$	dim $\mathcal{P}$	$\Omega'(\mathcal{P})$
A_n, $n\neq 1,3$	$2(n+1$	$\{\pm\theta_i; i=1,\dots,(n+1)\}$
$B_n, n\geq 1$	$2n+1$	$\{\pm\theta_i; i=1,\dots,n\}$
$C_n, n\geq 3$	$4n$	$2\cdot\{\pm\theta_i; i=1,\dots,n\}$
$D_n, n\geq 3$	$2n$	$\{\pm\theta_i; i=1,\dots,n\}$
E_6	54	$\{\pm(\lambda+\theta_i),\ \pm(\theta_i+\theta_j); i<j\}$
E_7	112	$2\cdot\{\pm(\theta_i+\theta_j);\ i<j\}$
E_8	248	$\Delta(E_8)$
G_2	7	$\{\pm\theta_i, i=1,2,3\}$
F_4	26	$\{\pm\theta_i, \frac{1}{2}(\pm\theta_1\pm\theta_2\pm\theta_3\pm\theta_4)\}$

Corollary: If $\mathfrak{g}$ is a simple compact Lie algebra, then it follows easily from Theorem 3 that

$$L_{top}(\mathfrak{g}) = L_{diff}(\mathfrak{g}) = L_{lin}(\mathfrak{g}).$$

Proof of Theorem 3: We divide the proof into the following five cases:

(i) The case $\mathfrak{g} = A_n$ $(n\geq 3)$, D_n $(n\geq 3)$ or E_6: In this case, the lowest dimensional representation $\mathcal{P}$ has no zero weights and $\Omega'(\mathcal{P})$ consists of a single orbit of $W(G)$ which is the unique orbit with smallest number of elements. Hence, it is clear that $\Omega'(\Psi) = \Omega'(\mathcal{P})$, and $F(G,X) = F(T,X)$ is also an acyclic cohomology submanifold with

$$\operatorname{codim} F(G,X) = \dim\mathcal{P} = L_{lin}(\mathfrak{g}).$$

(ii) $\mathfrak{g} = B_n$, $(n \geq 1)$; E_8: In this case, $\Omega'(\varphi)$ again consists a single orbit of $W(G)$ which is the unique orbit with smallest number of elements. However $\Omega'(\varphi) \subseteq \Delta(\mathfrak{g})$ and $\Omega(\varphi)$ does have zero weights with multiplicity 1 and 8 respectively. Hence, it is easy to show that $\Omega'(\Psi) = \Omega'(\varphi)$ and then, it follows from Theorem 1 that $F(G,X) = F(W(G),F(T,X))$ is also an acyclic cohomology manifold with codim $F(G,X) = \dim \varphi = L_{lin}(\mathfrak{g})$.

(iii) $\mathfrak{g} = C_n$, $(n \geq 3)$, E_7: In this case, $\Omega'(\varphi)$ consists of the smallest orbit of $W(G)$ with multiplicity 2. Since all the other orbits of $W(G)$ consists of too many weights, it is clear that

$$\Omega'(\Psi) = \begin{cases} \ell \cdot \{\pm\theta_i\} & \text{if } \mathfrak{g} = C_n, \quad n \geq 3 \\ \ell \cdot \{\pm(\theta_i + \theta_j)\} & \text{if } \mathfrak{g} = E_7 \end{cases}, \quad \ell = 1,2.$$

We shall show that ℓ must be equal to 2. Since the proofs are C_n and E_7 are essentially the same, we shall only prove the C_n case. It follows from the fact that

$$\Delta(G_y) \supseteq \Delta(G) \setminus \Omega'(\Psi) = \{\pm\theta_i \pm \theta_j\}$$

for any point $y \in F(T,X)$, G_y must be in fact equal to G itself. Let $x \in X$ be a point fixed under the corank one subtorus $\theta_1^{\perp}$. Then, it follows easily from (7) that $G_x = Sp(n-1)$ and ℓ must be equal to 2.

(iv) $\mathfrak{g} = G_2$, F_4: In this case $\Omega'(\varphi)$ consists of the short roots of $\mathfrak{g}$, and $\Delta(\mathfrak{g})$ splits into two orbits of equal size, i.e., the long roots and the short roots. Hence, it follows easily from the condition $\dim \Omega'(\Psi) \leq L_{lin}(\mathfrak{g})$ that either

$$\Omega'(\Psi) = \{\text{long roots}\} \quad \text{or} \quad \Omega'(\Psi) = \{\text{short roots}\} = \Omega'(\varphi).$$

Similar reason as that of (iii) will show that the case $\Omega'(\Psi) =$ {long roots} is impossible. Then, again, it follows from Theorem 1 that $F(G,X) = F(W(G),F(T,X))$ is also an acyclic cohomology manifold and codim $F(G,X) = \dim \mathcal{P} = L_{lin}(\mathfrak{g})$.

(v) $\mathfrak{g} = A_2$: In this case $L_{lin}(\mathfrak{g}) = 6$ and there are only two orbits with only three pairs of weights, namely, $\Delta(\mathfrak{g})$ and $\{\pm\theta_i\}$. As the proceeding cases, it is not difficult to show that $F(G,X)$ is also acyclic and codim $F(G,X) = 8$ and 6 respectively.

Q.E.D.

§4. Topological actions on acyclic cohomology manifolds with simple weight pattern

Generally speaking, the weight system $\Omega(\Psi)$ is an invariant of primary importance which dominant the orbit structure of the given action Ψ on acyclic manifold. It is then rather natural to begin with an investigation of those action Ψ on acyclic manifold with simple pattern of weight system $\Omega(\Psi)$. As one may expect, those actions with simpler weight patterns should be exactly those actions with simpler orbit structures.

(A). Topological actions on acyclic manifolds with $\Omega'(\Psi) = \Delta(G)$.

Theorem 4: Let G be a compact connected Lie groups and Ψ be a topological G-action on an acyclic cohomology manifold X. If the system of non-zero weights, $\Omega'(\Psi)$, equals to the root system of G, $\Delta(G)$, then

(i) the principal orbit type of Ψ, (H_Ψ), is the conjugacy class of maximal tori, i.e. $(H_\Psi) = (T)$,

(ii) The Weyl group $W(G)$ acts as a group generated by (topological) reflections on $Y = F(T,X)$,

(iii) the natural map $Y/W \to X/G$ induced by the inclusion $Y \subseteq X$ is one-to-one and onto.

Proof: We shall prove the above theorem by "induction" on compact Lie groups. Observe that Theorem 5 is trivial if G is torus. Hence we may assume that G is non-commutative, i.e. $\Delta(G) \neq \phi$. Let $Y = F(T,X)$, $\alpha \in \Delta(G)$, $T_\alpha = \alpha^\perp$ be the corank one subtorus perpendicular to α, $G_\alpha = N(T_\alpha)$ and $\tilde{G}_\alpha = G_\alpha/T_\alpha$. Let $Y_\alpha = F(T_\alpha,X)$ and Ψ_α be the induced action of $\tilde{G}_\alpha$ on Y_α. Since $\Omega'(\Psi) = \Delta(G)$, it is clear that $\Omega'(\Psi_\alpha) = \{\pm\alpha\}$. Hence, proposition 3 applies to show that

$$F(G_\alpha,X) = F(\tilde{G}_\alpha,Y_\alpha) = F(W(G_\alpha),Y) = H_\alpha$$

where $W(G_\alpha) \simeq Z_2$ acts as a topological reflection on Y with codimension one "hyperplane" H_α. Therefore, for those point $y \in (Y-H_\alpha)$, $G_y \not\supseteq G_\alpha$ and consequently $G_y \neq G$ is a proper maximal rank subgroup of G. Then, it follows from (7) of proposition 1 that the weight system of the slice at y, $\Omega'(S_y) = \Delta(G_y)$. Hence, by induction assumption, the principal isotropy subgroups of the G_y-action on S_y are maximal tori of G_y. Since principal orbits are everywhere dense, it is clear that $(H_\Psi) = (T)$. Finally, assertion (iii) follows easily from the fact $F(K,X) = F(W(K),Y)$ for all maximal rank connected subgroups K of G.

(B) Actions of classical groups with simple weight patterns.

Let $G = SO(n)$ (resp. $SU(n)$, $Sp(n)$) be a classical group and $= \rho_n$ (resp. μ_n, ν_n) be the standard linear action of G on R^n (resp. C^n, H^n). We shall study the orbit structures of topological G-actions Ψ on acyclic cohomology manifolds with their weight patterns "model" after k-copies of the standard linear action, namely

$$\Omega'(\Psi) = \Omega'(k\varphi).$$

Let $(\theta_1,\theta_2,\ldots,\theta_m)$ be the usual coordinate of a maximal torus T of G; where

$$m = \begin{cases} \left[\frac{n}{2}\right] & \text{if } G = SO(n) \\ n \text{ and } \theta_1+\dots+\theta_n = 0 & \text{if } G = SU(n) \\ n & \text{if } G = Sp(n). \end{cases}$$

Then

$$\Omega'(\Psi) = \begin{cases} k\cdot\{\pm\theta_i\} & \text{if } G = SO(n),\ SU(n) \\ 2k\cdot\{\pm\theta_i\} & \text{if } G = Sp(n). \end{cases}$$

Theorem 5: Let $G = SO(n)$ (resp. $SU(n)$, $Sp(n)$) and Ψ be a topological G-action on an acyclic cohomology manifold X. Let $T \subseteq G$ be a maximal torus with the usual coordinate $(\theta_1,\dots\theta_m)$. If

$$\Omega'(\Psi) = k\cdot\{\pm\theta_i\}$$

then all connected isotropy subgroups G_x^0 are conjugate to the standard $SO(j)$ (resp. $SU(j)$) for a suitable $j \leq n$. Furthermore, the connected principal isotropy subgroup type (H_Ψ^0) is non-trivial when and only when $k \leq (n-2)$ (resp. $k \leq (n-2)$, $k \leq 2(n-1)$); and in fact, $H_\Psi^0 = SO(n-k)$ (resp. $SU(n-k)$, $Sp(n-\frac{k}{2})$, k must be even when $G = Sp(n)$ and $k < 2n$).

Proof: Let $x \in X$ be an arbitrary point of X and G_x^0 be the connected isotropy subgroup of x. Up to conjugation, we may assume $(G_x \cap T)^0$ is a maximal torus of G_x^0. It follows from (6) of Proposition 1, §3, that

$$T_1 = T_x^0 = (G_x^0 \cap T)^0 = \theta_{i_1} \cap\dots\cap \theta_{i_s}$$

for a suitable collection of θ_i. On the other hand, there exists an element $w \in W(G)$ which maps $\theta_{i_1},\dots,\theta_{i_s}$ to $\theta_1,\theta_2,\dots,\theta_s$ respectively. Hence, we may assume that the maximal torus of G_x to to be $\theta_1^{\perp} \cap \ldots \cap \theta_s^{\perp}$. Now, it follows from (7) of Proposition 1 that

$$\Delta(G_x^0) \supseteq \{\Delta(G)|T_1\} \setminus \{\Omega'(\Psi)|T_1\} .$$

Hence, it follows easily from Lie theory that

$$G_x^0 = SO(j), \left[\frac{j}{2}\right] = (m-s) \text{ (resp. } SU(n-2),\ Sp(n-s)).$$

Similarly, it follows from (6), (7) and (8) that the connected principal isotropy subgroup $(H_{\Psi}^0) = (SO(n-k))$ (resp. $SU(n-k)$, $Sp(n-\frac{k}{2})$, k even if $G = Sp(n)$, $k < 2n$).

Q.E.D.

Remark: The above theorem is the "topological version" of Theorem of [18] for differentiable actions of classical groups on acyclic manifolds. However, due to the fact that topological weight system only retains the direction of the weight vectors, the above theorem only determines the connected component of the isotropy subgroups. In fact, it is possible to prove G_x are actually connected if one similarly introduce the concept of p-weights system $\Omega_p'(\Psi)$ and assume that $\Omega_p'(\Psi) = \Omega_p'(k\varphi)$ for all prime p.

Next we consider the following two cases:

(i) $G = SU(n)$, $\Omega'(\Psi) = \Omega'(\Lambda^2\mu_n) = \{\pm(\theta_i+\theta_j);\ i < j\}$

(ii) $G = Sp(n)$, $\Omega'(\Psi) = \Omega'(\Lambda^2\nu_n) = \{\pm\theta_i\pm\theta_j;\ i < j\}$.

Theorem 6: Let $G = SU(n)$ (resp. $Sp(n)$) and Ψ be a topological G-action on an acyclic manifold X. If

$$\Omega'(\Psi) = \Omega'(\Lambda^2\mu_n) \quad (\text{resp. } \Omega'(\Lambda^2\nu_n)),$$

then the connected orbit types of Ψ are the same as those of

$\Lambda^2\mu_n$ (resp. $\Lambda^2\nu_n$). In particular, the connected principal orbit type $(H_\Psi^0) = \left(SU(2)^{\left[\frac{n}{2}\right]}\right)$ (resp. $(Sp(1)^n)$). Moreover, the F-variety of a point $x \in X$, $F(x) = F(G_x^0, X)$ is always an acyclic cohomology submanifold and codim $F(x)$ = codim $F(y)$ for a suitable y in the representation space of $\Lambda^2\mu_n$ (resp. $\Lambda^2\nu_n$) with $G_x^0 = G_y$.

<u>Proof</u>: In the case $G = Sp(n)$, $\Omega'(\Psi) = \Omega'(\Lambda^2\nu_n)$, since $\Omega'(\Lambda^2\nu_n)$ consists of exactly those short roots, the proof is essentially the same as that of Theorem 4. Let $Y = F(T,X)$. Since $\Delta(G)\backslash\Omega'(\Lambda^2\nu_n) = \Delta(Sp(1)^n)$, it follows from (7) that $F(T,X) = F(Sp(1)^n, X)$. The same reason as that of Theorem 4 shows that the symmetric group of order n

$$\mathfrak{S}_n = N(Sp(1)^n)/Sp(1)^n$$

acts as a group generated by topological reflections on Y. Furthermore, let K be a subgroup containing $H = Sp(1)^n$ and $\widetilde{K} = N(H,K)/H \subseteq N(H,G)/H = \mathfrak{S}_n$. Then one has $F(K,X) = F(\widetilde{K},Y)$.

Next we consider the case $G = SU(n)$, $\Omega'(\Psi) = \Omega'(\Lambda^2\mu_n)$. Since $\Delta(G)\cap\Omega'(\Psi) = \phi$, $F(G,X) = F(T,X)$, which is clearly acyclic. Let $\omega = \theta_1+\theta_2$, $T_\omega = \omega^\perp$, and $x \in \{F(T_\omega X)-F(T,X)\} \neq \phi$. It is clear that T_ω is a maximal torus of G_x. Furthermore, it is not difficult to show that $G_x^0 = SU(2) \times SU(n-2)$ by (7). It follows also from (7) that $\Omega'(\Psi) = \Omega'(\Lambda^2\nu_{(n-2)})$. Hence, by induction on n, we conclude that the principal orbit type

$$(H_\Psi^0) = \left(SU(2)^{\left[\frac{n}{2}\right]}\right).$$

Again, one may show that $F(K,X) = F(\widetilde{K},X)$ for all closed subgroups $K \supseteq H = SU(2)^{\left[\frac{n}{2}\right]}$.

Q.E.D.

(C) Determination of $L_{top}(\mathfrak{g})$ for general compact Lie algebras:

Observe that

$$L_{lin}(C_1 \times C_n) = L_{lin}(C_n) \qquad (n \geq 2)$$

$$L_{lin}(C_1 \times E_7) = L_{lin}(E_7)$$

$$L_{lin}(C_1 \times C_1) = L_{lin}(R^1 \times A_1) = 4$$

$$L_{lin}(R^1 \times A_n) = L_{lin}(A_n) \qquad (n \neq 1,3)$$

For a given compact Lie algebra $\mathfrak{g}$ in general, let

k_0 = dim of the center of $\mathfrak{g}$ = the number of $\mathbb{R}^1$ factors in $\mathfrak{g}$

k_1 = the number of C_1 factors in $\mathfrak{g}$

ℓ_0 = the number of factors of A_n-type $(n \neq 1,3)$ in $\mathfrak{g}$

ℓ_1 = the number of factors of E_7-type or C_n-type $(n \geq 2)$ and $\mathfrak{g} = R^{k_0} + C_1^{k_1} + \tilde{\mathfrak{g}}$

where $\tilde{\mathfrak{g}} = \Sigma \mathfrak{g}_j$ and $\mathfrak{g}_j \neq C_1$. Then, it is not difficult to see that $L_{lin}(\mathfrak{g})$ can be computed as follows:

(i) $L_{lin}(\tilde{\mathfrak{g}}) = \Sigma L_{lin}(\mathfrak{g}_j)$

(ii) If $k_0 \leq \ell_0$ and $k_1 \leq \ell_1$, then $L_{lin}(\mathfrak{g}) = L_{lin}(\tilde{\mathfrak{g}}) = \Sigma L_{lin}(\mathfrak{g}_j)$

(iii) If we define $m_0 = \text{Max}(0, k_0 - \ell_0)$, $m_1 = \text{Max}(0, k_1 - \ell_1)$, then $L_{lin}(\mathfrak{g}) = L_{lin}(R^{m_0} + C_1^{m_1}) + L_{lin}(\tilde{\mathfrak{g}})$

where $L_{lin}(R^{m_0} + C_1^{m_1}) = \begin{cases} 2m_1 + 1 & \text{if } m_0 = 0 \text{ and } m_1 \text{ is odd} \\ 2(m_0 + m_1) & \text{otherwise.} \end{cases}$

<u>Theorem 7</u>: For any compact Lie algebra $\mathfrak{g}$, $L_{top}(\mathfrak{g}) = L_{lin}(\mathfrak{g})$. Let us first consider the case that $\mathfrak{g}$ is semi-simple and the rank of every simple factor is at least 2.

<u>Lemma</u>: Let $\mathfrak{g} = \mathfrak{g}_1 + \ldots + \mathfrak{g}_\ell$ be the sum of simple compact Lie algebras of rank ≥ 2 and Ψ be an almost effective topological G-action on an ayclic cohomology manifold X with $\mathfrak{g}$ = the Lie algebra of G. If

$$\dim X = \dim \Psi \leq L_{lin}(\mathfrak{g}) + 1 = \sum_{j=1}^{\ell} L_{lin}(\mathfrak{g}_j) + 1$$

then the system of non-zero weights of Ψ, $\Omega'(\Psi)$, must be completely splitting, i.e., $\Omega'(\Psi) = \sum_{j=1}^{\ell} \Omega'(\Psi|G_j)$, where G_j is the normal subgroup with $\mathfrak{g}_j$ as its Lie algebra. Moreoever, $\Omega'(\Psi|G_j)$ must be exactly the one given in the table of Theorem 3 and

$$\left(\dim X - \dim F(G)\right) = \sum_{j=1}^{\ell} L_{lin}(\mathfrak{g}_j) = L_{lin}(\mathfrak{g}).$$

Hence, in particular, $L_{top}(\mathfrak{g}) = L_{lin}(\mathfrak{g})$.

<u>Proof</u>: We shall prove by induction on the number of factors ℓ. The starting case $\ell = 1$ is exactly Theorem 3. Now, let us assume that the lemma holds for $\ell \leq k$ and proceed to show that the lemma also holds for $\ell = (k + 1)$.

We claim that there must exist at least one simple factor say $\mathfrak{g}_{k+1}$, such that $\Omega'(\Psi)$ is splitting with respect to the decomposition $G = G' \times G_{k+1}$. For otherwise, it is not difficult to estimate the number of weights in $\Omega'(\Psi)$, by using the invariance (with respect to Weyl group) property, to show that $\dim \Psi \geq \dim \Omega'(\Psi) > L_{lin}(\mathfrak{g}) + 1$, which contradicts to the assumption. Hence, there exists a suitable simple factor, say G_{k+1},

$$\Omega'(\Psi) = \Omega'(\Psi|G') + \Omega'(\Psi|G_{k+1})$$

where $G = G' \times G_{k+1} = (G_1 \times \ldots \times G_k) \times G_{k+1}$. Let T_{k+1} be a maximal torus of G_{k+1}, it is easy to see that $G' = G_1 \times \ldots \times G_k$ acts almost effectively on $X' = F(T_{k+1}, X)$. Then, it follows from the induction assumption that $\dim X' \geq L_{top}(\mathfrak{g}') = \sum_{j=1}^{k} L_{lin}(\mathfrak{g}_j)$, and consequently, the number of weights in $\Omega'(\Psi|G_{k+1})$ = codim of

$$F(T_{k+1}, X) \text{ is } \leq \left[\dim X - \sum_{j=1}^{k} L_{lin}(\mathfrak{g}_j)\right] \leq L_{lin}(\mathfrak{g}_{k+1}) + 1.$$

Then, it is quite simple to determine the possibility of $\Omega'(\Psi|G_{k+1})$, and show that $X'' = F(G_{k+1}, X)$ is also an acyclic cohomology submanifold. Obviously, $G' = G_1 \times \ldots \times G_k$ also acts almost effectively on X'' and again follows from induction assumption that

$$\dim X'' \geq L_{top}(\mathfrak{g}') = L_{lin}(\mathfrak{g}') = \sum_{j=1}^{k} L_{lin}(\mathfrak{g}_j),$$

$$\Omega'(\Psi|G') = \sum_{j=1}^{k} \Omega'(\Psi|G_j).$$

where $\Omega'(\Psi|G_j)$, $j = 1,2,\ldots,k$, are exactly those given by the table of Theorem 3. Therefore, by Theorem 3, $F(G_1, X)$ is acyclic and $\dim F(G_1, X) = \dim X - L_{lin}(\mathfrak{g}_1)$. Hence, the lemma follows by induction on $F(G_1, X)$ with respect to the induced $G_2 \times \ldots \times G_{k+1}$-action.

Q.E.D.

Proof of Theorem 7:

Let $\mathfrak{g}$ be a given compact Lie algebra and

k_0 = dim of the center of $\mathfrak{g}$

k_1 = the number of C_1 factors in $\mathfrak{g}$

ℓ_0 = the number of factors of A_n-type $(n \neq 1,3)$ in $\mathfrak{g}$

ℓ_1 = the number of factors of E_7-type or C_n-type $(n \geq 2)$.

Hence,

$$\mathfrak{g} = \mathbb{R}^{k_0} + C_1^{k_1} + \tilde{\mathfrak{g}} = \mathbb{R}^{k_0} + C_1^{k_1} + \sum_{j=1}^{\ell} \mathfrak{g}_j$$

where $rk(\mathfrak{g}_j) \geq 2$ for $j = 1,2,\ldots,\ell$, and corresponding by

$$G = T^{k_0} \times Sp(1)^{k_1} \times G' = T^{k_0} \times Sp(1)^{k_1} \times \left\{ \prod_{j=1}^{\ell} G_j \right\}$$

where the Lie algebra of G_j is $\mathfrak{g}_j$, $j = 1,2,\ldots,\ell$. By definition, it is obvious that

$$L_{top}(\mathfrak{g}) \geq L_{top}(\mathfrak{g}') \quad \text{for any } \mathfrak{g}' \subseteq \mathfrak{g}.$$

Hence, in case $k_0 \leq \ell_0$ and $k_1 \leq \ell_1$, Theorem 7 follows from the above lemma, namely,

$$k_0 \leq \ell_0 \text{ and } k_1 \leq \ell_1 \Rightarrow L_{lin}(\mathfrak{g}) = L_{lin}(\tilde{\mathfrak{g}})$$

$$\Rightarrow L_{lin}(\mathfrak{g}) = L_{lin}(\tilde{\mathfrak{g}}) = L_{top}(\tilde{\mathfrak{g}}) \leq L_{top}\mathfrak{g} \leq L_{lin}(\mathfrak{g})$$

$$\Rightarrow L_{lin}(\mathfrak{g}) = L_{top}(\mathfrak{g}).$$

The general case can again be proved by induction on the total number of simple (or $\mathbb{R}^1$) factors. Roughly, one first show by simple estimate of the number of weights in $\Omega'(\Psi)$ that the assumption $\dim \Psi = \dim X \leq L_{lin}(\mathfrak{g})$ implies one of the following cases must occur, namely,

(i) There exists a simple factor G_1 of rank ≥ 2, such that

$$\Omega'(\Psi) = \Omega'(\Psi|G_1) + \Omega'(\Psi|(G/_{G_1}))$$

is splitting with respect to $G = G_1 \times (G/_{G_1})$.

(ii) There exists a factor G_1 of $C_1 \times C_n$-type or $C_1 \times E_7$-type or $T' \times A_n$-type $(n \neq 3)$ such that

$$\Omega'(\Psi) = \Omega'(\Psi|G_1) + \Omega'(G/_{G_1})$$

is splitting with respect to $G = G_1 \times (G/_{G_1})$.

(iii) There exists a circle subgroup T^1 in the center such that

$$\Omega'(\Psi) = \Omega'(\Psi|T^1) + \Omega'(\Psi|(G/_{T}1))$$

is splitting with respect to $G = T^1 \times (G/_{T}1)$.

Then, it is not difficult to proceed the proof of Theorem 7 by induction on the number of factors in almost the same way as in the proof of the lemma. We shall leave the detail to the reader.

§5. The relationship between the weight system $\Omega(\Psi)$ and the principal orbit type (H_Ψ).

(A) Let G be a compact connected Lie group, X be an acyclic cohomology manifold and Ψ be a given topological G-action on X.

In this section, we shall investigate the relationship between its dominant algebraic invariant, the weight system $\Omega(\Psi)$, and its dominant geometric characteristic, the principal orbit type (H_Ψ). The main result of this section is the following algorithm which enables us to compute the connected principal orbit type (H^0_Ψ) from the weight system $\Omega(\Psi)$. Namely,

<u>Theorem 8</u>: Let $\omega_1, \omega_2, \ldots, \omega_k \in \Omega'(\Psi)$ be a sequence of non-zero weights and $T = S_0 \supsetneq S_1 \supsetneq \cdots \supsetneq S_k$ be a sequence of subtori of an arbitrarily chosen maximal torus T of G satisfying the following recursive conditions:

$$S_0 = T \,, \quad \omega_1 \in \{\Omega'(\Psi) - \Omega'(Ad_G)\} \neq \phi$$

$$S_1 = \omega_1^\perp, \quad \omega_2 \in \{\Omega'(\Psi|S_i) - \Omega'(Ad_G|S_i)\} \neq \phi$$

- - - - - - - - - - - - - - - - - - -

$$S_i = S_{i-1} \cap \omega_i^\perp, \quad \omega_{i+1} \in \{\Omega'(\Psi|S_i) - \Omega'(Ad_G|S_i)\} \neq \phi$$

- - - - - - - - - - - - - - - - - - -

$$S_k = S_{k-1} \cap \omega_k^\perp, \quad \{\Omega'(\Psi|S_k) - \Omega'(Ad_G|S_k)\} = \phi \text{ (empty)} .$$

Then S_k is a maximal torus of a suitable principal isotropy subgroup H^0_Ψ and the root system of H^0_Ψ is given by the following equation

$$\Delta(H^0_\Psi) = \{\Omega'(Ad_G|S_k) - \Omega'(\Psi|S_k)\} .$$

Conversely, suppose $S = (H^0_\Psi \cap T)$ is a maximal torus of a principal isotropy subgroup H_Ψ. Then there exist such a sequence of weights

$\omega_1, \omega_2, \ldots, \omega_k \in \Omega'(\Psi)$ and a sequence of subtori

$T = S_0 \supsetneqq S_1 \supsetneqq \cdots \supsetneqq S_k = S$ with $S_k = S$.

Remark: If $\{\Omega'(\Psi) - \Omega'(Ad_G)\} = \phi$ is empty, then $k = 0$, $S_k = T$. The difference $\{\Omega'(\Psi|S_i) - \Omega'(Ad_G|\ S_i)\}$ is a difference of sets with multiplicities, namely the multiplicity of a given weight ω in the difference set is zero or is equal to the difference of multiplicity if it is positive, i.e.

$$m_\omega = \begin{cases} (m'_\omega - m''_\omega) & \text{if } (m'_\omega - m''_\omega) > 0 \\ 0 & \text{otherwise.} \end{cases}$$

The above Theorem 8 has the following important corollaries:

Corollary 1: Let Ψ_1, Ψ_2 be topological G-actions on acyclic cohomology manifolds X_1, X_2 respectively. If they have the same system of non-zero weights, i.e. $\Omega'(\Psi_1) = \Omega'(\Psi_2)$ then they also have the same connected principal orbit type, i.e., $(H^0_{\Psi_1}) = (H^0_{\Psi_2})$.

Corollary 2: Let Ψ, Ψ_1, Ψ_2 be topological G-actions on acyclic cohomology manifolds X, X_1, X_2 respectively. If

$$\Omega'(\Psi) = \Omega'(\Psi_1) + \Omega'(\Psi_2) \text{ (sum with multiplicity)}$$

then $(H^0_\Psi) = (H^0_{\Psi_1}) \mathring{\cap} (H^0_{\Psi_2})$ is the intersection in general position of $(H^0_{\Psi_1})$ and $(H^0_{\Psi_2})$, namely, the conjugacy class of those smallest possible intersections among $\{g_1 H^0_{\Psi_1} g_1^{-1} \cap g_2 H^0_{\Psi_2} g_2^{-1}\}$.

Proof of Corollary 2: Let $\Psi_1 \times \Psi_2$ be the diagonal G-action on $X_1 \times X_2$, i.e., $g(x_1, x_2) = (gx_1, gx_2)$ for all $x_1 \in X_1$, $x_2 \in X_2$. Then it is clear that

$$\Omega'(\Psi_1 \times \Psi_2) = \Omega'(\Psi_1) + \Omega(\Psi_2)$$

and $(H^0_{\Psi_1 \times \Psi_2}) = (H^0_{\Psi_1}) \mathring{\cap} (H^0_{\Psi_2})$, (intersection in general position).

Hence, $\Omega'(\Psi) = \Omega'(\Psi_1) + \Omega'(\Psi_2) = \Omega'(\Psi_1 \times \Psi_2)$ and Corollary 2 follows from Corollary 1.

Corollary 3: Suppose Ψ_1, Ψ_2 are respectively topological G-actions on acyclic cohomology manifold X_1, X_2 and

$$\Omega'(\Psi_1) \supseteq \Omega'(\Psi_2) \quad \text{(with multiplicity)}.$$

Then

$$(H^0_{\Psi_1}) \leqslant (H^0_{\Psi_2})$$

in the sense that $H^0_{\Psi_1}$ is conjugate to a subgroup of $H^0_{\Psi_2}$.

Proof of Corollary 3: Chosen a maximal torus T of G. After suitable conjugation, we may assume that $(H^0_{\Psi_2}) \cap T) = S''$ is a maximal torus of $H^0_{\Psi_2}$. By Theorem 8, there exist a sequence of weights $\omega_1, \omega_2, \ldots, \omega_{k''} \in \Omega'(\Psi_2)$ and a sequence of subtori

$T = S_0 \supsetneqq S_1 \supsetneqq \cdots \supsetneqq S_{k''} = S''$ satisfying the recursive conditions:

$$S_i = S_{i-1} \cap \omega_i^{\perp} \; ; \; \omega_{i+1} \in \{\Omega'(\Psi_2|S_i) - \Omega'(Ad_G|S_i)\}.$$

Since we assume that $\Omega'(\Psi_1) \supseteq \Omega'(\Psi_2)$, it is clear that

$$\{\Omega'(\Psi_1|S) - \Omega'(Ad_G|S)\} \supseteq \{\Omega'(\Psi_2|S) - \Omega'(Ad_G|S)\}$$

holds for any subtorus $S \subseteq T$. Hence, if

$$\{\Omega'(\Psi_1|S'') - \Omega'(Ad_G|S'')\} = \phi$$

is also empty, then S'' is also a maximal torus of a suitable $H^0_{\Psi_1}$; otherwise, we may continue the sequences

$$\{\omega_1, \omega_2, \ldots, \omega_{k''}\}\ ,\quad \{T = S_0 \supsetneq S_1 \supsetneq \cdots \supsetneq S_{k''}\}$$

to obtain the corresponding sequences for $\Omega'(\Psi_1)$, i.e.,

$$\{\omega_1, \omega_2, \ldots, \omega_{k''}, \ldots, \omega_{k'}\}\ ,\ \{T = S_0 \supsetneq S_1 \supsetneq \cdots \supsetneq S_{k''} \supsetneq \cdots \supsetneq S_{k'} = S'\}$$

such that $S' = S_{k''}$ is a maximal torus of a suitable $(H^0_{\Psi_1})$. Notice that $S'' \supseteq S'$. Furthermore, we have

$$\Delta(H^0_{\Psi_2}) = \Omega'(Ad_G|S'') - \Omega'(\Psi_2|S'')$$

$$\Delta(H^0_{\Psi_1}) = \{\Omega'(Ad_G|S') - \Omega'(\Psi_1|S')\}$$

$$\Omega'(\Psi_2|S') \subseteq \Omega'(\Psi_1|S')\}\ .$$

Hence,

$$\Delta(H^0_{\Psi_1}) = \{\Omega'(Ad_G|S') - \Omega'(\Psi_1|S')\} \subseteq \{\Omega'(Ad_G|S') - \Omega'(\Psi_2|S')\}$$

$$= \Omega'(Ad_{H_{\Psi_2}}|S')$$

and consequently $(H^0_{\Psi_1}) \leqslant (H^0_{\Psi_2})$.

(B) Proof of Theorem 8:

Lemma 1: If $\{\Omega'(\Psi) - \Delta(G)\} = \phi$ is empty, i.e.

$\Omega'(\Psi) \subseteq \Delta(G) = \Omega'(Ad_G)$, then the connected principal isotropy subgroups (H^0_Ψ) are of maximal rank, i.e. $rk(H^0_\Psi) = rk(G)$.

Since $\Omega'(\Psi) \subseteq \Delta(G)$ implies that $\Omega'(\Psi)$ is a completely splitting, it is easy to use Proposition 2 of §3 to reduce the proof of Lemma 1 to the special case that G is simple. However, in the case G is simple, we have the following more precise result:

Lemma 1': Let G be a simple compact connected Lie group and Ψ be a non-trivial topological G-action on an acyclic cohomology manifold X with $\Omega'(\Psi) \subseteq \Delta(G)$. Then, either

(i) $\Omega'(\Psi) = \Delta(G)$ and $(H^0_\Psi) = (T)$, [maximal tori]

or (ii) $\Omega'(\Psi) = \{$the set of short roots of G$\}$ and respectively

$$(H^0_\Psi) = \begin{cases} (D_n) \\ (Sp(1)^n) \\ (Spin(8)) \\ (SU(3)) \end{cases} \text{when } G = \begin{cases} B_n \\ C_n \\ F_4 \\ G_2 \end{cases}.$$

Proof of Lemma 1': The case $\Omega'(\Psi) = \Delta(G)$ is proved in Theorem 4, §4. Hence, we shall only consider the case $\phi \neq \Omega'(\Psi) \subsetneq \Delta(G)$. Since $\Omega'(\Psi)$ and $\Delta(G)$ are both invariant under the Weyl group $W(G)$, $\Delta(G)$ must consist of more than one "orbit" under the action of $W(G)$. On the other hand, it is a well-known fact that the root system $\Delta(G)$ of a simple compact connected Lie group G consists of at most two orbits of $W(G)$, namely, "the set of long roots" and "the set of short roots". And B_n, C_n, F_4, G_2 are those simple compact Lie groups with roots of different length.

Let $\alpha \in \Omega'(\Psi) \subseteq \Delta(G)$ be such a root which is also in $\Omega'(\Psi)$, and G_α be the connected normalizer of $T_\alpha = \alpha^\perp$. Then it follows from Proposition 3 of §3 that

$$F(G_\alpha, X) = F(\sigma_\alpha, F(T,X)) \subsetneq F(T,X)$$

is a codimension one submanifold of $F(T,X)$, where σ_α is the order 2 element of $W(G_\alpha)$. Hence there exists $y \in F(T,X)$ with

$$T \subseteq G_y \subsetneq G.$$

On the other hand, it follows from Proposition 1 of §3 that

$$\Omega'(\Psi) = \{\Delta(G) - \Omega(G_y)\} + \Omega'(S_y). \tag{7}$$

We claim that $\{\Delta(G) - \Delta(K)\}$ consists of at least one short root for any proper maximal rank subgroup $K \subsetneq G$. Of course, we need only to check the above assertion for those maximal, maximal rank subgroups of G. Since the Dynkin diagram of such subgroups K are subdiagrams of the extended diagram of G, i.e., the diagram of the Cartan polyhedra of G. Namely,

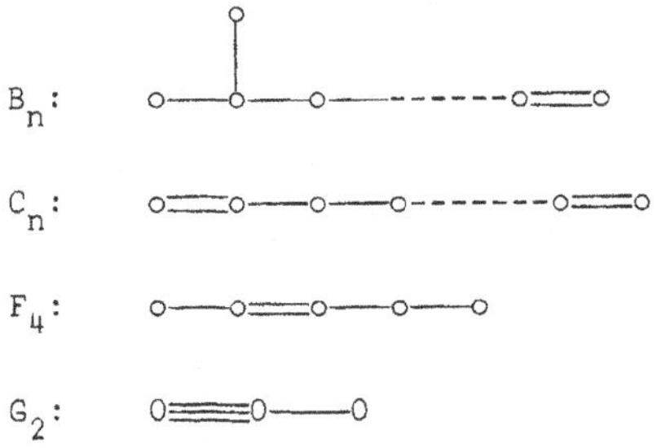

It is not difficult to check that $\Delta(K)$ does not contain all short roots of $\Delta(G)$. Hence it follows from equation (7) that $\Omega'(\Psi)$ must contain some short roots and consequently

$$\Omega'(\Psi) = \{\text{the set of short roots}\}$$

if $\Omega'(\Psi) \subsetneq \Delta(G)$. And the same reason as in the proof of Theorem 4 will show that the connected principal isotropy subgroups $(H^0_\Psi) = (D_n)$ or $(Sp(1)^n)$, or $(Spin(8))$, or $SU(3)$ for the case $G = B_n, C_n, F_4$, or G_2 respectively.

Lemma 2: Suppose $\{\Omega'(\Psi) - \Delta(G)\}$ is non-empty. Then to any weight $\omega \in \{\Omega'(\Psi) - \Delta(G)\}$, there exists a point $x \in X$ such that

$$T^0_x = \omega^\perp = G^0_x \cap T$$

and $\omega^\perp$ is a maximal torus of G^0_x.

Proof of Lemma 2: Let $K = G_y$, $y \in F(T,X)$ be minimal among the set of all maximal rank isotropy subgroups, and S_y be the K-action on the slice at x. Then it follows from the following equation (7) (of Proposition 1 in §3),

$$\Omega'(\Psi) = \Omega'(S_y) + \{\Delta(G) - \Delta(K)\}$$

that

$$\{\Omega'(\Psi) - \Delta(G)\} = \{\Omega'(S_y) - \Delta(K)\} .$$

On the other hand, it follows from Borel Formula that, to any $\omega \in \{\Omega'(S_y) - \Delta(K)\} = \{\Omega'(\Psi) - \Delta(G)\}$, there exists a suitable point x in the slice at x such that $T^0_x = \omega^\perp$. Since $K = G_y$ is the minimal among all maximal rank isotropy subgroups, we have

$$(rk(G) - 1) = rk(T^0_x) \leqslant rk(K_x = G_x) < rk(K) = rk(G).$$

Hence $rk(G_x) = rk(G) - 1 = rk(T^0_x)$ and $T^0_x = \omega^\perp$ is a maximal torus of G_x. The proof of Lemma 2 is complete.

Proof of Theorem 8:

We shall prove Theorem 8 by induction on compact connected Lie groups.

(1) <u>In the case $G = T$ is commutative</u>: Then $\Delta(G) = \phi$ and $H^0 = \ker^0(\Psi)$ and it follows from Corollary of §3 that

$H^0 = \ker^0(\Psi) = \cap\{\omega^\perp; \omega \in \Omega'(\Psi)\}$. Hence Theorem 8 holds for the case $G = T$.

(2) Now, we assume that Theorem 8 holds for all proper connected subgroups of G and proceed to show that Theorem 8 also holds for G itself.

If $\{\Omega'(\Psi) - \Delta(G)\} = \phi$ is empty, then it follows from Lemma 1 that $(H^0{}_\Psi)$ are maximal rank subgroups and it follows from the equation

$$\Omega'(\Psi|T) = \Omega'(Ad_G|T) - \Omega'(Ad_{H_\Psi}|T)$$

that $\Delta(H_\Psi) = \{\Delta(G) - \Omega'(\Psi)\}$. Hence Theorem 8 holds when $\{\Omega'(\Psi) - \Delta(G)\} = \phi$.

If $\{\Omega'(\Psi) - \Delta(G)\} \neq \phi$ is <u>non-empty</u>, then for every weight $\omega_1 \in \{\Omega'(\Psi) - \Delta(G)\}$, there exists $y \in X$ such that $S_1 = \omega_1^\perp$ is a maximal torus of G_y^0, (Lemma 2). Let Ψ_y be the G_y-action on a <u>slice</u> S_y at y. Then it follows from the following equation of Proposition 1 of §3

$$\Omega'(\Psi|S_1) = \{\Omega'(Ad_G|S_1) - \Omega'(Ad_{G_y}|S_1)\} + \Omega'(\Psi_y)$$

that

$$\{\Omega'(\Psi|S) - \Omega'(Ad_G|S)\} = \{\Omega'(\Psi_y|S) - \Omega'(Ad_{G_y}|S)\}$$

holds for any subtorus S of S_1. Hence, the sequence of weights

$$(\omega_2|S_1), (\omega_3|S_1), \ldots, (\omega_k|S_1) \in \Omega'(\Psi_y)$$

and the sequence of subtori

$$S_1 \supsetneq S_2 \supsetneq \cdots \supsetneq S_k$$

satisfy the recursive condition of Theorem 8 for the G_y-action, Ψ_y, on the slice, S_y, at y. Therefore, by the induction assumption, S_k is a maximal torus of a suitable principal isotropy subgroup, H_{Ψ_y}, of the G_y-action Ψ_y. However, the principal orbit type theorem of Montgomery-Samelson-Yang [21] assert that principal orbits are everywhere dense. Hence, it is clear that the principal isotropy subgroups (H_{Ψ_y}) of the G_y-action, Ψ_y, on the slice are also principal isotropy subgroups of the original G-action Ψ, i.e. $(H_{\Psi_y}) = (H_\Psi)$.

(3) Conversely, let $S = H_\Psi^0 \cap T$ be a maximal torus of a connected principal isotropy subgroup H_Ψ. If H_Ψ^0 is a maximal rank subgroup, i.e. $S = T$, then

$$\Omega'(\Psi) = \Omega'(\Psi|S) = \Omega'(Ad_G|S) - \Omega'(Ad_{H_\Psi}|\ S) = \Delta(G) - \Delta(H_\Psi)$$

implies that

$$\{\Omega'(\Psi) - \Delta(G)\} = \phi \text{ and } \Delta(H_\Psi) = \Delta(G) - \Omega'(\Psi).$$

If H_Ψ^0 is not of maximal rank, then it follows from Lemma 1 that $\{\Omega'(\Psi) - \Delta(G)\} \neq \phi$ is non-empty. Again, by Lemma 2, to every $\omega_1 \in \{\Omega'(\Psi) - \Delta(G)\}$, there exists $y \in X$ such that $\omega_1^\perp = G_y \cap T$ is a maximal torus of G_y. Since principal orbits are everywhere dense, one may assume after suitable conjugation that H_Ψ^0 is also a connected principal isotropy subgroup of the G_y-action on the slice S_y. Hence, by induction, Theorem 8 also holds for G itself.

§6. Classification of principal orbit types for actions of simple compact Lie groups on acyclic cohomology manifolds

In view of the principal orbit type theorem of Montgomery-Samelson-Yang [21], the principal orbits are population-wise dominating everywhere and hence the type of principal orbits is a geometric characteristic of fundamental importance. On the other hand, for spaces of given specific type, the possibilities of principal orbit types are usually rather limited. Hence, in the study of topological actions of a given compact connected Lie group G on spaces of a certain type (such as acyclic cohomology manifolds as we do in the present paper), one of the natural problems of primary importance is the classification of principal orbit types for all topological G-actions on spaces of a given type. In this section, we shall first work out the classification of principal orbit types for topological actions of those simple compact connected Lie groups on acyclic manifolds.

Since the family of linear actions usually offer one of the most valuable class of typical examples for such a purpose, it is natural to work out the much easier problem of classification of principal orbit types for linear actions of simple compact connected Lie groups.

(A) Classification of connected principal isotropy subgroups for linear actions of simple compact connected Lie groups:

The classification of linear actions of simple compact connected Lie groups with non-trivial connected principal isotropy subgroups, i.e. $H_\psi^0 \neq \{id\}$, has already been carried out in [15] and [32] independently. We list the result as follows:

Table A: Real irreducible representations of simple compact connected Lie groups with non-trivial connected principal isotropy subgroups:

Notations: We shall use the usual Lie algebra terminology. Let $\mathfrak{g}$ be a simple compact Lie algebra of rank r and $\Pi(G) = \{\alpha_1, \alpha_2, \ldots, \alpha_r\}$ be a system of primitive roots of $\mathfrak{g}$. We shall identify a real representation Ψ with its complexification and denote the r basic representation corresponding to $\alpha_1, \alpha_2, \ldots, \alpha_r$ by $\phi_1, \phi_2, \ldots, \phi_r$ respectively.

I. $\mathfrak{g} = A_r$:

rank r	Ψ	$\Omega'(\Psi)$	(H_Ψ^0)
$r \geq 1$	Ad	$\Delta(\mathfrak{g})$	(T); maximal tori
$r \geq 2$	$\phi_1 + \phi_r$	$\{\pm\theta_i,\ i = 1, \ldots, r+1\}$	$(A_{(r-1)})$
$r \geq 4$	$\phi_2 + \phi_{r-1}$	$\{\pm(\theta_i + \theta_j),\ i<j\}$	$(SU(2)^{\left[\frac{r+1}{2}\right]})$
$r = 3$	ϕ_2	$\{(\theta_i + \theta_j),\ i<j\}$	$(B_2 = C_2)$
$r = 5$	$2\phi_3$	$2\cdot\{(\pm\theta_i \pm\theta_j \pm\theta_k),\ i<j<k\}$	$T^2 \subseteq SU(3) \times SU(3) \subseteq SU(6)$

II. $\mathfrak{g} = B_r,\ r \geq 3$:

rank r	Ψ	$\Omega'(\Psi)$	(H_Ψ^0)
$r \geq 3$	Ad	$\Delta(\mathfrak{g})$	(T); maximal tori
$r \geq 3$	ϕ_1	$\{\pm\theta_i;\ i = 1, \ldots, r\}$	(D_r)
$r = 3$	ϕ_3	$\{\frac{1}{2}(\pm\theta_1 \pm\theta_2 \pm\theta_3)\}$	(G_2); $\dfrac{Spin\ (7)}{G_2} = S^7$
$r = 4$	ϕ_4	$\{\frac{1}{2}(\pm\theta_1 \pm\theta_2 \pm\theta_3 \pm\theta_4)\}$	$\big(Spin\ (7)\big)$; $\dfrac{Spin\ (9)}{Spin\ (7)} = S^{15}$

III. $\mathfrak{g} = C_r,\ r \geq 2$:

rank r	Ψ	$\Omega'(\Psi)$	(H_Ψ^0)
$r \geq 2$	Ad	$\Delta(\mathfrak{g})$	(T); maximal tori
$r \geq 2$	$2\phi_1$	$2\cdot\{\pm\theta_i, i = 1,\dots,r\}$	(C_{r-1})
$r \geq 2$	ϕ_2	$\{\pm\theta_i\pm\theta_j,\ i < j\}$	$(Sp(1)^r)$

IV. $\mathfrak{g} = D_r,\ r \geq 4$:

rank r	Ψ	$\Omega'(\Psi)$	(H_Ψ^0)
$r \geq 4$	Ad	$\Delta(\mathfrak{g})$	(T); maximal tori
$r \geq 4$	ϕ_1	$\{\pm\theta_i,\ 1 = 1,\dots,r\}$	(B_{r-1})
$r = 5$	$\phi_4 + \phi_5$	$\{\frac{1}{2}(\pm\theta_1\pm\theta_2\pm\theta_3\pm\theta_4\pm\theta_5)\}$	$(SU(4))$
$r = 6$	$2\phi_5$ or $2\phi_6$	$\{\frac{1}{2}(\pm\theta_1\pm\dots\pm\theta_6)\}$ even with -1 or odd	$(SU(2)\times SU(2)\times SU(2))$

V. Exceptional Lie Groups

For each of the five exceptional Lie groups, we have the adjoint representation whose principal orbit type is the maximal tori (T). Besides that, we have the following irreducible representations with non-trivial principal isotropy subgroups.

$\mathfrak{g}$	Ψ	$\Omega'(\Psi)$	(H^0_Ψ)
G_2: 	ϕ_1	$\{\pm\theta_1, \pm\theta_2, \pm\theta_3\}$	$(SU(3))$
F_4: 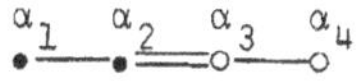	ϕ_1	$\{\frac{1}{2}(\pm\theta_1\pm\theta_2\pm\theta_3\pm\theta_4), \pm\theta_i\}$	$(Spin(8))$
E_6: 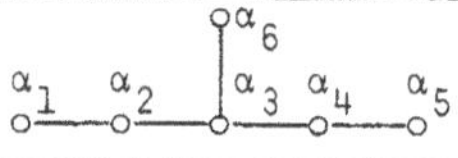	$\phi_1 + \phi_5$		$(Spin(8))$
E_7: 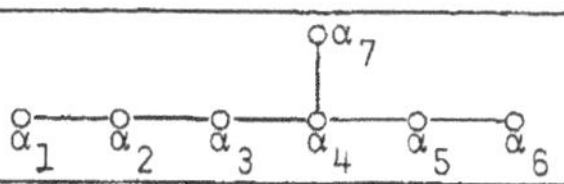	$2\phi_1$		$(Spin(8))$

Remark: Observe that if $\Psi = \Psi_1 + \Psi_2$, then the respective principal isotropy subgroups of Ψ, Ψ_1 and Ψ_2 are related as follows:

$$(H_\Psi) = (H_{\Psi_1}) \mathring{\cap} (H_{\Psi_2}) \quad \text{(intersection in general position).}$$

Hence, it is not difficult to extend the above list for irreducible linear representations to include all linear actions of simple Lie groups.

(B) G-admissible and G-indecomposable systems of weights:

Definition: Let G be a compact connected Lie group of rank k and Ω' be a system of non-zero weights defined over a torus group T of rank k. Ω' is called g-admissible if there exists a topological G-action Ψ on an acyclic cohomology manifold X with $\Omega'(\Psi) = \Omega'$. Furthermore Ω' is called G-decomposable if there exist two non-trivial topological G-actions Ψ_1, Ψ_2 on acyclic cohomology manifolds X_1 and X_2 such that $\Omega' = \Omega'(\Psi_1) + \Omega'(\Psi_2)$.

A G-admissible system of weights Ω' is called G-indecomposable if it is impossible to decompose Ω' into the sum of two non-trivial G-admissible systems of weights.

An obvious necessary condition for a system of weights Ω' to be G-admissible is that Ω' is invariant under the action of Weyl group $W(G)$. However, it is, in general, far from being sufficient. The following lemma is a slight improvement of such a necessary condition.

Lemma 1: If Ω' is G-admissible, then Ω' is invariant under the action of $W(G)$ and moreover, to any $\omega \in \{\Omega' - \Delta(G)\}$ and $\alpha \in \Delta(G)$ with $(\omega,\alpha) \neq 0$, there exists at least one $\omega' \in \Omega'$ such

$$(\alpha - \omega') \equiv 0 \pmod{\omega}.$$

Proof: Let Ψ be a topological G-action on acyclic manifold X with $\Omega'(\Psi) = \Omega'$. (Such an action Ψ exists by the assumption that Ω' is G-admissible.) Let ω be an arbitrary weights of $\{\Omega' - \Delta(G)\}$. It follows from Lemma 2 of §5 that there exists $x \in X$ such that

$$T_x^0 = G_x^0 \cap T = \omega^{\perp} \quad \text{is a maximal torus of } G_x^0.$$

Let $\alpha \in \Delta(G)$ be a root with $(\omega,\alpha) \neq 0$ (non-perpendicular). We claim that $(\alpha|\omega^{\perp}) \in \{\Delta(G)|\omega^{\perp} - \Delta(G_x)\}$. Since the case $(\alpha|\omega^{\perp}) = 0$ is obvious, we shall prove the above claim for the case $(\alpha|\omega^{\perp}) \neq 0$ as follows. If the multiplicity of $(\alpha|\omega^{\perp})$ in $\{\Delta(G)|\omega^{\perp}\}$ is > 1, then it is obvious that $(\alpha|\omega^{\perp})$ also belongs to $\{\Delta(G)|\omega^{\perp} - \Delta(G_x)\}$ since every root in $\Delta(G_x)$ has multiplicity one. If the multiplicity of $(\alpha|\omega^{\perp})$ in $\{\Delta(G)|\omega^{\perp}\}$ is one, then $(\alpha|\omega^{\perp}) \in \Delta(G_x)$ if and only if the Lie algebra of G_x, $\mathfrak{g}_x$, containing the eigen-space of

α, which is obviously impossible because $(\omega,\alpha) \neq 0$. Hence $(\alpha|\omega^{\perp}) \notin \Delta(G_x)$, (or $(\alpha|\omega^{\perp}) \in \{\Delta(G)\omega^{\perp} - \Delta(G_x)$.

Now, it follows from the following equation (7) of Proposition 1 in §3

$$\Omega'|\omega^{\perp} = \Omega'(S_x) + \{\Delta(G)\omega^{\perp} - \Delta(G_x)\}$$

that there exists at least one $\omega' \in \Omega'$ such that

$$\omega'|\omega^{\perp} = \alpha|\omega^{\perp} \in \{\Delta(G)|\omega^{\perp} - \Delta(G_x)\}$$

or equivalently, $(\omega' - \alpha) \equiv 0 \pmod{\omega}$.

Examples:

(1) Let $G = SU(r+1)$ and $\{\theta_1,\theta_2,\ldots,\theta_{r+1}\}$ be the weight system of the usual representation of $SU(r+1)$ on $\mathbb{C}^{r+1}$, $(r \geqslant 2)$. Suppose Ω' is a G-admissible system of weights containing $\pm(a\theta_1 + b\theta_2)$, $a > 1$ and $(a,b) = 1$. Then, by the above lemma and $(a\theta_1 + b\theta_2, \theta_1 - \theta_3) = a \neq 0$, there exists $\omega' \in \Omega'$ such that

$$(\omega' - \alpha) \equiv 0 \pmod{\omega}$$

or equivalently, there exists a suitable integer k, such that

$$\omega' = \alpha + k\cdot\omega = (\theta_1 - \theta_3) + k(a\theta_1 + b\theta_2)$$

$$= (k\cdot a + 1)\cdot\theta_1 + (k\cdot b)\cdot\theta_2 - \theta_3.$$

Hence, in particular $\{\pm(a\theta_i + b\theta_j)\}$ does not form a G-admissible system of weights.

(2) Let $G = B_r$ (resp. C_r, D_r), $r \geqslant 3$, and $\{\theta_1,\theta_2,\ldots,\theta_r\}$ be the usual orthonormal basis in the Cartan subalgebra of G.

Suppose Ω' is a G-admissible system of weights containing $\pm(a\theta_1 + b\theta_2)$, $a > b > 0$ and $(a,b) = 1$. Then, again by the above lemma and $(a\theta_1 + b\theta_2, \theta_1 - \theta_3) \neq 0$, there exists k, such that, $\omega' = (ka + 1)\theta_1 + k\cdot b\theta_2 - \theta_3$ also belongs to Ω'.

(C) Classification of connected principal orbit types for topological actions of simple compact Lie groups on acyclic cohomology manifolds:

We state the main result of §6 as follows:

Theorem 9': Let G be a simple compact connected Lie group. Let Ψ be a given topological G-action on an acyclic cohomology manifold X with indecomposable system of non-zero weights $\Omega'(\Psi)$. If the connected principal orbit type, $(H^0{}_\Psi)$, of Ψ is non-trivial, then there exists a unique irreducible linear G-action ψ with the same non-zero weight system and the same connected principal orbit type, i.e. $\Omega'(\psi) = \Omega'(\Psi)$ and $(H^0{}_\psi) = (H^0{}_\Psi)$, except the following undecided possibilities:

(i) $G = \mathrm{Spin}(11)$, $\Omega'(\Psi) = \{\frac{1}{2}(\pm\theta_1 \pm \dots \pm\theta_5)\} + m\{\pm\theta_i\}$ $2 \leq m \leq 3$

(ii) $G = \mathrm{Spin}(12)$, $\Omega'(\Psi) = \Omega'(\phi_5) + m\{\pm\theta_i\}$ $2 \leq m \leq 4$.

(D) Proof of Theorem 9':

Lemma 2: Let G be a simple compact connected Lie group and Ψ be a topological G-action on an acyclic cohomology manifold. If

$$\Omega'(\Psi) \supsetneq \Delta(G)$$

then the connected principal isotropy subgroups is trivial, i.e., $(H^0{}_\Psi) = \{id\}$.

Proof of Lemma 2:

Let $S \subseteq H_\Psi^0$ be a maximal torus of a principal isotropy subgroup H^0_Ψ. Then it follows from equation (8) of Corollary 1 in §3 that

$$\Omega'(\Psi)|S \equiv \Delta(G)|S - \Delta(H_\Psi) \quad \text{(mod zero weights)}.$$

On the other hand, we assume that $\Omega'(\Psi) \supsetneq \Delta(G)$, hence

$$\{\Omega'(\Psi) - \Delta(G)\}|S \equiv 0 \quad \text{(mod zero weights)}.$$

However, $\{\Omega'(\Psi) - \Delta(G)\}$ is, by assumption, non-empty and invariant under the Weyl group $W(G)$, it is not difficult to see that $\{\Omega'(\Psi) - \Delta(G)\}$ spans the Cartan subalgebra of G. Therefore, $\{\Omega'(\Psi) - \Delta(G)\}|S \equiv 0$ (mod 0) holds only when $S = \{id\}$, hence H_Ψ^0 must be trivial.

An outline of the proof of theorem 9':

The basic idea of the proof of Theorem 9' is rather straightforward, it consists of the following steps:

(i) We may assume that $\Omega'(\Psi)$ does not contain $\Delta(G)$, for otherwise, it follows from the above lemma 2 that either $H^0_\Psi = \{id\}$ or $\Omega'(\Psi) = \Delta(G)$ and $(H^0_\Psi) = (T)$. On the other hand, it follows from the equation (8) of Corollary 1 in §3, i.e.,

$$\Omega'(\Psi|H^0_\Psi) = \Omega'(Ad_G|H^0_\Psi) - \Delta(H^0_\Psi) \tag{8}$$

that the following condition is a necessary condition for the non-triviality of (H^0_Ψ), namely,

(*): There exists a circle subgroup $S \subseteq T$, such that

$$\Omega'(\Psi|S) \subseteq \Omega'(Ad_G|S)$$

or, a slightly weaker version which is also easier to check,

(*'): There exists a circle subgroup $S \subseteq T$ such that

$$\dim(\Omega'(\Psi|S)) \leq \dim(\Omega'(Ad_G|S)).$$

Since the order of Weyl group $W(G)$ is usually much larger than the number of roots of G, (for example $ord(W(A_n)) = (n+1)!$ as compared to $\#(\Delta(A_n)) = n(n+1))$. Hence, it is not difficult to show by lemma 1 that almost all indecomposable weight systems, $\Omega'(\Psi)$, consist of too many weights to satisfy condition (*), or even (*'), except a few simple possibilities. Therefore, one needs only to examine the remaining few simple possibilities.

(ii) Among those remaining few possibilities of indecomposable weight patterns $\Omega'(\Psi)$, at most two or three of them are not realizable by linear actions which deserve special treatment. For such non-linear possibilities of indecomposable weight patterns, one may apply the algorithm of Theorem 8 to compute their connected principal isotropy subgroups type (H^0_Ψ). If some of them turn out to be trivial, then one may again rule them out.

(iii) After the above two steps of elimination, there are altogether only 15 remaining possibilities of non-linear weight patterns that cannot be eliminated solely by means of weights. For these 15 cases, we proceed to study their orbit structure in detail and then try to examine cohomologically whether it is indeed possible to build an acyclic cohomology manifold with those specific orbit structures. So far, the only undecided cases are the possibilities stated in Theorem 9'.

Proof of Theorem 9' for $G = A_n$:

In the case $G = A_n$, we usually parametrize the Cartan subalgebra by $(n+1)$ coordinates $(\theta_1, \theta_2, \ldots, \theta_{n+1})$ with the relation $\theta_1 + \theta_2 + \ldots + \theta_{n+1} = 0$. Then, its Weyl group $W(A_n)$ acts as the full permutation group of the $(n+1)$ coordinates and every weight vector is an integral linear combination of $\{\theta_j\}$. Suppose $\Omega'(\Psi)$ is an indecomposable system of weights. Since $\Omega'(\Psi)$ is invariant under the Weyl group W, we can write $\Omega'(\Psi)$ as the sum of orbits of W as follows

$$\Omega'(\Psi) = W(\pm\omega_1) + W(\pm\omega_2) + \ldots \quad .$$

We may assume that $W(\pm\omega_1)$ is one of the orbits with the largest cardinality and ω_1 lies in the Weyl chamber. Namely

$$\omega_1 = a_1\theta_1 + \ldots + a_k\theta_k \ , \qquad a_1 \geqslant \ldots \geqslant a_k$$

Furthermore, since the topological weights only concern about their perpendicular hyperplanes, we may assume that

$$(a_1, \ldots, a_k) = 1 \quad \text{and} \quad a_1 \geqslant |a_k| \quad .$$

By lemma 2, we may assume that $\Omega'(\Psi) \cap \Delta(G) = \phi$, for otherwise, $\Omega'(\Psi) \supseteq \Delta(G)$ and then $(H^0{}_\Psi)$ is either trivial or equal to (T). Suppose $a_1 > 1$ and $k \neq (n+1)$. Then $(\omega_1, (\theta_1 - \theta_{k+1})) \neq 0$ and it follows from lemma 1 that there exists $\omega_1' \in \Omega'(\Psi)$ such that

$$\omega_1' = (\theta_1 - \theta_{k+1}) + \ell \cdot \omega_1$$

for a suitable integer ℓ. For most cases, for example, if $k \leqslant \left[\frac{n+1}{2}\right]$, the cardinality of $W(\omega_1')$ is larger than that of $W(\omega_1)$, which is a contradiction to the choice of ω_1. Hence, either $W(\omega_1)$ already consists of too many weights which makes $\Omega'(\Psi)$ impossible to satisfy condition (*'), or $a_1 = |a_2| = \ldots = |a_k| = 1$. Therefore $(H^0_\psi) \neq \{id\}$ implies that $a_1 = |a_2| = \ldots = |a_k| = 1$.

(i) If $a_1 = a_2 = \ldots = a_k = 1$, then one may assume that $k \leqslant \left[\frac{n+1}{2}\right]$. (by using the relation $\theta_1 + \theta_2 + \ldots + \theta_{n+1} = 0$). Notice that the weight system of real basic representations are as follows:

$$\Omega'(\varphi_k + \varphi_k^*) = W\{\pm(\theta_1 + \ldots + \theta_k)\} \quad \text{if} \quad k < \left[\frac{n+1}{2}\right]$$

$$\Omega'(2\varphi_k) = W\{\pm(\theta_1 + \ldots + \theta_k)\} \quad \text{if} \quad k = \frac{n+1}{2} \not\equiv 0 \pmod 2$$

$$\Omega'(\varphi_k) = W\{(\theta_1 + \ldots + \theta_k)\} \quad \text{if} \quad k = \frac{n+1}{2} \equiv 0 \pmod 2.$$

Hence, it follows from Theorem 8, Corollary 3 of Theorem 8 and I of Table A, that $(H^0_\Psi) \neq \{id\}$ only when $\Omega'(\Psi)$ is, in fact, the same as one of those listed in Table A-I with possibly the following exception:

$$G = A_5 \quad \text{and} \quad \Omega'(\Psi) = W\cdot\{(\theta_1+\theta_2+\theta_3)\} + m\cdot\{\pm\theta_i\} \ ; \ (m = 0 \text{ or } 1).$$

However, in either cases of $m = 0$ or 1, a detail computation of their orbit structures will imply that the homogeneous space $\frac{SU(6)}{SU(3) \times SU(3)}$ has the same rational cohomology as that of S^{19}, which is a contradiction. Hence $\Omega' = W\{(\theta_1+\theta_2+\theta_3)\} + m\{\pm\theta_i\}$, $m = 0, 1$ are, in fact, not admissible for A_5.

(ii) The remaining possible cases are $\omega_1 = (\theta_1 + \ldots + \theta_j - \theta_{j+1} \ldots \theta_k)$. Since $\Omega'(\Psi) \cap \Delta(G) = \phi$, we may assume that $k > 2$ and $j \geqslant \frac{k}{2}$. Again, it is not difficult to show that $W\cdot(\pm\omega_1)$ consists of too many weights which makes $\Omega'(\Psi)$ impossible to satisfy condition (*').

Proof of Theorem 9' for $G = B_n, C_n, D_n$ and exceptional Lie groups:

The proof of Theorem 9' for simple Lie groups other than A_n is essentially the same as that of A_n-case. The first step is to use condition * and Lemma 1 to reduce the possible candidates of indecomposable weight system $\Omega'(\Psi)$ with non-trivial connected principal isotropy subgroup type $(H^0_\Psi) \neq \{id\}$ to a handful distinguished ones. Among the few remaining candidates of weight patterns, there are the following three kinds:

(i) Those weight patterns that can be realized by linear actions, then it follows directly from Corollary 1 of Theorem 8 that their connected principal orbit types (H^0_Ψ) are the same of those of the corresponding linear actions.

(ii) Those weight patterns which cannot be realized by linear actions, however, the algorithm of Theorem 8, applying to them, will yield a trivial connected principal isotropy subgroups type. Hence, as far as the proof of Theorem 9' is concerned, they will not cause any trouble even if some of them turn out to be admissible.

(iii) Finally, there remains the following possible candidates of weight patterns which are non-linear and will yield non-trivial connected principal isotropy subgroup type, i.e. $(H^0_\Psi) \neq \{id\}$, if some of them happen to be admissible.

(1) $G = SU(6)$, $\Omega' = \{(\theta_i+\theta_j+\theta_k)\} + m\cdot\{\pm\theta_i\}$, $m = 0, 1$

(2) $G = Spin(11)$, $\Omega' = \{\frac{1}{2}(\pm\theta_1\pm\ldots\pm\theta_5)\} + m\cdot\{\pm\theta_i\}$, $m = 0,1,2,3$

(3) $G = Spin(13)$, $\Omega' = \{\frac{1}{2}(\pm\theta_1\pm\ldots\pm\theta_6)\} + m\cdot\{\pm\theta_i\}$, $m = 0,1$

(4) $G = Sp(3)$, $\Omega' = \{(\pm\theta_1\pm\theta_2\pm\theta_3)\} + (m+1)\ \{\pm\theta_i\}$, $m = 0,1$

(5) $G = Spin(12)$, $\Omega' = \Omega'(\varphi_5) + m\,\{\pm\theta_i\}$ $0 \leqslant m \leqslant 4$

(iv) In the above five types of non-linear weight patterns with m = 0 it is not difficult to determine the cohomological aspect of the detail "orbit structure" if some of them happen to be admissible. For example, in the case $m = 0$, the principal orbit types are respectively the following, which are, in fact, topologically contradictory to the assumption that X is acyclic.

(1) $G = SU(6)$, $\Omega' = \{(\theta_i+\theta_l+\theta_k)\} \Rightarrow (H^0_\Psi) = (SU(3) \times SU(3))$, $F(G) = F(T)$ is acyclic and $\dim \frac{SU(6)}{SU(3)\times SU(3)} = 19 = \dim X - \dim F(G) - 1$, which implies that $\frac{SU(6)}{SU(3)\times SU(3)}$ is a rational cohomology sphere, a contradiction.

(2) $G = Spin(11)$, $\Omega' = \{\frac{1}{2}(\pm\theta_1\pm\ldots\pm\theta_5)\} \Rightarrow H^0_\Psi = SU(5)$, $F(G) = F(T)$ is acyclic and $\dim \left(\frac{Spin(11)}{SU(5)}\right) = 31 = \dim X - \dim F(G) - 1$, which implies that $\frac{Spin(11)}{SU(5)}$ is a rational cohomology sphere, again a contradiction.

(3) $G = Spin(13)$, $\Omega' = \{\frac{1}{2}(\pm\theta_1\pm\ldots\pm\theta_6)\} \Rightarrow$ there exists an orbit of the type $Spin(13)/SU(6)$ and the weight system of the $SU(6)$ action on the slice has $\Omega'(S_x) = \{(\theta_i+\theta_j+\theta_k)\}$, which is proved in (1) to be impossible. Hence $\{\frac{1}{2}(\pm\theta_1\pm\ldots\pm\theta_6)\}$ is not Spin(13)-admissible.

(4) $G = S_p(3)$ $\Omega' = \{(\pm\theta_1\pm\theta_2\pm\theta_3)\} + \{\pm\theta_i\} \Rightarrow H^0_\Psi = SU(3)$, $F(G) = F(T)$ is acyclic, and $\dim \frac{Sp(3)}{SU(3)} = 13 = \dim X - \dim F(G) - 1$ which again implies that $\frac{Sp(3)}{SU(3)}$ is a rational cohomology sphere, an obvious contradiction.

(5) Spin(12), $\Omega' = \Omega'(\varphi_5) \Rightarrow H^0_\Psi = SU(6)$, $F(G) = F(T)$ is acyclic and $\dim \frac{Spin(12)}{SU(6)} = 31 = \dim X - \dim F(G) - 1$ which again is impossible because $\frac{Spin(12)}{SU(6)}$ is not a rational cohomology sphere.

(v) More detail but basically the same method will show that the above five types of non-linear weight patterns with $m = 1$ are not admissible either. Hence, the only remaining undecided cases are the following:

$$G = Spin(11),\ \Omega' = \{\tfrac{1}{2}(\pm\theta_1 \pm \dots \pm\theta_5)\} + m\cdot\{\pm\theta_i\} \quad m = 2 \text{ or } 3$$

$$G = Spin(12),\ \Omega' = \Omega'(\varphi_5) + m\cdot\{\pm\theta_i\} \quad 2 \leqslant m \leqslant 4 .$$

The proof of Theorem 9' is thus complete.

It follows from the above Theorem 9' and Corollary 2 of Theorem 8, we have the following classification Theorem.

Theorem 9: Let G be a simple compact connected Lie group and X be a given topological G-action on an acyclic cohomology manifold X. If the connected principal orbit type of Ψ, (H^0_Ψ), is non-trivial in the sense $H^0_\Psi \neq \{id\}$, then there exists a unique linear G-action φ with the same weight system, i.e., $\Omega(\Psi) = \Omega(\varphi)$, and the same connected principal orbit type, i.e., $(H^0_\Psi) = (H^0_\varphi)$, except possibly the following undecided cases (if they happen to be admissible):

(i) $G = Spin(11)$, $\Omega'(\Psi) = \{\tfrac{1}{2}(\pm\theta_1 \pm \dots \pm\theta_5)\} + m\{\pm\theta_i\}$, $2 \leqslant m \leqslant 3$

(ii) $G = Spin(12)$, $\Omega'(\Psi) = \Omega'(\varphi_5) + m\{\pm\theta_i\}$ $2 \leqslant m \leqslant 4$.

Conjecture: The above two possibilities are not admissible.

§7. Classification of connected principal orbit types for actions of (general) compact connected Lie groups on acyclic cohomology manifolds

Let G be a (general) compact connected Lie group and $\mathfrak{g}$ be the Lie algebra of G. It follows from a well-known structural theorem for compact Lie algebras that $\mathfrak{g}$ decomposes uniquely into the direct sum of its center $\mathfrak{g}_0$ and its simple normal factors $\mathfrak{g}_1, \mathfrak{g}_2, \ldots, \mathfrak{g}_\ell$, namely

$$\mathfrak{g} = \mathfrak{g}_0 \oplus \mathfrak{g}_1 \oplus \ldots \oplus \mathfrak{g}_\ell, \quad (\mathfrak{g}_0 \text{ may be trivial}).$$

Hence, there exists a suitable finite covering group $\tilde{G}$ of G such that

$$\tilde{G} = G_0 \times G_1 \times \ldots \times G_\ell, \quad (G_0 \text{ may be trivial, i.e., } G_0 = \{id\})$$

where G_0 is a torus group and $G_1, \ldots, G_\ell$ are simple compact Lie groups with $\mathfrak{g}_1, \ldots, \mathfrak{g}_\ell$ as their Lie algebra respectively. Hence, in the study of connected principal orbit types, (H^0_Ψ), we may assume without loss of generality that G is itself a product of its connected center G_0 and its simple normal factors $G_1, \ldots, G_\ell$, i.e.,

$$G = G_0 \times G_1 \times \ldots \times G_\ell .$$

(A) Several reductions:

(i) Suppose that the connected center of G, G_0, is non-trivial, i.e., G is non-semi-simple, and Ψ is an almost effective G-action on an acyclic cohomology manifold X. Let $\Omega_0 = \Omega(\Psi|G_0)$ be the weight system of the restriction of Ψ to G_0 and X_0, X_ω, $\omega \in \Omega'_0$ be the following subspaces:

$$X_0 = F(G_0, X), \quad X_\omega = F(G_0^\omega, X)$$

where G_0^ω is the kernel of $\omega \in \Omega_0'$. Then X_0, X_ω are obviously acyclic cohomology submanifolds of X invariant under G and moreover, the original G-action Ψ is, to a large extent, determined by those restricted G-action on X_0, and X_ω, $\omega \in \Omega_0'$ respectively. For example, if $\dim X_0 = 0$, then

$$\Omega(\Psi) = \sum_{\omega \in \Omega_0'} \Omega(\Psi | X_\omega)$$

and it follows directly from Corollary 2 of Theorem 8 that

$$(H^0_\Psi) = \dot{\bigcap} \{H_\omega ; \ \omega \in \Omega_0'\}$$

where (H_ω) is the connected principal orbit type of $\Psi | X_\omega$.

The general situation, i.e., $\dim X_0 > 0$, can also be taken care of by the following generalization of Corollary 2 of Theorem 8.

<u>Lemma</u>: Let G be a given compact connected Lie group and Ψ, Ψ_1, Ψ_2 be topological G-actions on pairs of acyclic cohomology manifolds (X,Y), (X_1,Y_1) and (X_2,Y_2) respectively (i.e., Y, Y_1, Y_2 are invariant acyclic submanifolds of X, X_1, X_2 respectively). If their respective systems of non-zero weights satisfy the following equations:

$$\Omega'(\Psi | Y) = \Omega'(\Psi_1 | Y_1) = \Omega'(\Psi_2 | Y_2)$$

and

$$\Omega'(\Psi) - \Omega'(\Psi | Y) = \{\Omega'(\Psi_1) - \Omega'(\Psi_1 | Y_1)\} + \{\Omega'(\Psi_2) - \Omega'(\Psi_2 | Y_2)\}$$

then their corresponding connected principal orbit types have the following relationship: (let (K) be the connected principal orbit type of $\Psi|Y$, or $\Psi_1|Y_1$, of $\Psi_2|Y_2$ which are, by Corollary 1 of Theorem 8, the same)

$$(H^0{}_{\Psi}) = (H^0{}_{\Psi_1}) \overset{(K)}{\cap} (H^0{}_{\Psi_2})$$

where $\overset{(K)}{\cap}$ means intersection in general position in K.

Proof: By the above assumption, there are points y, y_1, y_2 in Y, Y_1, and Y_2 respectively such that

$$G^0_y = G^0_{y_1} = G^0_{y_2} = K.$$

Let $\Psi_g, \Psi_{y_1}, \Psi_{y_2}$ be the K-action on the slices S_y, S_{y_1}, S_{y_2}, at y, y_1, y_2 respectively. Then, it follows from the above equation that $\Omega'(\Psi_y) = \Omega'(\Psi_{y_1}) + \Omega'(\Psi_{y_2})$. Hence, the above Lemma follows from Corollary 2 of Theorem 8. Now, suppose $(H_0), (H_\omega)$ be the connected principal orbit types of $(\Psi|X_0)$ and $(\Psi|X_\omega)$ respectively. Then

$$(H^0{}_{\Psi}) = \overset{(H_0)}{\cap} \{(H_\omega); \omega \in \Omega_0'\}.$$

Hence, as far as the principal orbit type is concerned, one may reduce the general cases to the cases $\dim (G_0) \leq 1$.

(ii) Suppose the connected principal isotropy subgroups of Ψ, $(H^0{}_{\Psi})$, are contained in a normal subgroup $K \subseteq G$, i.e.,

$$H^0{}_{\Psi} \subseteq K \subseteq G \text{ , and } K \text{ is normal.}$$

Then the connected principal isotropy subgroups of the restriction $\Psi|K$ are the same as that of Ψ, namely

$$\{H^0_{(\Psi|K)}\} = \{H^0_\Psi\} .$$

Hence, in the study of principal isotropy subgroups types of topological G-actions, one may reduce to the study of those cases whose connected principal isotropy subgroups, (H^0_Ψ), are not contained in any proper normal subgroups of G. Namely

$$"H^0_\Psi \subseteq K \subseteq G \quad \text{and} \quad K \quad \text{normal} \Rightarrow K = G".$$

(iii) Finally, in view of Corollary 2 of Theorem 8, one may assume that the system of non-zero weights, $\Omega'(\Psi)$, is indecomposable. Furthermore, in case that the group G is non-simple, it follows easily from the definition that a splitting weight system is automatically decomposable. Hence, an indecomposable system of weights is necessary non-splittable.

(B) The case $G = G_1 \times G_2$ is the product of two simple Lie groups.

Let $G = G_1 \times G_2$ be the product of two simple Lie groups G_1, G_2 and $\mathfrak{h} = \mathfrak{h}_1 \oplus \mathfrak{h}_2$ be a Cartan subalgebra of G and $\mathfrak{h}_1, \mathfrak{h}_2$ be respectively Cartan subalgebras of G_1, G_2. It is clear that the classification of possibilities of connected principal orbit types for general topological G-actions on acyclic manifolds can be reduced to the classification of connected principal orbit types for those G-actions with indecomposable system of weights, $\Omega'(\Psi)$, and non-splitting connected principal isotropy subgroups (H^0_Ψ).

Theorem 10': Let $G = G_1 \times G_2$ be the product of two simple Lie groups G_1, G_2 and Ψ be a topological G-action on an acyclic cohomology manifold X. If the weight system of Ψ, $\Omega'(\Psi)$, is indecomposable and the connected principal isotropy subgroup of Ψ,

(H^0_Ψ), is not contained in G_1 or G_2, then either

(i) $G = SU(n) \times SU(m)$, $\Omega'(\psi) = \Omega'([\mu_n \otimes_{\mathbb{C}} \nu_m]_{\mathbb{R}}), (H^0_\psi) = SU(n-m) \times T^{(m-1)}$;

or

(ii) $G = Sp(n) \times Sp(m)$, $\Omega'(\psi) = \Omega'(\nu_n \otimes_{\mathbb{Q}} \nu_m), (H^0_\psi) = Sp(n-m) \times Sp(1)^m$,

where μ_n, ν_n are the standard representation of $SU(n)$, $Sp(n)$ on $\mathbb{C}^n$ and $\mathbb{Q}^n$ (quaternion n-space) respectively.

As a straightforward consequence of Theorem 10', Corollary 2 of Theorem 8 and Proposition 2 of §3, we have the following classification theorem for the possibilities of principal orbit types of topological G-action on acyclic manifolds where G is the product of two simple Lie groups.

<u>Theorem 10</u>: Let $G = G_1 \times G_2$ be the product of two simple Lie groups G_1, G_2 and Ψ be a topological G-action on an acyclic cohomology manifold. If the connected principal isotropy subgroups type of Ψ, (H^0_Ψ), is <u>non-trivial</u>, then there are only the following possibilities:

(i) The weight system $\Omega'(\Psi)$ is splitting, i.e.

$$\Omega'(\Psi) = \Omega'(\Psi|G_1) + \Omega'(\Psi|G_2)$$

hence it follows from Proposition 2 of §3 that

$$(H^0_\Psi) = (H^0_\Psi \cap G_1) \times (H^0_\Psi \cap G_2) = (H^0_{\Psi|G_1}) \times (H^0_{\Psi|G_2})$$

which was classified in Theorem 9.

(ii) $H^0_\Psi \subseteq G_1$ or $H^0_\Psi \subseteq G_2$, then

$$(H^0_\Psi) = (H^0_{\Psi|G_1}) \quad \text{or} \quad (H^0_\Psi) = (H^0_{\Psi|G_2})$$

which was classified in Theorem 9.

(iii) $G = SU(n) \times SU(m)$, (Resp. $Sp(n) \times Sp(m)$) and

$$\Omega'(\Psi) = \Omega'([\mu_n \otimes_{\mathbb{C}} \mu_m]_{\mathbb{R}}), \quad (\text{resp. } \Omega'(\Psi) = \Omega'(\nu_n \otimes_Q \nu_m))$$

$$(H^0_\Psi) = (SU(n-m) \times T^{m-1}), \ (\text{resp.} (H^0_\Psi) = (Sp(n-m) \times Sp(1)^m)).$$

Proof of Theorem 10':

For convenience, we shall assume that $rk(G_1) \geqslant rk(G_2)$ and write the weight system $\Omega'(\Psi)$ as the sum of the following three parts

$$\Omega'(\Psi) = \Omega_1 + \Omega_2 + \tilde{\Omega}$$

where Ω_1, Ω_2 are the subset of those weights line in $\mathfrak{h}_1$ and $\mathfrak{h}_2$ respectively and $\tilde{\Omega}$ is the subset of those weights of mixed form. Since $\Omega'(\Psi)$ is assumed to be indecomposable, it must be also non-splittable, namely, $\tilde{\Omega} \neq \phi$. Suppose $\omega_1 + \omega_2 \in \tilde{\Omega}$ is such a mixed weight, $\omega_1 \in \mathfrak{h}_1$ and $\omega_2 \in \mathfrak{h}_2$. Then the whole "orbit" of $(\omega_1 + \omega_2)$ are also weights in $\tilde{\Omega}$, namely

$$W(G)\cdot(\omega_1 + \omega_2) = \{(\sigma_1(\omega_1) + \sigma_2(\omega_2)),$$

$$\sigma_1 \in W(G_1), \ \sigma_2 \in W(G_2)\} \subseteq \tilde{\Omega} \quad .$$

We claim that except the two possibilities mentioned in Theorem 10', (i.e., $G = SU(n) \times SU(m)$, $\Omega'(\Psi) = \tilde{\Omega} = \Omega'([\mu_n \otimes_{\mathbb{C}} \mu_m]_{\mathbb{R}})$ and $G = Sp(n) \times Sp(m)$, $\Omega'(\psi) = \tilde{\Omega} = \Omega'(\nu_n \otimes_Q \nu_m)$), the connected principal isotropy subgroups of ψ, H^0_Ψ, lie in G_1 and hence contradicts to the assumption that H^0_Ψ is not contained in G_1 or G_2. A detail proof of the above assertion is rather tedious and it seems to be inevitable to do somewhat case by case checking. However, in principle, it is simply a straightforward application of the algorithm of Theorem 8.

Notice that, for a given simple Lie group, there are only a few distinguished orbits (under the action of Weyl group) whose number of weights is not larger than the number of positive roots. Hence, except a few particularly simple cases which can easily be checked by the algorithm of Theorem 8, $\tilde{\Omega}$ contains at least an orbit $W(G)\cdot\{(\omega_1 + \omega_2)\}$ such that either the number of weights in $\{W(G_1)\cdot\omega_1\}$ is more than that of positive roots of G_1 or the number of weights in $\{W(G_2)\cdot\omega_2\}$ is more than that of positive roots of G_2. Let us show, as a typical example, that, in the later case, $H^0{}_\psi$ must be contained in G_1. Suppose $k = rk(G_2)$. Then it is possible to choose the proceeding $2k$ weights of the algorithm of Theorem 8 among those weights in $\{W(G)\cdot(\omega_1 + \omega_2)\}$ as follows: $\alpha_i \in W(G_1)$, $\beta_i, \beta_i' \in W(G_2)$

$$\gamma_1 = \alpha_1(\omega_1) + \beta_1(\omega_2), \quad \gamma_1' = \alpha_1(\omega_1) + \beta_1'(\omega_2)$$

$$\gamma_2 = \alpha_2(\omega_1) + \beta_2(\omega_2), \quad \gamma_2' = \alpha_2(\omega_1) + \beta_2'(\omega_2)$$

- - - - - - - - - - - - - - - - - - - -

$$\gamma_k = \alpha_k(\omega_1) + \beta_k(\omega_2), \quad \gamma_k' = \alpha_k(\omega_1) + \beta_k'(\omega_2)$$

satisfying the following conditions:

$$\{\alpha_1(\omega_1), \alpha_2(\omega_1), \ldots, \alpha_k(\omega_1)\} \text{ linearly independent}$$

$$\{\beta_1(\omega_2) - \beta_1'(\omega_2), \beta_2(\omega_2) - \beta_2'(\omega_2), \ldots, \beta_k(\omega_2) - \beta_k'(\omega_2)\}$$

also linearly independent. Then, it is not difficult to see that the maximal torus of $H^0{}_\psi$ determined by the algorithm $S \subseteq H^0{}_\psi$ is contained in $T_1 \subseteq G_1$, namely,

$$S \subseteq \gamma_1^{\perp} \cap \gamma_1'^{\perp} \cap \ldots \cap \gamma_k^{\perp} \cap \gamma_k'^{\perp} \subseteq (\beta_1(\omega_2) - \beta_1'(\omega_2))^{\perp} \cap \ldots \cap (\beta_k(\omega_2) - \beta_k'(\omega_2))^{\perp}$$

$$= \mathfrak{h}_2^{\perp} = \mathfrak{h}_1 .$$

We leave the detail proof of Theorem 10' to the reader.

(C) <u>The case that G is a general semi-simple compact Lie group</u>:

For the case that G is a general semi-simple compact Lie group, we state the following generalization of Theorem 10' without proof. In fact, its proof is a slight modification of that of Theorem 10'.

<u>Theorem 10"</u>: Let G be a semi-simple compact connected Lie group and Ψ be an almost effective topological G-action on an acyclic cohomology manifold. If the weight system of Ψ, $\Omega'(\Psi)$, is indecomposable and the connected principal isotropy subgroups of Φ, H^0_{Ψ}, are <u>not</u> contained in any <u>proper</u> normal subgroups of G, then there are only the following two possibilities:

(i) $G = SU(n) \times SU(m)$, $\Omega'(\Psi) = \Omega'([\mu_n \otimes_{\mathbb{C}} \mu_m]_{\mathbb{R}})$, $(H^0_{\Psi}) = (SU(n-m) \times T^{m-1})$

(ii) $G = Sp(n) \times Sp(m)$, $\Omega'(\Psi) = \Omega'(\nu_n \otimes_{\mathbb{Q}} \nu_m)$, $(H^0_{\Psi}) = (Sp(n-m) \times Sp(1)^m)$.

<u>Remark</u>: With the above neat strong theorem for indecommposable weight system and Corollary 2 of Theorem 8, it is not difficult to write down the complete classification of principal orbit types of all possible G-actions on acyclic manifolds for a <u>given</u> compact connected semi-simple Lie group G. However, such a statement for all compact connected semi-simple Lie groups <u>in general</u> is not very neat and seems unnecessary to state it as a theorem.

§8. Concluding remarks

Since the principal orbit type is the dominanting orbit type, the classification results of §6 and §7 are rather useful in the study of other geometric behaviors of topological actions of compact connected Lie groups. However, we shall wait until the next paper to give a more systematic account of such applications. In concluding this paper, we would like to remark on two immediate applications of the results of this paper:

(A) Local Theorems:

Almost all the theorems and lemmas proved in this paper for global G-actions on acyclic cohomology manifolds can be localized to get their corresponding local theorems for the action of G_x on the slice at x, S_x. For example, we state the localized version of Theorem 2, Theorem 5 and Theorem 8 respectively as follows:

Theorem $\overline{2}$: Let M^m be a cohomology manifold of dimension m with a given effective topological G-action. If there exists a point $x \in M$ such that $G_x \cong \mathrm{Spin}(k)$, then the dimension of M is bounded below by the following estimate:

$$m = \dim M \geqslant \begin{cases} \{\dim(G) - \dim(\mathrm{Spin}(k)\} + 2^{\left[\frac{k}{2}\right]} & \text{if } k \not\equiv 0 \pmod 4 \\ \{\dim(G) - \dim(\mathrm{Spin}(k)\} + 2^{\left[\frac{k}{2}\right]} + k & \text{if } k \equiv 0 \pmod 4). \end{cases}$$

Theorem $\overline{5}$: Let M be a cohomology manifold with a given topological G-action and $x_0 \in M$ be a point with $G^0_{x_0} \cong SO(n)$ (resp. $SU(n)$, $Sp(n)$). If the weight system of the $G^0_{x_0}$-action on a slice at x_0, S_{x_0}, is as follows:

$$\Omega('\Psi_{x_0}) = k\cdot\{\pm\theta_i\} ,$$

then there exists a neighborhood of the orbit $G(x_0)$ such that all connected isotropy subgroups, G_x^0, of the neighboring points are conjugate to the standard SO(j) (resp. SU(j), Sp(j)), $j \leqslant n$.

Theorem $\overline{8}$: Suppose M is a connected cohomology manifold with a given topological G-action Ψ, and G_x is the isotropy subgroup of $x \in M$. If the local weight system of the G_x-action on a slice at x is $\Omega'(\Psi_x)$, then the connected principal orbit type of Ψ, $(H^0{}_\Psi)$, can be computed by the algorithm of Theorem 8. (Cf. Theorem 8 of §5.)

(B) Differentiable actions on manifolds M with $H^*(M,\mathbb{Q}) \cong H^*(S^m,\mathbb{Q})$.

Let M be a differentiable G-manifold with $H^*(M,\mathbb{Q}) \cong H^*(S^m,\mathbb{Q})$. Then the cone over M, CM, is naturally a topological G-space which is obviously an acyclic cohomology manifold. Hence, the results of this paper apply directly to obtain interesting theorems which are mostly new. Hence, the idea of geometric weight system for topological actions not only open up new approach to a systematic investigation of topological actions, it also provides new results as well as better proofs of many interesting theorems for differentiable actions. Of course, in the case of differentiable actions on homology spheres, it is usually possible to sharpen the results by further using the differentiability. For example, it is not difficult to improve Theorem 5 for differentiable actions on $\mathbb{Z}$-homology spheres so that the isotropy subgroups themselves, G_x, are conjugate to SO(j) (resp. SU(j), Sp(j)). [See 14, p. 745-750 for a technique to prove such an improvement.]

(C): It is also possible to define p-weights to topological actions by using p-tori and p-primary subgroups. They can be used to detect the p-components of G_x/G_x^0 . However, they are usually not as important as the torus-weight in the study of actions of <u>connected</u> compact Lie groups. On the other hand, if one is interested in the study of actions of complicated finite groups, then such p-weights are obviously of importance.

REFERENCES

[1] R. Bing, A homeomorphism between the 3-sphere and the sum of two solid horned spheres, Ann. of Math. 56 (1952), pp. 354-362.

[2] S. Bochner, Compact groups of differentiable transformations, Ann. of Math. 46 (1945), pp. 372-381.

[3] A. Borel, Fixed point of elementary commutative groups, Bull. of A.M.S. 65 (1959), pp. 322-326.

[4] A. Borel, Sous-groupes commutatifs et torsion des groupes de Lie compacts connexes, Tohoku Math. J. 13 (1961), pp. 216-240.

[5] A. Borel et al., Seminar on transformation groups, Ann. of Math. Studies 46, Princeton University Press (1961).

[6] A. Borel and J. de Sibenthal, Sur les sous-groupes fermes de rang maximum des groupes de Lie compacts connexes, Comm. Math. Helv. 23 (1949), pp. 200-221.

[7] A. Borel and J. P. Serre, Sur certain sous-groupes de Lie compacts, Comm. Math. Helv. 27 (1953), pp. 128-139.

[8] G. Bredon, Exotic actions on spheres, Proc. of the Conference on Transf. groups, New Orleans (1967), Springer (1968).

[9] P. Conner and D. Montgomery, An example for SO(3) action, Proc. of the Nat. Acad. of Sci. U.S.A. 48 (1962), pp. 1918-1922.

[10] P. Conner and E. Floyd, On the construction of periodic maps without fixed points, Proc. of A.M.S. 10 (1959), pp. 354-360.

[11] E. Floyd, Examples of fixed point sets of periodic maps I, Ann. of Math. 55 (1952), pp. 167-171, II, Ann. of Math. 64, pp. 396-398.

[12] E. Floyd and R. Richardson, An action of a finite group on an n-cell without stationary points, Bull. of A.M.S. 65 (1959), pp. 73-76.

[13] W. C. Hsiang and W. Y. Hsiang, Classification of differentiable actions on S^n, R^n, and D^n with S^k as the principal orbit type, Ann. of Math. 82 (1965), pp. 421-433.

[14] W. C. Hsiang and W. Y. Hsiang, Differentiable actions of compact connected classical groups I, Amer. J. Math. 89 (1967), pp. 705-786.

[15] W. C. Hsiang and W. Y. Hsiang, Differentiable actions of compact connected classical groups II, (to appear in Ann. of Math.) (mimeo at Chicago University 1968).

[16] W. C. Hsiany and W. Y. Hsiang, Differentiable actions of compact connected Lie groups III (to appear).

[17] W. Y. Hsiang, On the principal orbit type and P.A. Smith theory of SU(p) actions, Topology 6 (1967), pp. 125-135.

[18] W. Y. Hsiang, On the geometric weight system of differentiable actions I, (to appear).

[19] W. Y. Hsiang, On the degree of symmetry and the structure of highly symmetric manifolds, (to appear).

[20] W. Y. Hsiang, A survey on regularity theorems in differentiable compact transformation groups, Proc. of the Conference on Transf. Groups, New Orleands (1967), Springer-Verlag (1968).

[21] D. Montgomery, H. Samelson and C. T. Yang, Exceptional orbits of highest dimension, Ann. of Math. 64 (1956), pp. 131-141.

[22] D. Montgomery and C. T. Yang, The existence of a slice, Ann. of Math. 64 (1957), pp. 108-116.

[23] D. Montgomery and C. T. Yang, Orbits of highest dimension, Trans. A.M.S. 87 (1958), pp. 284-293.

[24] G. D. Mostow, Equivariant embedding in euclidean spaces, Ann. of Math. 65 (1957), pp. 432-446.

[25] G. D. Mostow, On a conjecture of Montgomery, Ann. of Math. 65 (1957), pp. 513-516.

[26] P. A. Smith, Fixed point theorems for periodic transformations, Amer. J. Math. 63 (1941), pp. 1-8.

[27] P. A. Smith, Fixed points of periodic transformations, App. B in Lefschetz's Algebraic Topology, 1942.

[28] M. Richardson and P. A. Smith, Periodic transformations on complexes, Ann. of Math. 39 (1938), pp. 611-633.

[29] N. Steenrod, The topology of fibre bundles, Princeton University Press, 1951.

[30] H. C. Wang, Homogeneous spaces with non-vanishing Euler characteristics, Ann. of Math. 50 (1949), pp. 925-953.

[31] H. Weyl, The classical groups, Princeton University Press, 1939.

[32] M. Kramer, Uber das verhalten endlicher Untergruppen bei Darstellungen kompaktes Liegruppen, (Dissertation, Bonn).

EQUIVARIANT SINGULAR HOMOLOGY AND COHOMOLOGY FOR ACTIONS OF COMPACT LIE GROUPS

Sören Illman

Princeton University

Introduction

This article constitutes a slightly extended version of my talk at the Amherst conference. It is a summary of the author's thesis [5]. I am grateful to my adviser, Professor William Browder, for his encouragement and interest in my work.

Let G be a compact Lie group. By a G-space we mean a topological space together with a left G-action. We have the category of all G-pairs and G-maps between G-pairs. Our main purpose is to construct an equivariant singular homology and cohomology theory with coefficients in an arbitrary given covariant coefficient system and contravariant coefficient system respectively on the category of all G-pairs and G-maps. Our construction is such that G besides being an arbitrary compact Lie group also can be a discrete group or an abelian locally compact group. For actions by discrete groups equivariant homology and cohomology theories of this type exist before, see G. Bredon [1], [2] and Th. Bröcker [3].

1. Equivariant singular theory

In this section G denotes a good locally compact group, by which we mean that G is either a compact Lie group, a discrete group, or an abelian locally compact group. Let R be a ring with unit. By an R-module we mean a left R-module.

Definition 1.1. A covariant coefficient system k for G, over the ring R, is a covariant functor from the category of G-spaces of the form G/H, where H is a closed subgroup (not fixed) of G, and G-homotopy classes of G-maps, to the category of R-modules.

A contravariant coefficient system ℓ is defined by the contravariant version of the above definition.

Theorem 1.2. Let G be a good locally compact group and k a covariant coefficient system for G over the ring R. There exists an equivariant homology theory $H_*^G(\ ;k)$, defined on the category of all G-pairs and G-maps, which satisfies all seven equivariant Eilenberg-Steenrod axioms and which has the given coefficient system k as coefficients.

Thus, if H is a closed subgroup of G we have

$$H_m^G(G/H;\, k) = 0 \qquad \text{for } m \neq 0$$

and there exists an isomorphism

$$\gamma\colon H_0^G(G/H;\, k) \xrightarrow{\cong} k(G/H)$$

which commutes with homomorphisms induced by G-maps $\alpha\colon G/H \longrightarrow G/K$. The meaning of the rest of Theorem 1.2 is clear. Let us point out that the excision axiom is satisfied in the following strong sense. An inclusion of the form

$$i\colon (X - U,\ A - U) \longrightarrow (X, A)$$

where $\overline{U} \subset A^{o}$ (U and A are G-subsets of the G-space X) induces isomorphisms

$$i_*\colon H_n^G(X\text{-}U, A\text{-}U;\, k) \xrightarrow{\cong} H_n^G(X, A;\, k)$$

for all n.

Theorem 1.3. Let G be a good locally compact group and ℓ a contravariant coefficient system for G over the ring R. Then there exists an equivariant cohomology theory $H_G^*(\ ;\ell)$, defined on the category of all G-pairs and G-maps, which satisfies all seven equivariant Eilenberg-Steenrod axioms and which has the given coefficient system ℓ as coefficients.

Construction of equivariant singular homology

Let Δ_n be the standard n-simplex, that is $\Delta_n = \left\{(x_0, \ldots, x_n) \in R^{n+1} \mid \sum_{i=0}^{n} x_i = 1,\ x_i \geq 0\right\}$. We consider Δ_m, $0 \leq m \leq n$, as a subset of Δ_n through the imbedding of

Δ_m into Δ_n which is given by $(x_0, \ldots, x_m) \mapsto (x_0, \ldots, x_m, 0, \ldots, 0)$.

Definition 1.4. Let $K_0, \ldots, K_n$ be a sequence of closed subgroups of G, such that $K_0 \supset K_1 \supset \cdots \supset K_n$. We define the standard equivariant n-simplex of type $(K_0, \ldots, K_n)$, denoted by

$$(\Delta_n; K_0, \ldots, K_n)$$

to be the G-space constructed in the following way. Consider the G-space $\Delta_n \times G$ and define a relation $\sim$ in $\Delta_n \times G$ as follows: $(x, g) \sim (x, g') \Leftrightarrow gK_m = g'K_m \in G/K_m$, for $x \in \Delta_m - \Delta_{m-1}$. Thus $\sim$ is an equivalence relation in $\Delta_n \times G$, and we define

$$(\Delta_n; K_0, \ldots, K_n) = (\Delta_n \times G)/\sim .$$

We denote by $p: \Delta_n \times G \longrightarrow (\Delta_n; K_0, \ldots, K_n)$ the natural projection and by $[x, g] \in (\Delta_n; K_0, \ldots, K_n)$ the image of $(x, g) \in \Delta_n \times G$ under this projection. The group G acts on $(\Delta_n; K_0, \ldots, K_n)$ by $(\bar{g}, [x, g]) \mapsto [x, \bar{g}g]$, $\bar{g} \in G$. Since G is locally compact, it follows that $\mathrm{id} \times p: G \times (\Delta_n \times G) \longrightarrow G \times (\Delta_n; K_0, \ldots, K_n)$ is a quotient map and hence the action by G on $(\Delta_n; K_0, \ldots, K_n)$ is continuous.

It is easy to show that $(\Delta_n; K_0, \ldots, K_n)$ is Hausdorff. The projection $p: \Delta_n \times G \longrightarrow (\Delta_n; K_0, \ldots, K_n)$ is not open in general and $(\Delta_n; K_0, \ldots, K_n)$ need not be a locally compact space. If G is a compact Lie group then $(\Delta_n; K_0, \ldots, K_n)$ is of course a compact Hausdorff space. If $K_0 = \cdots = K_n = K$ then $(\Delta_n; K_0, \ldots, K_n) = \Delta_n \times G/K$.

The orbit space of the G-space $(\Delta_n; K_0, \ldots, K_n)$ is Δ_n, and if $x \in \Delta_m - \Delta_{m-1}$ then the orbit over x is G/K_m. Let $\pi: (\Delta_n; K_0, \ldots, K_n) \longrightarrow \Delta_n$ be the projection onto the orbit space.

Denote by $e^i: \Delta_{n-1} \longrightarrow \Delta_n$, $0 \leq i \leq n$, the face map defined by $e^i(x_0, \ldots, x_{n-1}) = (x_0, \ldots, x_{i-1}, 0, x_i, \ldots, x_{n-1})$. We define equivariant face maps

$$\bar{e}^i: (\Delta_{n-1}; K_0, \ldots, \hat{K}_i, \ldots, K_n) \longrightarrow (\Delta_n; K_0, \ldots, K_n)$$

by $\bar{e}^i([x, g]) = [e^i(x), g]$. Each $\bar{e}^i$ is a G-homeomorphism onto its image and $\bar{e}^i$

covers e^i.

The following consideration and simple lemma play an essential role in our construction. Consider the equivariant n-simplexes $(\Delta_n; K_0, \ldots, K_n)$ and $(\Delta_n; K'_0, \ldots, K'_n)$ and let

$$h: (\Delta_n; K_0, \ldots, K_n) \longrightarrow (\Delta_n; K'_0, \ldots, K'_n)$$

be a G-map which covers $id: \Delta_n \longrightarrow \Delta_n$. Let $x \in \Delta_m - \Delta_{m-1} \subset \Delta_n$ and restrict h to the orbits over x. This gives us a G-map

$$h_x: G/K_m \longrightarrow G/K'_m .$$

Lemma 1.5. Let k be a covariant coefficient system for G and let h be as above. Then h determines for each m, $0 \leq m \leq n$, a unique homomorphism

$$(h_m)_*: k(G/K_m) \longrightarrow k(G/K'_m),$$

and we have $(h_m)_* = (h_x)_*$ for any $x \in \Delta_m - \Delta_{m-1}$. Moreover, for any m and q, such that $0 \leq q \leq m \leq n$, the diagram

$$\begin{array}{ccc} k(G/K_m) & \xrightarrow{(h_m)_*} & k(G/K'_m) \\ \downarrow p_* & & \downarrow p'_* \\ k(G/K_q) & \xrightarrow{(h_q)_*} & k(G/K'_q) \end{array}$$

commutes. Here $p: G/K_m \longrightarrow G/K_q$ is the natural projection, that is $p(gK_m) = gK_q$, and correspondingly for p'.

If h is a G-homeomorphism then $(h_m)_*$ is an isomorphism and we have $(h_m)_*^{-1} = \left((h^{-1})_m\right)_*$. The corresponding contravariant version is valid.

Proof. Let $x \in \Delta_m - \Delta_{m-1}$ and $y \in \Delta_q - \Delta_{q-1}$, where $0 \leq q \leq m \leq n$. It is easy to show that the diagram

$$\begin{array}{ccc} G/K_m & \xrightarrow{h_x} & G/K'_m \\ \downarrow p & & \downarrow p' \\ G/K_q & \xrightarrow{h_y} & G/K'_q \end{array}$$

is G-homotopy commutative. The lemma follows from this.

Definition 1.6. Let X be a G-space. A G-map

$$T: (\Delta_n; K_0, \dots, K_n) \longrightarrow X$$

is called an equivariant singular n-simplex of type $(K_0, \dots, K_n)$ in X. We call K_n the main type of T and denote

$$t(T) = K_n$$

The equivariant singular (n-1)-simplex of type $(K_0, \dots, \hat{K}_i, \dots, K_n)$

$$T^{(i)} = T\bar{e}_n^{\,i}: (\Delta_{n-1}; K_0, \dots, \hat{K}_i, \dots, K_n) \longrightarrow X$$

is called the i:th face of T, i = 0, ..., n.

Observe that $t(T^{(i)}) = t(T) = K_n$ for $i = 0, \dots, n-1$ and $t(T^{(n)}) = K_{n-1}$.

Given an equivariant singular n-simplex

$$T: (\Delta_n; K_0, \dots, K_n) \longrightarrow X$$

we form

$$Z_T \otimes k(G/t(T)) = Z_T \otimes k(G/K_n).$$

Here Z_T denotes the infinite cyclic group on the generator, T, and the tensor product is over the integers. The R-module structure on (k(G/t(T)) makes $Z_T \otimes k(G/t(T))$ into an R-module, isomorphic with k(G/t(T)).

Definition 1.7. We define

$$\hat{C}_n^G(X; k) = \sum_T \oplus\, k(G/t(T))$$

where the direct sum is over all equivariant singular n-simplexes in X.

The boundary homomorphism

$$\hat{\partial}_n: \hat{C}_n^G(X; k) \longrightarrow \hat{C}_{n-1}^G(X, k)$$

is defined as follows. Let T be an equivariant singular n-simplex and $a \in k(G/t(T))$. Then we define

$$\hat{\partial}_n(T \otimes a) = \sum_{i=0}^{n} (-1)^i T^{(i)} \otimes (p_i)_*(a).$$

Here $(p_i)_*: k(G/t(T)) \longrightarrow k(G/t(T^{(i)}))$, $i = 0, \ldots, n$, is the homomorphism induced by the natural projection $p_i: G/t(T) \longrightarrow G/t(T^{(i)})$. Thus we have

$$\hat{\partial}_n(T \otimes a) = \sum_{i=0}^{n-1} (-1)^i T^{(i)} \otimes a + (-1)^n T^{(n)} \otimes (p_n)_*(a).$$

This defines the homomorphism $\hat{\partial}_n$.

It is immediately seen that $\hat{\partial}_{n-1} \hat{\partial}_n = 0$. Thus we have the chain complex

$$\hat{S}^G(X; k) = \{\hat{C}_n^G(X; k), \hat{\partial}_n\}.$$

Our main interest is not in $\hat{S}^G(X; k)$, but in a quotient of $\hat{S}^G(X, k)$. We shall define this quotient now.

Let $T: (\Delta_n; K_0, \ldots, K_n) \longrightarrow X$, $T': (\Delta_n; K'_0, \ldots, K'_n) \longrightarrow X$ be equivariant singular n-simplexes in X, and $a \in k(G/K_n)$, $a' \in k(G/K'_n)$. We define $T \otimes a \sim T' \otimes a' \Longleftrightarrow$ there exists a G-homeomorphism $h: (\Delta_n; K_0, \ldots, K_n) \longrightarrow (\Delta_n; K'_0, \ldots, K'_n)$ which covers $id: \Delta_n \longrightarrow \Delta_n$, such that $T = T'h$ and $(h_n)_*(a) = a'$. Here $(h_n)_*: k(G/K_n) \longrightarrow k(G/K'_n)$ is as described in Lemma 1.5.

Definition 1.8. Let

$$\overline{C}_n^G(X; k) \subset \hat{C}_n^G(X; k)$$

be the submodule of $\hat{C}_n^G(X; k)$ consisting of all elements of the form

$$\sum_{i=1}^{s} (T_i \otimes a_i - T'_i \otimes a'_i)$$

where $T_i \otimes a_i \sim T'_i \otimes a'_i$ for $i = 1, \ldots, s$.

We then define

$$C_n^G(X, k) = \hat{C}_n^G(X; k) / \overline{C}_n^G(X; k).$$

Lemma 1.9. The boundary homomorphism $\hat{\partial}_n$ induces

$$\partial_n: C_n^G(X, k) \longrightarrow C_{n-1}^G(X, k).$$

Proof. This is easily verified using Lemma 1.5.

Since $\hat{\partial}_{n-1}\hat{\partial}_n = 0$ it follows that $\partial_{n-1}\partial_n = 0$. Thus we have the chain complex

$$S^G(X, k) = \{C_n^G(X, k), \partial_n\}.$$

Definition 1.10. We define

$H_n^G(X, k)$ = n:th homology of the chain complex $S^G(X; k)$.

The relative groups $H_n^G(X,A; k)$ for a G-pair (X,A),

the boundary $\partial: H_n^G(X,A; k) \longrightarrow H_{n-1}^G(A; k)$, and

induced homomorphisms $f_*: H_n^G(X,A; k) \longrightarrow H_n^G(Y,B; k)$

by a G-map $f: (X,A,) \longrightarrow (Y,B)$, are now defined in a standard way.

The homotopy and excision axioms are proved by imitating the proofs of the homotopy and excision axioms for ordinary singular homology given in Eilenberg-Steenrod [4]. The details are too long to be given here. Complete details can be found in [5].

The dimension axiom

Let H be a closed subgroup of G. We wish to show that

$$H_m^G(G/H; k) \cong \begin{cases} k(G/H) & m = 0 \\ 0 & m \neq 0 \end{cases}$$

Define $\hat{C}_n^G \operatorname{Iso}(G/H; k)$ to be the submodule of $\hat{C}_n^G(G/H; k)$ generated by all elements of the form $V \otimes a$, where the equivariant singular n-simplex V is of the type

$$V: (\Delta_n; K, \ldots, K) = \Delta_n \times G/K \longrightarrow G/H$$

and moreover V is such that the restriction

$$V|: \{x\} \times G/K \longrightarrow G/H$$

is a G-homeomorphism for every $x \in \Delta_n$. As usual $a \in k(G/t(V))$.

We have the R-module $C_n^G \operatorname{Iso}(G/H; k)$ and the chain complex $S^G \operatorname{Iso}(G/H; k) = \{C_n^G(G/H; k), \partial_n\}$. It is not difficult to show that

$$H_m(S^G \operatorname{Iso}(G/H; k)) \cong \begin{cases} k(G/H) & m = 0 \\ 0 & m \neq 0 \end{cases}$$

The main part of the proof of the dimension axiom now consists of showing that the inclusion

$$\eta: S^G \mathrm{Iso}(G/H; k) \longrightarrow S^G(G/H; k)$$

is a chain homotopy equivalence. It is in proving this that we at one point use the assumption that G is a good locally compact group. Use is made of the Covering homotopy theorem by Palais, see [6], Theorem 2.4.1. on page 51, in the case G is a compact Lie group. The details are quite long and can be found in [5].

Construction of equivariant singular cohomology

To construct equivariant singular cohomology we take the "dual" in an appropriate sense of the chain complex which gave us equivariant singular homology. Let k_0 be the covariant coefficient system for which $k_0(G/H) = Z$ (the integers) for each closed subgroup H of G and all the induced homomorphisms are the identity on Z. We denote

$$\hat{S}^G(X; k_0) = \hat{S}^G(X) = \{\hat{C}^G_n(X), \hat{\partial}_n\}$$

Thus $\hat{C}^G_n(X)$ is the free abelian group on all equivariant singular n-simplexes in X. Now let ℓ be an arbitrary contravariant coefficient system for G over the ring R. Denote

$$L = \sum_H \otimes\, \ell(G/H)$$

where the direct sum is over all closed subgroups H of G.

Definition 1.11. We define

$$\hat{C}^n_G(X; \ell) = \mathrm{Hom}_t(\hat{C}^G_n(X), L)$$

Here $\mathrm{Hom}_t(\hat{C}^G_n(X), L)$ consists of all homomorphisms of abelian groups

$$c: \hat{C}^G_n(X) = \sum_T \otimes Z_T \longrightarrow \sum_H \otimes\, \ell(G/H) = L$$

which satisfy the condition

$$c(T) \in \ell(G/t(T))$$

for every equivariant singular n-simplex T in X. The R-module structure in L makes $\hat{C}^n_G(X; \ell)$ into an R-module.

The coboundary homomorphism

$$\hat{\delta}^{n-1}: \hat{C}^{n-1}_G(X; \ell) \longrightarrow \hat{C}^n_G(X; \ell)$$

is defined as follows. Let $c \in \hat{C}^{n-1}_G(X; \ell)$ then δc is defined by

$$(\delta c)(T) = \sum_{i=0}^{n} (-1)^i (p_i)^* c(T^{(i)})$$

where $(p_i)^*: \ell(G/t(T^{(i)})) \longrightarrow \ell(G/t(T))$ is induced by the natural projection $p_i: G/g(T) \longrightarrow G/t(T^{(i)})$. Then $\hat{\delta}^n \hat{\delta}^{n-1} = 0$ and we have the cochain complex

$$\hat{S}^*_G(X; \ell) = \{\hat{C}^n_G(X; \ell), \hat{\delta}^n\}.$$

Our main interest is in a subcomplex of $\hat{S}^*_G(X; \ell)$. We define this subcomplex now.

Definition 1.12. Let

$$C^n_G(X; \ell) \subset \hat{C}^n_G(X; \ell)$$

be the submodule of $\hat{C}^n_G(X; \ell)$ consisting of all the homomorphisms $c \in \mathrm{Hom}_t(\hat{C}^G_n(X), L)$ which satisfy the following condition.

Let $T': (\Delta_n; K'_0, \ldots, K'_n) \longrightarrow X$ be an equivariant singular n-simplex in X and let $h: (\Delta_n; K_0, \ldots, K_n) \longrightarrow (\Delta_n; K'_0, \ldots, K'_n)$ be a G-homeomorphism which covers $\mathrm{id}: \Delta_n \longrightarrow \Delta_n$. Denote $T = T'h$. Then

$$c(T) = (h_n)^* c(T') \in \ell(G/K_n).$$

Here $(h_n)^*: \ell(G/K'_n) \longrightarrow \ell(G/K_n)$ is as described in Lemma 1.5.

Lemma 1.13. The coboundary homomorphism $\hat{\delta}^n$ restricts to

$$\delta^n: C^n_G(X; \ell) \longrightarrow C^{n+1}_G(X; \ell).$$

Proof. Use Lemma 1.5.

Thus $\delta^n \delta^{n-1} = 0$, and we have the cochain complex

$$S^*_G(X; \ell) = \{C^n_G(X; \ell), \delta^n\}.$$

Definition 1.14. We define

$$H^n_G(X; \mathcal{L}) = \text{n:th homology of the cochain complex } S^*_G(X; \mathcal{L}).$$

For the remaining details we again refer to [5], where also constructions of a transfer homomorphism, a "Kronecker index," and a cup-product in cohomology are given.

2. Equivariant CW complexes

In this section G denotes a compact Lie group. The definition of an equivariant CW complex is obtained from the definition of an ordinary CW complex simply by instead of adjoining cells E^n by a map from S^{n-1} one adjoins G-spaces of the form $E^n \times G/H$, where $H \subset G$ is some closed subgroup (not fixed) of G, by a G-map from $S^{n-1} \times G/H$. The precise definitions are as follows.

Definition 2.1. Let X be a Hausdorff G-space and A a closed G-subset of X, and n a non-negative integer. We say that X is obtainable from A by adjoining equivariant n-cells if the following is true. There exists a collection $\{c^n_j\}_{j\in J}$ of closed G-subsets of X such that

1. $X = A \cup (\bigcup_{j\in J} c^n_j)$, and X has the topology coherent with $\{A, c^n_j\}_{j\in J}$.

2. Denote $\dot{c}^n_j = c^n_j \cap A$, then

$$(c^n_j - \dot{c}^n_j) \cap (c^n_i - \dot{c}^n_i) = \emptyset \qquad \text{if } j \neq i.$$

3. For each $j \in J$ there exists a closed subgroup H_j of G and a G-map

$$f_j: (E^n \times G/H_j, S^{n-1} \times G/H_j) \longrightarrow (c^n_j, \dot{c}^n_j)$$

such that $f_j(E^n \times G/H_j) = c^n_j$, and f_j maps $E^n \times G/H_j - S^{n-1} \times G/H_j$ homeomorphically onto $c^n_j - \dot{c}^n_j$.

Definition 2.2. An equivariant relative CW complex (X, A) consists of a Hausdorff G-space X, a closed G-subset A of X, and an increasing filtration of X by closed G-subsets $(X,A)^k$ $k = 0, 1, \ldots,$ such that the following conditions are satisfied.

1. $(X,A)^0$ is obtainable from A by adjoining equivariant 0-cells, and for $k \geq 1$ $(X,A)^k$ is obtainable from $(X,A)^{k-1}$ by adjoining equivariant k-cells.

2. $X = \bigcup_{k=0} (X,A)^k$, and X has the topology coherent with $\{(X,A)^k\}_{k \geq 0}$.

The closed G-subset $(X,A)^k$ is called the k-skeleton of (X,A). If $A = \emptyset$ we call X an equivariant CW complex and denote the k-skeleton by X^k.

Let G' be another compact Lie group. Let the G-pair (X,A) be a G-equivariant relative CW complex and the G'-pair (Y,B) be a G'-equivariant relative CW complex. Assume that both X and Y are locally compact or that one of them, say X, is compact. Then the $G \times G'$-pair $(X,A) \times (Y,B)$ is a $G \times G'$-equivariant relative CW complex.

The Propositions 2.3-2.5 below are proved in the same way as the corresponding results for ordinary CW-complexes.

Proposition 2.3. Let (X,A) be an equivariant relative CW complex. Then (X,A) has the G-homotopy extension property.

Let $\varphi: G \longrightarrow G'$ be a continuous homomorphism, and let X be a G-space and Y a G'-space. We call a map $f: X \longrightarrow Y$ a φ-map if $f(gx) = \varphi(g)f(x)$, for every $g \in G$ and $x \in X$.

Proposition 2.4. Assume that X is a G-equivariant CW complex and Y is a G'-equivariant CW complex. Then any φ-map $f: X \longrightarrow Y$ is φ-homotopic to a skeletal φ-map $f: X \longrightarrow Y$. If the skeletal φ-maps $\bar{f}_1, \bar{f}_2: X \longrightarrow Y$ are φ-homotopic there exists a skeletal φ-homotopy from $\bar{f}_1$ to $\bar{f}_2$.

Let the G-space X be a locally compact equivariant CW complex. Then $X \times X$ is a $G \times G$-equivariant CW complex and the diagonal $d: X \longrightarrow X \times X$ is a φ-map with $\varphi: G \longrightarrow G \times G$ the diagonal. Thus the diagonal d is φ-homotopic to a skeletal φ-map.

Proposition 2.5. Let X and Y be G-equivariant CW complexes. Then a G-map $f: X \longrightarrow Y$ is a G-homotopy equivalence if and only if for each closed subgroup H of G the restriction $f^H: X^H \longrightarrow Y^H$ induces a one-to-one correspondence between the path components of X^H and Y^H, and isomorphisms $f^H_*: \pi_k(X^H, x) \longrightarrow \pi_k(Y^H, f(x))$, for all $k \geq 1$ and every $x \in X^H$.

Here X^H denotes the fixed point set of X. Our main result about equivariant CW complexes is:

Theorem 2.6. Every differentiable G-manifold M is an equivariant CW complex.

By a theorem of C. T. Yang, see [7], the orbit space $G\backslash M$ can be triangulated. Moreover the triangulation is such that all points in the interior of a simplex belong to the same orbit type. Take the first barycentric subdivision of this triangulation. We claim that the G-space over an n-simplex in this "new" triangulation is G-homeomorphic with some standard equivariant n-simplex $(\Delta_n; K_0, \ldots, K_n)$. Our proof of this is somewhat long. It makes repeated use of the differentiable slice theorem and of the covering homotopy theorem by Palais, [6], Theorem 2.4.1. This gives us a stronger result than the fact that M is an equivariant CW complex. The details are given in [5].

Now Proposition 2.5 gives a necessary and sufficient condition for a G-map between differentiable G-manifolds to be a G-homotopy equivalence.

If M is a compact differentiable G-manifold then M is a finite equivariant CW complex. This fact can be used to give a partially new proof of the result by Atiyah and Segal that $K^*_G(M)$ is a finitely generated R(G)-module.

An equivariant CW complex X is called finite dimensional if $X = X^m$ for some m.

Theorem 2.7. Let X be a finite dimensional equivariant CW complex. Then the n:th homology of the chain complex

$$\cdots \longleftarrow H^G \;(X^{n-1}, X^{n-2}; k) \xleftarrow{\partial} H^G_n(X^n, X^{n-1}; k) \longleftarrow \cdots$$

is isomorphic to $H_n^G(X; k)$.

The corresponding cohomology version is valid.

Corollary 2.8. Let M be a compact differentiable G-manifold. Assume that k is a finitely generated covariant coefficient system (i.e.each $k(G/H)$ is f.g.) over a noetherian ring R. Then $H_m^G(M; k) = 0$ for $m > \dim M$ and each $H_n^G(M; k)$ is a finitely generated R-module.

The analogous result for cohomology is true.

REFERENCES

[1] G. Bredon, Equivariant cohomology theories, Bull. Amer. Math. Soc. 73 (1967), 269-273.

[2] __________, Equivariant cohomology theories, Lecture Notes in Mathematics, Vol. 34, Springer-Verlag, 1967.

[3] Th. Bröcker, Singuläre Definition der Äquivarianten Bredon Homologie, Manuscripta Matematica 5 (1971), 91-102.

[4] S. Silenberg and N. Steenrod, Foundations of Algebraic Topology, Princeton University Press, 1952.

[5] S. Illman, Equivariant singular homology and cohomology for actions of compact Lie groups, Thesis, Princeton University, 1971.

[6] R. Palais, The classification of G-spaces, Memoirs of Amer. Math. Soc. 36 (1960).

[7] C. T. Yang, The triangulability of the orbit space of a differentiable transformation group, Bull. Amer. Math. Soc. 69 (1963), 405-408.

CYCLIC BRANCHED COVERS AND O(n)-MANIFOLDS

Louis Kauffman*
University of Illinois
at Chicago Circle

I) INTRODUCTION

This article is a second look at standard O(n)-actions. We show that some Brieskorn varieties have a natural interpretation as pullbacks and hence may be constructed non-algebraically. This is closely tied with the structure of these varieties as branched coverings of spheres, branching over other Brieskorn varieties.

In the case where the O(n)-manifold has a one-dimensional fixed point set and orbit space D^4, we show how the equivariant classification problem is connected with the three-dimensional problem of finding the symmetries of a link. In the case of torus links this involves Brieskorn examples once again.

We conclude with a calculation of the number of equivariant diffeomorphism classes of O(n)-manifolds corresponding to a given torus link.

The author would like to thank Glen Bredon for kindly pointing out his independent discovery of the pullback constructions.

II) ALGEBRAIC VARIETIES

Let $\mathbb{C}^m$ denote complex m-space, $f(z) = f(z_1,\ldots,z_m)$ a polynomial in m complex variables. Let $V = V(f) = \{z \in \mathbb{C}^m | f(z) = 0\}$. Milnor studied the topology of V in the neighborhood of a point $x \in \mathbb{C}^m$. Let $S_\varepsilon = S_\varepsilon^{2m-1}$ be a small sphere centered at x. Consider $K = V \cap S_\varepsilon$. Let $\phi : S_\varepsilon - K \longrightarrow S^1$, $\phi(z) = f(z)/|f(z)|$. Then ϕ is the projection map of a smooth fiber bundle [see 4].

*Supported in part by NSF Grant No. GP 28487.

The point x is <u>singular</u> if all partial derivatives $\frac{\partial f}{\partial z_i}$ vanish at x. A point is said to be an isolated singularity if it has a neighborhood in which all other points are non-singular. When x is an isolated singularity, K is itself a smooth manifold.

Given f with an isolated singularity at $0 = (0,\ldots,0)$, we wish to study $F = x^k+f(z)$ regarded as a polynomial in $(m+1)$ complex variables. Let $\mathbb{V} = \{(x,z) \in \mathbb{C}^{m+1} | F(x,z) = 0\}$, $\mathbb{K} = \mathbb{V} \cap S_\varepsilon^{2m+1}$. We wish to show that $\mathbb{K}$ is a branched covering space of S_ε^{2m-1} with branch set K. Actually, the argument given here will only cover the case of weighted homogeneous polynomials. The result may be obtained for a general $f(z)$ by a similar but slightly more involved argument.

The polynomial $f(z)$ is said to be weighted homogeneous of type $(w_1,\ldots,w_m)$ if it can be expressed as a linear combination of monomials $z_1^{i_1}\cdots z_m^{i_m}$ for which $i_1/w_1 + i_2/w_2 + \cdots + i_m/w_m = 1$, where $w_1,\ldots,w_m$ are positive rational numbers. Thus the Brieskorn polynomials $z_1^{a_1} + z_2^{a_2} + \cdots + z_m^{a_m}$ are weighted homogeneous of type $(a_1,\ldots,a_n)$.

Given f weighted homogeneous of type $(w_1,\ldots,w_m)$ define $\rho*z = (\rho^{1/w_1}z_1,\ldots,\rho^{1/w_m}z_m)$ for ρ real and positive. Clearly $f(\rho*z) = \rho f(z)$.

<u>Proposition</u>. Let $f(z) = f(z_1,\ldots,z_m)$ be weighted homogeneous and suppose that f has an isolated singularity at $0 \in \mathbb{C}^m$. Let $F(x,z) = x^k+f(z)$, $K = V(f) \cap S_\varepsilon^{2m+1}$, $\mathbb{K} = V(F) \cap S_\varepsilon^{2m+1}$. Define $p: \mathbb{K} \longrightarrow S_\varepsilon^{2m-1}$ via $p(x,z) = \rho*z$ such that $|\rho*z| = \varepsilon$. Then p is a k-fold branched covering with branch set K.

<u>Proof</u>: First note that given $(x,z) \in \mathbb{K}$, $|x|^2+|z|^2 = \varepsilon^2$ whence $|z| \leq \varepsilon$. Since $|\rho*z|$ is strictly increasing as a function of ρ, there is a unique $\rho > 0$ such that $|\rho*z| = \varepsilon$. Hence p is well-defined. (Also $z \neq 0$ since $z = 0 \Rightarrow f(z) = 0 \Rightarrow x = 0$.)

Suppose $p(x,z) = p(x',z')$, $(x,z),(x',z') \in \mathbb{K}$. Then $\rho * z = \rho' * z'$ for some $\rho, \rho' > 0$. This implies that $z' = \rho'' * z$ and may assume $\rho'' \geq 1$ (otherwise choose $z = \rho'' * z'$). Now $x'^k = f(z') = f(\rho'' * z) = \rho'' f(z)$ and $x^k = f(z)$. Thus $x'^k = \rho'' x^k$. $\varepsilon^2 = |x'|^2 + |z'|^2 = |\rho''^{1/k} x|^2 + |\rho'' * z|^2 \geq |x|^2 + |z|^2 = \varepsilon^2$. Hence $\rho'' = 1$ and therefore $z' = z$, $x^k = x'^k$. Thus $p(x,z) = p(x',z') \Rightarrow z' = z$, $x^k = x'^k$.

Given $z \in S_\varepsilon^{2m-1}$ there is certainly a unique ρ, $1 \geq \rho \geq 0$ such that $|\rho f(z)|^2 + |\rho * z|^2 = \varepsilon^2$. Then $p^{-1}(z) = \{(x, \rho * z) \mid x^k + f(\rho * z) = 0\}$. Hence $p^{-1}(z)$ consists of k distinct points except when $z \in K$ and then $p^{-1}(z) = \{(0,z)\}$.

<u>Corollary</u>. Let $\phi : S_\varepsilon^{2m-1} - K \longrightarrow S^1$ be the fibration $\phi(z) = f(z)/|f(z)|$. $\hat{\phi} : \mathbb{K} - K \longrightarrow S^1$, $\hat{\phi}(x,z) = x/|x|$, where $K \equiv \{(0,z) \in \mathbb{K}\}$. $\lambda_k : S^1 \longrightarrow S^1$, $\lambda_k(\mu) = -\mu^k$. Then the following diagram commutes:

$$\begin{array}{ccc} \mathbb{K} - K & \xrightarrow{\ p\ } & S_\varepsilon^{2m-1} - K \\ \Big\downarrow \hat{\phi} & & \Big\downarrow \phi \\ S^1 & \xrightarrow{\ \lambda_k\ } & S^1 \end{array}$$

Hence $\mathbb{K} - K \xrightarrow{\hat{\phi}} S^1$ is also a smooth fibration.

<u>Proof</u>:
$$\begin{aligned} \lambda_k \hat{\phi}(x,z) &= \lambda_k(x/|x|) = -x^k/|x^k| \\ &= f(z)/|f(z)| = f(\rho * z)/|f(\rho * z)| \\ &= \phi p(x,z) \end{aligned}$$
$$\therefore \lambda_k \hat{\phi} = \phi p.$$

<u>Corollary</u>. Let $\mathbb{K}_d = V(x^d + f(z)) \cap S_\varepsilon^{2m+1}$. $p_d : \mathbb{K}_d \longrightarrow S_\varepsilon^{2m-1}$ as above. Suppose $k > 1$ and define $p : \mathbb{K}_{kd} \longrightarrow \mathbb{K}_d$ by $p(x,z) = (\rho^{1/d} x^k, \rho * z)$ where ρ is chosen so that $|p(x,z)| = \varepsilon$. Then p is a k-fold

branched cover with branch set $K \subset \mathbb{K}_d$. The following diagram commutes

$$\begin{array}{ccc} \mathbb{K}_{kd} & \xrightarrow{\ p\ } & \mathbb{K}_d \\ {\scriptstyle p_{kd}}\searrow & & \swarrow{\scriptstyle p_d} \\ & S^{2m-1}_{\varepsilon} & \end{array}$$

Proof: Clear.

III) O(n)-ACTIONS

A) Recall the outline of Jänich's classification theory for group actions [see 3]. Roughly, one has a manifold X with $O(n)$ acting on it such that all isotropy groups are conjugate to either $O(n-1)$ or $O(n-2)$. Thus, in the case where the orbit space is D^2, the interior points correspond to orbits of type $O(n)/O(n-2)$ and the boundary points to orbits of type $O(n)/O(n-1)$. Inasmuch as $X \xrightarrow{\pi} D^2$ can be viewed as a pasting together of the bundle of orbits over D^2 and the bundle of orbits over $S^1 = \partial D^2$, the "pasting data" is given by a certain reduction of structural group. This turns out to correspond to a map $\sigma: \partial D^2 \longrightarrow S^1$. Under appropriate conditions $O(n)$-manifolds with two orbit types and orbit space D^2 are classified by degree (σ) and, ignoring orientation, by $|\deg(\sigma)|$.

If, along with orbits homeomorphic to $O(n)/O(n-2)$, $O(n)/O(n-1)$ one allows fixed points, then under appropriate conditions the classification can be reduced to the above two-orbit case by removing a tubular neighborhood of the fixed point set. For example, if the orbit space is D^4 with fixed points corresponding to a knot $K \subset \partial D^4 = S^3$ then it turns out that sufficient pasting data is given by a map $\sigma: S^3 - N(K) \longrightarrow S^1$ where $N(K)$ = tubular neighborhood of K in S^3 and σ is of degree ± 1 when restricted to a meridian circle on the boundary of $N(K)$.

The technical conditions under which the above remarks hold are restrictions on the slice representations, the representations of the isotropy group normal to an orbit. One must require that the representations of $O(n)$, $O(n-1)$, $O(n-2)$ are respectively, one-dimensional trivial direct sum two copies of the standard representation of $O(n)$, three-dimensional trivial direct sum one standard representation of $O(n-1)$, four-dimensional trivial. Manifolds satisfying these conditions for either two or three orbit types will be referred to as $O(n)$-manifolds.

B) Consider the case of $O(n)$-manifolds with orbit types $O(n-1)$, $O(n-2)$ and orbit space D^2. The simplest example is $S^{2n-1} = \{(x,y) \in \mathbb{R}^n \times \mathbb{R}^n \mid |x|^2+|y|^2 = 1\}$. $O(n)$ acts via $g(x,y) = (gx,gy)$. Regard D^2 as the disk of radius 1 in the complex plane and define $\lambda_d : D^2 \longrightarrow D^2$ by $\lambda_d(z) = z^d$, $d \in \mathbb{Z}$, $d \neq 0$. Let X_d be the pullback

$$\begin{array}{ccc} X_d & \longrightarrow & S^{2n-1} \\ \downarrow & & \downarrow \pi \\ D^2 & \xrightarrow{\lambda_d} & D^2 \end{array}$$

$$X_d = \{(z,(x,y)) \in D^2 \times S^{2n-1} \mid z^d = \pi(x,y)\}$$

$$G \times X_d \longrightarrow X_d$$

$$g,(z,(x,y)) \longmapsto (z,(gx,gy)).$$

<u>Proposition</u>. X_d has pasting data $\sigma_d : \partial D^2 \longrightarrow S^1$ of degree d.

<u>Proof</u>: X_d is an $O(n)$-manifold. In fact, the construction shows that it is a cyclic branched cover of S^{2n-1} branching along the Stiefel manifold $\pi^{-1}(0) \approx O(n)/O(n-2)$. The pasting data for S^{2n-1} is a map $\sigma_1 : \partial D^2 \longrightarrow S^1$ of degree 1. It is easily seen, by looking at the source of σ_1 as a certain bundle cross section, that the corresponding cross section σ_d for X_n is given by $\sigma_d = \lambda_d \circ \sigma_1$. This certainly is of degree d.

<u>Corollary</u>. The set of manifolds $\{X_d \mid d \in \mathbb{Z} \setminus \{0\}\}$ are representatives of the distinct equivariant diffeom classes of $O(n)$-manifolds with the

above orbit types and orbit space D^2.

Proof: By the classification theory these manifolds are distinguished by degree (σ).

Let $K_d = V(z_0^d+z_1^2+z_2^2+\cdots+z_n^2) \cap S^{2n+1}$. Regard $z_j = x_j+iy_j$, $x = (x_j)$, $y = (y_j) \in \mathbb{R}^n$,
$K_d = \{(z_0,x,y) \mid z_0^d+\langle x,x\rangle-\langle y,y\rangle+2i\langle x,y\rangle = 0\} \cap S^{2n+1}$ where $\langle \ , \ \rangle$ is the inner product on $\mathbb{R}^n$. Then $O(n)$ acts on K_d via $g(z_0,x,y) = (z_0,gx,gy)$.

Since
$$K_d: z_0^d+\langle x,x\rangle-\langle y,y\rangle+2i\langle x,y\rangle = 0$$
$$|z_0|^2 + \langle x,x\rangle + \langle y,y\rangle = 1$$
z_0 determines the orbit. If $\pi: K_d \longrightarrow \mathbb{C}$, $\pi(z_0,x,y) = z_0$ then it is easy to see that $\pi(K_d)$ is a disk of positive radius depending on d (see [2, p. 31]).

Proposition. The map $p: K_d \longrightarrow S^{2n-1}$ is $O(n)$-equivariant. Models for the orbit spaces may be chosen so that the following diagram commutes:

$$\begin{array}{ccc} K_d & \xrightarrow{\ p\ } & S^{2n-1} \\ \downarrow & & \downarrow \\ D^2 & \xrightarrow{\ \lambda_d\ } & D^2 \end{array}$$

Hence K_d is equivariantly diffeomorphic to the pullback construction X_d.

Proof: $p(z,x,y) = \rho(x,y)$ such that $|\rho(x,y)| = 1$ and is thus obviously equivariant. To see the second statement first consider

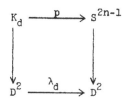

$$\begin{array}{ccc} K_d & \xrightarrow{q} & K_1 \\ \downarrow & & \downarrow \\ D^2 & \xrightarrow{\lambda_d} & D^2 \end{array}$$

The above discussion shows that this is commutative since $q(z_0,x,y) = \rho(z_0^k,x,y)$.

Now note that $K_1 \longrightarrow S^{2n-1}$ is an equivariant diffeomorphism. Since p factors $K_d \longrightarrow K_1 \longrightarrow S^{2n-1}$ we are done.

C) Now look at the three orbit type case. We wish to discuss the equivariant classification of such manifolds. Let $S_n(D^4,L)$ denote the set of unoriented equivariant diffeomorphism classes of $O(n)$-manifolds with orbit space D^4 and fixed point set $L \subset \partial D^4$; L is a link in S^3. The classification proceeds by a reduction to the two orbit type case via removal of a tubular neighborhood of the fixed point set. Suppose L has r components. Let T be a tubular neighborhood of L in S^3. Then T is a fiber bundle over L with fiber D^2. Choose a point from each component of L and let S_i^1, $i = 1,\ldots,r$ be the boundary of the fiber of T over that point. Let $\mathcal{L} = [\sigma : S^3 - L \longrightarrow S^1 \mid \deg \sigma | S_i^1 = \pm 1,\ i = 1,\ldots,r]$ where $[\ ]$ denotes homotopy classes of maps. Suppose $g \in \mathrm{Diff}(D^4,L)$. Then $\sigma \in \mathcal{L} \Rightarrow g \circ \sigma \in \mathcal{L}$. Hence $\mathrm{Diff}(D^4,L)$ acts on $\mathcal{L}$. Also $\mathbb{Z}_2$ acts on $\mathcal{L}$ via composition with $S^1 \longrightarrow S^1$ of degree -1. Jänich proves the

Theorem. $S_n(D^4,L)$ is in bijective correspondence with $\mathcal{L}/\mathbb{Z}_2 \times \mathrm{Diff}(D^4,L)$.

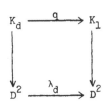

It is interesting to observe that this result has a mild reformulation which turns the classification for a given link into a purely three-dimensional problem and sometimes lets one find the

number of distinct manifolds in $S_n(D^4,L)$. The observations are:

1. By Cerf [0] any element of $\mathrm{Diff}(S^3,L)$ extends to an element of $\mathrm{Diff}(D^4,L)$.

2. Since elements of $\mathcal{L}$ are determined by their restrictions to the meridian circles S_i^1 they really correspond to the 2^r possible choices of orientation for these circles. Also, orienting a meridian circle is equivalent to choosing an orientation for the corresponding link component. Since any $g \in \mathrm{Diff}(S^3,L)$ may, for our purpose, be assumed to carry a tubular neighborhood of L to itself, an orientation for L will go to a new one under g, and the orientations on meridian circles will be correspondingly altered.

3. Hence, let $\mathcal{O}(L) = \{(\varepsilon_1,\ldots,\varepsilon_r) \mid \varepsilon_i = \pm 1\}$ be the set of possible orientations for L. $\mathbb{Z}_2$ operates on $\mathcal{O}(L)$ via $T(\varepsilon_1,\ldots,\varepsilon_r) = (-\varepsilon_1,\ldots,-\varepsilon_r)$. $\mathrm{Diff}(S^3,L)$ operates via $g(\varepsilon_1,\ldots,\varepsilon_r)$ = orientation on $g(L)$. Jänich's result becomes:

<u>Proposition</u>. $S_n(D^4,L)$ is in bijective correspondence with $\mathcal{O}(L)/\mathbb{Z}_2 \times \mathrm{Diff}(S^3,L)$.

<u>Corollary</u>. There are at most 2^{r-1} elements in $S_n(D^4,L)$. Each manifold corresponds to a given orientation for the link L and two manifolds are diffeomorphic if there is a diffeomorphism of S^3 carrying the link with first orientation to the link with second orientation.

<u>Examples</u>: a) L = two circles with linking number ± 1.

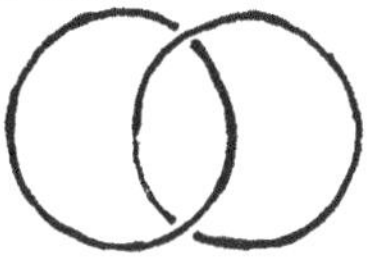

There are two possible orientations:

L_1:  L_2: 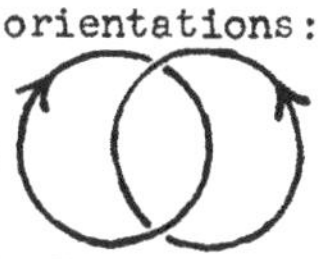

<u>Claim</u>: If $M_1, M_2 \in S_n(D^4,L)$ correspond to L_1 and L_2 respectively, then $M_1 \approx M_2$.

<u>Proof</u>: Let $g:(S^3,L) \longrightarrow (S^3,L)$ be the composition $g = g_1 g_2$ where g_2 reverses orientation of S^3, and g_1 turns the right-hand circle over. Then $g(L_1) = L_2$.

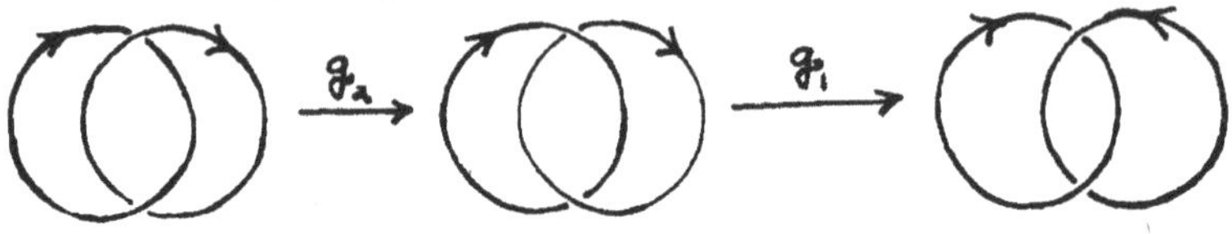

Thus there is only one element in $S_n(D^4,L)$. This may be described variously as the tangent sphere bundle to S^{n+1}, the Brieskorn variety $\mathbb{V}(x_0^2+x_1^2+\cdots x_{n+1}^2) \cap S^{2n+3}$, etc. The above provides an amusing way to prove that these seemingly different examples are really the same.

b) L = two circles linked k-times, $k > 1$.

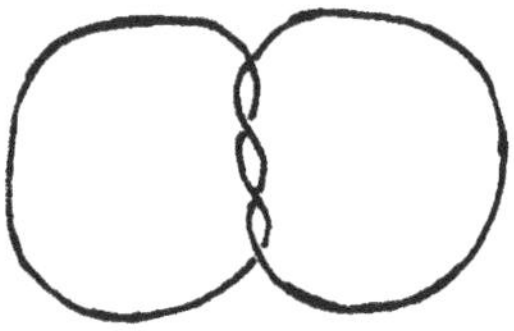

Here the two choices of orientation are in fact different as a computation of the homology of the corresponding manifolds in $S_1(S^3,L)$ shows.

<u>Remark</u>. One can go further and explicitly construct the manifold corresponding to a given oriented link by surgery following the procedure for knots in [1]. In the above case one manifold is the Brieskorn variety $\Sigma(k,2,\ldots,2)$; the other is obtained by one surgery from a sphere.

c) The simplest example is L = single circle. Its corresponding $O(n)$-manifold is S^{2n+1} where
$S^{2n+1} = \{(z,x,y) \in \mathbb{C} \times \mathbb{R}^n \times \mathbb{R}^n \mid |z|^2+|x|^2+|y|^2 = 1\}$,
$g(z,x,y) = (z,gx,gy)$. Let $\pi:S^{2n+1} \longrightarrow D^4$ be the projection to the orbit space. Consider pullback constructions for this example. Letting a,b be integers ≥ 1 and regarding
$D^4 = \{(z_1,z_2) \mid |z_1|^2+|z_2|^2 = 1, \ z_1,z_2 \in \mathbb{C}\}$ define $f:D^4 \longrightarrow D^4$ by

$f(z_1,z_2) = \rho(z_1^a,z_2^b)$ where ρ is chosen so that $|(\rho z_1^a,\rho z_2^b)| = |(z_1,z_2)|$. Let $X_{a,b}$ be the pullback

$$\begin{array}{ccc} X_{a,b} & \longrightarrow & S^{2n+1} \\ \downarrow & & \downarrow \pi \\ D^4 & \xrightarrow{f_{a,b}} & D^4 \end{array}$$

Then the fixed point set of the induced action on $X_{a,b}$ will correspond in D^4 to the inverse image of the fixed point circle under $f_{a,b}$. We can assume that the circle is embedded in S^3 as $S^1 = \{(\lambda\sqrt{2}/2,-\lambda\sqrt{2}/2) \mid |\lambda| = 1,\ \lambda \in \mathbb{C}\}$. Then $f_{a,b}^{-1}(S^1) = \{\frac{\sqrt{2}}{2}(\lambda,\mu) \mid |\lambda| = |\mu| = 1,\ \lambda^a+\mu^b = 0\}$. This is an (a,b) torus link. Letting $d = \gcd(a,b)$, $a = d\alpha$, $b = d\beta$, it consists of d torus knots of type (α,β). Each pair of knots are linked with linking number $\alpha\beta$.

<u>Proposition</u>. Let $\mathbb{K}_{a,b} = V(x^a+y^b+\Sigma_{i=1}^n z_i^2) \cap S^{2n+3}$

$p = p_1p_2 : \mathbb{K}_{a,b} \longrightarrow \mathbb{K}_{1,1}$ where

$p_2 : \mathbb{K}_{a,b} \longrightarrow \mathbb{K}_{a,1}$

$p_1 : \mathbb{K}_{a,1} \longrightarrow \mathbb{K}_{1,1}$ are the branched covering maps constructed in section II.

Then one can choose models for orbit spaces so that the following diagram commutes:

$$\begin{array}{ccc} \mathbb{K}_{a,b} & \xrightarrow{p} & \mathbb{K}_{1,1} \\ \downarrow & & \downarrow \\ D^4 & \xrightarrow{f_{a,b}} & D^4 \end{array}$$

<u>Proof</u>: As before $\mathbb{K}_{a,b}$ is an $O(n)$-manifold via action on the coordinates $z_1,\ldots,z_n$. (x,y) determines the orbit and the set of

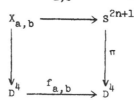

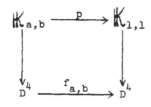

(x,y) in the image of $\pi: \mathbb{K}_{a,b} \longrightarrow \mathbb{C}^2$, $\pi(x,y,z_1,\ldots,z_n) = (x,y)$ form a 4-disk and hence may be regarded as the orbit space. The rest is clear.

Corollary. $\mathbb{K}_{a,b}$ is equivariantly diffeomorphic to $X_{a,b}$.

Proof: Since $\mathbb{K}_{1,1}$ is equivariantly diffeomorphic to S^{2n+1} we may use the map $\pi: \mathbb{K}_{1,1} \longrightarrow D^4$ in place of $S^{2n+1} \longrightarrow D^4$. The fixed point set $F \subset S^{2n+1}$ is $F = \{(x,-x,0) \mid 2|x|^2 = 1\}$, $\pi(F) = \{(\frac{\sqrt{2}}{2}\lambda, \frac{\sqrt{2}}{2}(-\lambda)) \mid |\lambda| = 1\}$. Hence the fixed point set is situated in D^4 as in the construction of $X_{a,b}$.

Remark. We could have identified $\mathbb{K}_{a,b}$ and $X_{a,b}$ by comparing the pasting data, i.e., the orientations for the links. It is clear that if $L_{a,b} = f_{a,b}^{-1}(S^1)$ then the choice of orientation for S^1 determines the orientation for $L_{a,b}$. This is equivalent to the statement that if $\tau: S^3 - S^1 \longrightarrow S^1$ is pasting data for S^{2n+1} then $f_{a,b} \circ \tau: S^3 - L_{a,b} \longrightarrow S^1$ is pasting data for $X_{a,b}$.

d) In general there are many elements in $S_n(D^4,L)$. The problem of determining just how many is closely related to what we will call the symmetries of a link [see 6]. When we wrote $\mathcal{O}(L) = \{(\varepsilon_1,\ldots,\varepsilon_d) \mid \varepsilon_i = \pm 1\}$ then $(1,1,\ldots,1)$ stood for some definite orientation for L from which all other orientations would be obtained by changing the orientations of one or more components. $\mathrm{Diff}(S^3,L)$ acts on $\mathcal{O}(L)$ as a subgroup of a possibly larger abstract group G_d of permutations and sign changes. G_d is a split extension $1 \longrightarrow \mathbb{Z}_2^d \longrightarrow G_d \longrightarrow S_d \longrightarrow 1$ where S_d = permutation group on d-letters, $\mathbb{Z}_2^d$ denotes the direct sum of d-copies of $\mathbb{Z}_2$, and S_d operates on $\mathbb{Z}_2^d$ by permutation. Define the symmetry group of L, Sym(L) to be the subgroup of G_d corresponding to the action of $\mathrm{Diff}(S^3,L)$ on $\mathcal{O}(L)$. Then we have the last reformulation:

<u>Proposition</u>. $S_n(D^4,L)$ is in bijective correspondence with

$$\mathcal{O}(L)/\mathbb{Z}_2 \times \mathrm{Sym}(L).$$

<u>Remark</u>. Sym(L) is not quite the usual symmetry group of L. It does not catalogue whether or not the diffeomorphism inducing the symmetry, preserves the orientation of S^3. Of course, this arises because we have ignored orientation of manifolds in $S_n(D^4,L)$.

<u>Proposition</u>. Let $L = L_{a,b}$, $d = \gcd(a,b)$, $a = d\alpha$, $b = d\beta$, $\alpha,\beta > 1$. Then $\mathrm{Sym}(L_{a,b}) \simeq \mathbb{Z}_2 \times S_d$.

<u>Proof</u>: Take for the chosen orientation $(1,1,\ldots,1)$ the orientation induced on L via $L_{a,b} = f^{-1}_{a,b}(S^1)$. Let the components of L be $K_1,\ldots,K_d$. These are nontrivial torus knots of type (α,β). Letting ℓ denote linking number in S^3 we know that $\ell(K_i,K_j) = \alpha\beta$ for $i \neq j$.

Note that a nontrivial torus knot is <u>not</u> amphicheiral (see 5, p. 31). This means that K_i can never be carried to K_j by any diffeomorphism which reverses the orientation of S^3. Hence we may restrict attention to those diffeomorphisms which preserve the orientation of S^3. However, such diffeomorphisms preserve linking numbers. Hence, given $g:(S^3,L) \longrightarrow (S^3,L)$, $\ell(g(K_i),g(K_j)) = \ell(K_i,K_j)$. Suppose $g(K_i) = \varepsilon_1 K_{i'}$, $g(K_j) = \varepsilon_2 K_{j'}$. Then $\varepsilon_1\varepsilon_2\alpha\beta = \ell(gK_i,gK_j) = \ell(K_i,K_j) = \alpha\beta$. Whence $\varepsilon_1\varepsilon_2 = +1$. Thus $\varepsilon_1 = \varepsilon_2$. The upshot is that, at best, a symmetry can only reverse all of the link orientations. In fact, each torus link has such a symmetry. It is obtained by turning the link around and then rotating it about its central axis by 180 degrees.

On the other hand, for each permutation $\tau \in S_d$ there is a diffeomorphism $g(\tau)$ such that $g(\tau)(K_i) = K_{\tau(i)}$. This is easily constructed by noting that the components of the link may be viewed as lying on concentric tori.

Hence $\mathrm{Sym}(L) \simeq \mathbb{Z}_2 \times S_d$.

<u>Corollary</u>. Under the above conditions $S_n(D^4, L_{a,b})$ is in bijective correspondence with $\mathcal{O}(L)/\mathbb{Z}_2 \times S_d$. It has $\frac{1}{2}(d+2)$ elements for d even and $\frac{1}{2}(d+1)$ elements for d odd.

<u>Proof</u>: The formulas follow from an easy counting argument.

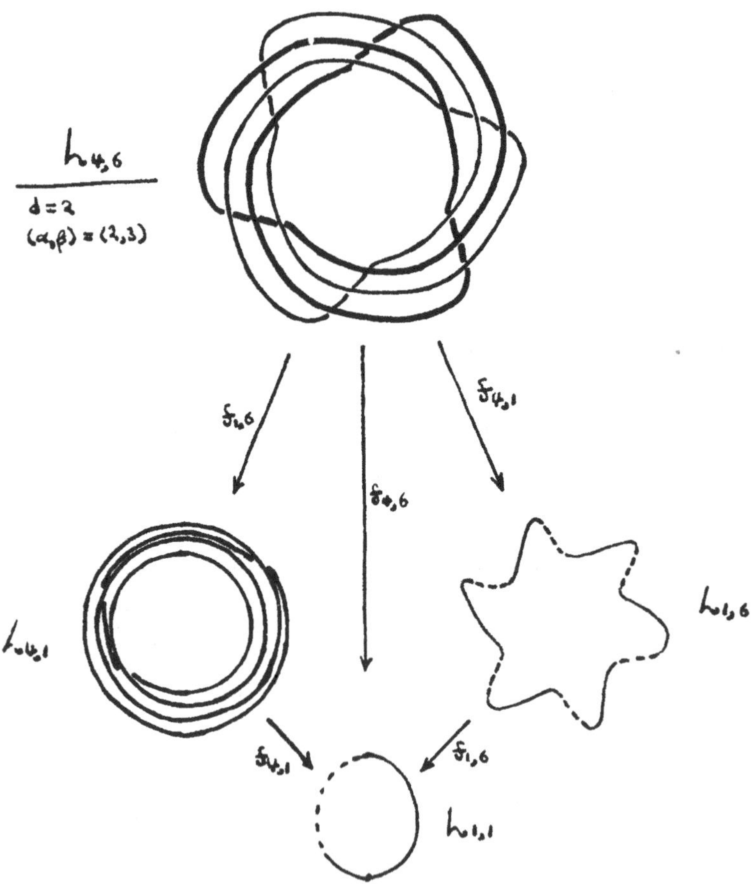

REFERENCES

0. J. Cerf, Sur les difféomorphismes de la sphere de dimension trois (Γ_4=0), Springer-Verlag, 1968.

1. D. Erle, Die Quadratische Form eines Knotens und ein Satz über Knoten mannigfaltigkeiten, Journal für Mathematik, Band 236. 174-217 (1969).

2. F. Hirzebruch and K. Mayer, O(n)-Mannigfaltigkeiten Exotische Sphären und Singularitäten, Springer-Verlag, 1968.

3. K. Jänich, Differenzierbare G-Mannigfaltigkeiten, Springer-Verlag, 1968.

4. J. Milnor, Singular Points of Complex Hypersurfaces, Princeton University Press, 1968.

5. K. Reidemeister, Knotentheorie, Chelsea, 1948.

6. W. Whitten, Symmetries of links, Trans. Amer. Math. Soc., 213-222 (1969).

DEGREE OF SYMMETRY OF CLOSED MANIFOLDS

by H.T. Ku, L.N. Mann, J.L. Sicks and J.C. Su*

The University of Massachusetts, Amherst

1. Introduction. In this note we will list some of the more important facts presently known concerning the degree of symmetry of manifolds. M^m will denote a closed connected differentiable m-dimensional manifold. The degree of symmetry of M, denoted by N(M), is defined as the supremum of the dimensions of all compact Lie groups which act effectively and differentiably on M. Given a differentiable action of G on M we can always find an invariant metric on M by averaging any Riemannian metric over the compact group G [3,VIII]. Therefore G acts as a group of isometries on M and, consequently, we may consider N(M) to be the supremum of the dimensions of the isometry groups of all possible Riemannian structures over M.

Although the formal notion of degree of symmetry was recently introduced by W. Y. Hsiang [11], investigations of this invariant were apparently made a good number of years ago in differential geometry. For example we have the following classical result.

THEOREM (Frobenius - Birkhoff [8], [17], [2]). $N(M^m) \le \frac{m(m+1)}{2}$ and $N(M^m) = \frac{m(m+1)}{2}$ *if and only if* M *is diffeomorphic to either the standard sphere* S^m *or the standard real projective space* RP^m.

---

*The authors were partially supported by the National Science Foundation.

2. *High degree of symmetry.* Using the techniques of differential geometry [26], [25] it was shown that there are very few manifolds with degree of symmetry in either of the following two ranges:

$$\frac{(m-1)m}{2} + 1 < N(M^m) < \frac{m(m+1)}{2}$$

$$\frac{(m-2)(m-1)}{2} + 3 < N(M^m) < \frac{(m-1)m}{2} .$$

In [16] it was demonstrated that these "gaps" in the degree of symmetry were part of a general pattern. In fact, if $N(M^m)$ is in any of the ranges,

$$\frac{(m-k)(m-k+1)}{2} + \frac{k(k+1)}{2} < N(M^m) < \frac{(m-k+1)(m-k+2)}{2}, \quad k = 1,2,3,...$$

then M is diffeomorphic to CP^2 or CP^5 [16].

This last result suggests a program of trying to classify those manifolds having a high degree of symmetry. The following seems to provide a good start.

LEMMA [14]. Suppose $N(M^m) \geq \frac{m^2}{4} + \frac{m}{2}$ and $m \geq 19$. *Then exactly one of the following holds*:

(α) $M = CP^k$, $m = 2k$.

(β) $M = CP^k \times S^1$, $m = 2k + 1$.

(γ) M *is a simple lens space finitely covered by* S^{2k+1}, $m = 2k + 1$.

(δ) *There exists an almost effective action of* Spin(n), $n \geq \frac{m}{2} + 1$, *on* M *with orbits which are some combination of fixed points, standard* (n-1) – *spheres and standard real projective* (n-1) – *spaces*.

By analyzing the possible orbit structures of case (δ), one establishes the following characterization of simply-connected highly symmetric manifolds.

THEOREM. *If* $\pi_1(M^m) = 0$ *and* $N(M^m) \geq m^2/4 + {}^m/_2$, $m \geq 19$, *exactly one of the following holds:*

(1) $M = CP^k$, $m = 2k$.

(2) $M = \partial(D^n \times X)$, $n \geq \frac{m}{2} + 1$, *where* X *is a compact manifold possibly with boundary.*

There are a number of immediate consequences of this result. We mention one below.

COROLLARY. *Suppose* $\pi_1(M^m) = 0$ *and* $N(M^m) \geq m^2/4 + {}^m/_2$, $m \geq 19$. *Then if the bordism class* $[M] \neq 0$ in Ω_m, $M = CP^k$.

3. *A generalization of the Frobenius-Birkhoff Theorem.* In [14] the following was established.

THEOREM (Ku, Mann, Sicks and Su). *Let* $M^m = M_1^{m_1} \times M_2^{m_2}$, $m \geq 19$. *Then*

$$N(M) \leq \frac{m_1(m_1+1)}{2} + \frac{m_2(m_2+1)}{2}$$

and if equality holds, M *is diffeomorphic to the product of two spheres, two real projective spaces or one of each.*

W. Y. Hsiang has conjectured that the above result holds for arbitrary products. In [12] he proves this conjecture under the additional (apparently technical) assumption that the first rational Pontrjagin class of M is zero.

It is not difficult to obtain a lower inequality for the degree of symmetry of a product manifold.

REMARK [14]. $N(M_1 \times M_2 \times \ldots \times M_k) \geq \sum_{i=1}^{k} N(M_i)$.

EXAMPLE. If an exotic sphere Σ^m bounds a π-manifold, it is known that $\Sigma^m \times S^2$ is diffeomorphic to $S^m \times S^2$. As we shall see in the next section, $N(\Sigma^m) < N(S^m)$.

Hence

$$N(\Sigma^m \times S^2) > N(\Sigma^m) + N(S^2).$$

4. Exotic structures. If Σ^m is an exotic sphere it is known that $N(\Sigma^m)$ is significantly smaller than $N(S^m)$.

THEOREM (W. Y. Hsiang [11]).

$$N(\Sigma^m) < \frac{1}{8} m^2 + 1 \qquad (m \geq 40).$$

This result is best possible in the sense that the exotic Kervaire sphere Σ_0^{8k+1} has degree of symmetry $\frac{1}{8} m^2 + \frac{7}{8}$ where $m = 8k + 1$. Moreover the Hsiang brothers have shown the following.

THEOREM (Hsiangs [10], [12]). If Σ^m is "very exotic," i.e. Σ^m does not bound a π-manifold,

$$N(\Sigma^m) < \frac{1}{16}(m+1)^2 + 5 \qquad (m \geq 35).$$

It is not known if the above bound is best possible. Continuing in this direction Schultz [22], [23] has exhibited a family $\{\Sigma^m\}$ of homotopy spheres of arbitrarily high dimension for which $N(\Sigma^m) \leq 30\ m/7$. The following question was independently raised by the Hsiang brothers and Glen Bredon at the Tulane Conference [19].

PROBLEM. Does there exist a Σ^m with $N(\Sigma^m) = 0$? In this connection it is interesting to note that Schultz [24] has recently shown that if Σ^m bounds a spin-manifold, $N(\Sigma^m) > 0$ for $m \leq 13$.

Of course one may consider similar questions for exotic structures on manifolds other than spheres. In [15], for example, it is proven that an exotic homotopy real projective m-space $(m \geq 72)$ has degree of symmetry smaller than $1/8\ m^2 + 1$.

5. Zero degree of symmetry. It follows from results of Mostert [18] that the only closed connected 2-manifolds which have non-zero degree of symmetry are S^2, T^2, RP^2 and the Klein bottle. Orlik and Raymond [20], [21] have classified the 3-manifolds with $N(M^3) \neq 0$.

Conner and Montgomery [4], [5] have shown that if M^m is a $K(\pi,1)$ with non-vanishing Euler characteristic, then $N(M^m) = 0$. This of course is consistent with Mostert's results in the 2-dimensional case. Recently Atiyah and Hirzebruch [1] have produced a large class of manifolds with zero degree of symmetry by showing that if M^{4k} is an orientable spin-manifold with $\hat{A}(M) \neq 0$, then $N(M) = 0$. As an application of the Atiyah-Hirzebruch result it is possible, using the results of [9], to show that for each $n \geq 2$, there exist infinitely many topologically distinct homotopy quaternionic projective n-spaces with zero degree of symmetry. For details see [13] of these proceedings.

Going beyond the notion of zero degree of symmetry, Conner and Raymond [6], [7] have exhibited closed manifolds which admit no effective finite group action.

PROBLEM. Do there exist manifolds whose homeomorphism groups (under the compact-open topology) contain no non-trivial compact subgroups?

REFERENCES

1. M. F. Atiyah and F. Hirzebruch, Spin-manifolds and group actions, Essays on Topology and Related Topics, Springer-Verlag (1969), 18-28.

2. Garrett Birkhoff, Extensions of Lie groups, Math-Zeit., 53 (1950), 226-235.

3. A. Borel et al., Seminar on Transformations Groups, Ann. of Math. Studies 46, Princeton Univ. Press, Princeton, N.J., 1960.

4. P. E. Conner and D. Montgomery, Transformation groups on a $K(\pi,1)$, I., Mich. Math. J. 6 (1959), 405-412.

5. P. E. Conner and F. Raymond, Actions of compact Lie groups on aspherical manifolds, Topology of Manifolds, Markham (1970), 227-264.

6. P. E. Conner and F. Raymond, Manifolds with few periodic homeomorphisms, these proceedings.

7. P. E. Conner, F. Raymond and P. Weinberger, Manifolds with no periodic maps, these proceedings.

8. L. P. Eisenhart, Riemannian Geometry, Princeton Univ. Press, Princeton, N. J. 1926.

9. W. C. Hsiang, A note on free differentiable actions of S^1 and S^3 on homotopy spheres, Ann. of Math. 83(1966), 266-272.

10. W. C. Hsiang and W. Y. Hsiang, The degree of symmetry of homotopy spheres, Ann. of Math. 89 (1969), 52-67.

11. W. Y. Hsiang, On the bound of the dimensions of the isometry groups of all possible Riemannian metrics on an exotic sphere, Ann. of Math. 85 (1967), 351-357.

12. W. Y. Hsiang, On the degree of symmetry and the structure of highly symmetric manifolds, mimeo., University of Cal., Berkeley.

13. H. T. Ku and M. C. Ku, Characteristic invariants of free differentiable actions of S^1 and S^3 on homotopy spheres, these proceedings.

14. H. T. Ku, L. N. Mann, J. L. Sicks and J. C. Su, Degree of symmetry of a product manifold, Trans. Amer. Math. Soc. 146 (1969), 133-149.

15. H. T. Ku, L. N. Mann, J. L. Sicks and J. C. Su, Degree of symmetry of a homotopy real projective space, Trans. Amer. Math. Soc., 161 (1971), 51-61.

16. L. N. Mann, Gaps in the dimensions of transformation groups, Ill. J. Math. 10 (1966), 532-546.

17. D. Montgomery and H. Samelson, Transformation groups of spheres, Ann. of Math. 44 (1943), 454-470.

18. P. S. Mostert, On a compact Lie group acting on a manifold, Ann. of Math. 65 (1957), 447-455.

19. P. S. Mostert, editor, Proceedings of the Conference on Transformation Groups, Springer-Verlag 1968.

20. P. Orlik and F. Raymond, Actions of SO(2) on 3-manifolds, Proceedings of the Conference on Transformation Groups, Springer-Verlag (1968), 297-318.

21. F. Raymond, Classification of the actions of the circle on 3-manifolds, Trans. Amer. Math. Soc. 131 (1968), 51-78.

22. R. Schultz, Improved estimates for the degree of symmetry of certain homotopy spheres, Topology 10 (1971), 227-235.

23. R. Schultz, Semifree circle actions and the degree of symmetry of homotopy spheres, Am. J. of Math. 93 (1971), 829-839.

24. R. Schultz, Circle actions on homotopy spheres bounding plumbing manifolds, to appear.

25. H. Wakakuwa, On n-dimensional Riemannian spaces admitting some groups of motions of order less than 1/2 n(n-1), Tohoku Math. J. (2), 6 (1954), 121-134.

26. H. C. Wang, On Finsler spaces with completely integrable equations of Killing, Journ. of the London Math. Soc. 22 (1947), 5-9.

TRANSFER HOMOMORPHISMS OF WHITEHEAD GROUPS OF SOME CYCLIC GROUPS, II[1)]

By Kyung Whan Kwun[2)]
Michigan State University

Consider the following

Assertion. Let M be a PL homotopy lens space of dimension ≥ 5 with $\pi_1 M$ of an odd order ≥ 5. Then given any PL free involution h and M' h-cobordant to M there exist infinitely many non-equivalent PL free involutions $h_1, h_2, \ldots$ of M and PL homotopy equivalences $f_i: M \to M'$ such that $f_i h = h_i f_i$.

A difficulty is not really getting infinitely many h_i but is getting one free involution of M' just knowing M admits a free involution.

Actually, the assertion above depends on a purely algebraic result which can be stated as follows.

Let Z_{2k+1} be the cyclic group of order $2k+1$ considered as a subgroup of Z_{4k+2} with $i: Z_{2k+1} \subset Z_{4k+2}$. Then the transfer homomorphism of Whitehead groups:

$$i^*: Wh(Z_{4k+2}) \to Wh(Z_{2k+1})$$

is an epimorphism for all k.

---

1) A summary of talk at this conference (a revised version).

2) Supported in part by NSF Grants GP-19462 and GP-29515X.

This result is a corollary to

THEOREM A. Let G be a finite abelian group of an odd order. Then

$$i^*: Wh(G \oplus Z_2) \to Wh(G)$$

is an epimorphism.

The proof of this theorem, its relation to the assertion and other related results are given in [1] and will not be repeated in this summary.

In the theorem, the oddity of the order of G is essential. In fact, we have

THEOREM B. The transfer homomorphism

$$i^*: Wh(Z_{2k} \oplus Z_2) \to Wh(Z_{2k})$$

is an epimorphism if and only if $k = 1, 2$ or 3.

We remark that the case $k = 1, 2$ or 3 is precisely the case where $Wh(Z_{2k}) = 0$.

We give below a proof. Let ϕ denote the Euler's Phi function. By checking several cases, one easily finds that $\phi(2k) \geq 4$ for $k \geq 4$. From here on assume $k \geq 4$. It then follows that there exist integers p and q such that $1 < p < k < q \leq 2k - p$ with $pq \equiv \pm 1$ mod $2k$. Now in general, if t is the generator of Z_{2k} and a unit $u \in ZZ_{2k}$ represents an element of $Wh(Z_{2k})$ in the image of i^*, $u = \pm t^i(\alpha^2 - \beta^2)$ for some i and $\alpha, \beta \in ZZ_{2k}$. This follows exactly as in [1]. That means that if $u = \sum_{j=0}^{2k-1} n_j t^j$, then the elements of $\{j \mid n_j \text{ odd}\}$ are all odd or all even. If $u = \Sigma n_j t^j$ where n_j is odd for some odd j and some even j, we will call u a mixed type. Hence in order to conclude the proof, it suffices to show that there exists a unit of mixed type. We start with a special case.

Special case.

There exist p,q such that $1 < p < k < q < 2k - p$, $pq \equiv \pm 1 \bmod 2k$. (This is equivalent to saying that there exists an integer p relatively prime to $2k$ such that $p^2 \not\equiv \pm 1 \bmod 2k$.) By [2, Lemma 12. 10, p. 408], there exists a unit $u \in ZZ_{2k}$ such that

$$u(1-t)(1-t) = (1-t^p)(1-t^q).$$

Hence $u = (1+t+\ldots+t^{p-1})(1+t+\ldots+t^{q-1}) - m(1+t+\ldots+t^{2k-1})$ for some unique integer m.

Now $p-1 < p < q-1 < q$ (p,q odd). t^{p-1}, t^p, t^{q-1} appear with coefficient p in $(1+t+\ldots+t^{p-1})(1+t+\ldots+t^{q-1})$. Hence if u is not of mixed type, m must be odd, in which case, t^{2k-1} and t^{2k-2} $(2k-2 > p+q-2)$ must appear with coefficient m. Hence u is of mixed type, no matter what.

General case.

We first take care of the case where $k > 5$. We show that if $k > 5$ then the situation satisfies the special case. If k is odd, let $p = k-2$. p is relatively prime to $2k$ and

$$p^2 \equiv (k-2)^2 \equiv k+4 \bmod 2k.$$

If $k > 5$, $k+4 \not\equiv \pm 1 \bmod 2k$.

If k is even, let $p = k-3$.

$$p^2 \equiv (k-3)^2 \equiv 9 \bmod 2k$$

and $9 \not\equiv \pm 1 \bmod 2k$ if $k > 5$.

Now if $k = 4$, from the congruence $3.3 \equiv 1$, we find a unit $u = (1+t+t^2)^2 - (1+t+\ldots+t^7)$ which is of mixed type. If $k = 5$, from the congruence $3.3 \equiv -1$, we obtained a unit

$$u = (1+t+t^2)^2 - (1+t+\ldots+t^9)$$

which is of mixed type.

As in [1], we obtain

COROLLARY. Let L be a PL homotopy lens space such that $\pi_1 L \simeq Z_{2k}$, $k \geq 4$. Let h be the obvious PL free involution of $L \times S^n$ ($\dim L + n \geq 5$) such that the corresponding orbit space is $L \times P_n$, where P_n, $n \geq 2$, is the real projective space. Then there exist infinitely many distinct PL h-cobordisms W starting with $L \times S^n$ such that h cannot be extended to a free PL involution of W.

REFERENCES

[1] K. W. Kwun, Transfer homomorphisms of Whitehead groups of some cyclic groups, to appear in Amer. Jour. Math. in 1971.

[2] J. Milnor, Whitehead torsion, Bull. Amer. Math. Soc. 72 (1966), 359-426.

SURGERY ON FOUR-MANIFOLDS AND TOPOLOGICAL TRANSFORMATION GROUPS

Julius L. Shaneson
Princeton University

Surgery theory in higher dimensions has been used by Browder [B1], Browder-Petrie [B5],[B6], and Rothenberg [R1] to study smooth and P. L. semi-free transformation groups. In this lecture we apply the results of [CS1] on four- and five-dimensional surgery to exhibit some topological actions with non-smoothable fixed point sets.

Let Z_p be the cyclic group of order p, p any natural number. By a _semi-free_ action of Z_p on the topological manifold M^n, we mean an action that has as isotropy subgroups only Z_p and the trivial subgroup. We will say that such an action is _flat_ if

(i) the set F of fixed points is a submanifold; and

(ii) the components F_i of F have disjoint product neighborhoods $F_i \times D^{k_i} \subset M$ (D^{k_i} = unit disk in $\mathbb{R}^{k_i}$) such that $\exists$ homomorphisms $\rho_i: Z_p \longrightarrow O(k_i)$ with

$$\xi\cdot(x,y) = (x,\rho_i(\xi)y)$$

for $\xi\in Z_p$, $x\in F_i$, $y\in D^{k_i}$.

Actions of Z_p on M and M' are called _equivalent_ if there is a homeomorphism $h: M \longrightarrow M'$ that is equivariant; i.e., for $\xi\in Z_p$ and $x\in M$, $h(\xi\cdot x) = \xi\cdot h(x)$. An action is said to be _smoothable_ if it is equivalent to a smooth action on a smooth manifold. Note that _smoothable_ semi-free actions on compact manifolds are always flat.

One can show that any flat semi-free action of Z_p on the topological n-sphere S^n, $n \geq 6$, with fixed points a topological sphere of codimension two is smoothable to a smooth action on a smoothing on S^n with fixed points a smooth homotopy sphere. This is proven by suitably smoothing the action on the complement of the fixed points using [K2] or [LR2], and [LR1]. (Actually, one can obtain a flat piecewise linear action for fixed points of higher codimension at least for p odd.) In this lecture I wish to discuss the following result coming from joint work with S. Cappell:

Theorem:

Let p be a natural number. Then there are infinitely many pairwise inequivalent non-smoothable flat semi-free actions of Z_p on S^5 with fixed points homeomorphic to S^3.

We cannot exclude the possibility that there may be smooth semi-free actions of Z_p on S^5 with fixed points a homotopy 3-sphere Σ^3 not diffeomorphic to S^3. However, these actions cannot be smoothings of the actions of the theorem, since, according to Moise, homeomorphic 3-manifolds are diffeomorphic.

Since the piecewise linear and smooth categories are equivalent in low dimensions (see [LR1], for example), the actions of the theorem will not be equivalent to P.L. actions.

For p=1, the theorem should be understood to assert that there is an infinite family of non-smoothable knotted flat 3-spheres in S^5. In fact the proof of the theorem gives a new construction of non-smoothable knots, based on the results of [CS1]. Indeed for all p, the non-smoothability will be seen to be implied by the non-smoothability of the fixed points as a flat knotted S^3 in S^5. A non-smoothable knot was first discovered by Lashof [L], also using the results of [CS1]. See also [CS2].

The actions of the theorem will be exhibited, roughly, by constructing the closed complements of the fixed points, with free Z_p actions, and then gluing in $S^3 \times D^2$ with a standard action. The construction of the complements would be more straightforward if we knew that surgery theory in dimension four was exactly analogous to higher dimensions.

Results on Surgery

Let $L_n(\pi)$ be the surgery obstruction group of Wall for the finitely presented group π with the trivial orientation character.

For dimensions ≥ 6, the obstruction to finding cobordisms, relative the boundary, of normal maps to simple homotopy equivalences lies in this group. Theorems 5.8 and 6.5 of [W] assert that for $n \geq 6$, the elements of $L_n(\pi)$ are all indeed the surgery obstructions of normal maps. We will discuss this result briefly for the case $n \equiv 1 \pmod 4$ (recall $L_n = L_{n+4}$). The construction involved can be described briefly as comparing the results of performing surgery in different ways on a trivial surgery problem in one dimension lower.

Let κ_r be the special hermitian kernel of dimension r over the integral group ring $Z\pi$ of the group π. [W, p. 47]. That is, κ_r consists of a free based $Z\pi$ module H_r with base $e_1,\ldots,e_r$, $f_1,\ldots,f_r$, and special hermitian* form [W, p. 47] $\lambda_r\colon H_r \times H_r \longrightarrow Z\pi$ associated with a form $\mu_r\colon H_r \longrightarrow Z\pi/\{\nu-\bar{\nu} \mid \nu\in Z\pi\}$ so that $\lambda_r(e_i,f_j) = \delta_{ij}$, $\lambda_r(e_i,e_j) = \lambda_r(f_i,f_j) = 0$, and $\mu_r(e_i) = \mu_r(f_j) = 0$. Let $SU_r(Z\pi)$ denote the (simple) automorphisms of κ_r. Then $SU_r(Z\pi)$ is contained in $SU_{r+1}(Z\pi)$ in a natural way, and $L_n(\pi)$ is

---

* $Z\pi$ has the involution given by $\bar{g} = g^{-1}$ for $g\in\pi$.

a quotient of $\lim_{r\to\infty} SU_r(Z\pi)$. (See §6 of [W].)

Now let M be an oriented closed smooth manifold of dimension $n-1 = 4\ell$, with $\ell > 1$, for the moment. Assume $\pi = \pi_1 M$. Given $\gamma\epsilon L_n(\pi)$, let $\alpha\epsilon SU_r(\Lambda)$ be a representative. Let "#" denote connected sum, and identify κ_r with the summand of the homology group $H_{2\ell}(M\#r(S^{2\ell}\times S^{2\ell}; Z\pi)$ with local coefficients generated by the second summand so that e_i and f_i are carried by the ith copy of $S^{2\ell}\times pt$ and $pt \times S^{2\ell}$, respectively. We may represent the classes $\alpha(e_1),\ldots,$ $\alpha(e_r)$ by disjointly embedded $S_i^{2\ell}\times D^{2\ell}\subset M\#r(S^2\times S^2)$; this uses the hypothesis $\ell > 1$. Then let

$$W_2 = (M\#r(S^2\times S^2)) \cup_{S_i\times D^{2\ell}} (D_i^{2\ell+1}\times D^{2\ell})$$

be the result of surgery on these classes. Then one can show that there is a normal map [B1][B2][W]

$$\begin{array}{ccc} \nu(W_2) & \xrightarrow{\ b_2\ } & \nu(M\times[\tfrac{1}{2},1]) \\ \downarrow & & \downarrow \\ (W_2;\ M\#r(S^2\times S^2),\partial_1 W_2) & \xrightarrow{\ f_2\ } & (M\times[\tfrac{1}{2},1];\ M\times\tfrac{1}{2},\ M\times 1), \end{array}$$

b_2 a map of normal bundles, so that $f_2 \mid M\#r(S^2\times S^2)$ is the natural collapsing map, and so that $f_2 \mid \partial_1 W_2$ is a simple homotopy equivalence. (Note that f_2 and $f_2 \mid \partial_1 W_2$, by general position and Van Kampen's theorem, induce isomorphisms of fundamental groups, so that it suffices, as in [W], to show that $f_2 \mid \partial_1 W_2$ induces isomorphisms of homology with local coefficients.)

On the other hand we can do surgery on the classes $e_1,\ldots,e_r$ themselves, and use this to give a normal map (f_1,b_1), where

$$f_1\colon (W_1;\ \partial_0 W_1,\ M\#r(S^2\times S^2)) \longrightarrow (M\times[0,\tfrac{1}{2}];\ M\times 0,\ M\times\tfrac{1}{2}),$$

where f_1 and f_2 agree on $M\#r(S^2\times S^2)$ and $f_1|\partial_0 W_1$ is a diffeomorphism. Let $P = M\#r(S^2\times S^2)$ and let $W = W_1 \cup_P W_2$. Define $(f,b) = (f_1,b_1) \cup_P (f_2,b_2)$, so that $f\colon W \longrightarrow M \times [0,1]$. Then (f,b) is a normal map which restricts to a homotopy equivalence of boundaries, and so has a surgery obstruction, $\sigma(f,b)$. In fact, it is not hard to show that $\sigma(f,b) = \gamma$.

For example, if α is the element Σ_r, whose matrix with respect to the base $e_1,\ldots,e_r$, $f_1,\ldots,f_r$ is $\begin{pmatrix} 0 & I \\ I & 0 \end{pmatrix}$, then $W = M\times I$.

Now suppose that $\dim M = 4$. Then we have only the following result, extracted from the proof of Theorem 3.1 of [CS1]:

Proposition. Let $\alpha \in SU_r(Z\pi)$, $r > 1$. Assume $M \cong Q\#r(S^2\times S^2)$, Q a compact 4-manifold. Identify the appropriate summand of $H_2(M\#r(S^2\times S^2); Z\pi)$ with κ_r, as above. Then $\exists$ disjointly embedded spheres with trivial normal bundle, $S_i^2 \times D^2 \subset M$, representing the classes $\alpha(e_i)$, $i = 1,\ldots,r$, so that the map

$$\pi_1(M - \cup_i(S_i\times 0)) \longrightarrow \pi_1(M)$$

induced by inclusion is an isomorphism. In particular, doing surgery using these embeddings gives a normal map (f,b), $f\colon(W;\partial_0 W, \partial_1 W) \longrightarrow (M\times I; M\times 0, M\times 1)$, such that $f|\partial_0 W$ is a diffeomorphism, $f|\partial_1 W$ is a homotopy equivalence, f induces an isomorphism of fundamental groups, and $\sigma(f,b)$ is the element represented in $L_5(\pi)$ by γ.

Note that the assertions about π_1 do not follow solely from general position.

Next, let $RU_r(Z\pi) \subset SU_r(Z\pi)$ be the subgroup generated by $\Sigma_1 \in SU_1(Z\pi) \subset SU_r(Z\pi)$ and by those elements which preserve the subspace of H_r generated by $e_1,\ldots,e_r$ (called a subkernel) and whose restriction to this subspace is simple with respect to this basis.

(By simple we can understand those automorphisms whose matrices with respect to the basis $\{e_1,\dots,e_r\}$ are products of elementary matrices and diagonal matrices of the form

$$\begin{pmatrix} \pm g & & & \\ & 1 & & \\ & & \ddots & \\ & & & 1 \end{pmatrix} \qquad g\in\pi.)$$

Then if $RU(Z\pi) = \lim_{r\to\infty} RU_r(Z\pi)$, $L_5(\pi) = SU(Z\pi)/RU(Z\pi)$.

If $g\colon \pi \longrightarrow \pi'$ is a homomorphism, let g_* denote the natural maps $SU_r(Z\pi) \longrightarrow SU_r(Z\pi')$ and $L_5(\pi) \longrightarrow L_5(\pi')$.

<u>Lemma 1</u>. <u>Assume</u> g <u>is an epimorphism</u>. <u>Then</u>

$$g_*(RU_r(Z\pi)) = RU_r(Z\pi').$$

<u>Proof</u>: The inclusion $\subseteq$ is clear. As in [W, p.57], $RU_r(Z\pi')$ is generated by Σ_1, and elements with the matrices

$$\begin{pmatrix} A & 0 \\ 0 & A^* \end{pmatrix}$$

and

$$\begin{pmatrix} I & 0 \\ C & I \end{pmatrix},$$

where A is elementary or diagonal as just above and A^* denotes its conjugate transpose; and where C is an $(r\times r)$ matrix of the form $D - D^*$. These are evidently in the image of g_*.

<u>Lemma 2</u>: <u>Let</u> $g\colon \pi = Z \longrightarrow Z_p$ <u>be the natural map</u>. (Z = integers.) <u>Then every element</u> $\gamma\in L_5(Z)$ <u>has a representative</u> $\alpha\in SU_r(Z\pi)$, <u>some</u> r, <u>so that</u> $g_*(\alpha) = \Sigma_r$ <u>in</u> $SU_r(Z[Z_p])$.

<u>Proof</u>: By [W, 14E.5b] the map $g_*\colon L_5(Z) \longrightarrow L_5(Z_p)$ is trivial. Let $\xi\in SU_r(Z\pi)$ represent γ. Then, after stabilizing if necessary, we may assume $g_*\xi$ is in $RU_r(Z[Z_p])$. Choose $\zeta\in RU_r(Z[\pi])$ with

$g_*\zeta = (g_*\xi)^{-1}\Sigma_r$, by lemma 1. Then $\alpha = \xi\zeta$ is the desired element.

Topological Surgery Obstructions

Using the work of Kirby and Siebenmann, one also has the notion of topological normal maps and surgery obstructions, at least in dimensions greater than five. (See §17B of [W] for example.) For a five-dimensional topological normal map, we may define the surgery obstruction by first taking products with CP^2 and then taking surgery obstructions. Of course, in case we already have a smooth or P.L. normal map, this agrees with the usual surgery obstruction (which is periodic under products with CP^2).

Fix the positive integer p. If X is a space with $\pi_1 X = Z$, let $\hat{X}$ denote the covering space associated to the subgroup $pZ \subset Z$.

Lemma 3: Let $\gamma \in L_5(Z)$. There is a topological normal map (f,b),

$$f: (W;\partial_0 W,\partial_1 W) \longrightarrow (S^3\times S^1\times I;\ S^3\times S^1\times 0,\ S^3\times S^1\times 1)$$

with the following properties

(i) $f|\partial_i W: \partial_i W \longrightarrow S^3\times S^1\times i$, $i=0,1$, are homeomorphisms;

(ii) $\sigma(f,b) = \gamma$;

(iii) f induces an isomorphism of fundamental groups; and

(iv) $\hat{f}: \hat{W} \longrightarrow (S^3\times S^1\times I)\hat{}\ \cong S^3\times S^1\times I$ induces isomorphisms of integral homology groups.

Notes: 1. It appears that (i) and (ii) can be proven using Theorem 5.8 of [W] in the topological category, topological transversality [K4], and the calculation in [S1] of $L_6(Z \oplus Z)$. The present proof uses none of these.

2. For $p=1$, (iv) is just the assertion that f induces homology isomorphisms.

To prove Lemma 3, let $\alpha\epsilon SU_r(Z[Z])$ be a representative of γ. By Lemma 2, we may assume the image of α in $SU_r(Z[Z_p])$ is precisely Σ_r.

Let

$$\delta: S^3 \times S^1 \times I \longrightarrow [-\tfrac{1}{2}, 4\tfrac{1}{2}]$$

be a self-indexing Morse function [S3] [M1] with precisely 2r critical points, r of index two and r of index 3. Then if

$$M = (S^3 \times S^1) \# r(S^2 \times S^2),$$

we may identify $M \times I$ with $f^{-1}[2\tfrac{1}{4}, 2\tfrac{3}{4}]$, so that we have

$$S^3 \times S^1 \times I = f^{-1}[-\tfrac{1}{2}, 2\tfrac{1}{4}] \cup M \times I \cup f^{-1}[2\tfrac{3}{4}, 4\tfrac{1}{2}],$$

a union of smooth manifolds. As above we identify x_r with the obvious summand of $H_2(M \# r(S^2 \times S^2); Z\pi)$. Then, as in the Proposition, we may represent the classes $\alpha(e_i)$ by embedded disjoint spheres with trivial normal bundles the complement of whose union has fundamental group Z. Then, as in the Proposition and preceding discussion, we may use these embeddings (and the obvious embeddings representing the classes e_i) to obtain a normal map

$$(g,c),\ g:(Q;\ \partial_0 Q, \partial_1 Q) \longrightarrow (M \times I;\ M \times 0,\ M \times 1),$$

with $g|\partial_0 Q$ a diffeomorphism and $g|\partial_1 Q$ a homotopy equivalence, and with $\sigma(g,c) = \gamma$. Further, g induces an isomorphism on fundamental groups.

The fact that α maps to Σ_r in $SU(Z[Z_p])$ implies that $\hat{g}: \hat{Q} \longrightarrow \hat{M} \times I$ induces isomorphisms of homology groups. In fact, $\hat{Q}$ and $\hat{g}$ can be constructed from $\hat{M}$ as in the discussion surrounding the Proposition, using an element $\hat{\alpha}$ that maps into Σ_r in $SU(Z)$ under the map induced by the natural map $pZ \longrightarrow \{e\}$. This in turn implies, by handlebody theory, our assertion about $\hat{g}$. We leave the details to the reader.

Now consider

$$g|\partial_1 Q: \partial_1 Q \longrightarrow M \times 1.$$

Then $(g|\partial_1 Q) \times \mathrm{id}_{S^1}$ represents an element of $hS(M\times S^1)$ (called $\mathscr{S}_{\mathrm{Diff}}(M\times S^1)$ in [W]) with vanishing normal invariant. (Compare [S2].) But it follows from the theory of Kirby and Siebenmann (essentially from the fact that $Z \cong \pi_4(G/PL) \longrightarrow \pi_4(G/Top) \cong Z$ is multiplication by two) and from surgery theory (and especially the calculation of $L_6(Z \oplus Z)$) that elements with vanishing normal invariant map trivially into $\mathscr{S}_{\mathrm{Top}}(M\times S^1)$. Hence $(g|\partial Q_1) \times \mathrm{id}_{S^1}$ is homotopic to a homeomorphism. (Compare [K2] [K4].) Now a standard argument shows that there is a topological h-cobordism $(U^5; \partial_1 Q, V)$ of $\partial_1 Q$ to V and an extension $\bar{g}: U \longrightarrow M \times 1$ of $g|\partial_1 Q$ so that $\bar{g}|V$ is a homeomorphism.

Now let W be obtained from the disjoint union $f^{-1}[-\frac{1}{2}, 2\frac{1}{4}] \cup (Q \cup_{\partial_1 Q} U) \cup f^{-1}[2\frac{3}{4}, 4\frac{1}{2}]$ by identifying x with

$$\begin{cases} (g|\partial_0 Q)^{-1}(x) & \text{for } x\in f^{-1}(2\frac{1}{4}) \\ (\bar{g}|V)^{-1}(x) & \text{for } x\in f^{-1}(2\frac{3}{4}) \end{cases}.$$

Let $f: W \longrightarrow S^3\times S^1\times I$ be induced by the union of maps $\mathrm{id} \cup g \cup \bar{g} \cup \mathrm{id}$. It is easy to see that there is a stable topological bundle map b covering f. Evidently, (i) of Lemma 3 is satisfied. That $\sigma(f,b) = \gamma$ follows from the facts that $\sigma(g,c) = \gamma$, that the other portions of f are homotopy equivalences, and the "addition theorem" for normal maps, as stated in 1.4 of [S1], for example. Properties (iii) and (iv) follow from the corresponding properties of (g,c) and Meyer-Vietoris sequences and Van Kampen's theorem, respectively.

Proof of Theorem (outline)

Recall that we have fixed an integer $p \geq 0$. Let $\lambda \in L_5(Z)$ be a generator. For each odd integer q, let (f_q, b_q), $f_q \colon W_q \longrightarrow S^3 \times S^1 \times I$, be a normal map satisfying the conclusion of Lemma 3, with $\gamma = q\lambda$. Then consider T_q obtained from the union $S^3 \times D^2 \cup \hat{W}_q \cup D^4 \times S^1$ by identifying $x \in \partial_0 \hat{W}_q$ with $(\hat{f}_q | \partial_0 \hat{W}_q)^{-1}(x)$ for x in $\partial(S^3 \times D^2)$ and $x \in \partial_1 \hat{W}_q$ with $(\hat{f}_q | \partial_1 \hat{W}_q)^{-1}(x)$ for $x \in \partial(D^4 \times S^1)$. The covering space $\hat{W}_g$ admits a free action of Z_p which on each boundary component corresponds under the above identification with the action on $S^3 \times S^1$ given by complex multiplication on the second factor. Thus we have a flat semi-free action of Z_p on T_q with fixed point set homeomorphic to S^3.

It follows from (iii) and (iv) of Lemma 3, Van Kampen's theorem, and Meyer-Vietoris sequences, that T is homotopy equivalent to S^5. By [K2] [LR2], T is smoothable; hence homeomorphic to S^5 [S3], (or see [H] for Stallings' version). So for each odd q we have a flat semi-free action ρ_q of Z_p on S^5.

Now $\sigma(\hat{f}_q, \hat{b}_q) = \sigma(f_q, b_q)$; this follows from known calculations (compare 10.5 of [HS]). Hence if F_q is the fixed point set of ρ_q, $S^3 - F_q$ and $S^3 - F_{q'}$ are not even of the same homotopy type, and so ρ_q and $\rho_{q'}$ are inequivalent.

Now, $\hat{f}_q \cup (\mathrm{id}) \colon d(T_q - S^3 \times D^2) \longrightarrow D^4 \times S^1$ is a normal map obtained from $\hat{f}_q$ by gluing on the identity map of $D^4 \times S^1$. So

$$\sigma(\hat{f}_q \cup \mathrm{id}, \hat{b}_q \cup \mathrm{id}) = \sigma(\hat{f}_q, \hat{b}_q) = q\lambda.$$

Now, as $q \equiv 1 \pmod 2$, $d(T_q - S^3 \times D^2)$ has no smooth structure extending the usual structure on the boundary, $S^3 \times S^1$. For if it did, we could make $\hat{f}_q$ transverse along $(D^4 \times pt)$ so as to obtain for $\hat{f}_q^{-1}(D^4 \times pt)$ a smooth, parallelizable 4-manifold with boundary a <u>smooth</u> S^3. Then, by the calculation of $L_5(Z) = L_9(Z)$ in terms of one-eighth the index of codimension-one submanifolds [B1][S1][W], and by periodicity under products with CP^2 (due to Sullivan in the simply-connected case; see [W]), the index of this four-manifold will be congruent to 8 modulo 16, contradicting Rohlin's theorem [R]. Therefore, any smoothing of S^5 induces an exotic smoothing on the product neighborhood $F_q \times \mathrm{Int}\, D^2$.

Finally, using engulfing one can show that S^3 can not be smoothly embedded, with respect to the exotic structure on $S^3 \times \mathrm{Int}\, D^2$, so as to be the zero section of $S^3 \times \mathrm{Int}\, D^2$ viewed as a topological microbundle. So by uniqueness of topological microbundle neighborhoods and 2.3 of [M2], the pair (F_q, S^5) is not homeomorphic to a smooth S^3 in S^5. So ρ_q is not smoothable. (This argument is essentially the same as one provided by Lashof. If one accepts the uniqueness portion of Theorem 1 of [K1], then the non-smoothability follows immediately from the preceding paragraph.)

References

[B1] W. Browder, Surgery and the theory of differentiable transformation groups, in Proceedings of the Conference on Transformation Groups, New Orleans, 1967, Springer 1968, 1-46.

[B2] ________, Surgery on simply-connected manifolds, to appear.

[B3] ________, Manifolds and homotopy theory, in Manifolds, Amsterdam 1970, Springer, 1971.

[B4] ________, Manifolds with $\pi_1 = Z$, Bull. Amer. Math. Soc. 72 (1966), 238-244.

[B5] W. Browder and T. Petrie, to appear. (See Bull. AMS 77 (1971) 160-163.)

[B6] ________, Semi-free and quasi-free S^1-actions on homotopy spheres, in Essays on Topology and Related Topics, Mémoires dédiés à Georges de Rham, 136-146, Springer, New York, 1970.

[CS1] S. Cappell and J. L. Shaneson, On four-dimensional surgery and applications, to appear.

[CS2] ________, Topological knot cobordism, to appear.

[HS] W. C. Hsiang and J. L. Shaneson, Fake tori, in Proceedings of the 1968 Georgia Conference (Topology of Manifolds), Markham Press, 1970, 19-50. (See also Proc. Nat. Acad. Sci. 62 (1969) 687-691.)

[K1] R. C. Kirby, Locally flat codimension two submanifolds have normal bundles, in Proceedings of the 1969 Georgia Conference (Topology of Manifolds), Markham Press, 1970.

[H] J. F. P. Hudson, PL Topology, Benjamin, 1969.

[K2] R. C. Kirby and L. C. Siebenmann, On the triangulation of manifolds and the Hauptvermatung, Bull. Amer. Math. Soc. 75 (1969), 742-749.

[K3] ________, Foundations of topology, Notices Amer. Math. Soc. 16 (1969), 848.

[K4] ________, to appear.

[L] R. K. Lashof, to appear.

[LR1] R. K. Lashof and M. Rothenberg, Microbundles and smoothing, Topology 3 (1965), 357-388.

[LR2] ________, Triangulation of Manifolds, I, II, Bull. AMS 75 (1969), 750-757.

[M1] J. Milnor, Lectures on the h-cobordism theorem, Princeton, 1965.

[M2] ________, Microbundles and differentiable structures, Notes, Princeton University, 1961.

[R] V. A. Rohlin, A new result in the theory of 4-manifolds, Doklady 8, 221-224 (1952).

[R1] M. G. Rothenberg, to appear. (See also Proc. Adv. Inst. on Alg. Top., Aarhus Univ., 1970, 455-475.)

[S1] J. L. Shaneson, Wall's Surgery Obstruction groups for Z × G, Ann. of Math. 90 (1969), 296-334. (See also Bull. AMS 74 (1968), 467-471.)

[S2] ________, On non-simply-connected manifolds, AMS Symposia in Pure Math, to appear.

[S3] S. Smale, Generalized Poincarés Conjecture in dimensions greater than four, Ann. of Math. 74 (1961), 391-406.

[S4] ________, On the structure of manifolds, Amer. J. of Math. 84 (1962), 387-399.

[S5] D. Sullivan, Triangulating and smoothing homotopy equivalences and homeomorphisms, Geometric Topology Seminar Notes, Princeton University, 1967.

[W] C. T. C. Wall, Surgery on compact manifolds, Academic Press, 1971.